Biopolymer Based Composites for Regenerative Medicines and Tissue Engineering Applications

Editor

Narendra Pal Singh Chauhan

Department of Chemistry
Faculty of Science
Bhupal Nobles' University, Udaipur
Rajasthan, India

CRC Press
Taylor & Francis Group
Boca Raton London New York

CRC Press is an imprint of the
Taylor & Francis Group, an **informa** business
A SCIENCE PUBLISHERS BOOK

Cover credit: Image reproduced by kind courtesy of Dr. Mazaher Gholipourmalekabadi.

First edition published 2026
by CRC Press
2385 NW Executive Center Drive, Suite 320, Boca Raton FL 33431

and by CRC Press
4 Park Square, Milton Park, Abingdon, Oxon, OX14 4RN

CRC Press is an imprint of Taylor & Francis Group, LLC

Library of Congress Cataloging-in-Publication Data (applied for)

ISBN: 978-1-032-85117-4 (hbk)
ISBN: 978-1-032-85576-9 (pbk)
ISBN: 978-1-003-51879-2 (ebk)

DOI: 10.1201/9781003518792

Typeset in Times New Roman
by Prime Publishing Services

Preface

In the rapidly evolving fields of regenerative medicine and tissue engineering, the quest for materials that can effectively mimic the complex environment of human tissues has led to an intense focus on biopolymer-based composites. These materials, which combine the inherent advantages of biopolymers with the functional benefits of composites, are emerging as vital tools in the development of next-generation medical therapies and technologies.

This book, Biopolymer-Based Composites for Regenerative Medicine and Tissue Engineering Applications, aims to capture the current landscape of research and development in this dynamic area. It brings together contributions from leading researchers worldwide, each offering a deep dive into specific biopolymers, their composites, and their applications in tissue engineering. As we delve into these chapters, we encounter a diverse array of materials cellulose, chitosan, collagen, gelatin, hyaluronic acid, and more each with unique properties that make them suitable for different tissue engineering applications. What unites these materials is their ability to be engineered into forms that not only support cell growth but also integrate seamlessly with human tissues, providing a foundation for true regeneration. However, this book is not just a collection of current practices; it also serves as a roadmap for future research. Each chapter addresses not only the successes but also the challenges and limitations of these materials, offering insights into areas where further innovation is needed. By exploring the intersection of material science, biology, and engineering, this book provides a comprehensive overview of where we are today and where we need to go next. One of the most compelling aspects of biopolymer-based composites is their potential to be both biocompatible and environmentally sustainable. As the world increasingly focuses on sustainability, the development of materials that are not only effective in medical applications but also derived from renewable sources are more critical than ever. This book highlights these aspects, offering a glimpse into a future where medicine and environment saviour approach go hand in hand. I would like to express my sincere gratitude to all the contributors who have dedicated their time and expertise to this volume.

Dr. Narendra Pal Singh Chauhan

Contents

CHAPTER 1

Keratin Nanofiber Composites for Tissue Engineering

Khushboo Jain,[1] *Dipti Soni,*[2] *Avinash Marwal,*[3] *Juan Pablo Fuentes,*[4] *Sapana Jadoun*[5] and *Narendra Pal Singh Chauhan*[6,*]

1. Introduction

The application of natural polymers in tissue engineering, drug delivery, and medical implants is recognized increasingly due to their biocompatibility, biodegradability, and low toxicity. Similar to the synthetic polymers, naturally derived polymers can be categorized into two groups based on their degradability in human tissues. The first group comprises of non-degradable or minimally degradable polymers, which includes glucans like cellulose, schizophyllan, dextran, and starch, as well as proteins such as pullulan, xylan, alginate, pectin, gum tragacanth, gum arabic, silk fibroin, and sericin. In contrast, the second group consists of polymers that are well-degradable in human tissues, including collagen and its derivative gelatin, elastin, keratin, and glycosaminoglycans such as hyaluronic acid, heparin, and chondroitin sulfate. This group also encompasses non-endogenous polymers like chitosan, gellan gum, zein, and poly(3-hydroxybutyrate-co-3-hydroxyvalerate) (PHBV). Notably, certain biodegradable polymers, including collagen, gelatin, elastin, keratin, and

[1] Department of Biotechnology, Faculty of Science, Janardan Rai Nagar Rajasthan Vidyapeeth (Deemed to be University), Udaipur-313001, Rajasthan, India.
[2] Department of Chemistry, Faculty of Science, Janardan Rai Nagar Rajasthan Vidyapeeth (Deemed to be University), Udaipur-313001, Rajasthan, India.
[3] Department of Biotechnology, Vigyan Bhanwan, Block-B, New Campus, Mohanlal Sukhadia University, Udaipur-313001, Rajasthan, India.
[4] Facultad de Ciencias, Universidad Santo Tomas, Puerto Mont, Chile.
[5] Sol-ARIS, Departamento de Química, Facultad de Ciencias, Universidad de Tarapacá, Avda. General Velásquez, 1775, Arica, Chile.
[6] Department of Chemistry, Faculty of Science, Bhupal Nobles' University, Udaipur-313001, Rajasthan, India.
* Corresponding author: narendrapalsingh14@gmail.com

DOI: 10.1201/9781003518792-1

glycosaminoglycans, contain specific cell-binding motifs, such as distinct amino acid sequences in proteins and oligosaccharide domains in glycosaminoglycans, which are recognized by cell adhesion receptors from both the integrin and non-integrin families (Bacakova and Svorcik, 2008; Bacakova et al., 2011).

Keratin, one of the most prevalent and underutilized non-food proteins, constitutes the primary structural component of hair, feathers, nails, and horns in mammals and birds. Despite its valuable properties, the disposal of keratin waste, presents significant challenges due to its high sulphur content (3–4% by weight),

Table 1. Application of keratin based material in tissue engineering.

S.No.	Source	Crosslink material	Application	Reference
1	Human hair	Polyvinyl alcohol	Tissue engineering	Aadil et al., 2023
2	Human hair	Poly(ethylene oxide)	Tissue engineering	Râpă et al., 2020
3	Wool fibre	Poly (ethylene oxide)	-	Sanchez Ramirez et al., 2021
4	Cattel hooves	Poly(ϵ-caprolactone) and polyethylene oxide	-	Su et al., 2020
5	Pure keratin (5–15%)	Poly (3-hydroxybutyrate)	Bone tissue engineering	Naderi et al., 2020
6	Iranian Afshari sheep Wool	Polyvinyl alcohol and poly (Ɛ-caproactone)	Skin regeneration	Mohammadi et al., 2022
7	Wool fibre	Polycaprolactone	Skin regeneration	Cruz-Maya et al., 2019
8	Keratin from animal byproduct	3-Glycidyloxypropyl) trimethoxy silane	Skin regeneration	Fortunato et al., 2019
9	Keratose (oxidative keratin from human hair)	Poly vinyl alcohol (PVA)	Neural Tissue engineering	Guo, 2018
10	Wool fiber	Poly (3-hydroxybutyric acid-co-3-hydroxyvaleric acid)	Wound healing	Ye et al., 2020
11	Human hair	Hydrophilic poly (acrylic acid), vascular endothelial growth factor (VEGF), Honey	Wound healing	Tavakoli et al., 2023
12	Pure keratin	Polyhydroxybutyrate and alumina	Bone tissue engineering	Ghafari et al., 2023
13	Feathers	Poly-ε-caprolactone + Bacillus velezensis P45	Antioxidant activity	Clerici et al., 2024
14	Pure keratin	Mupicorin, Pluronic, pectin	Angiogenic activity and wound healing	Mirhaj et al., 2023
15	Merino wool yarn	-	-	Tissera et al., 2024

which renders combustion inefficiency and also is environmentally harmful. Globally, the annual production of keratin, which includes fibrous by-products from the wool textile industry, substandard raw wools from livestock, and by-products from the butchery sector, is estimated to exceed 5 million tonnes (Barone et al., 2005). While ground horn and nails are utilized as nitrogen-rich fertilizers in gardening, and more recently have been subjected to caustic hydrolysis to create biodegradable surfactants for fire extinguisher foams, a substantial portion of keratin waste—particularly from unusable wool and poultry feathers—remains undervalued and is often discarded (Martínez-Hernández et al., 2007). A more effective approach to harness this extensive protein biomass could involve pooling and processing these keratin wastes into biopolymers. Keratin-based materials demonstrate considerable potential in biotechnological and biomedical applications, particularly in tissue engineering and the fabrication of affinity membranes, owing to their biocompatibility, capacity to promote fibroblast proliferation, and ability to absorb heavy metal ions and volatile organic compounds (VOCs) (Table 1).

This chapter primarily examines the diverse sources of keratin, highlighting its prevalence in biological materials such as hair, feathers, nails, and horns from various mammals and birds. It will delve into the methodologies employed for the extraction and processing of keratin to produce nanofiber composites, including techniques such as electrospinning. These methods facilitate the transformation of keratin into nanoscale structures, which enhances its mechanical properties and surface area, making it suitable for various applications. Furthermore, the chapter will explore the potential applications of keratin-based nanofiber composites in tissue engineering, emphasizing their biocompatibility, biodegradability, and capacity to support cellular attachment and proliferation.

2. About Keratin

Keratin is uniquely characterized by its significant proportion of cysteine residues, which constitute 7–20% of the total amino acid composition. These cysteine residues are crucial as they form both inter- and intra-molecular disulfide bonds, resulting in a compact three-dimensional structure that imparts enhanced stability to the protein (Dowling et al., 1986). However, keratin-based materials typically exhibit poor mechanical properties due to their relatively low molecular weight, which ranges from 9 to 60 kDa.

Keratin is classified into two main categories: epithelial keratins and trichocyte keratins. Epithelial keratins, also referred to as soft keratins, possess a low sulfur content (approximately 1%) and are primarily located in the stratum corneum of the skin. They typically contain less than 3% cysteine and play a vital role in stabilizing epithelial cells. In contrast, trichocyte keratins, known as hard keratins, are found in structures associated with the epidermis, including wool, hair, feathers, nails, horns, and hooves of mammals, reptiles, and birds (Figure 1). These keratins contain a higher cysteine content (ranging from 4% to 17%) and serve as structural scaffolding. The oxidation of cysteine residues leads to the formation of disulphide bonds, which stabilizes and crosslinks keratin filaments, contributing to the protein's overall stability. Additionally, keratins are abundant in other amino acids, such as

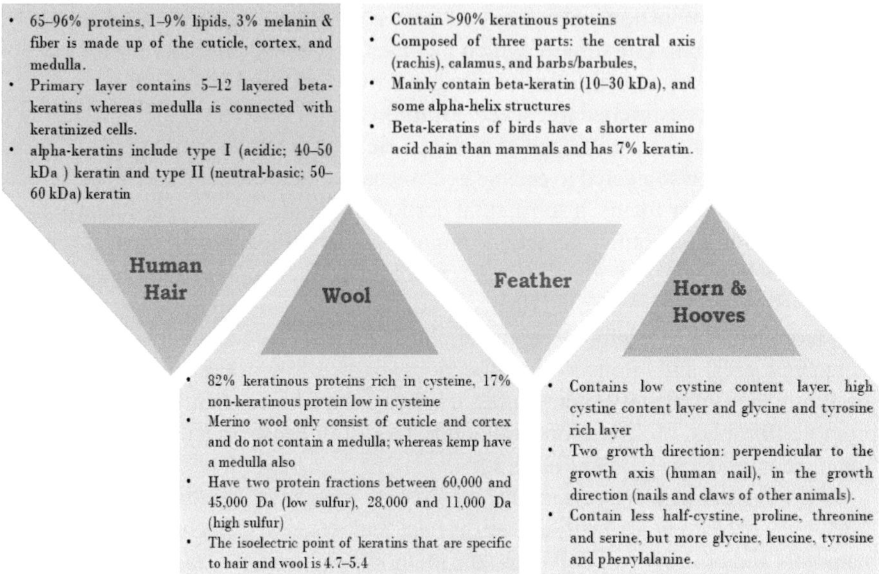

Figure 1. Different Sources of Keratin.

glutamic acid, aspartic acid, threonine, leucine, proline, serine, and glycine, further influencing their structural properties and functionalities (Rajabi et al., 2020).

Keratin-based materials exhibit suboptimal mechanical properties primarily due to their relatively low molecular weight. Additionally, unlike many synthetic polymers, keratin is not thermoplastic, limiting its processability under standard conditions. To enhance the mechanical performance of keratin for applications such as electrospinning, it is essential to blend it with compatible polymers. This blending process typically employs volatile and easy-to-handle solvents, which facilitates the uniform dispersion of keratin within the polymer matrix. Such an approach is particularly advantageous when aiming to produce nanofibers with improved structural integrity and functionality (Aluigi et al., 2008; Zoccola et al., 2008). By preparing keratin-based nanofibers through this method, researchers can leverage the biocompatibility and unique properties of keratin while overcoming its intrinsic mechanical limitations, making these materials suitable for various biomedical and biotechnological applications.

3. Method for Development of Keratin-based Nanofibers

A range of techniques have been developed to produce small-diameter fibers for high-volume manufacturing, including methods such as fibrillation, the island-in-sea technique and an innovative melt-blowing system. Additionally, highly precise methods like nanolithography and self-assembly are available (Tonin et al., 2010). However, these approaches are often limited by factors such as narrow material compatibility, high costs, and low production rates. In contrast, electrospinning is a cost-effective and straightforward process that offers a moderate production rate. It is classified as a physical phenomenon within the realm of electro-fluid dynamics and operates based on the electrostatic repulsion of charges, governed by Coulomb's law. Utilizing an electrostatic field, an electrospinning apparatus generates electrically driven polymer jets that subsequently solidify into sub-micron-scale polymer filaments, commonly referred to as nanofibers.

3.1 Electrospinning

Electrospinning can operate with both polymer melts and solutions, although there is significantly more literature on electrospinning polymer solutions compared to the less-explored melt-electrospinning. A standard electrospinning apparatus consists of three primary components:

1. **High Voltage Supply:** Provides the electrical energy necessary for the process.
2. **Capillary (Jet Source):** Acts as the conduit for the polymer solution, delivering it to the electrospinning region.
3. **Metal Collector (Target):** Receives the electrospun nanofibers, often grounded to facilitate charge accumulation.

In typical configurations, the solution is charged by the generator while the collector remains grounded, though it is possible to reverse this setup by charging the collector instead (Kilic et al., 2008).

3.2 Working Mechanism

The electrospinning process can be outlined as follows:

- A pump pushes the polymer solution through the capillary at a constant flow rate using a syringe.
- The high voltage generator, connected via a wire to the capillary's metal tip, applies voltage to the solution.
- A continuous jet of the solution is ejected from the capillary tip towards the grounded collector.
- The production rate of a single jet typically remains below 10 g h^{-1} of nanofibers, influenced by factors such as polymer concentration and flow rate.
- Upon activation of the voltage generator, the electrical potential at the droplet surface at the capillary tip increases to a level sufficient to transform the droplet from a spherical shape into a stable conical structure, known as the Taylor cone.
- Electrostatic forces counteract surface tension, with Coulomb repulsion promoting the formation of the cone shape. Once the electrostatic forces surpass the surface tension, a charged fluid jet is ejected from the cone's apex.

- The jet initially follows a straight trajectory, with a diameter significantly smaller than that of the capillary. As it travels towards the collector, the diameter progressively decreases due to:
  a) Electrostatic repulsion, which extends the jet.
  b) Solvent evaporation, which solidifies the polymer content, enhancing the evaporation rate as the diameter shrinks.
- As the solvent evaporates, changes in the concentration and viscoelastic properties of the jet occur.
- Following the straight section, the jet experiences bending instability, leading to a disordered trajectory characterized by oscillations that eventually evolve into a spiral motion, resembling whipping.
- Ultimately, the jet solidifies into nanofibers that are deposited onto the collector, forming a non-woven structure.

Despite the apparent simplicity of electrospinning, the onset of bending instability remains a complex phenomenon, with various theories and mechanisms proposed to elucidate it. The fibre formation is influenced by multiple factors, as presented below. Given these factors, the electrospun jet may undergo various phenomena, including branching, splitting, buckling, jet breakup (spraying), and the formation of flat ribbon or beaded nanofibers, leading to diverse morphologies (Reneker and Yarin, 2008).

- **Operating Parameters:** Applied voltage, distance between the tip and collector, flow rate of the solution, and tip diameter.
- **Polymer Solution Properties:** Molecular weight, concentration, relaxation time, viscosity, electrical conductivity, surface tension, and solvent vapor pressure.
- **Environmental Conditions:** Temperature, humidity, and ventilation.

4. Future Directions and Opportunities

Selection and design of the biomaterial should consider not only the structural and physicochemical properties of the constituents but also the desired bioactivity, biodegradability, and mechanical strength (Joyce et al., 2021). The latter properties are established by a combined use of traditional material science and the latest advancements in nanofiber, composite, and porous materials. Often, this ability to direct cell-dependent growth and function and extracellular matrix formation for tissue engineering applications has relied on the controlled delivery of active biomolecules, interaction with supporting cells, or simulated extrinsic mechanical cues in concert with the support framework offered by biomaterials, such as chitin, silk, collagen, and gelatin (Hutmacher et al., 2001; Sheikh et al., 2015). Another important functional features, surely essential for successful tissue regeneration *in vivo*, are biodegradation and removal of biomaterials. As such, one would expect that the engineering of biodegradable nanofibers and porous materials with optional biocomposites or supported substrates can provide flexibility not easily offered by traditional techniques and materials. No doubt, not all these features would be required within a tissue engineering construct, and even challenges exist for merging such features into one composite, yet benefits gained from such efforts are evident (Khajavi et al., 2016; Vasita and Katti, 2006). A couple of future directions and opportunities are listed below:

A. Enhancing Mechanical Properties

Tissue engineering is an active interdisciplinary research area that aims at repairing or replacing damaged or diseased tissues through the use of cell or tissue transplantation with biomedical engineering. Wound healing is an important part of the process of tissue repair and consists of several stages including inflammation, cell migration, and new tissue deposition. Actually, this is a very complex sequence of events involving quite a lot of cell types, as well as several growth factors and cytokines (Sharma et al., 2021). However, extracellular matrix accumulation can become excessive and provoke hypertrophic scars and keloids, a few raised, thickened areas of fibrous tissue deposited at the wound site, essentially composed of collagen I and III (Cen et al., 2008; Kijeńska et al., 2012). That's the reason for an urgent need of a wound dressing which not only provides functionalities that promote wound healing such as cell migration and blood coagulation but which also can be used to prevent the formation of hypertrophic scarring and even treat established hypertrophic scars. This report will cover the demand for advanced wound dressings and then review recent studies based on keratin nanofiber composites that have displayed excellent hemostatic performance and provide great potential to speed up healing without the scarring or hypertrophy characteristic of many currently used dressings (Ji et al., 2022; Serag et al., 2022). Lastly, the future outlook in the area of keratin nanofiber-based wound dressing, focusing mainly on the development of effective, biocompatible, and advantageous wound dressing for tissue engineering, will be provided.

B. Biocompatibility and Immunogenicity

Biocompatibility means it is the degree to which a material does not evoke an unacceptable therapeutic response. After treatment and interaction with body fluids or tissues, the artificial materials accept the prosthetic devices. The foreign body reaction of inflammation, and the coordinated host response is an unavoidable sequence of events round an implantation of artificial materials in animals, and damaged tissues are made of non-living material (Helmus et al., 2008). Depending on the severity and the duration of the inflammatory response, the degree of fibrosis, and the tissue loss, a poor therapeutic outcome could entail disability of the organs and death of the organ. Especially, the development and introduction of promising biocompatible materials can rescue the disease from deterioration by the progressing malformations or even repair tissue injuries. Among them, keratin-chitosan composite nanofibers are a biocompatible fibrous protein, which shows low immunogenicity because once applied to the wound site, it incites no or only a mild local immune response (Ishii et al., 2009; Capanema et al., 2019; Adi et al., 2020; Raut et al., 2023).

C. Innovations and Research Directions

In the last few years, the field of keratin nanofibers-based tissue engineering has highly advanced. Recently, by fabricating the nanofibers with diverse morphologies such as strong hydrogels with dynamic covalent crosslinking and three-dimensional nanofibers substrates that play a role in the control of cell behaviour, this area of research has experienced notable developments (Varesano et al., 2014; Yang et al., 2023). Today, along with microelectrode arrays, the potential applications of keratin nanofibers, in general, are mainly specialized in the field of skin wound healing. Keratin nanofibers still face numerous challenges in most applications for tissue repair, for example, weak mechanical properties, difficult controllability of hydrogel microstructure, and single modalities in signal transduction pathways (Bacakova et al., 2019). Nowadays, with the incorporation of new techniques for manufacturing, new materials, and even new technologies, the entry barriers of tissue engineering development remain weak for keratin nanofibers. This has established the proof that functional keratin nanofibers with varied properties could be produced in the future based on specific particular applications and varying precursor characteristics, such as structure, biological activity, and new biodegradable properties (Edwards et al., 2015). If the biological activity and mechanical properties of the keratin nanofibers are helping to aim for new applications, then a large number of functional parts are individually prepared and assembled to form complex tissue structures. This may mimic the cellular distribution in tissues of complex organs and closely revitalize their functions. Importing various material substances into keratin nanofibers and guiding their specific location, inherent connectivity, as well as the prepared 3D structure simultaneously within the hydrogel construction process may be one of the main approaches for this purpose (Choi et al., 2021; Serag et al., 2022).

5. Conclusion

Keratin is a natural biopolymer, which forms the cells' structure in the human body, such as skin, hair, and nails. Keratin nanofibers possess an excellent biocompatibility and immunogenicity and may be a good candidate for being applied to use in tissue engineering materials or drug delivery systems. This review introduces advanced development of nanofibers from keratin for application as tissue engineering scaffold material. Since keratins are a composition of the human body, they have better biocompatibility and immunological properties. Hence, it is industrially used as a raw cosmetic material; there are numerous studies on the same, ranging from medical applications to solubilizations or regenerations as nanofibers. For many years, keratin scaffold nanofibers have also been studied and applied in tissue engineering for the regeneration of human skin, blood vessel intima, and liver tissue. Nanofibers have received a tremendous amount of attention in recent years as matrix materials for the emerging discipline of tissue engineering. A multitude of techniques have been developed to produce nanofibers. In this respect, tissue reconstruction scaffold development is highly important and must be safe for implantation inside the body. Various composites have been developed based on the high mechanical strength, high water content, and the long-considered biological cellular response brought by the keratin fibrous structure. Among the many classes of composite materials, it has been shown that inorganic nanomaterials and organic natural materials can be used synergistically to enhance the mechanical properties of the constructed composite while offering novel functionality. Therefore, these keratin nanofiber composites possess inherent merits for use in tissue engineering and other potential biological applications.

References

Aadil, Keshaw, Nathani, Akash, Rajendran, Archana, Sharma, Chandra, Lenka, Nibedita and Gupta, Pratima. 2023. Investigation of human hair keratin based nanofiberous scaffold for skin tissue engineering application. 10.21203/rs.3.rs-2567435/v1.

Adi, Y. W., S. Ekatiwi and S. Suharti. 2020, April. Preliminary study on preparing carboxymethyl cellulose–keratin biofilm. In IOP Conference Series: Earth and Environmental Science (Vol. 475, No. 1, p. 012074). IOP Publishing.

Aluigi, A., C. Vineis, A. Varesano, G. Mazzuchetti, F. Ferrero and C. Tonin. 2008. Structure and properties of keratin/PEO blend nanofibres. Eur. Polym. J. 44: 2465–2475.

Bacakova, L. and V. Svorcik. 2008. Cell colonization control by physical and chemical modification of materials. pp. 5–56. *In*: Kimura, D. (ed.). Cell Growth Processes: New Research. Hauppauge, New York, USA: Nova Science Publishers, Inc. ISBN 978-1-60456-123-6.

Bacakova, L., E. Filova, M. Parizek, T. Ruml and V. Svorcik. 2011. Modulation of cell adhesion, proliferation and differentiation on materials designed for body implants. Biotechnol. Adv. 29(6): 739–767. DOI: 10.1016/j.biotechadv.2011.06.004.

Bacakova, L., J. Pajorova, M. Zikmundova, E. Filova, P. Mikes, V. Jencova et al. 2019. Nanofibrous scaffolds for skin tissue engineering and wound healing based on nature-derived polymers. Current and Future Aspects of Nanomedicine 1–30.

Barone, J. R., W. F. Schmidt and C. F. Liebner. 2005. Thermally processed keratin films. J. Appl. Polym. Sci. 97: 1644–1651.

Capanema, N. S., I. C. Carvalho, A. A. Mansur, S. M. Carvalho, A. P. Lage and H. S. Mansur. 2019. Hybrid hydrogel composed of carboxymethylcellulose–silver nanoparticles–doxorubicin for anticancer and antibacterial therapies against melanoma skin cancer cells. ACS Applied Nano Materials 2(11): 7393–7408.

Cen, L., W. E. I. Liu, L. E. I. Cui, W. Zhang and Y. Cao. 2008. Collagen tissue engineering: development of novel biomaterials and applications. Pediatric Research 63(5): 492–496.

Choi, W. S., J. H. Kim, C. B. Ahn, J. H. Lee, Y. J. Kim, K. H. Son et al. 2021. Development of a multi-layer skin substitute using human hair keratinic extract-based hybrid 3D printing. Polymers 13(16): 2584.

Clerici, N. J., A. A. Vencato, R. Helm Júnior, D. J. Daroit and A. Brandelli. 2024. Electrospun Poly-ε-Caprolactone nanofibers incorporating keratin hydrolysates as innovative antioxidant scaffolds. Pharmaceuticals 17(8): 1016. https://doi.org/10.3390/ph17081016.

Cruz-Maya, I., V. Guarino, A. Almaguer-Flores, M. A. Alvarez-Perez, A. Varesano and C. Vineis. 2019. Highly polydisperse keratin rich nanofibers: Scaffold design and *in vitro* characterization. J. Biomed. Mater. Res. Part A 107(8): 1803–1813. https://doi.org/10.1002/jbm.a.36699.

Dowling, L. M., W. G. Crewther and D. A. D. Parry. 1986. Secondary structure of component 8c-1 of α-keratin. Biochem. J. 236: 705–712.

Edwards, A., D. Jarvis, T. Hopkins, S. Pixley and N. Bhattarai. 2015. Poly (ε-caprolactone)/keratin-based composite nanofibers for biomedical applications. Journal of Biomedical Materials Research Part B: Applied Biomaterials 103(1): 21–30.

Fortunato, G. M., F. Da Ros, S. Bisconti, A. De Acutis, F. Biagini, A. Lapomarda et al. 2019. Electrospun structures made of a hydrolyzed keratin-based biomaterial for development of *in vitro* tissue models. Frontiers in Bioengineering and Biotechnology 7: 174.

Ghafari, F., S. Karbasi and M. B. Eslaminejad. 2023. Investigating of physical, mechanical, and biological properties of polyhydroxybutyrate-keratin/alumina electrospun scaffold utilized in bone tissue engineering. Materials Chemistry and Physics 297: 127340.

Guo, T., X. Yang, J. Deng, L. Zhu, B. Wang and S. Hao. 2018. Keratin nanoparticles-coating electrospun PVA nanofibers for potential neural tissue applications. J. Mater. Sci.: Mater. Med. 30(1). doi:10.1007/s10856-018-6207-5.

Helmus, M. N., D. F. Gibbons and D. Cebon. 2008. Biocompatibility: meeting a key functional requirement of next-generation medical devices. Toxicologic Pathology 36(1): 70–80.

Hutmacher, D. W., J. C. H. Goh and S. H. Teoh. 2001. An introduction to biodegradable materials for tissue engineering applications. Annals-academy of Medicine Singapore 30(2): 183–191.

Ishii, D., T. H. Ying, T. Yamaoka and T. Iwata. 2009. Characterization and biocompatibility of biopolyester nanofibers. Materials 2(4): 1520–1546.

Ji, Y., W. Song, L. Xu, D. G. Yu and S. W. Annie Bligh. 2022. A review on electrospun poly (amino acid) nanofibers and their applications of hemostasis and wound healing. Biomolecules 12(6): 794.

Joyce, K., G. T. Fabra, Y. Bozkurt and A. Pandit. 2021. Bioactive potential of natural biomaterials: Identification, retention and assessment of biological properties. Signal Transduction and Targeted Therapy 6(1): 122.

Khajavi, R., M. Abbasipour and A. Bahador. 2016. Electrospun biodegradable nanofibers scaffolds for bone tissue engineering. Journal of Applied Polymer Science 133(3).

Kijeńska, E., M. P. Prabhakaran, W. Swieszkowski, K. J. Kurzydlowski and S. Ramakrishna. 2012. Electrospun bio-composite P (LLA-CL)/collagen I/collagen III scaffolds for nerve tissue engineering. Journal of Biomedical Materials Research Part B: Applied Biomaterials 100(4): 1093–1102.

Kilic, A., F. Oruc and A. Demir. 2008. Effects of polarity on electrospinning process. Text. Res. J. 78: 532–539.

Liu, Xinkuan, Xu, Haixia, Zhang, Mingxin and Deng-Guang Yu. 2021. Electrospun medicated nanofibers for wound healing: review. Membranes 11(10): 770. DOI:10.3390/membranes11100770.

Martínez-Hernández, A. L., C. Velasco-Santos, M. De-Icaz and V. M. Castaño. 2007. Dynamical–mechanical and thermal analysis of polymeric composites reinforced with keratin biofibers from chicken feathers. Compos. Part B: Eng. 38: 405–410.

Mirhaj, M., J. Varshosaz, S. Labbaf, R. Emadi, A. M. Seifalian, F. Sharifianjazi et al. 2023. Mupirocin loaded core-shell pluronic-pectin-keratin nanofibers improve human keratinocytes behavior, angiogenic activity and wound healing. International Journal of Biological Macromolecules 253: 126700.

Mohammadi, R. M., Z. Arab-Bafrani, F. Karimi and N. Javid. 2022. Designing hybrid nanofibers based on keratin-poly (vinyl alcohol) and poly (Ɛ-caprolactone) for application as wound dressing. J. Ind. Text. 51(1_suppl): 1729S–1949S. doi:10.1177/1528083721988978.

Naderi, P., M. Zarei, S. Karbasi and H. Salehi. 2020. Evaluation of the effects of keratin on physical, mechanical and biological properties of poly (3-hydroxybutyrate) electrospun scaffold: Potential application in bone tissue engineering. Eur. Polym. J. 124: 109502. https://doi.org/10.1016/j.eurpolymj.2020.109502.

Rajabi, M., A. Ali, M. McConnell and J. Cabral. 2020. Keratinous materials: Structures and functions in biomedical applications. Mater. Sci. Eng. C 110: 110612.

Râpă, M., C. Gaidău, L. M. Stefan, E. Matei, M. Niculescu, M. D. Berechet et al. 2020. New nanofibers based on protein by-products with bioactive potential for tissue engineering. Materials 13(14): 3149. https://doi.org/10.3390/ma13143149.

Raut, M. P., E. Asare, S. M. D. Syed Mohamed, E. N. Amadi and I. Roy. 2023. Bacterial cellulose-based blends and composites: versatile biomaterials for tissue engineering applications. International Journal of Molecular Sciences 24(2): 986.

Reneker, D. H. and A. L. Yarin. 2008. Electrospinning jets and polymer nanofibers. Polymer 49: 2387–2425.

Sanchez Ramirez, D. O., I. Cruz-Maya, C. Vineis, V. Guarino, C. Tonetti and A. Varesano. 2021. Wool Keratin-based nanofibres—*in vitro* validation. Bioengineering 8: 224. https://doi.org/10.3390/bioengineering8120224.

Serag, E., A. M. A. El-Aziz, A. El-Maghraby and N. A. Taha. 2022. Electrospun non-wovens potential wound dressing material based on polyacrylonitrile/chicken feathers keratin nanofiber. Scientific Reports 12(1): 15460.

Sharma, S., P. Sudhakara, J. Singh, R. A. Ilyas, M. R. M. Asyraf and M. R. Razman. 2021. Critical review of biodegradable and bioactive polymer composites for bone tissue engineering and drug delivery applications. Polymers 13(16): 2623.

Sheikh, Z., S. Najeeb, Z. Khurshid, V. Verma, H. Rashid and M. Glogauer. 2015. Biodegradable materials for bone repair and tissue engineering applications. Materials 8(9): 5744–5794.

Su, S., T. Bedir, C. Kalkandelen, A. O. Başar, H. T. Şaşmazel, C. B. Ustundag et al. 2021. Coaxial and emulsion electrospinning of extracted hyaluronic acid and keratin based nanofibers for wound healing applications. European Polymer Journal 142: 110158.

Tavakoli, M., M. Mirhaj, J. Varshosaz, M. H. Al-Musawi, Y. Q. Almajidi, A. M. Danesh Pajooh et al. 2023. Keratin-and VEGF-incorporated honey-based sponge–nanofiber dressing: an ideal construct for wound healing. ACS Applied Materials & Interfaces 15(48): 55276–55286.

Tissera, N. D., R. N. Wijesena, N. Ludowyke, G. Priyadarshana, D. Dahanayake, R. M. de Silva et al. 2024. Keratin protein nanofibers from merino wool yarn: a top-down approach for the disintegration of hierarchical wool architecture to extract α-keratin protein nanofibers. RSC Adv. 14(10): 6793–6804. https://doi.org/10.1039/d3ra07063h.

Tonin, C., A. Aluigi, A. Varesano and C. Vineis. 2010. Keratin-based nanofibres. *In*: Nanofibers: 139–158.

Varesano, A., C. Vineis, C. Tonetti, D. O. S. Ramírez and G. Mazzuchetti. 2014. Chemical and physical modifications of electrospun keratin nanofibers induced by heating treatments. Journal of Applied Polymer Science 131(15).

Vasita, R. and D. S. Katti. 2006. Nanofibers and their applications in tissue engineering. International Journal of Nanomedicine 1(1): 15–30.

Yang, J., T. Lei, X. Yang, J. Fan, H. Wu, S. He et al. 2023. Investigation of keratin/poly (ethylene oxide) nanofiber membrane prepared with different post-crosslinking method. Fibers and Polymers 24(2): 715–727.

Ye, J. P., J. S. Gong, C. Su, Y. G. Liu, M. Jiang, H. Pan et al. 2020. Fabrication and characterization of high molecular keratin based nanofibrous membranes for wound healing. Colloids and Surfaces B: Biointerfaces 194: 111158.

Zoccola, M., A. Aluigi, C. Vineis, F. Ferrero, M. Piacentino and C. Tonin. 2008. Study on cast membranes and electrospun nanofibers made from keratin/fibroin blends. Biomacromolecules 9: 2819–2825.

CHAPTER 2

Cellulose Composites for Tissue Engineering

Fatemeh Koeini,[1,2] *Mitra Gharivi,*[2,3] *Fatemeh Rafati,*[2,4] *Mehri Nazeri,*[5,6] *Jila Majidi,*[2,7] *Narendra Pal Singh Chauhan,*[8,*] *Sanaz Alizadeh*[9,*] and *Mazaher Gholipourmalekabadi*[2,5,10,*]

1. Introduction

Traditional tissue degeneration therapies primarily rely on tissue transplantation, encompassing autografts, allografts, and xenografts. However, drawbacks such as donor site complications, the risk of pathogen transfer, and the potential for significant immune rejection hampered these methods (Mirtaghavi et al., 2020). Tissue engineering, a rapidly evolving field, aims to regenerate and replace damaged tissues by utilizing temporary scaffolds. These scaffolds serve as a substrate for

[1] Biomedical Engineering Department, Amirkabir University of Technology (Tehran Polytechnic), Tehran, Iran.

[2] Cellular and Molecular Research Center, Iran University of Medical Sciences, Tehran, Iran.

[3] Department of Materials Science and Engineering, Sharif University of Technology, Tehran, Iran.

[4] Department of Materials engineering and metallurgy, Amirkabir University of Technology (Tehran Polytechnic), Tehran, Iran.

[5] Department of Medical Biotechnology, Faculty of Allied Medicine, Iran University of Medical Sciences, Tehran, Iran.

[6] Research Center for Molecular Medicine, Hamadan University of Medical Sciences, Hamadan, Iran.

[7] Department of Tissue Engineering & Regenerative Medicine, Faculty of Advanced Technologies in Medicine, Tehran, Iran.

[8] Department of Chemistry, Faculty of Science, Bhupal Nobles' University, Udaipur-313002, Rajasthan, India.

[9] Burn Research Center, Iran University of Medical Science, Tehran, Iran.

[10] NanoBiotechnology and Regenerative Medicine Innovation Group, Noavarn Salamat ZHINO (PHC), Tehran, Iran.

* Corresponding authors: narendrapalsingh14@gmail.com, s.alizadeh@rsct.ir; alizadehsannaz@gmail.com, mazaher.gholipour@iums.ac.ir; mazaher.gholipour@gmail.com

DOI: 10.1201/9781003518792-2

cell attachment and guide three-dimensional tissue formation (Rahmati et al., 2021). Typically, cells are cultured on biodegradable scaffolds designed to mimic the hierarchical, nano-to-macro scale architecture of the ECM (Aziz et al., 2022). The cells should receive all the signals from the scaffolds for cell and tissue differentiation, cell proliferation, cell adhesion, cell migration as well as tissue regeneration and repair. These scaffolds must provide the necessary cues for cell and tissue differentiation, proliferation, adhesion, migration, and ultimately, tissue regeneration and repair. Crucially, scaffolds must exhibit biocompatibility to avoid immune rejection, be sterilizable for safe implantation, be biodegradable to allow for eventual resorption, and be bioactive to stimulate appropriate tissue responses (Eldeebaziz et al., 2022). With the rapid advancements in science and technology, so many kinds of biomaterials of both synthetic and natural origin have become prominent subjects of study for their potential biomedical applications, such as tissue-engineered scaffolds (Choi et al., 2022). Among these, polysaccharides, long-chain biopolymeric carbohydrates composed of monosaccharide units, have emerged as promising candidates for advanced tissue engineering and regenerative medicine due to their advantageous properties over synthetic polymers (Sunasee et al., 2016). Notably, cellulose and chitin, two abundantly available natural polysaccharides, are being intensively researched for their applications in tissue engineering in the form of fibrous mats, hydrogels, films, and fillers (Seddiqi et al., 2021).

Cellulose is a bio-based, hydrophilic, renewable, and environmentally friendly polysaccharide that holds significant promise for meeting the growing demand for sustainable and biocompatible materials. This polymer is a linear syndiotactic homopolymer made up of D-anhydro glucopyranose units that are linked by β-(1→4)-glycosidic bonds (Popa et al., 2022). As the most abundant organic substance and polysaccharide on Earth, cellulose is widely available in various biomass sources, including trees, tunicates, and bacteria. Its non-toxic nature, thermal and chemical stability, biodegradability, and white color in its pure form, combined with a molecular mass ranging from 1.44×10^6 to 1.8×10^6 g/mol, make it an attractive material for diverse applications (Gopi et al., 2019).

Previous studies have confirmed that cellulose has the potential to serve as a feasible biomaterial for both *in vivo* and *in vitro* applications because of its mechanical strength, flexibility, durability, low density, damping performance, and the ability to provide hierarchical three-dimensional support (Hasanin, 2022). Furthermore, the tunable morphological dimensions and functional groups (-OH, -CHO, -COOH, and $-SO_3H$) of cellulose have contributed to its increasing popularity in recent research (Bangar and Whiteside, 2021). The presence of hydroxyl groups and their capacity to form hydrogen bonds significantly influence the crystalline structure and other physical properties of cellulose. These strong intra- and inter-chain hydrogen bonds contribute to the formation of tightly packed cellulose molecular chains (Mali and Sherje, 2022).

Cellulose exhibits a diverse range of morphological forms, including fiber, microfibrils/nanofibrils, and micro/nanocrystalline cellulose, each offering distinct physical properties, dimensions, and shapes depending on the source and processing methods employed (Khalid et al., 2023; Bangar and Whiteside, 2021). These variations arise from inherent differences in organic material sources, as well as specific biosynthesis and processing conditions that dictate their geometry and dimensions (Chen et al., 2022). As building blocks, these diverse forms of cellulose yield materials with unique microstructures and properties, rendering them suitable for a wide array of biomedical applications. To further enhance its properties, cellulose can be converted into various derivatives, broadly categorized as cellulose esters and cellulose ethers, each possessing distinct characteristics and applications. Cellulose esters, in particular, are well-suited for biomedical applications due to their stability, high water permeability, and biocompatibility (Edgar et al., 2001; Liu et al., 2021). Cellulose ethers are also ideal for biomedical applications due to their hydrophilicity, mechanical strength, and stability against oxidative, thermal, and biological degradation (Hasanin, 2022).

The aforementioned attributes of cellulose materials have rendered them outstanding candidates for reinforcing polymeric structures such as poly-lactic acid (PLA), polycaprolactone (PCL), polyhydroxybutyrate (PHB), alginate, gelatin, collagen, silk fibroin, and chitosan, targeting scaffold fabrication (Cakmak et al., 2020; Burger et al., 2020; Mohammadalipour et al., 2022; Ghorbani et al., 2020; Dutta

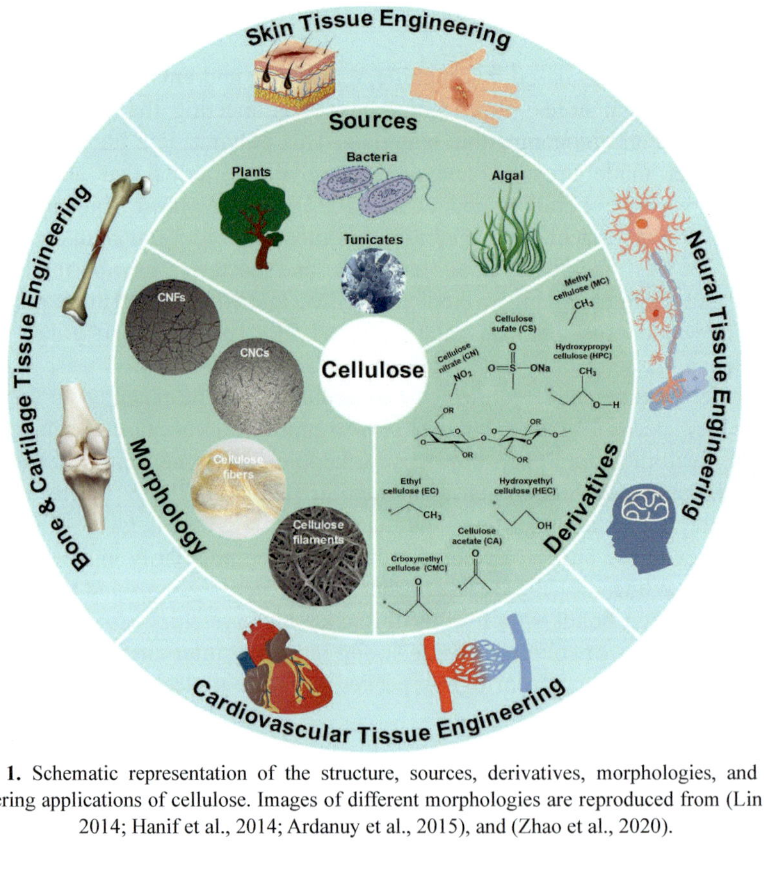

Figure 1. Schematic representation of the structure, sources, derivatives, morphologies, and tissue engineering applications of cellulose. Images of different morphologies are reproduced from (Lin et al., 2014; Hanif et al., 2014; Ardanuy et al., 2015), and (Zhao et al., 2020).

et al., 2021; Lohrasbi et al., 2020; Chen et al., 2020). Moreover, numerous studies have investigated the application of cellulose matrices combined with organic and inorganic nanomaterials, including ceramic nanoparticles, carbon-based nanosheets, metallic nanoparticles, and their oxides, for the artificial promotion of new tissues (Ashraf et al., 2020; Umar Aslam Khan et al., 2021; Saxena et al., 2021). Overall, cellulose and its composites have garnered substantial research interest due to their unique properties, which confer high effectiveness in various biological applications (Pita-Vilar et al., 2023; Yuan et al., 2021). This chapter explores the various sources, types, and morphological features of cellulose, along with the advantages and impacts of creating composites from these materials on the final properties of tissue-engineered constructs tailored to specific application cases. Finally, we summarized recent developments of cellulose-based composites in tissue engineering, with particular emphasis on their applications in wound dressing, bone tissue engineering, nerve tissue repair, and cardiovascular treatments (Figure 1).

2. Cellulose Sources

Cellulose is the most abundant naturally occurring biopolymer found in plants (e.g., cotton, hemp, and wood), marine animals (e.g., tunicates), algae (e.g., *Valonia*), and bacteria (e.g., *Gluconacetobacter xylinus*) (Lv et al., 2023; Cheng et al., 2020). Nevertheless, plant fiber is the primary source of cellulose, which serves as a structural component within the complex architecture of plant cell walls. While cellulose from different sources has the same chemical composition, its crystalline structure varies significantly, leading to diverse physical and chemical properties. Given cellulose's abundant availability, renewability, sustainability, and biocompatibility, there is growing interest in exploring various sources and techniques to harness this valuable material. The following sections will review the different sources of cellulose and discuss their potential in more detail.

2.1 Wood and Plant Cellulose

Plants are by far the primary and most numerous sources of cellulose, and in the present day, a well-developed framework is available for the cultivation, re-expression and extraction of cellulose from the plant sources (Zinge and Kandasubramanian, 2020). Today, the distinct components of different plants are used for cellulose harvesting, extraction, and processing. Cellulose is the main structural component of the primary cell wall of most plants, which includes stems, fruits, seeds, leaves and other elements (De France et al., 2021). Plants such as cotton (Dominic et al., 2020), jute (Correia and Valera, 2019), hemp (Panthapulakkal and Sain, 2013), bamboo (Liew et al., 2015), sisal, flax, kenaf, ramie, etc. are fundamental sources of cellulose, while industrial and agricultural residues such as wheat straw (Siró and Plackett, 2010), rice husk (Dominic et al., 2020), lotus leaf stem (Chen et al., n.d.), sunflower stem (Fortunati et al., 2016), corn husk (Akinjokun et al., 2021), soybean stock (Siró and Plackett, 2010), orange bagasse (Mishra et al., 2018), plum seed husk (Frone et al., 2017), and sugarcane bagasse (Mandal and Chakrabarty, 2011), are among the secondary sources of cellulose.

Table 1. Cellulose, hemicellulose, and lignin content as the main constituents in the chemical makeup of plant fibers and wood (Madsen and Gamstedt, 2013; Liitiä et al., 2003; Thirumala et al., 2013).

Source	Type	Cellulose (wt.%)	Hemicellulose (wt.%)	Lignin (wt.%)
Plant fibers (except cotton)		30–75	10–35	0–20
Wood fibers	Softwood	45–50	18–35	23–35
	Hardwood	40–50	24–40	23–35

Wood and plant-derived cellulose, despite having a similar chemical structure to other cellulose sources, possess a complex and multi-level structure due to the presence of lignin, a complex organic polymer that is essential to the growth of plant cell walls (Table 1) (Moraïs et al., 2010). Together with hemicellulose and cellulose, lignin forms a strong, complex structure from which cellulose is extracted and pulped before being extensively cleaned with caustic chemicals (Kargarzadeh et al., 2018). Lignin's non-biodegradability under physiological conditions, coupled with the harsh chemicals required to remove lignin and hemicellulose, may pose potential risks in biomedical applications. The presence of residual chemicals and their byproducts from the extraction process may have detrimental impacts on the body, reducing the potential of this type of cellulose for biomedical applications (Kalia et al., 2009).

2.2 Bacterial Cellulose

Although plant-based cellulosic materials are considered as the most common sources of cellulose, other living species including some types of bacteria can also be employed to produce cellulose (Gregory et al., 2021). Certain types of bacteria, including Gram-negative bacteria species such as *Achromobacter*, *Rhizobium*, *Agrobacterium*, Acetobacter, Pseudomonas, Salmonella, *Enterobacter*, *Komagataeibacter, Coronobacter, Gluconobacter,* and *Gluconacetobacter*, as well as Gram-positive bacterial species such as *Sarcina ventriculi*, Synthesise a type of extracellular polysaccharide called bacterial cellulose (BC). They do this as a protection mechanism against ultraviolet radiation, unfavorable pH conditions, and harsh chemical environments (de Amorim et al., 2020). However, among the mentioned genera, the Gram-negative aerobic bacterium *Gluconacetobacter xylinus* (formerly known as *Acetobacter xylinus*) stands out as the main producer of BC. This strain of bacteria can effectively metabolize a variety of carbon and nitrogen sources and produces BC fibers with a diameter less than 100 nm through oxidative fermentation in both synthetic and non-synthetic media (Jung et al., 2007). There are two types of cultures used in BC preparation today: static and agitated/dynamic. A higher oxygen concentration promotes the formation of a dense BC membrane at the liquid-air interface in the static state. This setting encourages superior aerobic growth (Kuo et al., 2016). Conversely, when oxygenated air is added immediately to the liquid growth media in the agitated culture method, cellulose takes the form of irregular pellets or suspended microfibers (Wang et al., 2019). Choosing between static and agitated culture depends on how the BC will be used. The physical, morphological, and mechanical characteristics of the resulting BC vary depending on the cultivation method. For example, cellulose obtained through stirred culture has a lower mechanical strength than cellulose obtained through static culture. Stirred culture also typically results in lower BC production levels and a higher chance of microorganism mutation, which can have a negative impact on BC production (Moniri et al., 2017). In contrast, static culture necessitates a bigger area for cultivation and a longer period for cultivation (Tyagi, 2016).

Apart from the particular growing techniques, a number of other significant variables affect the yield and structure of bacterial cellulose in addition to its mechanical and physical properties. Regardless of whether static or agitated fermentation processes are used, one of these crucial variables is the makeup of the culture medium. At least one carbon source, one nitrogen source, and a few key nutrients including phosphorus, potassium, sulfur, and magnesium are usually present in a normal fermentation medium (Manan et al., 2022). The commonly employed synthetic complex culture medium for BC synthesis is the Hestrin–Schramm medium, which consists of 2.0 wt.% glucose as a carbon source, peptone, and 0.5 wt.% yeast extract as nitrogen sources. Furthermore, the medium may incorporate other substances, such as 0.5% peptone, 0.27% sodium phosphate, and 0.115% citric acid, alongside its fundamental constituents. The pH of the medium is a significant parameter that is usually regulated to around 6.0 through the use of hydrochloric acid or sodium hydroxide (Of et al., 1954). Adjusting the cultivation medium can significantly impact the bacterial growth as well as the characteristics of the resulting cellulose. Researchers can enhance the process of BC production by methodically adjusting parameters such as temperature, pH, nutrition content, and other additives (Sperotto et al., 2021).

Although the glucose in the growth medium for bacteria cultivation is the preferred carbon source for BC synthesis, other sugars, such as xylose, lactose, galactose, mannose, arabinose, fructose, and sucrose present in different agricultural and food waste materials, and industrial byproducts can also be used (Jozala et al., 2015). The biosynthesis of BC from glucose is a multi-step, intricate process. Many specialized enzymes and regulatory proteins are used in this process to carefully regulate the synthesis. However, the precise specifics pertaining to BC's supramolecular structure are still unknown. There are four phases in the enzymatic process used to produce BC. To make glucose-6-phosphate (Glc-6-P), glucose must first be phosphorylated by glucokinase. This glucose-6-phosphate is then isomerized to glucose-1-phosphate (Glc-1-P) by phosphoglucomutase. Uridine diphosphate glucose (UDP-Glc) is the result of Glc-1-P being converted by the enzyme UDP-glucose pyrophosphorylase. In the end, cellulose synthase transforms UDP-glucose into cellulose (Buldum and Mantalaris, 2021).

BC is frequently recognized for its extraordinary purity. Microbially generated cellulose is devoid of lignin, pectin, hemicelluloses, and other biogenic compounds, in contrast to plant cellulose (Heinze, 2015). The purification process for plant cellulose involves intensive mechanical and chemical separation steps, resulting in a significant consumption of energy and making it ecologically detrimental. In contrast, BC merely necessitates simple purification following fermentation, such

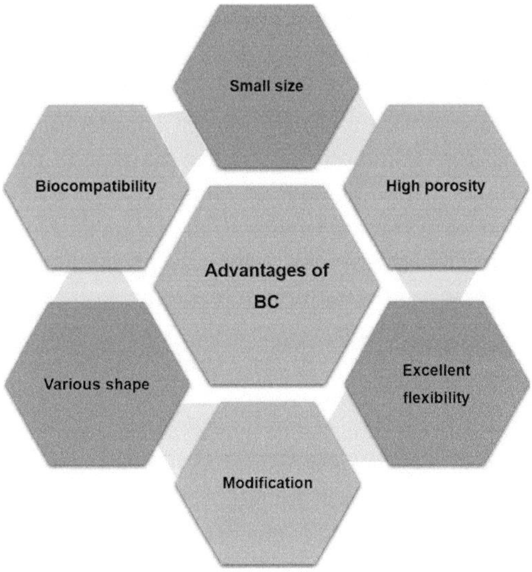

Figure 2. Schematic diagram of the advantages of BC for regenerative medicine applications (Choi and Shin, 2020; Moniri et al., 2017).

as the use of alkali, organic acids, or repeated washing with reverse osmosis water to remove contaminants such as cells and medium components (Andriani et al., 2020). A porous structure, high mechanical strength, a high degree of polymerization between 2000 and 6000, a high crystallinity index (above 60%), and a larger capacity to absorb and hold water are also provided by the three-dimensional arrangement of BC microfibrils (Mohite and Patil, 2014; Aziz et al., 2022). BC has become a viable biomaterial for a range of applications due to its unique features, including its nonporous structure, chemical insensitivity, transparency, and capacity to create dynamic fibers (Figure 2) (El-Gendi et al., 2023). BC acquires new characteristics when it is composited or altered with polymers, reinforcing agents, and active chemicals. Then, its use is increased in a number of industries, including the food, pharmaceutical, and medical domains (Fernandes et al., 2020).

2.3 Algal Cellulose

The demand for cellulose and its derivatives in the market is continuously increasing, with an expected annual production of 180 billion tons. Lignocellulose biomass sources, such as wood, cotton, flax, hemp, and jute, play a vital role in the cellulose industry (Birgersson et al., 2023). However, the removal of lignin from lignocellulose biomass presents a significant challenge. This process often requires strong chemical treatments, which can result in the partial degradation of cellulose and the release of harmful effluents that pose environmental risks. As a result, quickly expanding and widely accessible biomass, like marine algae, also known as macroalgae or seaweed, is drawing interest as a substitute supply of cellulose on a global scale (Bogolitsyn et al., 2020). Seaweeds are complex organisms that serve as the foundation of most aquatic food chains. They consume light, nutrients, and dissolved carbon dioxide from the water (Machado et al., 2024). Various types of algae such as red (*Rhodophyta*), green (*Chlorophyta*), and brown (*Phaeophyceae*) are known for cellulose production; however, green algae are the most favored for cellulose extraction (Johnsy George and Sabapathi, 2015). Native cellulose, along with other polysaccharides like agar, alginates, galactans, carrageenans, laminarans, fucoidans, and ulvans, is present in the cell walls of red, green, and brown algae. It is composed of elementary fibrils measuring between 10 and 30 nm and exhibits a remarkably high level of crystallinity (Zanchetta et al., 2021; Moreira et al., 2023; Hua et al., 2013). In general, algal macromolecules are mostly composed of different monosaccharides joined by glucosidic linkages; some also include linear backbones with repeating disaccharide units (Wang et al., 2017). Seaweed-derived cellulose has the potential to be used in several biomedical applications, such as polymer nanocomposites, drug delivery systems, wound dressings, tissue engineering (Yu et al., 2023), and controlling the inflammatory response in monocytes and macrophages (Liu et al., 2024; Birgersson et al., 2023).

2.4 Tunicate Cellulose

The importance of marine resources has increased recently (Zhao and Li, 2014). Among these are aquatic invertebrate animals called tunicates, which are members of the class Ascidiacea and subphylum Tunicata. They are capable of manufacturing cellulose. These animals' epidermal tissues contain enzyme complexes that catalyze the creation of cellulose (Cheng et al., 2020). The cellulose generated by tunicates is noteworthy for its environmental sustainability and its ability to sequester carbon on the ocean floor. With over 3000 species discovered too far, tunicates have a great deal of diversity (Lv et al., 2023).

Cellulose produced by tunicates interacts with proteins to form their protective outer layer, also known as the tunic (Lv et al., 2023). Within this structure, cellulose nanofibrils are organized in a multilayered arrangement, oriented parallel to the surface of the tunic. Notably, the morphology and dimensions of these nanofibrils exhibit interspecies variation. The highly reactive surface of tunicate cellulose renders it an exceptional candidate for cross-linking with other polymers, thereby expanding its potential for various biological applications (Lv et al., 2024). However, the inherent high crystallinity of this cellulose presents a challenge, as it impedes dissolution in water, resulting in only swelling. This limitation significantly restricts its practical applications. Consequently, extracted tunicate cellulose is typically processed into nanocellulose to enhance its utility (Lv et al., 2023). Tunicate nanocellulose not only exhibits exceptional characteristics such as high tensile strength, purity, and Young's modulus but also possesses noteworthy optical, electrical, magneto-mechanical, and rheological properties (Lv et al., 2024). Moreover, the abundance of hydroxyl groups on the surface of nanocellulose offers significant potential for further functional modification (Xu et al., 2024).

Tunicate cellulose has garnered significant attention across various sectors, including food packaging, biomedicine, paper production, functional materials, and environmental applications. This interest stems from its remarkable physical and chemical properties, notably its high aspect ratio and elastic modulus. Furthermore, the versatility of tunicate cellulose allows for its conversion into nanocellulose or its modification for integration with other polymeric materials. This facilitates the fabrication of diverse structures, such as gels, films, and nano-complexes. Recent advancements have even explored the potential of tunicate cellulose-based nano-fiber bio-ink in the burgeoning field of 3D bioprinting for organ and tissue regeneration (Sreedharan et al., 2024; Šturcová et al., 2005).

3. Types of Cellulose

Cellulose's backbone mostly consists of hydroxyl groups, allowing the substance to form hydrogen bonds both internally and between molecules. As a result, this leads to the creation of a metastable, semicrystalline structure that consists of crystalline and amorphous regions (Ko et al., 2018). The presence of diverse molecular orientations and hydrogen bonding networks in the crystalline area leads to the formation of different cellulose polymorphs (Kovalenko, 2010). Cellulose exists in four distinct polymorphs, namely cellulose I, cellulose II, cellulose III, and cellulose IV. Although the patterns and molecular form may seem similar, the arrangement of the chains within these groups varies significantly (French and Johnson, 2007). Cellulose I also known as native cellulose, is packed in a parallel chain arrangement without inter-sheet hydrogen bonding (Mahmud et al., 2019). When cellulose I is subjected to a mercerization process, its bundled fibers break down into smaller microfibrils. When these microfibrils are then reassembled, it results in the formation of a more thermodynamically stable structure called cellulose II, or regenerated cellulose (Wada et al., 2004). Cellulose III is regained by exposing cellulose I or cellulose II to liquid ammonia or organic amine treatment, with subsequent removal of the swelling agents ("Cellulose", 2008). Cellulose IV can be prepared from both cellulose II and III by heat treatment (Isogai et al., 1989).

Cellulose can be modified into different cellulose derivatives by destroying the ordered crystallization region to enhance its properties, which has created new possibilities and expanded the applications of cellulosic. Cellulose derivatives are classified into cellulose ester and cellulose ether according to the specific chemical reaction performed on the free hydroxyl groups in the anhydro glucose units. Derivatives may differ in their basic properties, such as chemical structure, hygroscopicity, water interaction, surface activity, and solubility (Oprea and Voicu, 2020a). Cellulose can be esterified by reaction with organic acids, anhydrides, or acid chlorides (Edgar, 2004). Stability, high water permeability, high strength, and low tissue non-toxicity make cellulose esters well-suited to the needs of biomedical applications (Edgar, 2007). On the other hand, cellulose ethers are commonly obtained by treating purified cellulose with alkylating reagents under heterogeneous conditions, usually in the presence of a base, typically sodium hydroxide, and an inert diluent (Burchard, 2003). This class of cellulose benefits from hydrophilicity, mechanical toughness, and good stability against oxidative, thermal, and biological degradation (Shuhui et al., n.d.). The etherification of cellulose produces methyl cellulose (MC), ethyl cellulose (EC), carboxymethyl cellulose (CMC), hydroxyethyl cellulose (HEC), and hydroxypropyl cellulose (HPC). whereas for the synthesis

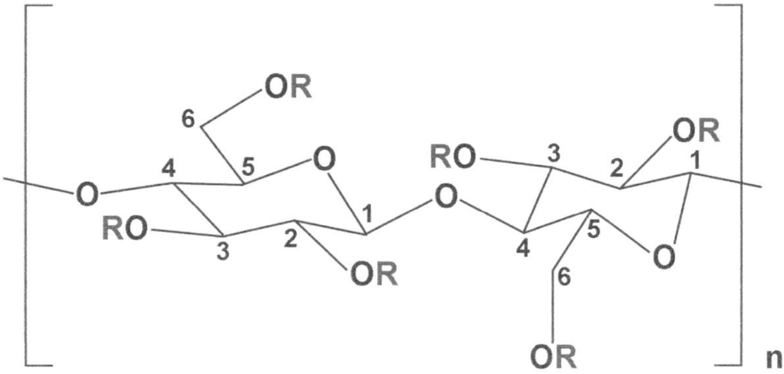

Figure 3. The overall chemical structure of a cellulose derivative can differ based on the type of R group present, leading to potential mono-, di-, or tri-substitution in the derivative.

Table 2. Cellulose derivatives and corresponding R groups.

Cellulose derivative	R groups
MC	H replaced by CH_3
EC	H replaced by CH_2CH_3
CMC	H replaced by CH_2COOH
HEC	H replaced by CH_2CH_2OH
HPC	H replaced by $CH_2CH(CH_3)$ OH
CA	H replaced by C_2H_3O
CN	H replaced by NO_2
CS	H replaced by SO_3H

of cellulose acetate (CA), cellulose nitrate (CN), and cellulose sulfate (CS), the esterification process needs to be done. The introduction of new cellulose ester and ether groups gives them to be used for a wide range of biomedical applications. This section will explore the properties and applications of cellulose derivatives (Figure 3, Table 2).

3.1 Methyl Cellulose (MC)

MC is a prominent cellulose ether that is produced by the reaction between alkali cellulose and methyl chloride (Oprea and Voicu, 2020a). This type of cellulose is widely used in commercial applications and is the most abundant biological polymer found in nature. In MC, ethyl groups ($-CH_3$) replace the hydroxyls at C-2, C-3 and/or C-6 positions of anhydro-D-glucose units. Depending on the degree of substitution, MC can dissolve in a wide range of organic solvents and has the potential to undergo thermogelling (Nasatto et al., 2015). MC is commonly used in various industrial applications because of its availability, reasonable price, and effortless processing capacity. The poor biodegradability behavior of MC limits its medical applications. Nevertheless, this problem can be resolved by employing different cross-linking methods, such as radiation or the use of chemical cross-linking agents like aldehydes or epoxy compounds (Rimdusit et al., 2008; Viera et al., 2007).

MC is frequently employed as an excipient in pharmaceutical formulations, serving as a coating agent or matrix for controlled drug delivery (Oprea and Voicu, 2020a). Furthermore, MC possesses the ability to enhance the flow performance of injectable pastes containing porous calcium phosphate glass microspheres by modifying their viscosity. However, MC-based pastes had strong characteristics, specifically at 20°C, but lost their properties when exposed to the higher body temperature of 37°C (Matamoros-Veloza et al., 2020). Research indicates that the physical characteristics and gelling temperature can be enhanced by blending polymers and adding various salts. The chitosan/methyl cellulose/Na_3PO_4 hydrogel demonstrated satisfactory cell viability and proliferation when used as a scaffold for chondrocytes. This indicates the potential to act as a 3D synthetic matrix for tissue engineering (Tang et al., 2010).

3.2 Ethyl Cellulose (EC)

EC is derived from cellulose by reacting alkali cellulose with ethyl chloride (Chen et al., 2022). In the EC structure, the hydroxyl (OH) end groups of the molecule are substituted with ethyl groups (C_2H_5) at carbon 2, carbon 3, or carbon 6 of the anhydrous glucose unit (Ahmadi et al., 2022a). EC is insoluble in water, although it can be dissolved in other organic solvents, including ethanol, methanol, toluene, chloroform, or ethyl acetate, depending on its level of substitution (Mahnaj et al., 2013; Oprea and Voicu, 2020b). EC is generally regarded as safe and included in the US Food and Drug Administration (FDA) Inactive Ingredients database (Wasilewska and Winnicka, 2019), and is commonly utilized as an encapsulating material for the controlled release of pharmaceuticals in tissue engineering applications (Mahnaj et al., 2013; Ahmadi et al., 2022b). In these applications, the release rate of drug in EC-coated formulations can be manipulated due to the low solubility of EC in water and its lower water uptake (Liu et al., 2010). This polymer is also recommended for reconstructing damaged tissue and coating medicinal devices because of its non-toxicity, stability, and biocompatibility (Chen et al., 2014; Mao et al., 2018). EC exhibits hydrophobic properties when compared to other cellulose derivatives, such as HPC, HEC, CMC, and MC. The hydrophobic nature of this material reduces its susceptibility to hydrolysis, making it advantageous for long-term uses, such as in bone implant scaffolds (Horvat et al., 2024) and biomimetic blood vessels (Aydogdu et al., 2019).

3.3 Carboxymethyl Cellulose

Among cellulose ethers, CMC is the first FDA approved cellulose derivative and has been extensively studied in biomedical field, ranging from tissue engineering to drug delivery (Camponeschi et al., 2015; Guo et al., 2015). CMC is a water-soluble derivative of cellulose that is produced by treating raw cellulose with a solution of sodium hydroxide, extracting the resulting alkali cellulose, and then reacting it with chloroacetic acid in an alcohol-based medium (Gelman, 1982). This modification involved substituting the hydroxyl group on glucopyranose chains of cellulose with carboxymethyl groups (-CH_2COOH), resulting in the formation of a polyelectrolyte capable of being responsive to changes in pH (Mallakpour et al., 2021). The negatively charged carboxymethyl groups present in CMC enable them to create a crosslinked matrix structure by complexing with materials that have an opposite charge. This property has the potential for use in tissue engineering (Gaihre and Jayasuriya, 2016). These groups also facilitate the ability of CMC to uptake water and create a hydrated environment analogous to the ECM (Singh et al., 2016). Notably, CMC is compatible with other polymeric compounds when used in composites. This property, along with its remarkable physical and chemical properties, makes CMC-based materials more efficient for the preparation of biomaterial scaffolds and drug-encapsulating nanoparticles (Zennifer et al., 2021).

In addition, CMC hybrid materials are primarily utilized in the biomedical and pharmaceutical areas, specifically in applications like skin tissue engineering (Ramli and Wong, 2011), pulp cells regeneration (Huangqin Chen and Mingwen Fan, 2007), biocompatible implants, vascular tissue engineering (Joy et al., 2018), and the diagnosis and treatment of different diseases (Fang et al., 2009). This is mostly due to its simple and cost-effective synthesis process, availability of raw materials, specific surface properties, strong mechanical strength, and the ability to control hydrophilicity, viscosity, and rheological properties (Rahman et al., 2021). It has been reported that CMC can modulate the scaffolds' pore architecture and swelling ratio, tailoring their physical and swelling properties to support cartilage formation (Namkaew et al., 2021). This type of cellulose is an ideal option for bone tissue engineering applications because of its low cost, biodegradability, biocompatibility, non-toxicity, low immunogenicity, and its ability to promote osteogenic induction *in vitro* and *in vivo* (Priya et al., 2021).

3.4 Hydroxyethyl Cellulose (HEC)

HEC is one of the leading commercial non-ionic cellulose derivatives that is synthesized by reacting alkali cellulose with ethylene oxide (Alessandro Sannino et al., 2009). Cellulose's hydrogen atom is converted into a hydroxyethyl group by the etherification treatment, thereby rendering the polymer water-soluble. A large amount of hydroxyl units in HEC makes it available to be functionalized, as these units can be attached by several functional groups (Ding et al., 2018). In this regard, graft copolymerization is shown to be an effective tool to modify the properties of HEC through different reagents (Noreen et al., 2020). This polymer has commonly been developed into films, microspheres, hydrogels, nanofibers, and scaffolds in tissue engineering applications due to its excellent properties in terms of high capacity for water absorption and retention (Kong et al., 2016), proven mechanical strength (Zulkifli et al., 2019), biocompatibility (Hussein et al., 2018), biodegradability, and low cytotoxicity (Tohamy et al., 2018). Studies conducted *in vitro* confirmed that the HEC-based scaffold is a desirable platform for human osteosarcoma cell attachment and growth (Chahal et al., 2017). HEC hydrogel membranes have proven to be efficient in exhibiting anti-inflammatory and antibacterial abilities, as well as improving the safety of tungsten oxide towards normal human cells (El Fawal et al.,

2018). Furthermore, HEC, as an absorbable hemostatic gauze surgical can be applied for hemostasis and further prevention of breast cancer after surgical tumor resection (Yan et al., 2020).

3.5 Hydroxypropyl Cellulose (HPC)

HPC, a cellulose derivative approved by the FDA, is produced through the chemical modification of cellulose fibers using an etherifying agent in an etherification reaction (Bhatt et al., 2011). This process introduces side chains of hydroxypropyl groups to the cellulose polymer backbone, forming $OCH_2CH(OH)CH_3$ groups that can transform insoluble cellulose into water-soluble derivatives (El-Newehy et al., 2017). HPC has become popular in the fields of pharmacy and medicine because of its transparency, high water content, physiological inertness, and good film-forming ability (Takeuchi et al., 2018). As a result, this polymer has been frequently utilized as a precursor for hydrogels (Miao et al., 2016), ophthalmic inserts (Nguyen and Latkany, 2011), and as a disintegrating and adhesive agent for pharmaceutical tablets (Picker-Freyer and Dürig, 2007). For instance, the mucoadhesion feature of HCP ensures better oral residence and drug release at the absorption site and could be effective in the repair or recovery of damaged cell membranes in control drug-delivery intestinal formulations (Filip et al., 2022). Other possible applications of HPC include its role as a steric stabilizer in the production of nanocrystalline ceramic oxide powders, as well as functioning as a sieve matrix for the separation of DNA through capillary and microchip electrophoresis (Espinha et al., 2018; Abdel-Halim and Al-Deyab, 2011).

Recent findings have shown that the HCP's surface, which has a strong affinity for water, may effectively hinder bacterial attachment. This characteristic makes HCP a promising candidate for advanced wound dressings with exceptional antibacterial properties (Lin et al., 2020). Yet, HPC may undergo modifications by adding stearyl groups, leading to the formation of hydrophobic HPC esters that exhibit resistance to moist environments (Saraiva et al., 2021). On account of being composed of both hydrophobic and hydrophilic groups, HPC is capable of experiencing a phase change in response to shifts in the temperature of a water solution, allowing for an adjustment of the mechanical characteristics of HPC (Ma et al., 2014). This allows for the mechanical characteristics of HPC to be adjusted, making it suitable for tissue-engineered constructs and thermosensitive matrices (Filip et al., 2022). In the light of functional scaffolds, the HPC fibers embedded within the thermoresponsive hydrogel have the ability to attract ions in solution, thereby improving the conductivity of the hydrogel (Gan et al., 2021). Additionally, polyether-grafted HPC materials can endow traditional HPC with multi-functional properties such as antibacterial activity and pH responsiveness (Deng et al., 2020).

3.6 Cellulose Acetate (CA)

CA is obtained from the chemical reaction between cellulose with acetic acid and acetic anhydride in the presence of sulphuric acid and is recognized as a biodegradable polymer. The solubility of CA depends on the degree of substitution (DS). DS of

2.5 is commonly used in the manufacturing of textiles, plastics, films, and cigarette filters because of its appropriate solubility and molecular weight (Puls et al., 2011; Fischer et al., 2008). CA is a great option for producing osseointegration membranes because of its outstanding biocompatibility (Oprea and Voicu, 2020a). CA fibers have multiple uses, such as wound healing mats, controlled drug delivery systems, tissue engineering scaffolds, protein immobilization membranes, and biosensors. For example, CA and silver-based compounds were used to create antibacterial electrospun wound healing mats (Konwarh et al., 2013). A different study used a layer-by-layer technique to assemble CA nanofibers with negatively charged sodium alginate and positively charged chitosan to create scaffolding materials for tissue engineering applications (Luo et al., 2013).

3.7 Cellulose Nitrate (CN)

CN, also known as nitrocellulose, is a polysaccharide consisting of pyranose rings, which is obtained by replacing three free hydroxyl groups of cellulose with nitrate groups in each pyranose ring (Berthumeyrie et al., 2014). CN is used as a propellant in the creation of explosive formulations in addition to being used in the manufacturing of dye, paint, and photographic film. The presence of nitrogenous chemicals causes serious toxicity and environmental problems, which restricts their use in other domains (Oprea and Voicu, 2020a). Because of its good physio-chemical stability and biocompatibility, the CN with less than 12.6% nitrogen is appropriate for use in biomedicine. Strong non-covalent interactions between the polymer's nitro groups and the amine functional groups in the protein structure are thought to be responsible for CN membranes' capacity to retain proteins (Li and Frey, 2010; Oprea and Voicu, 2020a). These properties make them appropriate for application in hemodialysis and biosensors (Eduok et al., 2021; Diagnosis, 2020).

3.8 Cellulose Sulfate (CS)

CS is a semi-synthesized cellulose derivative that has a similar structure to polysaccharide sulfates. It consists of β-1,4-glucan as a main chain, with sulfate groups replacing hydroxyl groups in the anhydroglucose unit (Wang et al., 2007). Cellulose is used in food packing, drug delivery systems, anticoagulants, and contraceptives due to its commercialization capability and biocompatibility. Furthermore, cellulose is highly valuable in the field of medicine due to its antiviral properties against the human immunodeficiency virus (Chen et al., 2014; Zhang et al., 2015). CS offers several advantages over pure cellulose, such as its ability to dissolve in water and its strong antibacterial capabilities at high concentrations. The advantageous molecular binding characteristics offered by the sulfate ligands in the CS structure were utilized in the cell culture-based production of influenza vaccines (Wolff et al., 2010).

4. Different Forms of Cellulose Based on Morphology

Cellulosic materials can be classified into categories based on morphology: (I) cellulose nanofibrils, (II) cellulose crystals, (III) cellulose fibers, and (IV) cellulose filaments. These particles have the same chemical composition, but they differ in

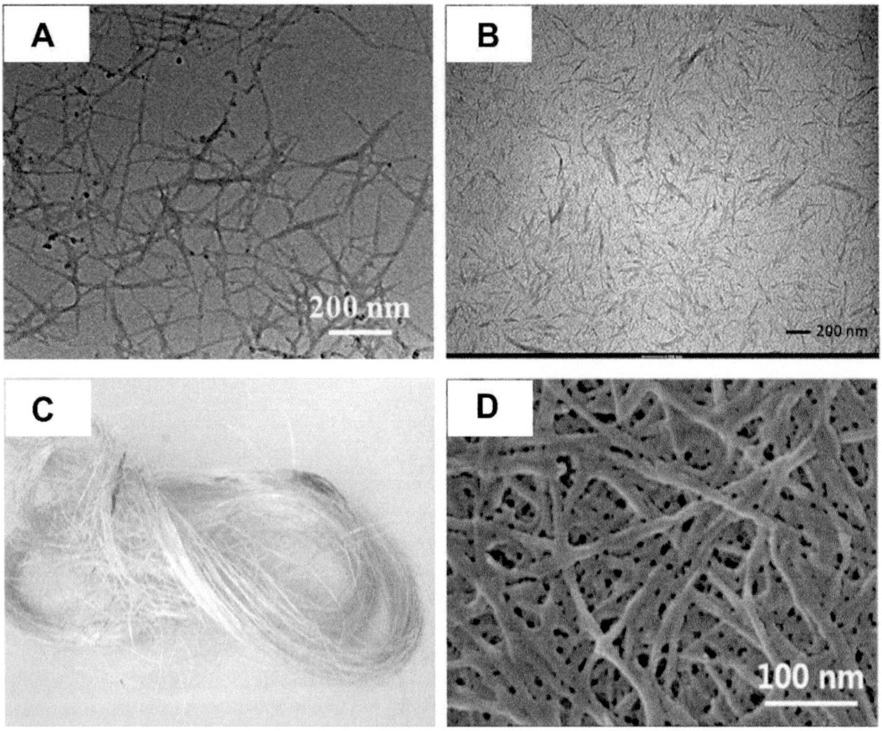

Figure 4. Different forms of cellulose based on morphology: (A) CNFs (Lin et al., 2014); (B) CNCs (Hanif et al., 2014); (C) cellulose fibers (Ardanuy et al., 2015); (D) cellulose filaments (Zhao et al., 2020).

terms of structure, particle size, purity, crystallinity, and other physicochemical characteristics (Varghese et al., 2023). In this section, each of these particles is discussed (Figure 4).

4.1 Cellulose Nanofibrils

Cellulose nanofibrils (CNF/NCF) are elongated and entangled fibers of nanocellulose, with diameters that vary from 1 to 100 nm and lengths that range from 100 nm to several micrometers (Dhali et al., 2021). The CNF consists of both amorphous and crystalline areas. These fibers exhibit a strong affinity for water, considerable flexibility with a high length-to-width ratio, and can be easily modified chemically (Xu et al., 2013). They are commonly produced by mechanical techniques such as ultrasonication, high-pressure homogenization, steam explosion, grinding, microfluidization, or cryocrushing. These techniques apply pressure to the fibers, leading to their breakdown from microfibers to nanofibers while preserving the amorphous and crystalline regions (Nechyporchuk et al., 2016). As the frequency of mechanical disintegration cycles increases, the size of the nanofibers decreases (Hanieh Kargarzadeh et al., 2017). One major drawback of mechanical disintegration is the significant energy consumption and the propensity for equipment to get obstructed (Dhali et al., 2021). Before performing the mechanical defibrillation procedure, different chemical approaches are used to make the extraction process easier and yield nanofibers (Hanieh Kargarzadeh et al., 2017). These methods include 2,2,6,6-tetramethylpiperidine-1-oxyl (TEMPO) mediated oxidation, organosolv treatments, partial carboxymethylation, and enzymatic hydrolysis (Pääkko et al., 2007). TEMPO-oxidized CNFs demonstrate excellent water dispersibility due to abundant carboxyl groups, along with enhanced morphological stability and crystallinity. The method is straightforward and cost-effective, making it a viable option for high-performance nanofibers (Gupta et al., 2019). Although TEMPO CNFs demonstrates a higher degree of fibrillation, their suitability varies and depends on the specific end-use application. A good example of a broad use is the improvement of the mechanical properties of paper and cardboard. In such situations, mechanical pretreatments demonstrate higher characteristics in comparison to TEMPO (Sanchez-Salvador et al., 2020). However, when it comes to optical applications, TEMPO produces a sample with a higher degree of fibrillation.

4.2 Crystalline Cellulose

Microcrystalline cellulose (MCC) and cellulose nanocrystal (CNC) are particles that are rigid and have a rod-like shape. They also have a highly organized crystalline structure. MCCs have a diameter that falls within the range of microns, whereas CNCs have a diameter that falls within the range of nanometers (Trache et al., 2017). Ranby created the MCC in 1949, and it was later released into the market under the brand name AvicelÂ®. Currently, MCC is manufactured by approximately 10 global sources, with wood cellulose being the predominant source (Thoorens et al., 2014). MCC, a modified form of partially degraded alpha-cellulose, is suitable for use as a filler in biocomposites due to its inherent biodegradability, heat resistance, and mechanical properties (Wardhono et al., 2020).

CNC particles are crystalline areas of cellulose microfibrils that are obtained by subjecting them to sulfuric acid hydrolysis (Al-Kaabi et al., 2021). These particles possess remarkable attributes such as a high aspect ratio, a large specific surface area, a high specific strength and modulus, biodegradability, reactive surfaces, and the ability to form stable suspensions in low-polarity liquids. The morphology of CNCs is influenced by the cellulose supply and the method used for extraction (Martínez-Sanz et al., 2011). They exhibit a wide range of lengths, often ranging from a few hundred nanometers to a few nanometers in width. CNCs, derived from several cellulose sources such as wood, plant, BC, algal, and tunicate, can be obtained by disintegrating MCC particles (Le Goff et al., 2015). Their elastic modulus ranges from 60 to 220 GPa, while their tensile strength ranges from 7500 to 7700 MPa (Wu et al., 2013). The utilization of CNC particle reinforcement provides advantages such as reduced weight, consistent distribution, long-lastingness, and the capacity to naturally break down over time. Due to its high strength and compact size, it is well-suited for biomedical uses such as medical implants, wound dressings, medication delivery, and tissue engineering scaffolds (Septevani et al., 2018).

4.3 Cellulose Fibers

Cellulose is formed as microfibrils at the site where wood or plants are produced. These microfibrils are then ordered to form cellulose fibers. Cellulosic fibers are commonly found in three different forms: strand fibers (long fibers measuring 20–100 cm in length), staple fibers (short fibers measuring less than 60 mm in length), and pulp fibers (very short fibers measuring 1–10 mm in length) (Ardanuy et al., 2015). A staple fiber consists of a single cell, whereas a strand fiber is made up of several cells. Cotton fibers are classified as staple fibers, measuring between 25 and 45 mm in length (Abbott et al., 2005). The length of cotton linters, which are the fibers that stay attached to the cotton seed, is only a few millimeters. Strand and staple fibers are derived from cultivated or wild plants immediately following a water retting procedure (Ardanuy et al., 2015). Pulp fibers are typically derived from individual cells found in wood and occasionally from plant sources such as sisal, banana, fique, and cotton linters. This extraction is achieved by a pulping process. Pulping is a thorough process including the use of solutions to extract lignin from the spaces between cells, resulting in individual plant cells that are only a few millimeters in length (Abbott et al., 2005).

4.4 Cellulose Filaments

Cellulose nanofibers can be fabricated into filamentary structures utilizing wet-extrusion (Walther et al., 2011), flow-focusing (Håkansson et al., 2014), or spinning (Shinichiro Iwamoto et al., 2011; Vuoriluoto et al., 2017) techniques. All cellulose filaments exhibit varying physicochemical characteristics, such as allomorph and crystallinity. However, they are uniformly mechanically robust, flexible, and possess extremely high aspect ratios. Filaments are intriguing due to their role as fundamental components of materials used in biological applications. Additionally, there exists an alternative form of cellulose filament that is not derived using the aforementioned methods but rather by extracting cellulose fibers from wood or plants (Tibolla et al., 2014). The peeling process is purely mechanical and does not include the use of chemicals or enzymes. Additionally, there is no discharge of waste, making the resulting filaments highly suitable for biomedical applications.

5. Classes of Cellulose-Based Composites

Biocomposite have recently gained significant attention and have been extensively studied due to their excellent properties, which often surpass those of single-polymer scaffolds. This is because biocomposite can address the limitations that may be inherent in individual polymers, making them more effective for successful tissue transplantation. Among the various types of biocomposite, cellulose-based composites have also attracted substantial interest. The structural characteristics of cellulosic materials have demonstrated their tremendous potential as both reinforcing agents and matrix components in the synthesis of diverse composite materials.

Cellulose-based composites have been synthesized to impart a wide range of desirable properties, including biological activities, physio-chemical characteristics, mechanical properties, conductivity, and magnetic capabilities, all of which have enabled their application in various biomedical fields (Moniri et al., 2017). The well-organized fibrous network structure of cellulose, coupled with its highly porous nature and high specific surface area, allows it to accommodate a broad spectrum of nanomaterials, such as metals (Ag (Raghavendra et al., 2013), Au (Arumugam et al., 2021), Mg (Zhang et al., 2020)), metal oxides (ZnO (Ul-Islam et al., 2014), TiO_2 (David et al., 2022), MgO (Shahzadi et al., 2022), CaO (Shahzadi et al., 2023)), mineral nanomaterials (SiO_2 (Angelova et al., 2019), $CaCO_3$ (Jia et al., 2012), montmorillonite (Ul-Islam et al., 2012), hydroxyapatite (Fu et al., 2018)) and carbonaceous nanomaterials (graphene oxide (Wei et al., 2022), reduced graphene oxide (rGO) (Esmaeili et al., 2020), carbon nanotubes (Ioniță et al., 2018)). Similarly, cellulose composites have been synthesized with natural polymers, including gelatin (Xing et al., 2010), collagen (Weichang Li et al., 2014), chitosan (Kim et al., 2011), as well as synthetic polymers like poly(3,4-ethylenedioxythiophene)–poly(styrene sulfonate) (PEDOT:PSS) (Khan et al., 2015), PLA (Rahman et al., 2014), and poly(vinyl alcohol) (PVA) (Millon and Wan, 2006), using the *in situ* composite synthetic strategy. The existing literature contains numerous publications that investigate these cellulose-based composites, and we provide a summary of a portion of the relevant studies.

6. Key Properties of Cellulose-Based Composites

6.1 Solubility

As discussed earlier, cellulose composite materials have attracted considerable attention because of their biocompatibility, mechanical properties, environmental adaptability, and lightweight nature. They have been extensively utilized in many biomedical applications. Undoubtedly, the solubility of compounds plays a crucial role in the design and manufacture of diverse bio compositions or formulations that consist of multiple components. This is because it can impact the miscibility or solubility between the solvent and polymer, or the components of the matrix in composite biomaterials.

Cellulose is known to be hardly dissolved in water and most organic solvents. Its hydrophilic nature with high inter- and intramolecular hydrogen bonds and Van der Waals forces within the integrated fibrils make it difficult for dissolution. With the strong interconnections between the intertwined chains of cellulose, it becomes insoluble in water and most commonly used solvents. Consequently, dissolving cellulose requires solvents with strong interactions as well as multi-step pretreatments (Burchard, 2003). The dissolution of cellulose has been investigated using a number of complex solvent solutions. Acetone/triethyloctylammonium chloride mixture and dimethyl sulfoxide (DMSO) and tetra-n-butyl ammonium fluoride ×3H and tetrabutylammonium fluoride × $3H_2O$/DMSO are typically used as conventional cellulose solvents. Yet, practical application of these compounds has been hindered by their limitations, including volatility, toxicity, high prices, and challenges in solvent recovery (Gericke et al., 2011).

Alternatively, ionic liquids, which are organic salts with low melting points and are considered environmentally friendly, can be used to dissolve cellulosic materials without compromising their molecular weight (Zhang et al., 2005). Due to their nonvolatile nature, low toxicity, high polarity, thermostability, high conductivity, recyclability, and solubility in organic solvents and water, these chemicals are beneficial for dissolving cellulose under homogeneous conditions. This is achieved by forming hydrogen bonds with hydroxyl oxygen and ether oxygen, which in turn increase the solubility of cellulose. An ionic liquid of N,N-dimethylacetamide/lithium chloride (DMAc/LiCl) (Potthast et al., 2002) is a classical and direct solvent system for cellulose and was found to be an effective solvent system. Moreover, a potent nonderivatizing ionic liquid solvent commonly employed for dissolving and regenerating cellulose is 1-allyl-methylimidazolium chloride (AMIMCl) (Stanton et al., 2018). Similarly, 1-butyl-3-methylimidazolium chloride (BMIMCl) dissolves cellulose by using $BMIM^+$ as an electron acceptor and Cl^- as an electron donor. With this association, the intermolecular hydrogel linkages break down, allowing the cellulose to dissolve in BMIMCl (Haq et al., 2019).

Studies have shown that imidazolium-based carboxylates and alkyl phosphate solvents, namely 1-ethyl-3-methylimidazolium alkyl phosphates [C2mim][(MeO)(R)PO_2], have a high degree of solubility for cellulose (Fukaya et al., 2008). The aqueous solution containing sodium hydroxide is an environmentally friendly solvent for cellulose that has the ability to break the intra-/intermolecular hydrogen bonds. In the case of NaOH/urea (Dormanns et al., 2016), cellulose creates a shell-like inclusion complex with alkali and urea by hydrogen bonding. In the NaOH/thiourea solvent solution, the intramolecular hydrogen link in cellulose is disrupted, allowing NaOH to readily bind with the cellulose chain and create a new hydrogen bond (Morgado et al., 2011). In conclusion, researchers are always seeking to produce new and enhanced cellulose solvents to meet the unique requirements for employing cellulose in biomedical applications. These novel solvent solutions must possess the appropriate characteristics to facilitate efficient processing and utilization of cellulose in the medical domain.

6.2 Mechanical Properties

An effective tissue-engineered scaffold requires a meticulously designed structure capable of resisting surgical operations and physiological stresses, where the scaffolds are used to contact culture medium *in vitro* or body fluid *in vivo* studies (Joshi et al., 2015). Scaffolds are designed to closely resemble the original tissue's mechanical properties, which are dependent on the specific anatomical site of application (Kumar et al., 2016). In this connection, the mechanical stability of biomaterials becomes important when designing an artificial scaffold.

The mechanical properties of cellulose make it an appealing component in many nanocomposites to be used for the regeneration of several types of tissues, including bone (Aravamudhan et al., 2013), skin (Esmaeili et al., 2020), cartilage (Nimeskern et al., 2013), blood vessels (Bäckdahl et al., 2006), etc. The outstanding mechanical properties of cellulose come from the structural hierarchy of its crystalline network. The cellulose chains are joined by hydrogen bonds during lateral growth, giving the material a rigid, rod-like form. The overall polymeric structure gains a great deal of strength and directional rigidity from this structural configuration (Azevedo et al., 2013). According to previous research, cellulose's mechanical properties can vary depending on a number of parameters, including the crystal structure, crystallinity percentage, anisotropy, and morphology (Zou et al., 2022). The elastic or Young's modulus of plant cellulose fibers typically falls between the 20–30 GPa range (Iwamoto et al., 2007), while this term is much higher (about 78 ± 17 GPa) for BC fibers with diameters ranging from 35 to 90 nm due to the higher crystallinity ratio (Guhados et al., 2005).

Reportedly, the conditions of processing applied for testing the polymer have an important effect on the mechanical characteristics of BC. For example, one of the earliest reports on the mechanical properties of air-dried BC films was done by Yamanaka et al. (Yamanaka et al., 1989), where they found that interfibrillar hydrogen bonds and a large contact area could result in Young's modulus and tensile strength being as high as > 15 GPa and 260 MPa, respectively. Later on, Nishi et al. (Nishi et al., 1990) observed significantly improved mechanical properties assigned to BC sheets of about 30 GPa as a result of the removal of impurities by alkaline and/or oxidative treatment. Wet BC demonstrates a tensile strength of 400 mPa, a maximum strain of around 20%, and a Young's modulus ranging from 130 to 145 GPa (Swingler et al., 2021). Other forms of cellulose exhibit superior mechanical behavior as well, have exceptional inherent specific modulus and strength as a result of their great crystallinity, organization, and intermolecular interactions (Yu et al., 2012). The Young's modulus of cellulose nanocrystals is close to 150 GPa, and their tensile strength is 10 GPa. Cellulose nanowhiskers (CNWs) have a modulus of 143 GPa because of their almost-perfect crystalline structure (Sturcova et al., 2005).

Considering the advantages mentioned above, cellulose is supposed to provide excellent mechanical properties through improved filler-matrix compatibility and the formation of filler networks if used as reinforcing components for manufacturing polymer nanocomposites (Hobzova et al., 2018; Nge et al., 2010). A number of researchers focused on the application of nanocellulose as reinforcing agents. For instance, crystalline cellulose of jute, when used as reinforcement in the PLA matrix, was found to enhance crystallinity, Vickers hardness, and yield strength more than that of neat PLA (Rahman et al., 2014). Strong interactions between BC and PVA resulted in the creation of an extremely durable nanocomposite material, according to Millon et al. (Millon and Wan, 2006). The outcome was an improvement in the mechanical characteristics and anisotropic behavior of BC/PVA. These composites have mechanical similarities to the porcine aorta and aortic valve, which suggests that they could be used to create prostheses that replace connective tissues in the cardiovascular system.

Elsewhere, in order to improve collagen's mechanical characteristics, rod-shaped CNCs were embedded into collagen films. The obtained composite film had a maximum tensile strength that was twice as high as the pure collagen film's at a CNC loading of 7 wt.% (Weichang Li et al., 2014). Similarly, the mechanical studies showed that the CNC incorporation into alginate/PCL nanofibers increased the tensile modulus by 65% (Rashtchian et al., 2020). It has been reported that

CNC uniformly dispersed in poly (oligoethylene glycol methacrylate) (POEGMA) hydrogel networks improved mechanical characteristics (up to 35-fold storage modulus) compared to POEGMA-only controls. POEGMA–CNC nanocomposite hydrogels are proposed as high-strength biodegradable scaffolds for stiffer tissues or cell growth platforms (De France et al., 2016). By temporarily attaching surrounding polymer chains and providing composite hydrogels with an effective load-carrying characteristic, CNCs can serve as multifunctional physical cross-linkers (Yang et al., 2014). Regarding this case, Zhang and colleagues integrated quaternized tunicate cellulose nanocrystals (Q-TCNCs) together with Fe^{3+} into a poly (acrylic acid-co-acrylamide) (PAAAM) nanocomposite network. In the nanocomposite hydrogels, Q-TCNCs functions as cross-linkers and compatible reinforcements at the interfaces, resulting in loose cross-linking; in contrast, compact cross-linking was achieved by ionic coordination between Fe^{3+} and -COO- of PAAAM. On the other hand, compact cross-linking was achieved by ionic coordination between Fe^{3+} and $-COO^{-}$ of PAAAM. The toughness of dual cross-linked hydrogel was 340-fold more greater than that of mono-cross-linked hydrogel, which was 10-fold greater than that of PAAAM hydrogel (Zhang et al., 2017).

Since CNWs tend to form strong networks through hydrogen-bonding and mechanical interlocking between whiskers, compounding them with both bioceramics and polymeric scaffolds is gaining a lot of attention and has proven to enhance the mechanical properties of different tissue-engineered matrixes, like 45S5 bioactive glass-based scaffold (Wei Li et al., 2014). This is because CNWs are biocompatible, have a low density, high specific strength, and modulus. Even at concentrations as low as 0.2 weight percent, it has been shown that the efficient dispersion of CNWs and their connectivity to form a robust percolating network within a cellulose acetate propionate scaffold may greatly improve the mechanical stability of the entire system (Pooyan et al., 2012).

Despite the fact that the cellulose-based materials are strong, robust, and bioaffinity-friendly, there are instances when they fall short when it comes to the mechanical qualities needed by particular tissues. As a result, inorganic material composites are becoming more popular for use in the development of functional and mechanical materials with exceptional performance. Introducing thermally stable ZnO can create a barrier that absorbs heat and prevents degradation of the BC skeleton at higher temperatures, reducing weight loss. Additionally, ZnO nanoparticles attached to cellulose microfibrils improve the toughness of cellulose sheets and limit the movement of polymer chains, leading to increased mechanical strength of the composites (Ul-Islam et al., 2014). Studies have demonstrated that the mechanical characteristics may be enhanced to some extent when the polymer matrix gets embedded with clay particles. For example, Ul-Islam et al. (Ul-Islam et al., 2012) *ex situ* impregnated montmorillonite particles inside BC sheets to create a biomaterial with improved physio-mechanical properties for various applications. The tensile strength of composites reached up to 210 MPa from 151.3 MPa (for pristine BC).

6.3 Biological Performance

6.3.1 Biocompatibility

When considering a material for use in biomedicine, the choice of material is among the most critical criteria to take into account. Biomaterials distinguished themselves by their ability to live and interact with specific tissues or physiological systems, including blood, interstitial fluids, immune cells, and chemicals, without causing excessive harm. Biocompatibility, in essence, refers to the ability of a biomedical material to perform with an appropriate host response when it is in contact with living tissue without provoking any toxic or allergic side-effects in biological systems (Torres et al., 2012). In general, a biocompatible implanted biomaterial has the ability to restore tissue with proper function and cellular phenotypic expression. It also prevents macrophage and foreign body giant cell responses, as well as oxidative stress. Additionally, it inhibits the formation of a fibrous capsule (scar) that can either surround the implant or infiltrate the porous material. Lastly, it inhibits immune responses, which could render the biomaterial ineffective (Naahidi et al., 2017). Consequently, it is necessary to assess a biomaterial in order to identify any biological reactions that may result in harm or undesirable consequences for the recipient. The ISO 10993-1-2011 standard states that in order to assess the biological activity of pharmaceuticals, both *in vitro* and *in vivo* investigations should be carried out. Determining the matrix's cytotoxic and adhesive properties is necessary for the *in vitro* evaluation of its suitability for use with cellular cultures. The incapacity of *in vitro* research to evaluate immune system activity and the body's overall reaction is one of its main drawbacks. For this reason, implantation testing on animals of various species is a standard component of preclinical biocompatibility testing (Ivanov et al., 2019).

Cellulose has been proven to be a biocompatible material that remains intact in the human digestive system due to the absence of the enzyme cellulase. Cellulose fiber possesses a substantial concentration of active hydroxyl groups on its surface, which aids in the attachment of cell adhesion proteins like fibronectin (Noiset et al., 1999). Numerous *in vitro* investigations assess the biocompatibility of cellulose-based scaffolds, focusing primarily on cell attachment and growth (Chiaoprakobkij et al., 2011; Zhijiang et al., 2012). For instance, bone-forming osteoblast cells and L929 fibroblasts proliferated normally on native BC (Xi et al., 2009). Additionally, Chinese Hamster Lung fibroblasts were found to adhere and proliferate on the surface of the Poly(3-hydroxubutyrate-co-4-hydroxubutyrate) (P(3HB-co-4HB))/BC scaffold in comparison to the pure P(3HB-co-4HB) scaffold. This can be attributed to the increased hydrophilicity of the composite scaffold, which facilitates protein adsorption in the presence of BC (Zhijiang et al., 2012).

To enhance positive interactions between cellulose implants and the surrounding tissues, researchers have attempted to impregnate various bioactive polymers and nanomaterials into the cellulose matrix to enhance its biological features for medical purposes, including tissue-engineered scaffolds (Khan et al., 2015; Chiulan et al., 2016; Xing et al., 2010). Xiang et al. (Xing et al., 2010) revealed that human mesenchymal stem cells (hMSCs) exhibited robust proliferation and displayed extensive F-actin and extracellular molecule networks on the gelatin-coated fiber surface, because gelatin makes the fiber surface suitable for cell culture and provides spacing for the cellulose fibers. Cai et al. (Cai and Kim, 2010) concluded that after incubating 3T3 fibroblast cells with BC/Poly(ethylene glycol) (PEG) scaffolds for 48 h, the cells presented stretched morphology on the surface, established pseudopodia, and showed increased proliferation. This indicates that the BC/PEG scaffolds have significantly superior biocompatibility compared to pure BC. Elsewhere, immersing the oxidized bacterial cellulose (OBC) in an acetic acid solution of chitosan and cross-linking it with 1-Ethyl-3-(3-dimethylaminopropyl)carbodiimide/N-Hydroxysuccinimide (EDC/NHS) led to an enhancement in the growth of L929 cells that were placed on the chitosan/OBC composite sponge compared to those on BC and OBC after 5 days of cultivation (Fu et al., 2021).

While the small pores (100–300 nm in diameter) of BC scaffolds may limit the infiltration of cells, Jing et al. (Jing et al., 2013) used a laser patterning technique to create uniform vertical pore arrays and introduced parallel-channel pores with a diameter of approximately 200–300 μm. They applied post-chemical modification to cover the surfaces of nanofibers with uniform coatings of gelatin and HAp. The study revealed that the composite scaffolds facilitated the proliferation and differentiation of chondrogenic rat cells on the scaffolds. Moreover, the introduction of bioactive glass (Cu-MBG) into CNF improved the angiogenesis capacity of the composite, as confirmed by upregulated angiogenic gene expression of 3T3 fibroblasts (e.g., Vegfa, Vegfc, Pdgf, and bFgf) in real-time quantitative polymerase chain reaction (qPCR) analysis. In addition to this, the incorporation of Cu-MBG into the CNF matrix significantly induced human umbilical vein endothelial cells (HUVECs) to sprout and promoted the fibroblast-endothelial cell interaction and ECM production, confirming the cytocompatibility and functionality of the scaffolds for use in biomedical applications (Wang et al., 2016). Therefore, materials derived from cellulose were found to be biocompatible *in vitro* when tested with various cell lines and stem cells; these materials can be utilized for cell growth *in vitro* and subsequent implantation *in vivo*.

In 1989, Miyamoto et al. (Miyamoto et al., 1989) conducted one of the first studies to use *in vivo* test methods to assess the tissue biocompatibility of cellulose and its derivatives. MC, EC, HEC, aminoethyl cellulose, and cellulosic polyion complexes were discovered to have relatively mild foreign body reactions. The amount of absorption by living tissue was discovered to be dependent on the sample's crystallinity and chemical composition. In a different study, Modulevsky et al. (Modulevsky et al., 2016) examined the biocompatibility of an apple-derived cellulose scaffold by subcutaneously implanting it in wild-type, immunocompetent mice for 1, 4, and 8 weeks. Despite observing a foreign body response 1 week post-surgery, all animal groups showed a gradual disappearance of the immune response by the end of 8 weeks. This result was concomitant with the migration of activated fibroblasts within the cellulose scaffold, along with the deposition of a new collagen ECM and the formation of blood vessels within the scaffold. Pértile et al. (Pértile et al., 2012) investigated the long-term fate of subcutaneously implanted breast cancer in mice, with implant durations of up to 1 year. The BC implants induced a modest and non-harmful inflammatory response, which gradually decreased over time and did not trigger a foreign body reaction. However, it was observed that the less porous BC implants tended to undergo calcification more frequently and resulted in stronger encapsulation. Similarly, Mendes et al. (Mendes et al., 2009) subcutaneously implanted BC membrane in mice and analyzed the tissue reaction at 7, 15, 30, 60, and 90 days post-implantation. During this time, they did not detect any signs of a foreign body reaction. A mild inflammatory response is suggested by the presence of polymorphonuclear cells and lymphocytes at 7-, 15-, and 30 days post-surgery. On the other hand, there was no sign of inflammatory infiltration 60 and 90 days after surgery. Based on these results, cellulose seems to be a perfect biomaterial for use in medicine because it is neither toxic nor immunogenic.

6.3.2 Degradability

The term "biodegradability" refers to a biomaterial's capacity to undergo breakdown by biological processes over a period of time. While some applications, such as biological implants, require stability, most applications, like tissue engineering scaffolds, are designed to progressively deteriorate until they are replaced by new tissue, are designed to progressively deteriorate until they are replaced by new tissue. When it comes to the concept of biodegradability, it is worth mentioning that cellulose-based materials are biodegradable utilizing enzyme-aided hydrolysis by cellulose-degrading microorganisms (Raut et al., 2023). These microorganisms possess the capability to produce cellulose-specific enzymes, known as cellulases, which can chemically hydrolyze the β-1,4-glucoside linkages of cellulose polymers, resulting in the release of glucose (Favi et al., 2016). However, cellulose cannot be easily biodegraded under the body's physiological conditions because the cellulase enzymes required for its degradation are not synthesized in human tissues (Sunasee et al., 2016). Hence, the limited ability of cellulose to biodegrade has emerged as a major restriction in its use as a scaffold material in tissue engineering.

To render cellulosic materials biodegradable and bioabsorbable, different approaches such as chemical and enzymatic modification have been proposed in recent years (Chen and Sun, 2000; Sannino et al., 2005; Hu and Catchmark, 2011). For example, cellulose-degrading enzymes (cellulases) and their combinations were incorporated into cellulose composites to induce degradation in pH conditions relevant to wound environments (Hu and Catchmark, 2011). In an experiment performed by Zhu et al. (Novotna et al., 2013) they used laser perforation and sulphonation treatment to obtain a double-modified BC. This modification has improved the mechanical characteristics of the BC scaffold while also greatly enhancing its biocompatibility and degradability. Furthermore, reports suggest that oxidizing chemicals can improve the biodegradability of cellulose.

So far, hydrogen peroxide (Yang et al., 2016), periodate (Li et al., 2009), and 2,2,6,6-tetramethylpyperidine-1-oxy radical (TEMPO) (Luo et al., 2013), and nitrogen dioxide (Peng et al., 2012), have been employed to oxidize the glucose unit of cellulose. Using hydrogen peroxide as the oxidation agent, Yang et al. (Yang et al., 2016) introduced large amounts of carbonyl and carboxyl groups on the surface of OBC. Further, gelatin and HA were integrated into the OBC scaffolds, resulting in the formation of OBC/gelatin/HAp composite scaffolds. It was found that these modified scaffolds could be completely degraded in simulated body fluid after approximately 90 days. Periodate oxidation causes the C2–C3 bond in the glucopyranoside ring to break, allowing dialdehyde groups to form per glucose unit. On the other hand, TEMPO-mediated oxidation transforms primary hydroxyl groups into aldehydes. Both processes lead to the hydrolytic degradation of cellulose under physiological conditions (Li et al., 2009; Luo et al., 2013). It was observed that the degradation process took place both *in vitro* and *in vivo* without triggering excessive tissue reactions (Singh et al., 1979; Singh et al., 1982). Adjusting the oxidizer content (Chimpibul et al., 2020) and reaction time (Hou et al., 2018) could control the oxidation degrees, which in turn influence the degradation rate. In general, the molecular weight, composition, crystallinity, water uptake capability, pH, and additives, as well as the material's morphology, are some of the factors that can affect the rate at which cellulosic materials degrade (Torgbo and Sukyai, 2020). This is of particular interest in the field of tissue engineering since different application scenarios can call for different rates of degradation.

The biodegradability of nanocellulose in an aqueous medium was initially observed by Kümmerer et al. (Kümmerer et al., 2011) whose work revealed that, due to its surface area, nanocellulose degraded more quickly than its macroscopic forms. Therefore, the addition of nanocellulose to polymer matrices is expected to enhance biodegradability performance, thereby meeting the required level of adjustability for tissue engineering scaffolds. The degradation rates of various collagen-HA-microfibrillated cellulose (MFC) scaffolds show that the rate of degradation rises as the MFC content increases. In contrast to conventional scaffolds, which degrade 50–70% in 6 weeks, the collagen-HA-MFC with 50% MFC (the highest level of MFC in the composite) breaks in week 5 (He et al., 2018). Shi et al. (Shi et al., 2012) recorded the rate at which PLA/CNC nanocomposites degrade throughout one month. When compared to plain PLA mats, the nanocomposite mats, particularly at high CNC-loading levels, saw faster degradation in a phosphate-buffered saline solution. The increased surface area of nanocomposite mats, possibly due to the smaller fiber diameter, and the presence of many well-defined pores on the surface of PLA/CNC nanocomposite fibers are the probable reasons for this occurrence. Huang et al. (Huang et al., 2018) examined the biodegradation behavior of the composite formed by reinforcing CNC in poly (butylene succinate) (PBS). After 28 days, neat PBS lost less than 5% of its weight, whereas PBS/CNC composite membrane lost 13.74%, showing that CNC dramatically accelerated the degradation of electrospun PBS fibrous membranes. The higher degradation rate of the PBS/CNC composite membrane was attributed to the improved hydrophilicity owed to the addition of CNCs, which promoted the permeation of water into the pores of electrospun PBS fibers.

7. Applications of Cellulose-Based Composites in Tissue Engineering

The development of new biomedical materials from natural polymers is a critical focus for scientists working to advance the field of biomedicine. Extensive research on cellulose and its applications has demonstrated its immense potential for a wide range of biomedical uses, owing to its unique and advantageous properties. In recent years, there has been a surge of interest in producing cellulose-based composites by combining them with other materials, such as polymers and nanoparticles. This innovative approach has enabled researchers to create a wealth of promising new biomedical products. In this section, we will explore the latest examples and research progress in the biomedical applications of cellulose, including its use in wound healing, bone tissue engineering, neural tissue repair, and cardiovascular therapy.

7.1 Cellulose Composites for Wound Repair

The skin serves as the first barrier between the human body and the external environment, and it is the largest covering of the body with a complex structure of tissues (Xia et al., 2020). It has a crucial role in various physiological functions, including maintaining homeostasis, regulating body temperature, controlling blood pressure and hydration levels, eliminating toxins, and preventing the invasion of pathogens (Tudoroiu et al., 2021). The skin typically consists of three layers: the epidermis, dermis, and hypodermis (Kilic Bektas et al., 2018). The epidermis, which lacks blood vessels, forms the outermost layer of the skin and acts as a major physical barrier to chemical transfer through the skin. Attached to both the mucous and epidermal membranes, the epidermis layer is composed of several cell sheets. Its outermost layer is a continuous shedding of epidermal dead cells, while the cells underneath it are continually proliferating (Kabashima et al., 2019). The dermis is responsible for the skin's thickness and lies between the hypodermis and the epidermis (Bian et al., 2022). Due to the presence of collagen and elastic fibers, the dermis mostly has a fibrous structure. Furthermore, the dermis has a crucial function in regulating body temperature, protecting the skin, and facilitating sensory perception (Bragazzi et al., 2019). The hypodermis located underneath the dermis, is the widest and thickest layer and is composed of elastin and loose connective tissue. The hypodermis contributes to thermal insulation and mechanical protection and supports the proliferation of keratinocytes and fibroblasts, as well as wound healing (Zimoch et al., 2021).

Skin wounds are a form of injury caused by external factors that result in the disruption of a sequence of skin cells. Wounds frequently fall into two categories: acute wounds and chronic wounds. The first may result from incisions, burns, or trauma. On the other hand, chronic wounds can be classified into severe burns, infectious wounds, diabetic foot ulcers, and so on (Kanikireddy et al., 2020). Acute wounds commonly heal within a relatively short period of four to six weeks; however, chronic wounds heal at a significantly slower pace, leading to severe patient discomfort and adversely affecting their quality of life (Nguyen et al., 2023).

Currently, wound management is a global challenge due to the complex and lengthy process of wound healing, which necessitates the use of appropriate wound dressings (Gharib Khajeh et al., 2022). Traditional passive wound dressings, such as gauze and tulle dressings, are often ineffective in treating wounds because they stick to the tissue, causing pain and dehydration when removed (Yu et al., 2021). Conversely, interactive modern dressings, including films, foams, hydrogels, and fibers, successfully promote the healing of wounds by creating a moist environment, controlling exudate, and offering improved capabilities such as antibacterial and antioxidant properties (Rahmani Del Bakhshayesh et al., 2018).

Cellulose and its derivatives are natural polymers that show high potential in skin tissue repair (Wahid et al., 2021). Due to its three-dimensional fiber structure, cellulose exhibits unique mechanical properties such as excellent tensile strength, hydrophilic nature, high water absorption capacity, good permeability, and moldability, especially in wet form (Ahmed et al., 2020; Zeng et al., 2023). In addition, for their nanofibrous and filter-like networks, cellulosic membranes prevent the penetration of bacteria into the wound site, minimizing the potential for wound infection by creating a physical barrier. With its high water retention and water vapor transmission capabilities, cellulose locks wound exudate inside its material, creating optimal moisture conditions for wound healing (Ahmed et al., 2020). In recent years, the above-mentioned properties have drawn attention to cellulose dressings for wound healing applications. BC scaffolds can flexibly adapt to any shape of the wound and prevent direct contact of the wound with the external environment to reduce the possibility of bacterial infection. Therefore, BC scaffolds can be attractive candidates for wound dressing materials (Del Valle et al., 2017). However, the lack of antibacterial ability in BC scaffolds has motivated researchers to incorporate antibacterial components into BC. Antimicrobial activity is essential to create an antibacterial barrier between the wound and the external environment.

So far, various antibiotics (e.g., tetracycline, amoxicillin, and fusidic acid) have been incorporated into BC scaffolds to improve their antibacterial properties (Shao et al., 2016; Ye et al., 2018). For instance, Ye et al. (Ye et al., 2019) made a sponge composite including BC and gelatin using glutaraldehyde as a cross-linker by the freeze-drying method. The fabricated BC/gelatin sponges have a regular honeycomb structure with a uniform porosity of about 94–95% and an excellent swelling property of 3000–3150%. BC/gelatin sponges show ampicillin release for up to 48 h, and ampicillin release from BC/gelatin sponges follows non-Fickian diffusion. Also, the obtained sponges show superior antibacterial activity, so they have high potential in various antibacterial applications, especially in wound dressings.

Nevertheless, the application of conventional antibiotics poses high health risks to humans as it promotes the development of antibiotic resistance in bacteria; hence, there is an increasing growth in the use of antimicrobial nanoparticles and polymers along with BC matrixes. For instance, Wu et al. (Wu et al., 2014) prepared BC/silver nanoparticles (AgNPs) composite hydrogels and systematically studied their wound healing performance. The composite hydrogels have been found to possess strong antibacterial activity against *S. aureus* while also exhibiting low toxicity towards rat fibroblasts and a high capability of wound healing when implanted in rat models. Khumrai et al. (Khamrai et al., 2019) developed an antibacterial mussel mimetic patch based on dopamine-modified BC (BC-DOPA) combined with AgNPs and rGO for wound healing applications. The composite patch had superior antibacterial efficacy against both gram-positive and gram-negative bacteria in comparison to pure BC and BC-DOPA. Furthermore, an enhancement in the proliferation and migration of NIH 3T3 fibroblast cells and A549 human lung epithelial cells was observed in the cell groups treated with the BC-DOPA/rGO/AgNPs composite patch, indicating excellent performance in wound healing.

Several researchers have shown that lignin and its derivatives exhibit good antibacterial properties and are considered good candidates for antibiotics. For example, Zemjkoski et al. (Zmejkoski et al., 2018) synthesized a type of lignin model compound called dehydrogenative polymer of coniferyl alcohol (DHP), from cone alcohol and then prepared BC/DHP composite hydrogels. The composite hydrogels were further investigated as a dressing material for the treatment of chronic wounds. The composite hydrogel demonstrated antibacterial activity against selected pathogenic bacteria, such as *P. aeruginosa*, *Serratia* sp., and *S. aureus*. The *in vitro* release results showed that DHP was released rapidly at the beginning (1 h), and a high concentration of antimicrobial compounds could be maintained for the next 72 h. Due to their high swelling and long-term antibacterial ability, BC/DHP composite hydrogels can be a promising candidate for chronic wound healing.

Along with displaying sufficient properties in favor of a wound dressing *in vitro*, cellulosic materials have been proven to show a promising wound healing rate *in vivo*. For instance, in Guamba and colleagues' study (Guamba et al., 2023), cellulose extracted from a pitahaya source was able to show improved antibacterial activity against Gram-negative bacteria as a model when applied over pigskin. Elsewhere, Jiji and co-workers (Jiji et al., 2019) reported that thymol-enriched BC hydrogels (BCT) possess low toxicity *in vitro* and can facilitate the growth of fibroblast cells. They finally conducted an *in vivo* study on female Wistar rats to evaluate the wound-healing performance of the obtained composite hydrogels. As shown in Figure 5A, compared to the control and pure BC groups, a significant increase in recovery (90.7%) was observed on day 20, and almost complete recovery occurred on day 25 in the BCT group. According to histopathological analysis (Figure 5B), they observed thicker granulation tissue and better wound healing through new skin formation. The results showed that BCT hydrogels are suitable for natural burn wound dressing materials. Lignin is considered the most abundant aromatic biopolymer on earth. Several researchers have shown that lignin and its derivatives exhibit good antibacterial properties and are considered good candidates for antibiotics.

Loh et al. (Loh et al., 2018) evaluated the ability of BC/poly acrylic acid (PAA) hydrogels to deliver human epidermal keratinocytes and dermal fibroblasts for the treatment of full-thickness skin lesions. *In vitro* studies found that BC/AA hydrogels exhibit excellent cell attachment and allow cell transfection. Furthermore, *in vivo* studies showed that BC/PAA hydrogels loaded with cells showed the best wound healing performance compared to the hydrogel alone and untreated groups. In addition, further histological analysis was performed using hematoxylin and eosin (H&E) and Masson's trichrome staining. The untreated group had many blood vessels, according to the results of the H&E staining assay, whereas the groups that received hydrogel alone and hydrogels filled with cells had nearly as many blood vessels. This suggests that whilst the wounds in the untreated group were still in the proliferation phase, the wounds in the hydrogels-containing cells and hydrogel-alone groups were in the regeneration phase. Furthermore, Masson's trichrome staining

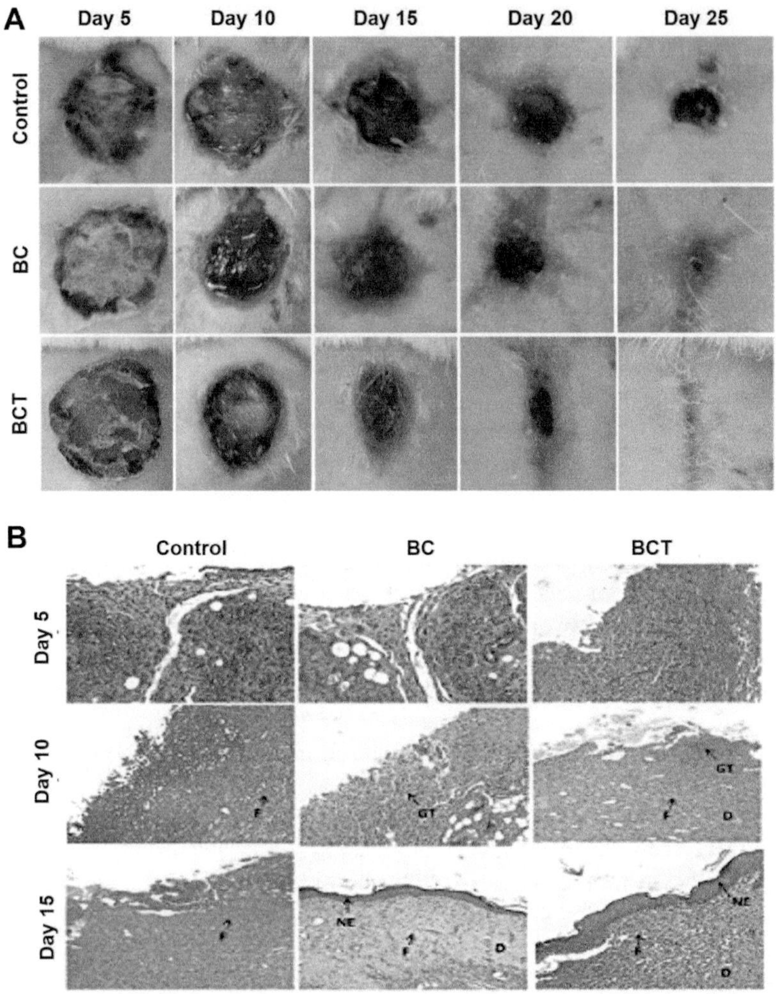

Figure 5. (A) Photographs of burn wounds, and (B) Histopathological micrographs of control, BC and BCT wound tissue at in different intervals (Jiji et al., 2019).

showed that the hydrogels loaded with cells group had more widespread collagen deposition than the hydrogel alone and untreated groups, suggesting that the HC group had a superior therapeutic effect. These findings demonstrated once more that BC/PAA composite hydrogels serve as both wound dressings and cell carriers to support full-thickness wounds.

BC wound dressings reinforced with ε-polylysine (ε-PL) and mussel-inspired polydopamine were investigated for the treatment of infected wounds by Fazli Wahid et al. (Wahid et al., 2021). As observed in the Sprague-Dawley rat model, functionalized membranes have strong antibacterial properties and significantly contribute to the wound healing process. Based on the histological studies, wound healing using modified membranes resulted in faster epithelialization, thicker skin and epidermal layers, and a smoother surface until day 14 compared to the control and pure BC groups. Also, these membranes can absorb tissue secretions, maintain environmental moisture, and reduce bacterial infection and inflammation, highlighting their potential as advanced wound dressings.

7.2 Cellulose Composite Scaffolds for Bone and Cartilage Regeneration

Bone and cartilage tissues play crucial roles in the human body. Hence, it is essential to preserve the health of these vital organs to have a healthy and dynamic lifestyle. Bones provide a protective and supportive structure for internal organs, whereas cartilage facilitates mobility by minimizing the friction between bones. Nowadays, the significant increase in bone and cartilage damage caused by aging, tumors, and traumas has prompted researchers to discover novel therapy approaches, surpassing the conventional techniques for addressing bone and cartilage defects (Zhang et al., 2019; Cheng et al., 2019; Koons et al., 2020). Conventional clinical treatment, such as allograft/autograft transplantation, encountered multiple drawbacks, including immune rejection, limited availability of donors, and the potential for microbial infections (Murizan et al., 2020). Otherwise, significant advancement in tissue engineering have provided an inexpensive, accessible, and safe strategy for the regeneration of bone and cartilage deficiencies. This is achieved by creating an optimal microstructure with interconnected pores, which facilitates cell migration and proliferation while also supporting the vascularization of ingrown tissue (Daugela et al., 2018). Various material categories and combinations have shown promise as candidates for tissue engineering applications. In general, a number of parameters, such as the desired production and implementation techniques, influence the choice of material for soft and hard tissue regeneration. Polymers, for example, are organic compounds that include extended chains of atoms connected by covalent bonds. Both natural and synthetic polymers serve as beneficial forms of materials for bone and cartilage tissue engineering (CTE) (Lu et al., 2019; Koons et al., 2020).

Cellulose and its derivatives are natural polymers that exhibit great potential in BTE due to their excellent biocompatibility, nontoxicity, and slow degradation rate, as well as their cost-effective production and accessibility through renewable sources (Wang et al., 2017). Also, the crystalline structure of cellulose nanofibers exhibits similarity to the collagen found in natural bone, hence offering advantageous mechanical characteristics (Torgbo and Sukyai, 2018). However, cellulose, in its pure form, may not be an ideal matrix for promoting osteoinduction, osteoconduction, and osteointegration owing to the limited interactive biological interaction between cells and the cellulose substrates. In order to improve the ability of this material to repair bone damage, different organic and inorganic components are frequently incorporated into cellulosic scaffolds. These new cellulose-based composites demonstrate better mechanical strength, biomineralization, and bone cell differentiation compared to pure cellulose (Janmohammadi et al., 2023; Zaman et al., 2020; Seddiqi et al., 2021).

The distinctive characteristics of BC sparked a growing interest in its application as a foundational matrix for biocomposite scaffolds synthesis. For instance, in order to simulate the composition and structure of natural bone, researchers introduced cellulose/HAp nanofiber composites as scaffolds to mimic natural bone with favorable mechanical properties, porous structure, and a higher level of biocompatibility (Hickey and Pelling, 2019). As previously stated, the Young's moduli of human trabecular bone range from 0.8 to 16.9 GPa (Wu et al., 2018), while, human cortical bone presents higher mechanical strength (~ 20 GPa in the Young's modulus) (Rho et al., 1993). Hence, it is essential to meticulously engineer the BTE scaffolds with suitable mechanical characteristics, as inadequate strength would cause immediate collapsing when subjected to external pressure, while a too high young modulus can lead to stress shielding, bone resorption, and poor osseointegration. In this regard, Ran and colleagues (Ran et al., 2017) constructed a multi-component organic/inorganic composite BC-gelatin/HAp double-network scaffold. The HAp particles reinforced the mechanical stability of the of the BC network and the gelatin network assisted in releasing stress that accumulated at the nodules. The combination of these elements led to significantly increased strength and stiffness, with a Young's modulus of 177 MPa and a fracture stress of 12.95 in dry conditions. In wet conditions, the elastic modulus was 0.27 MPa, and the fracture stress was 0.28 MPa. Despite the fact that pure BC has significant potential as a scaffold for bone tissue, it lacks an appropriate pore structure that restricts the infiltration of cells. Feng et al. (Fang et al., 2009) discovered that the incorporation of HAp into BC can achieve optimal pore size required for bone scaffolds, thereby enhancing bioactivity and promoting bone formation. The HAp/BC nanocomposite scaffolds revealed desirable mechanical characteristics and pore structure, making them very appropriate for this particular application. Furthermore, the obtained nanocomposites exhibited a much greater affinity for serum proteins and had a positive impact on subsequent level of proliferation.

Aside from BC, CA can be used as a multifunctional material for simulating bone tissue. CA-based scaffolds serve a promising hydrophilicity and flexibility, encouraging the attachment, proliferation, migration, and differentiation of osteoblasts (Gouma et al., 2012). Medicinal plant extracts such as Cissus quadrangularis (CQ) and Spinacae olareacea (SO) have been used to improve the bioactivity and mechanical properties of Na-CMC-based composite scaffolds. For example, the addition of CQ led to osteoblasts proliferation and increased mineralization. As anticipated, there was a lack of significant biomineralization detected in the control group when the osteogenic supplement was absent. Cell biomineralization was observed on scaffolds that contained CQ, even when they were cultured in a medium that did not contain osteogenic additives. Additionally, the results indicated that the activity of ALP of SaOS-2 cells on scaffolds, improved as the concentration of CQ on chitosan/Na-CMC scaffolds increased (Tamburaci et al., 2018).

Chahal et al. (Chahal et al., 2016) prepared an electrospun HEC-PVA nanofibrous composite for non-load bearing bone regeneration. The MTS assay was employed to measure cell proliferation on the surface of HEC/PVA electrospun nanofibers at various time intervals during cell culture. After 3 days of cell culture, there was a significant increase in cell proliferation on the surface of HEC/PVA scaffolds compared to the control sample. On the other hand, the rise in HEC concentration led to an increase in the tensile strength of HEC/PVA mats. This could be attributed to the fact that HEC membrane, which is 70–80 μm thick, has a high tensile strength of 30 MPa. The elastic behavior of PVA and HEC/PVA mats shows a similar trend in terms of tensile strength. The HEC/PVA (60:40) mats exhibited the highest E-modulus, measuring approximately 349.25 ± 69 MPa, whereas the lowest E-modulus was seen for PVA, measuring approximately 188 ± 50 MPa. Therefore, HEC not only improved the biocompatibility, but also enhanced the elastic modulus, and tensile strength of this scaffold.

Cartilage has a limited capacity to self-healing and regenerate after injury due to the total lack of innervation and vascular supply. It also contains only a sparse population of chondrocytes, and the mature form of this cell type does not exhibit mitotic activity and has a low metabolism (Lindahl et al., 2015). Alternative treatments, including tissue engineering, offer a promising alternative approach by introducing chondrogenic cells onto or into biodegradable scaffolds to create cartilage for repairing defects. Different types of cellulose materials are used in cartilage repair and regeneration. For instance, by combining CNF with thermoplastic starch (SC) by salt leaching technique, it is possible to increase the biocompatibility of scaffolds, and the ratio of starch to cellulose affects both the water absorption capacity and porosity. When incorporating 10 wt.% of CNF into salt-leached scaffolds containing 70% salt, the water absorption at equilibrium decreased by approximately 18.4% compared to the pure starch film. The presence of a rigid network of hydrogen bonds between cellulose and starch may be the cause. Nevertheless, when the cellulose nanofibers were present in higher quantities (15 wt.%), the samples exhibited an elevated water absorption capacity, which could potentially boost the swelling capability of the starch-based scaffolds. By increasing the concentration of nanofibers, the clustering of nanofibers results in an improvement in the capillary action within the composite structure. When 5 wt.% nanofibers are added to scaffolds with 90% salt content, it is shown that the water resistance is significantly improved. Besides, the proliferation rate of the cells increases after 4 days when exposed to SC15/70 and SC15/90. This demonstrates the inductive function of cellulose nanofibers in promoting cell adhesion and proliferation (Nasri-Nasrabadi et al., 2014).

The research conducted by Muller and his colleagues (Müller et al., 2006) has shown that CaP-coated cellulose exhibits favorable characteristics for cartilage tissue engineering. Physiological fluids easily wet CaP-coated cellulose, leading to a significant enhancement in cell adhesion compared to untreated cellulose. Based on Chondrocyte adhesion tests, both CaP-coated and Ca $(OH)_2$-treated scaffolds can be categorized as biocompatible, since they exhibit a cell vitality of 495% (Figure 6A). The formation of cartilage was examined in all samples using primary bovine cartilage cells. The specimens underwent a 6-week incubation period. The histological sections were treated with Safranine-O stain to assess the production of proteoglycans. According to the Figure 6B, the scaffolds coated with CaP were fully covered by cartilaginous tissue. In another experiment, HEC with a solution of chitosan and glycerol phosphate was used to make a hydrogel scaffold, and the results showed that this gel can maintain the full thickness of cartilage in living rabbits for up to 1 week (Hoemann et al., 2005). In conclusion, cellulose-based materials have emerged as promising candidates for bone and cartilage tissue engineering

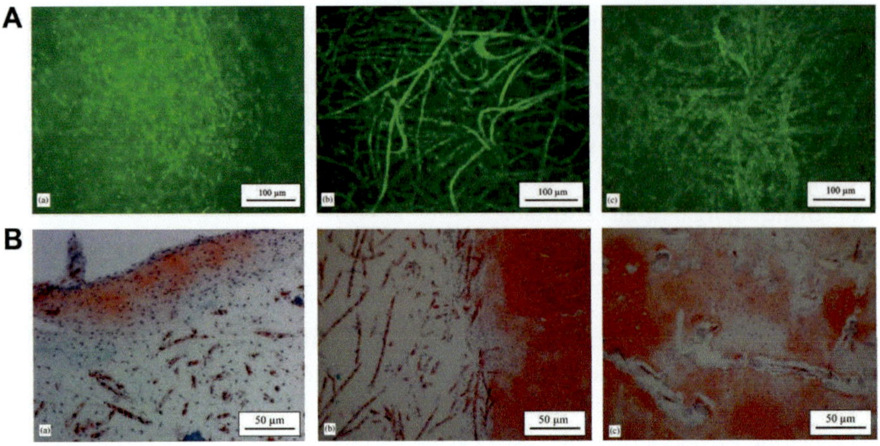

Figure 6. (A) FDA+PI staining images of various cellulose scaffolds (a) untreated, (b) $Ca(OH)_2$-treated and (c) CaP-coated cellulose, seeded with primary chondrocytes after 1 week, and (B) Safranine-O staining images of cellulose scaffolds seeded with primary bovine chondrocytes, after 6 week (Müller et al., 2006).

due to their desirable properties. By incorporating other components, researchers have been able to enhance the bioactivity, degradation, and mechanical strength of cellulose-based scaffolds for bone regeneration. Moreover, cellulose materials have also shown potential for mimicking cartilage. Continued research and advances in this field hold great promise for the future of tissue engineering and regenerative medicine.

7.3 Cellulose-based Composite Scaffolds for Neural Tissue Engineering

Nerve injuries have been recognized as the leading cause of disability and mortality, accompanied by a large cost burden on society (Feigin et al., 2019). Damaged axons in the nervous system are often irreparable or semi-reparable, given the formation of glial scars that limit the few remaining neurons from regenerating and restoring tissue function adequately (Doblado et al., 2021). Currently, autologous nerve grafting is the gold standard for treating patients with extensive nerve loss. Yet, this surgery presents significant limitations such as the donor's restricted nerve supply, the loss of donor nerve function, and the risk of neuroma development. Therefore, seeking alternate treatments for autologous nerve transplants is critically important (Schmidt and Leach, 2003). Neural tissue engineering (NTE) provides a glimmer of hope for the future of regenerative medicine by providing novel medical strategies to restore nerve functionality as alternatives to the traditional transplantation approach (Julian George et al., 2020).

In NTE, scaffold biomaterials serve as structural frameworks that support cell growth, cell migration, and cell remodeling while also promoting angiogenesis and tissue regeneration. The scaffolds designed for NTE must have sufficient rigidity to withstand physiological stress and supply mechanical support to the growing tissue. Simultaneously, it is crucial for the scaffolds to avoid triggering any undesirable immune response or toxicity upon interaction with living cells. To date, numerous researchers have explored the potential applications of cellulose-based composites in neuroregeneration. Among various scaffolding methods, the electrospinning technique has emerged as a promising candidate to create an artificial cellular scaffold with controllable size and a manipulated porous structure that has a similar architecture to the natural ECM. Moreover, the electrospinning process is known for its highly controllability, stability, and versatility (Frost et al., 2018).

Several synthetic polymers, such as PLA, PCL, poly(lactide-coglycoside) (PLGA), and polyurethane (PU), have already been successfully applied to fabricate nanofibers by electrospinning (Lee and Livingston Arinzeh, 2011). However, its applicability is severely limited due to weak hydrophilicity, slow degradation kinetics, and the lack of cell recognition sites (Mohammadalizadeh et al., 2022). Utilizing cellulose derivatives in conjunction with synthetic polymers has been proposed to overcome these shortcomings. For instance, Altun et al. (Altun et al., 2019) prepared BC/PCL blend nanofibrous scaffolds (50:50 wt.%) by electrospinning to enhance peripheral nervous system neural tissue regeneration. It was found that the obtained composite polymer blends, compared with neat PCL, exhibited improved L2929 murine fibroblast adhesion and proliferation. In addition, the representative laser scanning confocal microscope and SEM images obtained from primary cell culture with dorsal root ganglia (DRG) cells on nanofibrous samples confirmed enhanced neural growth and neurite outgrowth when using BC/PCL nanofibrous scaffold. In another study, Yang et al. (Yang et al., 2019) reported of the drop in water contact angle of PLA/CNF composite nanofiber membranes, which indicates hydrophilicity, with the increase of the CNF loading. Elsewhere, Goudarzi et al. (Moazzami Goudarzi et al., 2021) demonstrated the fabrication and characterization of electrospun composite membranes based on PCL/gelatin and acetylated cellulose nanofibers (ACNF). They displayed that the homogeneous dispersion of the ACNF within the matrix raised the ultimate tensile strength (UTS) of the PCL/gelatin from 2.5 ± 0.1 MPa to 4.3 ± 0.1 MPa while reducing its degradation rate by 1.66 times. It was discovered that the addition of ACNF to scaffold at a concentration of 2 wt.% could promote cell proliferation. The authors anticipated that this composite could be a promising candidate for soft tissue engineering, such as nerves and similar tissues.

Neural cell signaling relies on electrical conductivity, which is not present in pure cellulose. Integrating cellulose with signal-transmitting and signal-receiving components shows potential for creating a material resembling the ECM of brain tissue (Korábková et al., 2024). These highly conductive structures enable the stimulation of cells cultivated on them and permit the direct transmission of electrical signals between cells. This ultimately enhances the interactions between cells and structures, promoting neurite outgrowth and neuronal differentiation (Koeini et al., 2024). Intrinsically conductive polymers, e.g., polypyrrole (PPy), polyaniline and polythiophene (PTh), and their derivatives, are a class of synthetic polymers that have been thoroughly utilized as electrically conductive compartments in cellulose-based composites for biological applications due to their advantages of being highly electrically conductive, biocompatible, suitable for chemical and physical stability, and flexible in processing (Khan et al., 2015). There have been several successful attempts at imparting electrical conductivity to cellulose materials. In one study, Muller et al. (Muller et al., 2013) by performing *in situ* oxidative chemical polymerization of PPy on the surface of BNC nanofibers using ammonium persulfate. The enhanced adhesion and proliferation of PC12 neuronal cells on the composite material demonstrates the potential of these materials as scaffolds for neural tissue applications. In another experiment, the fabrication of electrically conductive composites by coating nanoporous cellulose gels (NCG) with PPy nanoparticles, which were synthesized via *in-situ* vapor-phase polymerization, resulted in the fine formation of particles on CNF nearby. The results demonstrated that the inclusion of PPy led to a significant improvement in electrical conductivity, reaching a value of 0.08 S cm^{-1}. This enhancement may be attributed to the establishment of a continuous PPy network. Remarkably, cell study results revealed that PC12 cells show a significant increase in cell proliferation on the surface of NCG/PPy composites with a higher PPy content of 3%. Additionally, the cells cultivated on the NCG/PPy-3 composite developed of multiple extended neurites. Therefore, the NCG/PPy composite scaffold appears to be a more suitable option for repairing nerve tissue. Similarly, Thunberg et al. (Thunberg et al., 2015) reported the synthesis of conductive PPy using *in-situ* PPy polymerization on electrospun CA nanofibers. The composite mats exhibited a 10^5-fold increase in conductivity compared to neat cellulose nanofibers, along with enhanced surface roughness. Not only cytotoxicity analysis demonstrated the composite had no cytotoxic potential, but also *in vitro* culturing using SH-SY5Y human neuroblastoma cells indicated enhanced adhesion and a more neuron-like phenotype of cells on the cellulose/PPy scaffolds.

By utilizing an electrical signal, conductive composites made from cellulose can enhance the effects of contact guiding in the outgrowth of neuritis. In this regard, Zha et al. (Zha et al., 2020) assessed the effect of electrical stimulation on PC12 cell proliferation cultured on polymeric nanofibrous mats comprising electrospun cellulose/poly(N-vinylpyrrole) (EC/PNVPY) and electrospun cellulose/poly(3-hexylthiophene) (EC/P3HT). When subjected to a consistent voltage of 5 V for 300 min per day, the PC12 cell density on EC/P3HT mats exhibited a significant increase compared to the electrospun pure EC and EC/PNVPY mats after 5 days of culturing. This suggests that the EC/P3HT composite is a suitable conductive substrate for delivering electrical stimulation to promote the proliferation of neuronal cells. More recently, Liu et al. (Liu et al., 2022) developed transparent BC/poly(3,4-ethylenedioxythiophene)-sulfonated nanofibers (BC/PEDOT-SNFs, BPS) composites through the *in-situ* polymerization of PEDOT with the assistance of an oxidation agent and the doping of SNF (Figure 7A). They claimed that self-assembly of SNF during the in-situ polymerization process of PEDOT could make PEDOT accumulate more on the surface of SNFs, and the final membrane formed a heterogenous layered structure as well as a complete conductive network, resulting in a conductivity of 1.8×10^2 S/cm. Moreover, they found that the BC/PEDOT-SNFs membrane nerve conduit, when implanted *in vivo* in Sprague-Dawley (SD) rats with sciatic nerve defects for 12 weeks, yielded a superior outcome in repairing peripheral nerves compared to a hollow nerve conduit. As

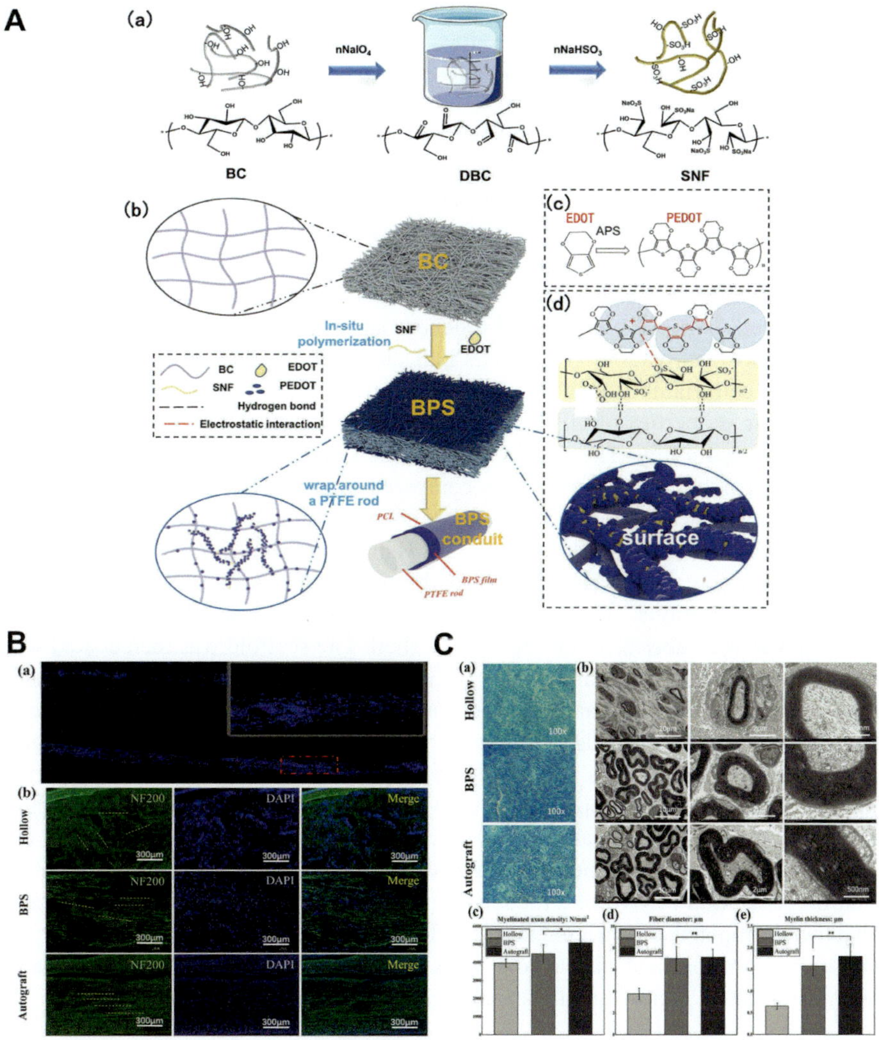

Figure 7. (A) Schematic of the preparation of SNFs, BPS membrane, and BPS conduit (a, b), the polymerization of PEDOT and illustration of the binding mechanism of the BPS membrane (c, d). (B) Representative images were taken of sections stained with NF200 (green) and DAPI (blue). (a) Shows a stained section of the BPS conduit at 2 weeks; (b) depicts the morphology of regenerated nerves in the Hollow, BPS, and Autograft groups at 12 weeks. (C) Includes information on the regenerated nerves: (a) Toluidine blue staining and (b) TEM images of the nerve cables in the Hollow, BPS, and Autograft groups at 12 weeks, along with (c) axon density, (d) fiber diameter, and (e) myelin thickness (Liu et al., 2022).

it has shown in Figure 7B, at 2 weeks, a large number of cells were observed on the inner wall of the BPS tube, likely due to Schwann cells forming supportive structures that promote nerve regeneration. The BPS group showed better alignment of regenerated nerve fibers compared to the Hollow group, resembling the Autograft group, suggesting that BPS materials enhance the microenvironment for nerve repair.

Similarly, Figure 7C demonstrates that fiber diameter and myelin thickness in the BPS group were comparable to the Autograft group and significantly greater than in the Hollow group, indicating that BPS significantly enhances nerve regeneration and produces mature nerve fibers similar to those in autografts.

Cellulose-based composites could also serve as a suitable platform for delivering different kinds of biological factors, such as neurotrophic factors and drugs, through a controlled release mechanism that can stimulate nerve regeneration. Farzamfar et al. (Farzamfar et al., 2018) demonstrated the fabrication of a CA/gelatin composite incorporating gabapentin (GBP), an anticonvulsant drug used as a signaling molecule, by the wet-electrospinning method and assessed its ability to support Schwann cells' behavior and neural regeneration using a rat's sciatic nerve defect model. *In vitro* cultivation on scaffolds indicated that all GBP-containing scaffolds had higher cell viability than the positive control due to the neuroprotective and antiapoptotic properties of GBP, showing their proliferative activity in addition to their cytocompatibility. The results showed the CA-based scaffolds enhanced the regeneration of the created nerve injury compared with the GBP-free CA/Gel scaffold, which demonstrates its applicability for NTE applications. In another experiment, Naseri-Nosar et al. (Naseri-Nosar et al., 2017) invented a biodegradable drug-loaded scaffold with core–shell structured fibrils by means of coaxial wet-electrospinning for NTE applications. They used wet-electrospun PLA as the core material, whereas CA was fabricated into the fibril's shell. The scaffold was then coated with citalopram-loaded gelatin nanocarriers (CGNs) through the nanoprecipitation method to enhance the attachment and proliferation of the neural cells. *In vitro* experiments revealed that Schwann cells cultivated on coated composite scaffolds had higher viability than uncoated ones. The main reasons for the enhanced biological properties are due to improved hydrophilicity, Arg-Gly-Asp sequences on the gelatin, which are cell-recognition sites, and the positive effect of citalopram on neuronal cell behavior. The authors further investigated the effects of the coated composite on neuroregeneration *in vivo* in a rat sciatic nerve defect model. The results indicated that when implanted as a guidance conduit, the conduits significantly prevented muscle weight loss and displayed a higher sciatic nerve function index compared with the negative control and uncoated composite scaffold.

7.4 Cellulose Based-Composite Scaffolds for Cardiovascular Diseases

Cardiovascular disease (CVD) is a group of conditions (including coronary heart disease, stroke, hypertension, and heart failure) that affect the structure and function of the heart and is considered the leading cause of death worldwide (Gisone et al., 2022). Approximately 30% of deaths worldwide are associated with CVDs, and in industrialized countries, heart failure affects 1–2% of the adult population (Miller et al., 2021; Camman et al., 2022). Current treatment options for CVDs mainly include traditional pharmacotherapy and surgery. These methods can only address symptoms and cannot initially regenerate the heart (Xu and Song, 2021). Heart transplants for end-stage heart failure are constrained by the absence of organ donors, immune rejection (Badylak and Nerem, 2010), surgical complications, the need for anticoagulants (Mhashilkar and Atala, 2012), and the limited durability of the transplants. Prosthetic valve devices pose the risk of thromboembolic complications and life-long bleeding (Pibarot and Dumesnil, 2009; Fioretta et al., 2021). In particular, xenografts of bovine or porcine heart valves are often associated with structural valve failure (Manji et al., 2012). Autografts used for coronary bypass transplants are limited by the availability of appropriate conduits and the additional risks associated with the operation (Desai et al., 2011). Given the above-mentioned treatment methods, it is urgently necessary to develop new, practical, and efficient technologies for CVD treatment.

In order to overcome the limitations of current CVD therapies, tissue engineering has emerged as a largely multidisciplinary field that integrates the principles of engineering, biology, and material science (Fooladi et al., 2022). Numerous studies have focused on creating myocardial tissue grafts using prefabricated, implantable scaffolds that mimic the biochemical, mechanical, and structural properties of cardiac ECM (Nguyen et al., 2019; Schussler et al., 2021). Cellulosic biomaterial possesses advantageous properties that make it an ideal material for utilization in cardiovascular tissue engineering. It is biocompatible, meaning that it is not causing an immunological or negative response in the body (da Silva et al., 2022). It supports cell attachment and proliferation in order to regenerate heart tissue (Lin et al., 2020); Furthermore, its adaptability in terms of shape and size allows it to closely mimic the structure and characteristics of the native ECM found in blood vessels (Lin et al., 2020); it also has excellent mechanical properties, including high tensile strength and flexibility, which make it resistant to the hemodynamic forces of the cardiovascular system. It has high permeability, which causes the diffusion of nutrients and oxygen throughout the scaffold (Lupașcu et al., 2022). The unique porous and interconnected network structure of cellulose fibers can improve angiogenesis; the structure of BC has channels that facilitate the infiltration of endothelial cells and other cells, resulting in the formation of functional blood vessels (Fusco et al., 2022).

There are various articles in the literature on cellulose-based composites for vascular replacement. For example, bacterial nanocellulose (BNC) modified with chitosan provided highly biocompatible vascular prostheses that showed no toxicity or allergic reactions in an animal model, showing potential for future use in tissue-engineered blood vessels for vascular surgery (Piasecka-Zelga et al., 2021). In another study, heparinized 3D porous BC/gelatin scaffolds were loaded with vascular endothelial growth factor. The incorporation of heparin, a known anticoagulant, into BC scaffolds could further increase their bioactivity and potential to promote angiogenesis. This modification strategy has been proven to overcome the challenge of poor cell adhesion and proliferation associated with BC scaffolds, ultimately improving their performance in tissue engineering applications. By combining the advantages of BC with the bioactivity of heparin, these scaffolds have the potential to significantly increase angiogenesis and overall tissue regeneration on a clinical scale (Wang et al., 2018).

In another paper, Bao et al. (Bao et al., 2022) fabricated 3D composite tubes of BC and fish gelatin treated with glutaraldehyde for small-caliber artificial blood vessels, enhancing their mechanical properties and biocompatibility. The *in vitro* findings demonstrated that the attachment, growth, and differentiation capacity of

HUVECs and human smooth muscle cells (HSMCs) were significantly enhanced when implanted on the inner surfaces of BNC/gelatin tubes as compared to pure BNC tubes. Additionally, the BC/gelatin tubes exhibited superior anticoagulant abilities. The optimal concentration of fish gelatin for the composite was determined based on the increased mechanical properties and biocompatibility of small caliber vessels of 0.1% weight/volume. In another study, small-diameter BC-based vascular grafting was performed for coronary artery bypass grafting, focusing on *in vivo* patency, endothelialization, and surgical-specific features in a porcine model. The aim of this research was to investigate the challenge of producing small-diameter grafts with an internal diameter of less than 6 mm and a length of more than 5 cm while combining BC with a cobalt-chromium mesh to potentially improve graft compliance and long-term patency. Coronary angiography, performed postoperatively at 4 weeks, confirmed complete patency of the grafts in all animals, with no signs of graft dissection or rupture, indicating successful *in vivo* function (Fusco et al., 2022).

In order to promote the adherence, proliferation, and metabolism of primary human saphenous vein endothelial cells (HSVECs) to BC, Fink et al. (Fink et al., 2011) employed a novel technique involving xyloglucan (XG) glycoconjugate and an adhesion peptide Arginylglycylaspartic acid (RGD). The XG approach can serve as a carrier molecule to introduce the RGD adhesion peptide to BC without causing chemical alterations or altering the hydrogel's shape. By modifying XG-RGD, it is possible to achieve an *in vivo* endothelialization process, eliminating the need for time-consuming preimplantation and cell expansion processes. This makes BC a promising material for artificial grafts in cardiovascular surgery. In addition, Tang et al. (Tang et al., 2015) investigated PVA/BNC composites to improve mechanical properties and biocompatibility for artificial blood vessels. Accordingly, PVA/BNC films showed improved tensile strength, reduced water absorption, increased oxygen resistance, and increased biodegradability compared to pure PVA films. Also, these composites increase cell adhesion, proliferation, and anticoagulant properties, which are very important for rapid endothelialization in small-caliber vessels.

A major challenge in vascular tissue engineering is the presence of a dense nanofibrous structure, which prevents the penetration of cells during implantation in the body. To address this challenge, a study reported BC/potato starch composites with a hierarchical structure and high biocompatibility to achieve vascular implants with a small diameter (> 6 mm). BC/potato starch (2 wt.%) tubes showed dense inner surfaces with peripheral macroporous outer layers. Also, *in vivo* evaluations confirmed the rapid regeneration of blood vessels in rabbits, with an intact endothelium monolayer, organized smooth muscle cells, rich new capillaries, and deposited ECM (Liu et al., 2022). On the other hand, Encheva et al. (Entcheva et al., 2004) used CA and regenerated cellulose to fabricate a cardiac patch with appropriate electromechanical functionality. In order to generate RC and increase the scaffold's nanoscale porosity, the researchers partially deacetylated and hydrolyzed CA. The scaffolds showed enhanced cell growth and connectivity due to the cytocompatibility of CA and the similarity of the polysaccharide scaffold microenvironment to native ECM, making them suitable candidates for cardiac tissue engineering.

8. Conclusion

In conclusion, cellulose has emerged as a promising class of materials for tissue engineering due to its unique characteristics and biocompatibility. The diverse sources of cellulose, ranging from wood and plant materials to bacterial, algal, and tunicate cellulose, provide a wide range for material selection tailored to specific tissue engineering needs. The various types of cellulose, including methyl cellulose, ethyl cellulose, and their derivatives, offer a spectrum of chemical modifications that can be exploited to enhance solubility, mechanical properties, and biological performance. The morphological diversity of cellulose, from nanofibrils to crystalline structures and fibers, allows for the creation of composites with customized architectures that can mimic the ECM of various tissues. The applications of cellulose-based composites in tissue engineering are vast and varied, spanning from wound repair to bone and cartilage regeneration, neural tissue engineering, and the treatment of cardiovascular diseases. These applications highlight the versatility and potential of cellulose composites to address a wide array of medical challenges.

As research in this field continues to advance, the integration of cellulose with other biomaterials and the development of advanced manufacturing techniques will further enhance their functionality and applicability. The future of cellulose composites in tissue engineering looks promising, with ongoing efforts to optimize their properties and expand their use in clinical settings. Ultimately, cellulose composites hold the potential to revolutionize the field of tissue engineering by providing sustainable, biocompatible, and highly functional materials for the regeneration of damaged or diseased tissues.

References

Abbott, Andrew P., Thomas J. Bell, Sandeep Handa and Barry Stoddart. 2005. O-acetylation of cellulose and monosaccharides using a zinc based ionic liquid. Green Chemistry 7(10): 705–7. https://doi.org/10.1039/B511691K.

Abdel-Halim, E. S. and Salem S. Al-Deyab. 2011. Utilization of hydroxypropyl cellulose for green and efficient synthesis of silver nanoparticles. Carbohydrate Polymers 86(4): 1615–22. https://doi.org/10.1016/j.carbpol.2011.06.072.

Ahmadi, Parisa, Ali Jahanban-Esfahlan, Azam Ahmadi, Mahnaz Tabibiazar and Mohammadamin Mohammadifar. 2022a. Development of ethyl cellulose-based formulations: a perspective on the novel technical methods. Food Reviews International 38(4): 685–732. https://doi.org/10.1080/87559129.2020.1741007.

Ahmadi, Parisa, Ali Jahanban-Esfahlan, Azam Ahmadi, Mahnaz Tabibiazar and Mohammadamin Mohammadifar. 2022b. Development of ethyl cellulose-based formulations: a perspective on the novel technical methods. Food Reviews International 38(4): 685–732. https://doi.org/10.1080/87559129.2020.1741007.

Ahmed, Jubair, Merve Gultekinoglu and Mohan Edirisinghe. 2020. Bacterial cellulose micro-nano fibres for wound healing applications. Biotechnology Advances 41(July). https://doi.org/10.1016/J.BIOTECHADV.2020.107549.

Akinjokun, Adebola Iyabode, Leslie Felicia Petrik, Aderemi Okunola Ogunfowokan, John Ajao and Tunde Victor Ojumu. 2021. Isolation and characterization of nanocrystalline cellulose from cocoa pod husk (CPH) biomass wastes. Heliyon 7(4). https://doi.org/10.1016/J.HELIYON.2021.E06680.

Al-Kaabi, Maryam A., Nabil Zouari, Dana Adel Da'na and Mohammad A. Al-Ghouti. 2021. Adsorptive batch and biological treatments of produced water: recent progresses, challenges, and potentials. Journal of Environmental Management 290(July): 112527. https://doi.org/10.1016/J.JENVMAN.2021.112527.

Altun, Esra, Mehmet O. Aydogdu, Sine O. Togay, Ahmet Z. Sengil, Nazmi Ekren, Merve E. Haskoylu, Ebru T. Oner et al. 2019. Bioinspired scaffold induced regeneration of neural tissue. European Polymer Journal 114(May): 98–108. https://doi.org/10.1016/J.EURPOLYMJ.2019.02.008.

Amorim, Julia Didier Pedrosa de, Karina Carvalho de Souza, Cybelle Rodrigues Duarte, Izarelle da Silva Duarte, Francisco de Assis Sales Ribeiro, Girlaine Santos Silva, Patrícia Maria Albuquerque de Farias et al. 2020. Plant and bacterial nanocellulose: production, properties and applications in medicine, food, cosmetics, electronics and engineering. A review. Environmental Chemistry Letters 18(3): 851–69. https://doi.org/10.1007/S10311-020-00989-9.

Andriani, Dian, Arina Yuthi Apriyana and Myrtha Karina. 2020. The optimization of bacterial cellulose production and its applications: a review. Cellulose 27(12): 6747–66. https://doi.org/10.1007/S10570-020-03273-9.

Angelova, Tsvetelina, Nadezhda Rangelova, Nelly Georgieva, Veronica Nemska, Tihomira Stoyanova, Veselina Uzunova, Lyubomir Aleksandrov and Rumiana Tzoneva. 2019. Study of potential biomedical application of sol-gel derived Zn-doped SiO_2-hydroxypropyl cellulose nanohybrids. Materials Science and Engineering: C 100(July): 608–15. https://doi.org/10.1016/J.MSEC.2019.03.018.

Aravamudhan, Aja, Daisy M. Ramos, Jonathan Nip, Matthew D. Harmon, Roshan James, Meng Deng, Cato T. Laurencin, Xiaojun Yu and Sangamesh G. Kumbar. 2013. Cellulose and collagen derived micro-nano structured scaffolds for bone tissue engineering. Journal of Biomedical Nanotechnology 9(4): 719–31. https://doi.org/10.1166/JBN.2013.1574.

Ardanuy, Mònica, Josep Claramunt and Romildo Dias Toledo Filho. 2015. Cellulosic fiber reinforced cement-based composites: a review of recent research. Construction and Building Materials 79(March): 115–28. https://doi.org/10.1016/J.CONBUILDMAT.2015.01.035.

Arumugam, Mayakrishnan, Balaji Murugesan, Nithya Pandiyan, Dhilip Kumar Chinnalagu, Gowri Rangasamy and Sundrarajan Mahalingam. 2021. Electrospinning cellulose acetate/silk fibroin/Au-Ag hybrid composite nanofiber for enhanced biocidal activity against MCF-7 breast cancer cell. Materials Science and Engineering: C 123(April): 112019. https://doi.org/10.1016/J.MSEC.2021.112019.

Ashraf, Roqia, Hasham S. Sofi, Towseef Akram, Hilal Ahmad Rather, Abdalla Abdal-hay, Nadeem Shabir, Rajesh Vasita et al. 2020. Fabrication of multifunctional cellulose/TiO_2/Ag composite nanofibers scaffold with antibacterial and bioactivity properties for future tissue engineering applications. Journal of Biomedical Materials Research Part A 108(4): 947–62. https://doi.org/10.1002/JBM.A.36872.

Aydogdu, Mehmet Onur, Joshua Chou, Esra Altun, Nazmi Ekren, Selami Cakmak, Mehmet Eroglu, Asila A. Osman et al. 2019. Production of the biomimetic small diameter blood vessels for cardiovascular tissue engineering. International Journal of Polymeric Materials and Polymeric Biomaterials 68(5): 243–55. https://doi.org/10.1080/00914037.2018.1443930.

Azevedo, Eduardo P., Rohini Retarekar, Madhavan L. Raghavan and Vijay Kumar. 2013. Mechanical properties of cellulose: chitosan blends for potential use as a coronary artery bypass graft. Journal of Biomaterials Science, Polymer Edition 24(3): 239–52. https://doi.org/10.1080/09205063.2012.690273.

Aziz, Tariq, Arshad Farid, Fazal Haq, Mehwish Kiran, Asmat Ullah, Kechun Zhang, Cheng Li et al. 2022. A review on the modification of cellulose and its applications. Polymers 14(15): 3206. https://doi.org/10.3390/POLYM14153206.

Bäckdahl, Henrik, Gisela Helenius, Aase Bodin, Ulf Nannmark, Bengt R. Johansson, Bo Risberg and Paul Gatenholm. 2006. Mechanical properties of bacterial cellulose and interactions with smooth muscle cells. Biomaterials 27(9): 2141–49. https://doi.org/10.1016/J.BIOMATERIALS.2005.10.026.

Badylak, Stephen F. and Robert M. Nerem. 2010. Progress in tissue engineering and regenerative medicine. Proceedings of the National Academy of Sciences of the United States of America 107(8): 3285. https://doi.org/10.1073/PNAS.1000256107.

Bangar, Sneh Punia and William Scott Whiteside. 2021. Nano-cellulose reinforced starch bio composite films—a review on green composites. International Journal of Biological Macromolecules 185(August): 849–60. https://doi.org/10.1016/J.IJBIOMAC.2021.07.017.

Bao, Luhan, Can Li, Man Tang, Lin Chen and Feng F. Hong. 2022. Potential of a composite conduit with bacterial nanocellulose and fish gelatin for application as small-diameter artificial blood vessel. Polymers 14(20): 4367. https://doi.org/10.3390/POLYM14204367.

Berthumeyrie, Sebastien, Steeve Collin, Pierre Olivier Bussiere and Sandrine Therias. 2014. Photooxidation of cellulose nitrate: new insights into degradation mechanisms. Journal of Hazardous Materials 272(2): 137–47. https://doi.org/10.1016/j.jhazmat.2014.02.039.

Bhatt, Neetu, P. K. Gupta and Sanjay Naithani. 2011. Hydroxypropyl cellulose from α-cellulose isolated from Lantana Camara with respect to DS and rheological behavior. Carbohydrate Polymers 86(4): 1519–24. https://doi.org/10.1016/J.CARBPOL.2011.06.054.

Bian, Donghui, Yan Wu, Guodong Song, Ramyar Azizi and Amir Zamani. 2022. The application of mesenchymal stromal cells (MSCs) and their derivative exosome in skin wound healing: a comprehensive review. Stem Cell Research & Therapy 13(1): 1–17. https://doi.org/10.1186/S13287-021-02697-9.

Birgersson, Paulina S., Maren Oftebro, Wenche I. Strand, Olav A. Aarstad, Gerd Inger Sætrom, Håvard Sletta, Øystein Arlov et al. 2023. Sequential extraction and fractionation of four polysaccharides from cultivated brown algae Saccharina Latissima and Alaria Esculenta. Algal Research 69(January): 102928. https://doi.org/10.1016/J.ALGAL.2022.102928.

Bogolitsyn, Konstantin, Anastasia Parshina and Lyudmila Aleshina. 2020. Structural features of brown algae cellulose. Cellulose 27(17): 9787–9800. https://doi.org/10.1007/S10570-020-03485-Z/METRICS.

Bragazzi, Nicola Luigi, Maha Sellami, Iman Salem, Rosalynn Conic, Mark Kimak, Paolo Daniele Maria Pigatto and Giovanni Damiani. 2019. Fasting and its impact on skin anatomy, physiology, and physiopathology: a comprehensive review of the literature. Nutrients 11(2). https://doi.org/10.3390/NU11020249.

Buldum, Gizem and Athanasios Mantalaris. 2021. Systematic understanding of recent developments in bacterial cellulose biosynthesis at genetic, bioprocess and product levels. International Journal of Molecular Sciences 22(13): 7192. https://doi.org/10.3390/IJMS22137192.

Burchard, Walther. 2003. Solubility and solution structure of cellulose derivatives. Cellulose 10(3): 213–25. https://doi.org/10.1023/A:1025160620576/METRICS.

Burger, Dennis, Marco Beaumont, Thomas Rosenau and Yasushi Tamada. 2020. Porous silk fibroin/cellulose hydrogels for bone tissue engineering via a novel combined process based on sequential regeneration and porogen leaching. Molecules 25(21): 5097. https://doi.org/10.3390/MOLECULES25215097.

Cai, Zhijiang and Jaehwan Kim. 2010. Bacterial cellulose/poly(ethylene glycol) composite: characterization and first evaluation of biocompatibility. Cellulose 17(1): 83–91. https://doi.org/10.1007/S10570-009-9362-5/FIGURES/6.

Cakmak, Abdullah M., Semra Unal, Ali Sahin, Faik N. Oktar, Mustafa Sengor, Nazmi Ekren, Oguzhan Gunduz and Deepak M. Kalaskar. 2020. 3D printed polycaprolactone/gelatin/bacterial cellulose/hydroxyapatite composite scaffold for bone tissue engineering. Polymers 12(9): 1962. https://doi.org/10.3390/POLYM12091962.

Camman, Marie, Pierre Joanne, Onnik Agbulut and Christophe Hélary. 2022. 3D models of dilated cardiomyopathy: shaping the chemical, physical and topographical properties of biomaterials to mimic the cardiac extracellular matrix. Bioactive Materials 7(January): 275–91. https://doi.org/10.1016/J.BIOACTMAT.2021.05.040.

Camponeschi, Francesca, Andrea Atrei, Giulia Rocchigiani, Lorenzo Mencuccini, Marianna Uva and Rolando Barbucci. 2015. New formulations of polysaccharide-based hydrogels for drug release and tissue engineering. Gels 1(1): 3. https://doi.org/10.3390/GELS1010003.

"Cellulose." 2008. In Crystalline Cellulose and Derivatives: Characterization and Structures, 101–74. Berlin, Heidelberg: Springer Berlin Heidelberg. https://doi.org/10.1007/978-3-540-73934-0_5.

Chahal, Sugandha, Fathima Shahitha Jahir Hussain, Anuj Kumar, Mohammad Syaiful Bahari Abdull Rasad and Mashitah Mohd Yusoff. 2016. Fabrication, characterization and *in vitro* biocompatibility of electrospun hydroxyethyl cellulose/poly (vinyl) alcohol nanofibrous composite biomaterial for bone tissue engineering. Chemical Engineering Science 144(April): 17–29. https://doi.org/10.1016/J.CES.2015.12.030.

Chahal, Sugandha, Fathima Shahitha Jahir Hussain, Mashitah Mohd Yusoff, Mohammad Syaiful Bahari Abdull Rasad and Anuj Kumar. 2017. Nanohydroxyapatite-coated hydroxyethyl cellulose/poly (vinyl) alcohol electrospun scaffolds and their cellular response. International Journal of Polymeric Materials and Polymeric Biomaterials 66(3): 115–22. https://doi.org/10.1080/00914037.2016.1190926.

Chen, Chao, Yuewei Xi and Yunxuan Weng. 2022. Recent advances in cellulose-based hydrogels for tissue engineering applications. Polymers 14(16): 3335. https://doi.org/10.3390/POLYM14163335.

Chen, Depu and Baoquan Sun. 2000. New tissue engineering material copolymers of derivatives of cellulose and lactide: their synthesis and characterization. Materials Science and Engineering: C 11(1): 57–60. https://doi.org/10.1016/S0928-4931(00)00131-4.

Chen, Guo, Bin Zhang, Jun Zhao and Hongwen Chen. 2014. Development and characterization of food packaging film fromcellulose sulfate. Food Hydrocolloids 35: 476–83. https://doi.org/10.1016/j.foodhyd.2013.07.003.

Chen, Juan, Tonghui Zhang, Weikang Hua, Peiyun Li and Xuefen Wang. 2020. 3D porous poly(lactic acid)/regenerated cellulose composite scaffolds based on electrospun nanofibers for biomineralization. Colloids and Surfaces A: Physicochemical and Engineering Aspects 585(January): 124048. https://doi.org/10.1016/J.COLSURFA.2019.124048.

Chen, Po Hsuen, Hsueh Chung Liao, Sheng Hao Hsu, Rung Shu Chen, Ming Chung Wu, Yi Fan Yang et al. 2014. A novel polyurethane/cellulose fibrous scaffold for cardiac tissue engineering. RSC Advances 5(9): 6932–39. https://doi.org/10.1039/C4RA12486C.

Chen, Yandan, Qiaomei Wu, Biao Huang, Mingjie Huang and Xiaolin Ai. n.d. Isolation and characteristics of cellulose and nanocellulose from lotus leaf stalk agro-wastes.

Cheng, Alice, Zvi Schwartz, Adrian Kahn, Xiyu Li, Zhenxing Shao, Muyang Sun, Yingfang Ao, Barbara D. Boyan and Haifeng Chen. 2019. Advances in porous scaffold design for bone and cartilage tissue engineering and regeneration. Tissue Engineering - Part B: Reviews 25(1): 14–29. https://doi.org/10.1089/ten.teb.2018.0119.

Cheng, Yanan, Ajoy Kanti Mondal, Shuai Wu, Dezhong Xu, Dengwen Ning, Yonghao Ni and Fang Huang. 2020. Study on the anti-biodegradation property of tunicate cellulose. Polymers 12(12): 3071. https://doi.org/10.3390/POLYM12123071.

Chiaoprakobkij, Nadda, Neeracha Sanchavanakit, Keskanya Subbalekha, Prasit Pavasant and Muenduen Phisalaphong. 2011. Characterization and biocompatibility of bacterial cellulose/alginate composite sponges with human keratinocytes and gingival fibroblasts. Carbohydrate Polymers 85(3): 548–53. https://doi.org/10.1016/J.CARBPOL.2011.03.011.

Chimpibul, Wichchulada, Tadashi Nakaji-Hirabayashi, Xida Yuan and Kazuaki Matsumura. 2020. Controlling the degradation of cellulose scaffolds with malaprade oxidation for tissue engineering. Journal of Materials Chemistry B 8(35): 7904–13. https://doi.org/10.1039/D0TB01015D.

Chiulan, Ioana, Denis Mihaela Panaitescu, Adriana Nicoleta Frone, Mircea Teodorescu, Cristian Andi Nicolae, Angela Căşărică, Vlad Tofan et al. 2016. Biocompatible polyhydroxyalkanoates/bacterial cellulose composites: preparation, characterization, and *in vitro* evaluation. Journal of Biomedical Materials Research Part A 104(10): 2576–84. https://doi.org/10.1002/JBM.A.35800.

Choi, Soon Mo and Eun Joo Shin. 2020. The nanofication and functionalization of bacterial cellulose and its applications. Nanomaterials 10(3): 406. https://doi.org/10.3390/NANO10030406.

Choi, Soon Mo, Kummara Madhusudana Rao, Sun Mi Zo, Eun Joo Shin and Sung Soo Han. 2022. Bacterial cellulose and its applications. Polymers 14(6): 1080. https://doi.org/10.3390/POLYM14061080.

Correia, Carla Almeda and Ticiane Sanches Valera. 2019. Cellulose nanocrystals and jute fiber-reinforced natural rubber composites: cure characteristics and mechanical properties. Materials Research 22(August): e20190192. https://doi.org/10.1590/1980-5373-MR-2019-0192.

Daugela, Povilas, Mindaugas Pranskunas, Gintaras Juodzbalys, Jolanta Liesiene, Odeta Baniukaitiene, Américo Afonso and Pedro Sousa Gomes. 2018. Novel cellulose/hydroxyapatite scaffolds for bone tissue regeneration: *in vitro* and *in vivo* study. Journal of Tissue Engineering and Regenerative Medicine 12(5): 1195–1208. https://doi.org/10.1002/TERM.2651.

David, Madalina Elena, Rodica Mariana Ion, Ramona Marina Grigorescu, Lorena Iancu, Alina Maria Holban, Florin Iordache et al. 2022. Biocompatible and antimicrobial cellulose acetate-collagen films containing MWCNTs decorated with TiO_2 nanoparticles for potential biomedical applications. Nanomaterials 12(2): 239. https://doi.org/10.3390/NANO12020239.

Deng, Jin Rui, Cong Lei Zhao and Yi Xian Wu. 2020. Antibacterial and pH-responsive quaternized hydroxypropyl cellulose-g-poly(thf-co-epichlorohydrin) graft copolymer: synthesis, characterization and properties. Chinese Journal of Polymer Science (English Edition) 38(7): 704–14. https://doi.org/10.1007/S10118-020-2372-3/METRICS.

Desai, Mital, Alexander M. Seifalian and George Hamilton. 2011. Role of prosthetic conduits in coronary artery bypass grafting. European Journal of Cardio-Thoracic Surgery 40(2): 394–98. https://doi.org/10.1016/J.EJCTS.2010.11.050.

Dhali, Kingshuk, Mehran Ghasemlou, Fugen Daver, Peter Cass and Benu Adhikari. 2021. A review of nanocellulose as a new material towards environmental sustainability. The Science of the Total Environment 775(June). https://doi.org/10.1016/J.SCITOTENV.2021.145871.

Diagnosis, Medical. 2020. Recent Advances in Cellulose-Based Biosensors For, 1–26.

Ding, Chaofan, Yue Li, Yanhui Wang, Jianbo Li, Yuanling Sun, Yanna Lin, Weiyan Sun and Chuannan Luo. 2018. Highly selective adsorption of hydroquinone by hydroxyethyl cellulose functionalized with magnetic/ionic liquid. International Journal of Biological Macromolecules 107(Pt A): 957–64. https://doi.org/10.1016/J.IJBIOMAC.2017.09.075.

Doblado, Laura Rodríguez, Cristina Martínez-Ramos and Manuel Monleón Pradas. 2021. Biomaterials for neural tissue engineering. Frontiers in Nanotechnology 3(April). https://doi.org/10.3389/fnano.2021.643507.

Dominic, Midhun, Rani Joseph, P. M. Sabura Begum, Bipinbal Parambath Kanoth, Julie Chandra and Sanmariya Thomas. 2020. Green tire technology: effect of rice husk derived nanocellulose (RHNC) in replacing carbon black (CB) in natural rubber (NR) compounding. Carbohydrate Polymers 230(February): 115620. https://doi.org/10.1016/J.CARBPOL.2019.115620.

Dormanns, Jan W., Jeremias Schuermann, Jörg Müssig, Benoît J.C. Duchemin and Mark P. Staiger. 2016. Solvent infusion processing of all-cellulose composite laminates using an aqueous NaOH/urea solvent system. Composites Part A: Applied Science and Manufacturing 82(March): 130–40. https://doi.org/10.1016/J.COMPOSITESA.2015.12.002.

Dutta, Sayan Deb, Jin Hexiu, Dinesh K. Patel, Keya Ganguly and Ki Taek Lim. 2021. 3D-printed bioactive and biodegradable hydrogel scaffolds of alginate/gelatin/cellulose nanocrystals for tissue engineering. International Journal of Biological Macromolecules 167(January): 644–58. https://doi.org/10.1016/J.IJBIOMAC.2020.12.011.

Edgar, Kevin J. 2004. Cellulose Esters, Organic. Encyclopedia of Polymer Science and Technology, July. https://doi.org/10.1002/0471440264.PST045.

Edgar, Kevin J. 2007. Cellulose esters in drug delivery. Cellulose 14(1): 49–64. https://doi.org/10.1007/S10570-006-9087-7/METRICS.

Edgar, Kevin J., Charles M. Buchanan, John S. Debenham, Paul A. Rundquist, Brian D. Seiler, Michael C. Shelton et al. 2001. Advances in cellulose ester performance and application. Progress in Polymer Science 9(26): 1605–88. https://doi.org/10.1016/S0079-6700(01)00027-2.

Eduok, Ubong, Amira Abdelrasoul, Ahmed Shoker and Huu Doan. 2021. Recent developments, current challenges and future perspectives on cellulosic hemodialysis membranes for highly efficient clearance of uremic toxins. Materials Today Communications 27(November 2020): 102183. https://doi.org/10.1016/j.mtcomm.2021.102183.

El-Gendi, Hamada, Ahmed Salama, Esmail M. El-Fakharany and Ahmed K. Saleh. 2023. Optimization of bacterial cellulose production from prickly pear peels and its *ex situ* impregnation with fruit byproducts for antimicrobial and strawberry packaging applications. Carbohydrate Polymers 302(February): 120383. https://doi.org/10.1016/J.CARBPOL.2022.120383.

El-Newehy, Mohamed H., Mehrez E. El-Naggar, Saleh Alotaiby, Hany El-Hamshary, Meera Moydeen and Salem Al-Deyab. 2017. Green electrospining of hydroxypropyl cellulose nanofibres for drug delivery applications. Journal of Nanoscience and Nanotechnology 18(2): 805–14. https://doi.org/10.1166/jnn.2018.13852.

Eldeeb, Alaa Emad, Salwa Salah and Nermeen A. Elkasabgy. 2022. Biomaterials for tissue engineering applications and current updates in the field: a comprehensive review. AAPS PharmSciTech 23(7): 1–25. https://doi.org/10.1208/S12249-022-02419-1.

Entcheva, Emilia, Harold Bien, Lihong Yin, Chiung Yin Chung, Melissa Farrell and Yordan Kostov. 2004. Functional cardiac cell constructs on cellulose-based scaffolding. Biomaterials 25(26): 5753–62. https://doi.org/10.1016/j.biomaterials.2004.01.024.

Esmaeili, Elaheh, Tarlan Eslami-Arshaghi, Simzar Hosseinzadeh, Elnaz Elahirad, Zahra Jamalpoor, Shadie Hatamie and Masoud Soleimani. 2020. The biomedical potential of cellulose acetate/polyurethane nanofibrous mats containing reduced graphene oxide/silver nanocomposites and curcumin: antimicrobial performance and cutaneous wound healing. International Journal of Biological Macromolecules 152(June): 418–27. https://doi.org/10.1016/J.IJBIOMAC.2020.02.295.

Espinha, André, Camilla Dore, Cristiano Matricardi, Maria Isabel Alonso, Alejandro R. Goñi and Agustín Mihi. 2018. Hydroxypropyl cellulose photonic architectures by soft nanoimprinting lithography. Nature Photonics 12(6): 343–48. https://doi.org/10.1038/s41566-018-0152-1.

Fang, Bo, Yi Zao Wan, Ting Ting Tang, Chuan Gao and Ke Rong Dai. 2009. Proliferation and osteoblastic differentiation of human bone marrow stromal cells on hydroxyapatite/bacterial cellulose nanocomposite scaffolds. Tissue Engineering - Part A 15(5): 1091–98. https://doi.org/10.1089/ten.tea.2008.0110.

Farzamfar, Saeed, Mahdi Naseri-Nosar, Ahmad Vaez, Farshid Esmaeilpour, Arian Ehterami, Hamed Sahrapeyma, Hadi Samadian et al. 2018. Neural tissue regeneration by a gabapentin-loaded cellulose acetate/gelatin wet-electrospun scaffold. Cellulose 25(2): 1229–38. https://doi.org/10.1007/S10570-017-1632-Z/FIGURES/7.

Favi, Pelagie Marlene, Sandra Patricia Ospina, Mukta Kachole, Ming Gao, Lucia Atehortua and Thomas Jay Webster. 2016. Preparation and characterization of biodegradable nano hydroxyapatite–bacterial cellulose composites with well-defined honeycomb pore arrays for bone tissue engineering applications. Cellulose 23(2): 1263–82. https://doi.org/10.1007/S10570-016-0867-4/FIGURES/13.

Fawal, Gomaa F. El, Marwa M. Abu-Serie, Mohamed A. Hassan and Mohamed S. Elnouby. 2018. Hydroxyethyl cellulose hydrogel for wound dressing: fabrication, characterization and *in vitro* evaluation. International Journal of Biological Macromolecules 111: 649–59. https://doi.org/10.1016/j.ijbiomac.2018.01.040.

Feigin, Valery L., Emma Nichols, Tahiya Alam, Marlena S. Bannick, Ettore Beghi, Natacha Blake, William J. Culpepper et al. 2019. Global, regional, and national burden of neurological disorders, 1990–2016: a systematic analysis for the global burden of disease study 2016. The Lancet Neurology 18(5): 459–80. https://doi.org/10.1016/S1474-4422(18)30499-X.

Fernandes, Isabela de Andrade Arruda, Alessandra Cristina Pedro, Valéria Rampazzo Ribeiro, Débora Gonçalves Bortolini, Mellany Sarah Cabral Ozaki, Giselle Maria Maciel et al. 2020. Bacterial cellulose: from production optimization to new applications. International Journal of Biological Macromolecules 164(December): 2598–2611. https://doi.org/10.1016/J.IJBIOMAC.2020.07.255.

Filip, Daniela, Doina Macocinschi, Mihaela Balan-Porcarasu, Cristian Dragos Varganici, Raluca Petronela Dumitriu, Dragos Peptanariu, Cristina Gabriela Tuchilus et al. 2022. Biocompatible self-assembled hydrogen-bonded gels based on natural deep eutectic solvents and hydroxypropyl cellulose with strong antimicrobial activity. Gels 8(10): 666. https://doi.org/10.3390/GELS8100666/S1.

Filip, Daniela, Doina Macocinschi, Mirela Fernanda Zaltariov, Bianca Iulia Ciubotaru, Alexandra Bargan, Cristian Dragos Varganici, Ana Lavinia Vasiliu et al. 2022. Hydroxypropyl cellulose/pluronic-based composite hydrogels as biodegradable mucoadhesive scaffolds for tissue engineering. Gels 8(8): 519. https://doi.org/10.3390/GELS8080519.

Fink, Helen, Lage Ahrenstedt, Aase Bodin, Harry Brumer, Paul Gatenholm, Alexandra Krettek and Bo Risberg. 2011. Bacterial cellulose modified with xyloglucan bearing the adhesion peptide RGD promotes endothelial cell adhesion and metabolism—a promising modification for vascular grafts. Journal of Tissue Engineering and Regenerative Medicine 5(6): 454–63. https://doi.org/10.1002/TERM.334.

Fioretta, Emanuela S., Sarah E. Motta, Valentina Lintas, Sandra Loerakker, Kevin K. Parker, Frank P.T. Baaijens et al. 2021. Next-generation tissue-engineered heart valves with repair, remodelling and regeneration capacity. Nature Reviews. Cardiology 18(2): 92–116. https://doi.org/10.1038/S41569-020-0422-8.

Fischer, Steffen, Katrin Thümmler, Bert Volkert, Kay Hettrich, Ingeborg Schmidt and Klaus Fischer. 2008. Properties and applications of cellulose acetate. Macromolecular Symposia 262(1): 89–96. https://doi.org/10.1002/masy.200850210.

Fooladi, Saba, Sanaz Faramarz, Shahriar Dabiri, Abdolmohammad Kajbafzadeh, Mohammad Hadi Nematollahi and Mehrnaz Mehrabani. 2022. An efficient strategy to recellularization of a rat aorta scaffold: an optimized decellularization, detergent removal, and Apelin-13 immobilization. Biomaterials Research 26(1): 1–17. https://doi.org/10.1186/S40824-022-00295-1/FIGURES/8.

Fortunati, E., F. Luzi, A. Jiménez, D. A. Gopakumar, D. Puglia, S. Thomas et al. 2016. Revalorization of sunflower stalks as novel sources of cellulose nanofibrils and nanocrystals and their effect on wheat gluten bionanocomposite properties. Carbohydrate Polymers 149(September): 357–68. https://doi.org/10.1016/J.CARBPOL.2016.04.120.

France, Kevin De, Zhihui Zeng, Tingting Wu and Gustav Nyström. 2021. Functional materials from nanocellulose: utilizing structure–property relationships in bottom-up fabrication. Advanced Materials 33(28): 2000657. https://doi.org/10.1002/ADMA.202000657.

France, Kevin J. De, Katelyn J. W. Chan, Emily D. Cranston and Todd Hoare. 2016. Enhanced mechanical properties in cellulose nanocrystal–poly(oligoethylene glycol methacrylate) injectable nanocomposite hydrogels through control of physical and chemical cross-linking. Biomacromolecules 17(2): 649–60. https://doi.org/10.1021/ACS.BIOMAC.5B01598.

French, Alfred D. and Glenn P. Johnson. 2007. Cellulose shapes. Cellulose: Molecular and Structural Biology, 257–84. https://doi.org/10.1007/978-1-4020-5380-1_15.

Frone, Adriana Nicoleta, Ioana Chiulan, Denis Mihaela Panaitescu, Cristian Andi Nicolae, Marius Ghiurea and Ana Maria Galan. 2017. Isolation of cellulose nanocrystals from plum seed shells, structural and morphological characterization. Materials Letters 194(May): 160–63. https://doi.org/10.1016/J.MATLET.2017.02.051.

Frost, Hanna K., Tomas Andersson, Sebastian Johansson, U. Englund-Johansson, Per Ekström, Lars B. Dahlin and Fredrik Johansson. 2018. Electrospun nerve guide conduits have the potential to bridge peripheral nerve injuries *in vivo*. Scientific Reports 8(1): 1–13. https://doi.org/10.1038/s41598-018-34699-8.

Fu, Lian Hua, Chao Qi, Yan Jun Liu, Wen Tao Cao and Ming Guo Ma. 2018. Sonochemical synthesis of cellulose/hydroxyapatite nanocomposites and their application in protein adsorption. Scientific Reports 8(1): 1–12. https://doi.org/10.1038/s41598-018-25566-7.

Fu, Ranran, Yanfei Ren, Kuanjun Fang, Yaning Sun, Zaicheng Zhang and Anqiao Luo. 2021. Preparation, characterization and biocompatibility of chitosan/TEMPO-oxidized bacterial cellulose composite film for potential wound dressing applications. Fibers and Polymers 22(7): 1790–99. https://doi.org/10.1007/S12221-021-0854-8/METRICS.

Fukaya, Yukinobu, Kensaku Hayashi, Masahisa Wada and Hiroyuki Ohno. 2008. Cellulose dissolution with polar ionic liquids under mild conditions: required factors for anions. Green Chemistry 10(1): 44–46. https://doi.org/10.1039/B713289A.

Fusco, Deborah, Florian Meissner, Bruno K. Podesser, Anna Marsano, Martin Grapow, Friedrich Eckstein et al. 2022. Small-diameter bacterial cellulose-based vascular grafts for coronary artery bypass grafting in a pig model. Frontiers in Cardiovascular Medicine 9(September): 881557. https://doi.org/10.3389/FCVM.2022.881557/BIBTEX.

Gaihre, Bipin and Ambalangodage C. Jayasuriya. 2016. Fabrication and characterization of carboxymethyl cellulose novel microparticles for bone tissue engineering. Materials Science and Engineering: C 69(December): 733–43. https://doi.org/10.1016/J.MSEC.2016.07.060.

Gan, Shuchun, Shihang Bai, Cheng Chen, Yongliang Zou, Yingjuan Sun, Jianhao Zhao and Jianhua Rong. 2021. Hydroxypropyl cellulose enhanced ionic conductive double-network hydrogels. International Journal of Biological Macromolecules 181(June): 418–25. https://doi.org/10.1016/J.IJBIOMAC.2021.03.068.

Gelman, Robert A. 1982. Characterization of carboxymethylcellulose: distribution of substituent groups along the chain. Journal of Applied Polymer Science 27(8): 2957–64. https://doi.org/10.1002/APP.1982.070270820.

George, Johnsy and S. N. Sabapathi. 2015. Cellulose nanocrystals: synthesis, functional properties, and applications. Nanotechnology, Science and Applications 8(November): 45–54. https://doi.org/10.2147/NSA.S64386.

George, Julian, Chia Chen Hsu, Linh Thuy Ba Nguyen, Hua Ye and Zhanfeng Cui. 2020. Neural tissue engineering with structured hydrogels in CNS models and therapies. Biotechnology Advances. Elsevier Inc. https://doi.org/10.1016/j.biotechadv.2019.03.009.

Gericke, Martin, Tim Liebert, Omar A.El Seoud and Thomas Heinze. 2011. Tailored media for homogeneous cellulose chemistry: ionic liquid/co-solvent mixtures. Macromolecular Materials and Engineering 296(6): 483–93. https://doi.org/10.1002/MAME.201000330.

Gharib Khajeh, Hossein, Mohammad Sabzi, Soghra Ramezani, Amir Abbas Jalili and Marjan Ghorbani. 2022. Fabrication of a wound dressing mat based on polyurethane/polyacrylic acid containing poloxamer for skin tissue engineering. Colloids and Surfaces A: Physicochemical and Engineering Aspects 633(January): 127891. https://doi.org/10.1016/J.COLSURFA.2021.127891.

Ghorbani, Marjan, Parinaz Nezhad-Mokhtari, Hessamaddin Sohrabi and Leila Roshangar. 2020. Electrospun chitosan/nanocrystalline cellulose-graft-poly(n-vinylcaprolactam) nanofibers as the reinforced scaffold for tissue engineering. Journal of Materials Science 55(5): 2176–85. https://doi.org/10.1007/S10853-019-04115-1/METRICS.

Gisone, Ilaria, Antonella Cecchettini, Elisa Ceccherini, Elisa Persiani, Maria Aurora Morales and Federico Vozzi. 2022. Cardiac tissue engineering: multiple approaches and potential applications. Frontiers in Bioengineering and Biotechnology 10(October). https://doi.org/10.3389/FBIOE.2022.980393.

Goff, Kevin J. Le, Cedric Gaillard, William Helbert, Catherine Garnier and Thierry Aubry. 2015. Rheological study of reinforcement of agarose hydrogels by cellulose nanowhiskers. Carbohydrate Polymers 116(February): 117–23. https://doi.org/10.1016/J.CARBPOL.2014.04.085.

Gopi, S., P. Balakrishnan, D. Chandradhara, D. Poovathankandy and S. Thomas. 2019. General scenarios of cellulose and its use in the biomedical field. Materials Today Chemistry 13(September): 59–78. https://doi.org/10.1016/J.MTCHEM.2019.04.012.

Gouma, P., R. Xue, C. P. Goldbeck, P. Perrotta and C. Balázsi. 2012. Nano-hydroxyapatite - cellulose acetate composites for growing of bone cells. Materials Science and Engineering C 32(3): 607–12. https://doi.org/10.1016/j.msec.2011.12.019.

Gregory, David A., Lakshmi Tripathi, Annabelle T. R. Fricker, Emmanuel Asare, Isabel Orlando, Vijayendran Raghavendran et al. 2021. Bacterial cellulose: a smart biomaterial with diverse applications. Materials Science and Engineering: R: Reports 145(July): 100623. https://doi.org/10.1016/J.MSER.2021.100623.

Guamba, Esteban, Nelson Santiago Vispo, Daniel C. Whitehead, Ajaya Kumar Singh, Ralph Santos-Oliveira, Dario Niebieskikwiat et al. 2023. Cellulose-based hydrogels towards an antibacterial wound dressing. Biomaterials Science 11(10): 3461–68. https://doi.org/10.1039/D2BM01369J.

Guhados, Ganesh, Wankei Wan and Jeffrey L. Hutter. 2005. Measurement of the elastic modulus of single bacterial cellulose fibers using atomic force microscopy. Langmuir 21(14): 6642–46. https://doi.org/10.1021/LA0504311/ASSET/IMAGES/MEDIUM/LA0504311N00001.GIF.

Guo, Xuejie, Li Xue, Weizhong Lv, Qi Liu, Rumin Li, Zhanshuang Li and Jun Wang. 2015. Facile synthesis of magnetic carboxymethylcellulose nanocarriers for pH-responsive delivery of doxorubicin. New Journal of Chemistry 39(9): 7340–47. https://doi.org/10.1039/C5NJ01190F.

Gupta, Praveen Kumar, Shreeya Sai Raghunath, Deepali Venkatesh Prasanna, Priyadharsini Venkat, Vidhya Shree, Chandrananthi Chithananthan et al. 2019. An update on overview of cellulose, its structure and applications. Cellulose, May. https://doi.org/10.5772/INTECHOPEN.84727.

Håkansson, Karl M. O., Andreas B. Fall, Fredrik Lundell, Shun Yu, Christina Krywka, Stephan V. Roth et al. 2014. Hydrodynamic alignment and assembly of nanofibrils resulting in strong cellulose filaments. Nature Communications 5(1): 1–10. https://doi.org/10.1038/ncomms5018.

Hanif, Zahid, Farrukh R. Ahmed, Seung Won Shin, Young Kee Kim and Soong Ho Um. 2014. Size- and dose-dependent toxicity of cellulose nanocrystals (CNC) on human fibroblasts and colon adenocarcinoma. Colloids and Surfaces B: Biointerfaces 119(July): 162–65. https://doi.org/10.1016/J.COLSURFB.2014.04.018.

Haq, Muhammad Abdul, Yasuhiro Habu, Kazuya Yamamoto, Akihiko Takada and Jun ichi Kadokawa. 2019. Ionic liquid induces flexibility and thermoplasticity in cellulose film. Carbohydrate Polymers 223(November): 115058. https://doi.org/10.1016/J.CARBPOL.2019.115058.

Hasanin, Mohamed S. 2022. Cellulose-based biomaterials: chemistry and biomedical applications. Starch - Stärke 74(7–8): 2200060. https://doi.org/10.1002/STAR.202200060.

He, Xichan, Xialian Fan, Wenpo Feng, Yifei Chen, Ting Guo, Fang Wang et al. 2018. Incorporation of microfibrillated cellulose into collagen-hydroxyapatite scaffold for bone tissue engineering. International Journal of Biological Macromolecules 115(August): 385–92. https://doi.org/10.1016/J.IJBIOMAC.2018.04.085.

Heinze, Thomas. 2015. Cellulose: structure and properties. Advances in Polymer Science 271: 1–52. https://doi.org/10.1007/12_2015_319.

Hickey, Ryan J. and Andrew E. Pelling. 2019. Cellulose biomaterials for tissue engineering. Frontiers in Bioengineering and Biotechnology 7(MAR): 1–15. https://doi.org/10.3389/fbioe.2019.00045.

Hobzova, Radka, Jakub Hrib, Jakub Sirc, Evgeny Karpushkin, Jiri Michalek, Olga Janouskova and Paul Gatenholm. 2018. Embedding of bacterial cellulose nanofibers within PHEMA hydrogel matrices: tunable stiffness composites with potential for biomedical applications. Journal of Nanomaterials 2018. https://doi.org/10.1155/2018/5217095.

Hoemann, Caroline D., J. Sun, A. Légaré, M. D. McKee and M. D. Buschmann. 2005. Tissue engineering of cartilage using an injectable and adhesive chitosan-based cell-delivery vehicle. Osteoarthritis and Cartilage 13(4): 318–29. https://doi.org/10.1016/j.joca.2004.12.001.

Horvat, Gabrijela, Jan Rožanc, Uroš Maver, Matjaž Finšgar, Željko Knez and Zoran Novak. 2024. Reinforcing ethyl cellulose aerogels with poly(lactic acid) for enhanced bone regeneration. Cellulose 31(7): 4421–39. https://doi.org/10.1007/S10570-024-05905-W/FIGURES/8.

Hou, Yuanjing, Xinyu Wang, Jing Yang, Rong Zhu, Zongrui Zhang and Yi Li. 2018. Development and biocompatibility evaluation of biodegradable bacterial cellulose as a novel peripheral nerve scaffold. Journal of Biomedical Materials Research Part A 106(5): 1288–98. https://doi.org/10.1002/JBM.A.36330.

Hu, Yang and Jeffrey M. Catchmark. 2011. Integration of cellulases into bacterial cellulose: toward bioabsorbable cellulose composites. Journal of Biomedical Materials Research. Part B, Applied Biomaterials 97(1): 114–23. https://doi.org/10.1002/JBM.B.31792.

Hua, Kai, Daniel O. Carlsson, Eva Ålander, Tom Lindström, Maria Strømme, Albert Mihranyan et al. 2013. Translational study between structure and biological response of nanocellulose from wood and green algae. RSC Advances 4(6): 2892–2903. https://doi.org/10.1039/C3RA45553J.

Huang, An, Xiangfang Peng, Lihong Geng, Lingli Zhang, Keqing Huang, Binyi Chen et al. 2018. Electrospun poly (butylene succinate)/cellulose nanocrystals bio-nanocomposite scaffolds for tissue engineering: preparation, characterization and *in vitro* evaluation. Polymer Testing 71(October): 101–9. https://doi.org/10.1016/J.POLYMERTESTING.2018.08.027.

Huangqin Chen and Mingwen Fan. 2007. Chitosan/carboxymethyl cellulose polyelectrolyte complex scaffolds for pulp cells regeneration. Http://Dx.Doi.Org/10.1177/0883911507081329 22(5): 475–91. https://doi.org/10.1177/0883911507081329.

Hussein, Jihan, Mona El-Banna, Tay Abdel Razik and Mehrez E. El-Naggar. 2018. Biocompatible zinc oxide nanocrystals stabilized via hydroxyethyl cellulose for mitigation of diabetic complications. International Journal of Biological Macromolecules 107(PartA): 748–54. https://doi.org/10.1016/J.IJBIOMAC.2017.09.056.

Ioniță, Mariana, Livia Elena Crică, Stefan Ioan Voicu, Sorina Dinescu, Florin Miculescu, Marieta Costache et al. 2018. Synergistic effect of carbon nanotubes and graphene for high performance cellulose acetate membranes in biomedical applications. Carbohydrate Polymers 183(March): 50–61. https://doi.org/10.1016/J.CARBPOL.2017.10.095.

Isogai, Akira, Makoto Usuda, Takashi Kato, Toshiyuki Uryu and Rajai H. Atalla. 1989. Solid-state CP/MAS 13C NMR study of cellulose polymorphs. Macromolecules 22(7): 3168–72. https://doi.org/10.1021/MA00197A045/ASSET/MA00197A045.FP.PNG_V03.

Ivanov, A. N., M. S. Saveleva, M. N. Kozadaev, O. V. Matveeva, Yu E. Sal'kovskiy, G. P. Lyubun et al. 2019. New approaches to scaffold biocompatibility assessment. BioNanoScience 9(2): 395–405. https://doi.org/10.1007/S12668-019-00613-3/METRICS.

Iwamoto, S., A. N. Nakagaito and H. Yano. 2007. Nano-fibrillation of pulp fibers for the processing of transparent nanocomposites. Applied Physics A: Materials Science and Processing 89(2): 461–66. https://doi.org/10.1007/S00339-007-4175-6/METRICS.

Iwamoto, Shinichiro, Akira Isogai and Tadahisa Iwata. 2011. Structure and mechanical properties of wet-spun fibers made from natural cellulose nanofibers. Biomacromolecules 12(3): 831–36. https://doi.org/10.1021/BM101510R.

Janmohammadi, Mahsa, Zahra Nazemi, Amin Orash Mahmoud Salehi, Amir Seyfoori, Johnson V. John, Mohammad Sadegh Nourbakhsh et al. 2023. Cellulose-based composite scaffolds for bone tissue engineering and localized drug delivery. Bioactive Materials 20(February 2022): 137–63. https://doi.org/10.1016/j.bioactmat.2022.05.018.

Jia, Ning, Shu Ming Li, Ming Guo Ma, Run Cang Sun and Jie Fang Zhu. 2012. Hydrothermal fabrication, characterization, and biological activity of cellulose/$CaCO_3$ bionanocomposites. Carbohydrate Polymers 88(1): 179–84. https://doi.org/10.1016/J.CARBPOL.2011.11.086.

Jiji, Swaminathan, Sivalingam Udhayakumar, Chellan Rose, Chellappa Muralidharan and Krishna Kadirvelu. 2019. Thymol enriched bacterial cellulose hydrogel as effective material for third degree burn wound repair. International Journal of Biological Macromolecules 122(February): 452–60. https://doi.org/10.1016/J.IJBIOMAC.2018.10.192.

Jing, Wang, Yang Chunxi, Wan Yizao, Luo Honglin, He Fang, Dai Kerong and Huang Yuan. 2013. Laser patterning of bacterial cellulose hydrogel and its modification with gelatin and hydroxyapatite for bone tissue engineering. Soft Materials 11(2): 173–80. https://doi.org/10.1080/1539445X.2011.611204.

Joshi, Mahesh Kumar, Arjun Prasad Tiwari, Hem Raj Pant, Bishnu Kumar Shrestha, Han Joo Kim, Chan Hee Park et al. 2015. *In situ* generation of cellulose nanocrystals in polycaprolactone nanofibers: effects on crystallinity, mechanical strength, biocompatibility, and biomimetic mineralization. ACS Applied Materials and Interfaces 7(35): 19672–83. https://doi.org/10.1021/ACSAMI.5B04682/SUPPL_FILE/AM5B04682_SI_001.PDF.

Joy, Jincy, Jessica Pereira, Rachida Aid-Launais, Graciela Pavon-Djavid, Alok R. Ray, Didier Letourneur et al. 2018. Gelatin—oxidized carboxymethyl cellulose blend based tubular electrospun scaffold for vascular tissue engineering. International Journal of Biological Macromolecules 107(February): 1922–35. https://doi.org/10.1016/J.IJBIOMAC.2017.10.071.

Jozala, Angela Faustino, Renata Aparecida Nedel Pértile, Carolina Alves dos Santos, Valéria de Carvalho Santos-Ebinuma, Marcelo Martins Seckler, Francisco Miguel Gama et al. 2015. Bacterial cellulose production by Gluconacetobacter xylinus by employing alternative culture media. Applied Microbiology and Biotechnology 99(3): 1181–90. https://doi.org/10.1007/S00253-014-6232-3/TABLES/3.

Jung, Jae Yong, Taous Khan, Joong Kon Park and Ho Nam Chang. 2007. Production of bacterial cellulose by gluconacetobacter hansenii using a novel bioreactor equipped with a spin filter. Korean Journal of Chemical Engineering 24(2): 265–71. https://doi.org/10.1007/S11814-007-5058-4/METRICS.

Kabashima, Kenji, Tetsuya Honda, Florent Ginhoux and Gyohei Egawa. 2019. The immunological anatomy of the skin. Nature Reviews. Immunology 19(1): 19–30. https://doi.org/10.1038/S41577-018-0084-5.

Kalia, Susheel, B. S. Kaith and Inderjeet Kaur. 2009. Pretreatments of natural fibers and their application as reinforcing material in polymer composites—a review. Polymer Engineering & Science 49(7): 1253–72. https://doi.org/10.1002/PEN.21328.

Kanikireddy, Vimala, Kokkarachedu Varaprasad, Tippabattini Jayaramudu, Chandrasekaran Karthikeyan and Rotimi Sadiku. 2020. Carboxymethyl cellulose-based materials for infection control and wound healing: a review. International Journal of Biological Macromolecules 164(December): 963–75. https://doi.org/10.1016/J.IJBIOMAC.2020.07.160.

Kargarzadeh, H., J. Huang, N. Lin, I. Ahmad, M. Mariano, A. Dufresne et al. 2018. Recent developments in nanocellulose-based biodegradable polymers, thermoplastic polymers, and porous nanocomposites. Progress in Polymer Science 87(December): 197–227. https://doi.org/10.1016/J.PROGPOLYMSCI.2018.07.008.

Kargarzadeh, Hanieh, Michael Ioelovich, Ishak Ahmad, Sabu Thomas and Alain Dufresne. 2017. Methods for extraction of nanocellulose from various sources. Handbook of Nanocellulose and Cellulose Nanocomposites, May, 1–49. https://doi.org/10.1002/9783527689972.CH1.

Khalid, Muhammad Yasir, Zia Ullah Arif, Reza Noroozi, Mokarram Hossain, Seeram Ramakrishna and Rehan Umer. 2023. 3D/4D printing of cellulose nanocrystals-based biomaterials: additives for sustainable applications. International Journal of Biological Macromolecules 251(November): 126287. https://doi.org/10.1016/J.IJBIOMAC.2023.126287.

Khamrai, Moumita, Sovan Lal Banerjee, Saikat Paul, Anup Kumar Ghosh, Priyatosh Sarkar and Patit Paban Kundu. 2019. A mussel mimetic, bioadhesive, antimicrobial patch based on dopamine-modified bacterial cellulose/RGO/Ag NPs: A green approach toward wound-healing applications. ACS Sustainable Chemistry and Engineering 7(14): 12083–97. https://doi.org/10.1021/ACSSUSCHEMENG.9B01163/SUPPL_FILE/SC9B01163_SI_001.PDF.

Khan, Shaukat, Mazhar Ul-Islam, Waleed Ahmad Khattak, Muhammad Wajid Ullah and Joong Kon Park. 2015. Bacterial cellulose-titanium dioxide nanocomposites: nanostructural characteristics, antibacterial mechanism, and biocompatibility. Cellulose 22(1): 565–79. https://doi.org/10.1007/S10570-014-0528-4/FIGURES/6.

Khan, Shaukat, Mazhar Ul-Islam, Muhammad Wajid Ullah, Yeji Kim and Joong Kon Park. 2015. Synthesis and characterization of a novel bacterial cellulose–poly(3,4-ethylenedioxythiophene)–poly(styrene sulfonate) composite for use in biomedical applications. Cellulose 22(4): 2141–48. https://doi.org/10.1007/S10570-015-0683-2/FIGURES/4.

Kilic Bektas, Cemile, Ilgin Kimiz, Aylin Sendemir, Vasif Hasirci and Nesrin Hasirci. 2018. A bilayer scaffold prepared from collagen and carboxymethyl cellulose for skin tissue engineering applications. Journal of Biomaterials Science. Polymer Edition 29(14): 1764–84. https://doi.org/10.1080/09205063.2018.1498718.

Kim, Jaehwan, Zhijiang Cai, Hyun Sook Lee, Gwang Seong Choi, Don Haeng Lee and Chulhee Jo. 2011. Preparation and characterization of a bacterial cellulose/chitosan composite for potential biomedical application. Journal of Polymer Research 18(4): 739–44. https://doi.org/10.1007/S10965-010-9470-9/FIGURES/6.

Ko, Sung Won, Juan Paolo E. Soriano, Afeesh Rajan Unnithan, Ji Yeon Lee, Chan Hee Park and Cheol Sang Kim. 2018. Development of bioactive cellulose nanocrystals derived from dominant cellulose polymorphs I and II from Capsosiphon fulvescens for biomedical applications. International Journal of Biological Macromolecules 110(April): 531–39. https://doi.org/10.1016/J.IJBIOMAC.2017.11.047.

Koeini, Fatemeh, Atefeh Solouk and Somaye Akbari. 2024. Graphene oxide-incorporated polylactic acid/polyamidoamine dendrimer electroconductive nanocomposite as a promising scaffold for guided tissue regeneration. Macromolecular Materials and Engineering, August, 2400100. https://doi.org/10.1002/MAME.202400100.

Kong, Bong Ju, Ayoung Kim and Soo Nam Park. 2016. Properties and in vitro drug release of hyaluronic acid-hydroxyethyl cellulose hydrogels for transdermal delivery of isoliquiritigenin. Carbohydrate Polymers 147(August): 473–81. https://doi.org/10.1016/J.CARBPOL.2016.04.021.

Konwarh, Rocktotpal, Niranjan Karak and Manjusri Misra. 2013. Electrospun cellulose acetate nano fibers: the present status and gamut of biotechnological applications. Biotechnology Advances 31(4): 421–37. https://doi.org/10.1016/j.biotechadv.2013.01.002.

Koons, Gerry L., Mani Diba and Antonios G. Mikos. 2020. Materials design for bone-tissue engineering. Nature Reviews Materials 5(8): 584–603. https://doi.org/10.1038/s41578-020-0204-2.

Korábková, Eva, Zhanna Boeva, Katarzyna Anna Radaszkiewicz, Kateřina Skopalová, Věra Kašpárková, Wenyang Xu et al. 2024. Stimuli-responsive thin film composites of conducting polymers and cellulose nanocrystals for tissue engineering. International Journal of Biological Macromolecules 265(April): 131036. https://doi.org/10.1016/J.IJBIOMAC.2024.131036.

Kovalenko, Valeri I. 2010. Crystalline cellulose: structure and hydrogen bonds. Russian Chemical Reviews 79(3): 231–41. https://doi.org/10.1070/RC2010V079N03ABEH004065/XML.

Kumar, A., K. C. Nune, L. E. Murr and R. D. K. Misra. 2016. Biocompatibility and mechanical behaviour of three-dimensional scaffolds for biomedical devices: process–structure–property paradigm. International Materials Reviews 61(1): 20–45. https://doi.org/10.1080/09506608.2015.1128310.

Kümmerer, Klaus, Jakob Menz, Thomas Schubert and Wim Thielemans. 2011. Biodegradability of organic nanoparticles in the aqueous environment. Chemosphere 82(10): 1387–92. https://doi.org/10.1016/J.CHEMOSPHERE.2010.11.069.

Kuo, Chia Hung, Jing Hua Chen, Bo Kang Liou and Cheng Kang Lee. 2016. Utilization of acetate buffer to improve bacterial cellulose production by Gluconacetobacter Xylinus. Food Hydrocolloids 53(February): 98–103. https://doi.org/10.1016/J.FOODHYD.2014.12.034.

Lee, Yee-Shuan and Treena Livingston Arinzeh. 2011. Electrospun nanofibrous materials for neural tissue engineering. Polymers 3(1): 413–26. https://doi.org/10.3390/polym3010413.

Li, Jian, Yizao Wan, Lianfeng Li, Hui Liang and Jiehua Wang. 2009. Preparation and characterization of 2,3-dialdehyde bacterial cellulose for potential biodegradable tissue engineering scaffolds. Materials Science and Engineering: C 29 (5): 1635–42. https://doi.org/10.1016/J.MSEC.2009.01.006.

Li, Lili and Margaret Frey. 2010. Preparation and characterization of cellulose nitrate-acetate mixed ester fibers. Polymer 51(16): 3774–83. https://doi.org/10.1016/j.polymer.2010.06.013.

Li, Wei, Nere Garmendia, Uxua Pérez De Larraya, Yaping Ding, Rainer Detsch, Alina Grünewald et al. 2014. 45S5 Bioactive glass-based scaffolds coated with cellulose nanowhiskers for bone tissue engineering. RSC Advances 4(99): 56156–64. https://doi.org/10.1039/C4RA07740G.

Li, Weichang, Rui Guo, Yong Lan, Yi Zhang, Wei Xue and Yuanming Zhang. 2014. Preparation and properties of cellulose nanocrystals reinforced collagen composite films. Journal of Biomedical Materials Research Part A 102(4): 1131–39. https://doi.org/10.1002/JBM.A.34792.

Liew, Fui Kiew, Sinin Hamdan, Md Rezaur Rahman, Mohamad Rusop, Josephine Chang Hui Lai, Md Faruk Hossen et al. 2015. Synthesis and characterization of cellulose from green bamboo by chemical treatment with mechanical process. Journal of Chemistry 2015(1): 212158. https://doi.org/10.1155/2015/212158.

Liitiä, Tiina, Sirkka Liisa Maunu, Bo Hortling, Tarja Tamminen, Osmo Pekkala and Antero Varhimo. 2003. Cellulose crystallinity and ordering of hemicelluloses in pine and birch pulps as revealed by solid-state NMR spectroscopic methods. Cellulose 10(4): 307–16. https://doi.org/10.1023/A:1027302526861/METRICS.

Lin, Che Min, Yung Chi Chang, Li Chang Cheng, Chao Hsien Liu, Shin C. Chang, Tzu Yang Hsien et al. 2020. Preparation of graphene-embedded hydroxypropyl cellulose/chitosan/polyethylene oxide nanofiber membranes as wound dressings with enhanced antibacterial properties. Cellulose 27(5): 2651–67. https://doi.org/10.1007/S10570-019-02940-W/METRICS.

Lin, Dehui, Zhe Liu, Rui Shen, Siqian Chen and Xingbin Yang. 2020. Bacterial cellulose in food industry: current research and future prospects. International Journal of Biological Macromolecules 158(September): 1007–19. https://doi.org/10.1016/J.IJBIOMAC.2020.04.230.

Lin, Jinyou, Liangbo Yu, Feng Tian, Nie Zhao, Xiuhong Li, Fenggang Bian et al. 2014. Cellulose nanofibrils aerogels generated from jute fibers. Carbohydrate Polymers 109(August): 35–43. https://doi.org/10.1016/J.CARBPOL.2014.03.045.

Lindahl, Anders, Mats Brittberg, David Gibbs, Jonathan I. Dawson, Janos Kanczler, Cameron Black et al. 2015. Cartilage and bone regeneration. Tissue Engineering, January, 529–82. https://doi.org/10.1016/B978-0-12-420145-3.00016-X.

Liu, Guodong, Mengjiao Ma, Haoye Meng, Jintao Liu, Yudong Zheng, Jiang Peng et al. 2022. *In-situ* self-assembly of bacterial cellulose/poly(3,4-ethylenedioxythiophene)-sulfonated nanofibers for peripheral nerve repair. Carbohydrate Polymers 281(April): 119044. https://doi.org/10.1016/J.CARBPOL.2021.119044.

Liu, Haohuai, Li Zhang, Pujiang Shi, Qin Zou, Yi Zuo and Yubao Li. 2010. Hydroxyapatite/polyurethane scaffold incorporated with drug-loaded ethyl cellulose microspheres for bone regeneration. Journal of Biomedical Materials Research Part B: Applied Biomaterials 95B(1): 36–46. https://doi.org/10.1002/JBM.B.31680.

Liu, Li, Xiongfa Ji, Lin Mao, Li Wang, Kun Chen, Zhijun Shi, Abeer Ahmed Qaed Ahmed et al. 2022. Hierarchical-structured bacterial cellulose/potato starch tubes as potential small-diameter vascular grafts. Carbohydrate Polymers 281(April). https://doi.org/10.1016/J.CARBPOL.2021.119034.

Liu, Suling, Qing Zhang, Shaheng Gou, Lili Zhang and Zhiguo Wang. 2021. Esterification of cellulose using carboxylic acid-based deep eutectic solvents to produce high-yield cellulose nanofibers. Carbohydrate Polymers 251(January): 117018. https://doi.org/10.1016/J.CARBPOL.2020.117018.

Liu, Yuansen, Yi Zhu, Quanxing Zheng, Hongliang Lu, Huafa Huang, Jianping Zhang et al. 2024. Recent progress in the efficacy of algal saccharides on skin repair. AlgRe 78(March): 103403. https://doi.org/10.1016/J.ALGAL.2024.103403.

Loh, Evelyn Yun Xi, Najwa Mohamad, Mh Busra Fauzi, Min Hwei Ng, Shiow Fern Ng and Mohd Cairul Iqbal Mohd Amin. 2018. Development of a bacterial cellulose-based hydrogel cell carrier containing keratinocytes and fibroblasts for full-thickness wound healing. Scientific Reports 8(1): 1–12. https://doi.org/10.1038/s41598-018-21174-7.

Lohrasbi, Sajedeh, Esmaeil Mirzaei, Ayoob Karimizade, Sara Takallu and Anita Rezaei. 2020. Collagen/cellulose nanofiber hydrogel scaffold: physical, mechanical and cell biocompatibility properties. Cellulose 27(2): 927–40. https://doi.org/10.1007/S10570-019-02841-Y/METRICS.

Lu, Yuezhi, Wenjie Zhang, Jie Wang, Guangzheng Yang, Shi Yin, Tingting Tang et al. 2019. Recent advances in cell sheet technology for bone and cartilage regeneration: from preparation to application. International Journal of Oral Science 11(2). https://doi.org/10.1038/s41368-019-0050-5.

Luo, Honglin, Guangyao Xiong, Da Hu, Kaijing Ren, Fanglian Yao, Yong Zhu et al. 2013. Characterization of TEMPO-oxidized bacterial cellulose scaffolds for tissue engineering applications. Materials Chemistry and Physics 143(1): 373–79. https://doi.org/10.1016/J.MATCHEMPHYS.2013.09.012.

Luo, Yu, Shige Wang, Mingwu Shen, Ruiling Qi, Yi Fang, Rui Guo et al. 2013. Carbon nanotube-incorporated multilayered cellulose acetate nanofibers for tissue engineering applications. Carbohydrate Polymers 91(1): 419–27. https://doi.org/10.1016/j.carbpol.2012.08.069.

Lupașcu, Raluca Elisabeta, Mihaela Violeta Ghica, Cristina Elena Dinu-Pîrvu, Lăcrămioara Popa, Bruno Ștefan Velescu and Andreea Letiția Arsene. 2022. An overview regarding microbial aspects of production and applications of bacterial cellulose. Materials 15(2): 676. https://doi.org/10.3390/MA15020676.

Lv, Xiaoyu, Jianing Han, Min Liu, Hui Yu, Kaihua Liu, Yifan Yang et al. 2023. Overview of preparation, modification, and application of tunicate-derived nanocellulose. Chemical Engineering Journal 452(January): 139439. https://doi.org/10.1016/J.CEJ.2022.139439.

Lv, Xiaoyu, Hui Yu, Jianing Han, Yage Hou, Yi Sun, Kaihua Liu et al. 2024. Tunicate cellulose nanocrystals reinforced modified calcium sulfate bone cement with enhanced mechanical properties for bone repair. Carbohydrate Polymers 323(January). https://doi.org/10.1016/J.CARBPOL.2023.121380.

Ma, Lin, Lili Wang, Lixin Wu, Dongxian Zhuo, Zixiang Weng and Rongrong Ren. 2014. Cellulosic nanocomposite membranes from hydroxypropyl cellulose reinforced by cellulose nanocrystals. Cellulose 21(6): 4443–54. https://doi.org/10.1007/S10570-014-0405-1/FIGURES/10.

Machado, Bárbara, Sofia M. Costa, Isabel Costa, Raul Fangueiro and Diana P. Ferreira. 2024. The potential of algae as a source of cellulose and its derivatives for biomedical applications. Cellulose 31(6): 3353–76. https://doi.org/10.1007/S10570-024-05816-W/METRICS.

Madsen, Bo and E. Kristofer Gamstedt. 2013. Wood versus plant fibers: similarities and differences in composite applications. Advances in Materials Science and Engineering 2013. https://doi.org/10.1155/2013/564346.

Mahmud, Md Musavvir, Asma Perveen, Rumana A. Jahan, Md Abdul Matin, Siew Yee Wong, Xu Li et al. 2019. Preparation of different polymorphs of cellulose from different acid hydrolysis medium. International Journal of Biological Macromolecules 130(June): 969–76. https://doi.org/10.1016/J.IJBIOMAC.2019.03.027.

Mahnaj, Tazin, Salah U. Ahmed and Fotios M. Plakogiannis. 2013. Characterization of ethyl cellulose polymer. Pharmaceutical Development and Technology 18(5): 982–89. https://doi.org/10.3109/10837450.2011.604781.

Mali, Prajakta and Atul P. Sherje. 2022. Cellulose nanocrystals: fundamentals and biomedical applications. Carbohydrate Polymers 275(January). https://doi.org/10.1016/J.CARBPOL.2021.118668.

Mallakpour, Shadpour, Maryam Tukhani and Chaudhery Mustansar Hussain. 2021. Recent advancements in 3D bioprinting technology of carboxymethyl cellulose-based hydrogels: utilization in tissue engineering. Advances in Colloid and Interface Science 292(June): 102415. https://doi.org/10.1016/J.CIS.2021.102415.

Manan, Sehrish, Muhammad Wajid Ullah, Mazhar Ul-Islam, Zhijun Shi, Mario Gauthier and Guang Yang. 2022. Bacterial cellulose: molecular regulation of biosynthesis, supramolecular assembly, and tailored structural and functional properties. Progress in Materials Science 129(August): 100972. https://doi.org/10.1016/J.PMATSCI.2022.100972.

Mandal, Arup and Debabrata Chakrabarty. 2011. Isolation of nanocellulose from waste sugarcane bagasse (SCB) and its characterization. Carbohydrate Polymers 86(3): 1291–99. https://doi.org/10.1016/J.CARBPOL.2011.06.030.

Manji, Rizwan A., Alan H. Menkis, Burcin Ekser and David K.C. Cooper. 2012. Porcine bioprosthetic heart valves: the next generation. American Heart Journal 164(2): 177–85. https://doi.org/10.1016/J.AHJ.2012.05.011.

Mao, Daoyong, Qing Li, Ningning Bai, Hongzhou Dong and Daikun Li. 2018. Porous stable poly(lactic acid)/ethyl cellulose/hydroxyapatite composite scaffolds prepared by a combined method for bone regeneration. Carbohydrate Polymers 180(January): 104–11. https://doi.org/10.1016/J.CARBPOL.2017.10.031.

Martínez-Sanz, Marta, Amparo Lopez-Rubio and Jose M. Lagaron. 2011. Optimization of the nanofabrication by acid hydrolysis of bacterial cellulose nanowhiskers. Carbohydrate Polymers 85(1): 228–36. https://doi.org/10.1016/J.CARBPOL.2011.02.021.

Matamoros-Veloza, Adriana, Kazi M. Zakir Hossain, Brigitte E. Scammell, Ifty Ahmed, Richard Hall and Nikil Kapur. 2020. Formulating injectable pastes of porous calcium phosphate glass microspheres for bone regeneration applications. Journal of the Mechanical Behavior of Biomedical Materials 102: 103489. https://doi.org/10.1016/j.jmbbm.2019.103489.

Mendes, Péricles Nóbrega, Sheila Canevese Rahal, Oduvaldo Câmara Marques Pereira-Junior, Viciany Erique Fabris, Sara Lais Rahal Lenharo, João Ferreira De Lima-Neto et al. 2009. In vivo and in vitro evaluation of an Acetobacter Xylinum synthesized microbial cellulose membrane intended for guided tissue repair. Acta Veterinaria Scandinavica 51(1): 1–8. https://doi.org/10.1186/1751-0147-51-12/TABLES/1.

Mhashilkar, Abner, M. and Anthony Atala. 2012. Advent and maturation of regenerative medicine. Current Stem Cell Research & Therapy 7(6): 430–45. https://doi.org/10.2174/157488812804484657.

Miao, Lei, Jiwen Hu, Mangeng Lu, Yuanyuan Tu, Xin Chen, Yinwen Li et al. 2016. Alkynyl-functionalization of hydroxypropyl cellulose and thermoresponsive hydrogel thereof prepared with P(NIPAAm-Co-HEMAPCL). Carbohydrate Polymers 137(February): 433–40. https://doi.org/10.1016/J.CARBPOL.2015.11.001.

Miller, Kathleen L., Yi Xiang, Claire Yu, Jacob Pustelnik, Jerry Wu, Xinyue Ma et al. 2021. Rapid 3D bioprinting of a human IPSC-derived cardiac micro-tissue for high-throughput drug testing. Organs-on-a-Chip 3(November): 100007. https://doi.org/10.1016/J.OOC.2021.100007.

Millon, L. E. and W. K. Wan. 2006. The polyvinyl alcohol–bacterial cellulose system as a new nanocomposite for biomedical applications. Journal of Biomedical Materials Research Part B: Applied Biomaterials 79B(2): 245–53. https://doi.org/10.1002/JBM.B.30535.

Mirtaghavi, Ali, Jikui Luo and Rajendran Muthuraj. 2020. Recent advances in porous 3D cellulose aerogels for tissue engineering applications: a review. Journal of Composites Science 4(4): 152. https://doi.org/10.3390/JCS4040152.

Mishra, Raghvendra Kumar, Arjun Sabu and Santosh K. Tiwari. 2018. Materials chemistry and the futurist eco-friendly applications of nanocellulose: status and prospect. Journal of Saudi Chemical Society 22(8): 949–78. https://doi.org/10.1016/J.JSCS.2018.02.005.

Miyamoto, Takeaki, Shin-ichi -i Takahashi, Hiraku Ito, Hiroshi Inagaki and Yasuhara Noishiki. 1989. Tissue biocompatibility of cellulose and its derivatives. Journal of Biomedical Materials Research 23(1): 125–33. https://doi.org/10.1002/JBM.820230110.

Moazzami Goudarzi, Zahra, Tayebeh Behzad, Laleh Ghasemi-Mobarakeh and Mahshid Kharaziha. 2021. An investigation into influence of acetylated cellulose nanofibers on properties of PCL/gelatin electrospun nanofibrous scaffold for soft tissue engineering. Polymer 213(January): 123313. https://doi.org/10.1016/J.POLYMER.2020.123313.

Modulevsky, Daniel J., Charles M. Cuerrier and Andrew E. Pelling. 2016. Biocompatibility of subcutaneously implanted plant-derived cellulose biomaterials. PLOS ONE 11(6): e0157894. https://doi.org/10.1371/JOURNAL.PONE.0157894.

Mohammadalipour, Mohammad, Saeed Karbasi, Tayebeh Behzad, Zahra Mohammadalipour and Maryam Zamani. 2022. Effect of cellulose nanofibers on polyhydroxybutyrate electrospun scaffold for bone tissue engineering applications. International Journal of Biological Macromolecules 220(November): 1402–14. https://doi.org/10.1016/J.IJBIOMAC.2022.09.118.

Mohammadalizadeh, Zahra, Elahe Bahremandi-Toloue and Saeed Karbasi. 2022. Synthetic-based blended electrospun scaffolds in tissue engineering applications. Journal of Materials Science 57(6): 4020–79. https://doi.org/10.1007/S10853-021-06826-W.

Mohite, Bhavna V. and Satish V. Patil. 2014. A novel biomaterial: bacterial cellulose and its new era applications. Biotechnology and Applied Biochemistry 61(2): 101–10. https://doi.org/10.1002/BAB.1148.

Moniri, Mona, Amin Boroumand Moghaddam, Susan Azizi, Raha Abdul Rahim, Arbakariya Bin Ariff, Wan Zuhainis Saad et al. 2017. Production and status of bacterial cellulose in biomedical engineering. Nanomaterials 7(9): 257. https://doi.org/10.3390/NANO7090257.

Moraïs, Sarah, Arnon Heyman, Yoav Barak, Jonathan Caspi, David B. Wilson, Raphael Lamed et al. 2010. Enhanced cellulose degradation by nano-complexed enzymes: synergism between a scaffold-linked exoglucanase and a free endoglucanase. Journal of Biotechnology 147(3–4): 205–11. https://doi.org/10.1016/J.JBIOTEC.2010.04.012.

Moreira, J. B., T. D. Santos, C. G. Cruz, J. T. D. Silveira, L. F. D. Carvalho, M. G. D. Morais et al. 2023. Algal polysaccharides-based nanomaterials: general aspects and potential applications in food and biomedical fields. Polysaccharides 4(4): 371–89. https://doi.org/10.3390/POLYSACCHARIDES4040022.

Morgado, D. L., E. Frollini, A. Castellan, D. S. Rosa and V. Coma. 2011. Biobased films prepared from NaOH/thiourea aqueous solution of chitosan and linter cellulose. Cellulose 18(3): 699–712. https://doi.org/10.1007/S10570-011-9516-0/FIGURES/7.

Muller, D., J. P. Silva, C. R. Rambo, G. M. O. Barra, F. Dourado and F. M. Gama. 2013. Neuronal cells' behavior on polypyrrole coated bacterial nanocellulose three-dimensional (3D) scaffolds. Journal of Biomaterials Science, Polymer Edition 24(11): 1368–77. https://doi.org/10.1080/09205063.2012.761058.

Müller, Frank A., Lenka Müller, Ingo Hofmann, Peter Greil, Magdalene M. Wenzel and Rainer Staudenmaier. 2006. Cellulose-based scaffold materials for cartilage tissue engineering. Biomaterials 27(21): 3955–63. https://doi.org/10.1016/j.biomaterials.2006.02.031.

Murizan, Nur Ilyana Sahira, Nur Syahirah Mustafa, Nor Hasrul Akhmal Ngadiman, Noordin Mohd Yusof and Ani Idris. 2020. Review on nanocrystalline cellulose in bone tissue engineering applications. Polymers 12(12): 2818. https://doi.org/10.3390/POLYM12122818.

Naahidi, Sheva, Mousa Jafari, Megan Logan, Yujie Wang, Yongfang Yuan, Hojae Bae, Brian Dixon et al. 2017. Biocompatibility of hydrogel-based scaffolds for tissue engineering applications. Biotechnology Advances 35(5): 530–44. https://doi.org/10.1016/J.BIOTECHADV.2017.05.006.

Namkaew, Jirapat, Panitporn Laowpanitchakorn, Nuttapong Sawaddee, Sirinee Jirajessada, Sittisak Honsawek and Supansa Yodmuang. 2021. Carboxymethyl cellulose entrapped in a poly(vinyl) alcohol network: plant-based scaffolds for cartilage tissue engineering. Molecules 26(3): 578. https://doi.org/10.3390/MOLECULES26030578.

Nasatto, Pauline L., Frédéric Pignon, Joana L. M. Silveira, Maria Eugênia R. Duarte, Miguel D. Noseda and Marguerite Rinaudo. 2015. Methylcellulose, a cellulose derivative with original physical properties and extended applications. Polymers 7(5): 777–803. https://doi.org/10.3390/polym7050777.

Naseri-Nosar, Mahdi, Majid Salehi and Shahriar Hojjati-Emami. 2017. Cellulose acetate/poly lactic acid coaxial wet-electrospun scaffold containing citalopram-loaded gelatin nanocarriers for neural tissue engineering applications. International Journal of Biological Macromolecules 103(October): 701–8. https://doi.org/10.1016/J.IJBIOMAC.2017.05.054.

Nasri-Nasrabadi, Bijan, Mohammad Mehrasa, Mohammad Rafienia, Shahin Bonakdar, Tayebeh Behzad and Shahin Gavanji. 2014. Porous starch/cellulose nanofibers composite prepared by salt leaching technique for tissue engineering. Carbohydrate Polymers 108(1): 232–38. https://doi.org/10.1016/j.carbpol.2014.02.075.

Nechyporchuk, Oleksandr, Mohamed Naceur Belgacem and Julien Bras. 2016. Production of cellulose nanofibrils: a review of recent advances. Industrial Crops and Products 93(December): 2–25. https://doi.org/10.1016/J.INDCROP.2016.02.016.

Nge, Thi Thi, Masaya Nogi, Hiroyuki Yano and Junji Sugiyama. 2010. Microstructure and mechanical properties of bacterial cellulose/chitosan porous scaffold. Cellulose 17(2): 349–63. https://doi.org/10.1007/S10570-009-9394-X/FIGURES/10.

Nguyen, Anh H., Paul Marsh, Lauren Schmiess-Heine, Peter J. Burke, Abraham Lee, Juhyun Lee et al. 2019. Cardiac tissue engineering: state-of-the-art methods and outlook. Journal of Biological Engineering 13(1): 1–21. https://doi.org/10.1186/S13036-019-0185-0.

Nguyen, Hien Minh, Tam Thi Ngoc Le, An Thanh Nguyen, Han Nguyen Thien Le and Thi Tan Pham. 2023. Biomedical materials for wound dressing: recent advances and applications. RSC Advances 13(8): 5509–28. https://doi.org/10.1039/D2RA07673J.

Nguyen, Theresa and Robert Latkany. 2011. Review of hydroxypropyl cellulose ophthalmic inserts for treatment of dry eye. Clinical Ophthalmology 5(1): 587–91. https://doi.org/10.2147/OPTH.S13889.

Nimeskern, Luc, Héctor Martínez Ávila, Johan Sundberg, Paul Gatenholm, Ralph Müller and Kathryn S. Stok. 2013. Mechanical evaluation of bacterial nanocellulose as an implant material for ear cartilage replacement. Journal of the Mechanical Behavior of Biomedical Materials 22(June): 12–21. https://doi.org/10.1016/J.JMBBM.2013.03.005.

Nishi, Y., M. Uryu, S. Yamanaka, K. Watanabe, N. Kitamura, M. Iguchi et al. 1990. The structure and mechanical properties of sheets prepared from bacterial cellulose - Part 2 Improvement of the mechanical properties of sheets and their applicability to diaphragms of electroacoustic transducers. Journal of Materials Science 25(6): 2997–3001. https://doi.org/10.1007/BF00584917/METRICS.

Noiset, Olivier, Yves Jacques Schneider and Jacqueline Marchand-Brynaert. 1999. Fibronectin adsorption or/and covalent grafting on chemically modified PEEK film surfaces. Journal of Biomaterials Science. Polymer Edition 10(6): 657–77. https://doi.org/10.1163/156856299X00865.

Noreen, Aqdas, Khalid Mahmood Zia, Shazia Tabasum, Sana Khalid and Rahila Shareef. 2020. A review on grafting of hydroxyethylcellulose for versatile applications. International Journal of Biological Macromolecules 150(May): 289–303. https://doi.org/10.1016/J.IJBIOMAC.2020.01.265.

Novotna, Katarina, Pavel Havelka, Tomas Sopuch, Katerina Kolarova, Vladimira Vosmanska, Vera Lisa et al. 2013. Cellulose-based materials as scaffolds for tissue engineering. Cellulose 20(5): 2263–78. https://doi.org/10.1007/S10570-013-0006-4/FIGURES/10.

Of, Stimulation, H. Hift, L. Ouellet, J. W. Littlefield, D. R. Sanadi, F. E. Hunter et al. 1954. Synthesis of cellulose by Acetobacter Xylinum. 2. Preparation of freeze-dried cells capable of polymerizing glucose to cellulose. Biochemical Journal 58(2): 345. https://doi.org/10.1042/BJ0580345.

Oprea, Madalina and Stefan Ioan Voicu. 2020a. Recent advances in composites based on cellulose derivatives for biomedical applications. Carbohydrate Polymers 247: 116683. https://doi.org/10.1016/j.carbpol.2020.116683.

Oprea, Madalina and Stefan Ioan Voicu. 2020b. Recent advances in composites based on cellulose derivatives for biomedical applications. Carbohydrate Polymers 247: 116683. https://doi.org/10.1016/j.carbpol.2020.116683.

Pääkko, M., M. Ankerfors, H. Kosonen, A. Nykänen, S. Ahola, M. Österberg et al. 2007. Enzymatic hydrolysis combined with mechanical shearing and high-pressure homogenization for nanoscale cellulose fibrils and strong gels. Biomacromolecules 8(6): 1934–41. https://doi.org/10.1021/BM061215P/ASSET/IMAGES/MEDIUM/BM061215PN00001.GIF.

Panthapulakkal, Suhara and Mohini Sain. 2013. Isolation of nano fibres from hemp and flax and their thermoplastic composites isolation of nano fibres from hemp and flax and their thermoplastic composites hemp and flax nanofibres and composites. Plastic and Polymer Technology PAPT 2 (1).

Peng, Shuai, Yudong Zheng, Jian Wu, Yaxun Wu, Yanxuan Ma, Wenhui Song et al. 2012. Preparation and characterization of degradable oxidized bacterial cellulose reacted with nitrogen dioxide. Polymer Bulletin 68(2): 415–23. https://doi.org/10.1007/S00289-011-0550-8/FIGURES/4.

Pértile, Renata A. N., Susana Moreira, Rui M. Gil Da Costa, Alexandra Correia, Luisa Guardão, Fátima Gartner et al. 2012. Bacterial cellulose: long-term biocompatibility studies. Journal of Biomaterials Science, Polymer Edition 23(10): 1339–54. https://doi.org/10.1163/092050611X581516.

Piasecka-Zelga, Joanna, Piotr Zelga, Karolina Gronkowska, Jakub Madalski, Joanna Szulc, Justyna Wietecha et al. 2021. Toxicological and sensitization studies of novel vascular prostheses made of bacterial nanocellulose modified with chitosan (MBC) for application as the tissue-engineered blood vessels. Regenerative Engineering and Translational Medicine 7(2): 218–33. https://doi.org/10.1007/S40883-021-00209-Y/TABLES/6.

Pibarot, Philippe and Jean G. Dumesnil. 2009. Prosthetic heart valves: selection of the optimal prosthesis and long-term management. Circulation 119(7): 1034–48. https://doi.org/10.1161/CIRCULATIONAHA.108.778886.

Picker-Freyer, Katharina M. and Thomas Dürig. 2007. Physical mechanical and tablet formation properties of hydroxypropylcellulose: in pure form and in mixtures. AAPS PharmSciTech 8(4): 1–9. https://doi.org/10.1208/PT0804092/METRICS.

Pita-Vilar, Maria, Angel Concheiro, Carmen Alvarez-Lorenzo and Luis Diaz-Gomez. 2023. Recent advances in 3D printed cellulose-based wound dressings: a review on *in vitro* and *in vivo* achievements. Carbohydrate Polymers 321(December): 121298. https://doi.org/10.1016/J.CARBPOL.2023.121298.

Pooyan, Parisa, Rina Tannenbaum and Hamid Garmestani. 2012. Mechanical behavior of a cellulose-reinforced scaffold in vascular tissue engineering. Journal of the Mechanical Behavior of Biomedical Materials 7(March): 50–59. https://doi.org/10.1016/J.JMBBM.2011.09.009.

Popa, Lăcrămioara, Mihaela Violeta Ghica, Elena Emilia Tudoroiu, Diana Georgiana Ionescu and Cristina Elena Dinu-Pîrvu. 2022. Bacterial cellulose—a remarkable polymer as a source for biomaterials tailoring. Materials 15(3): 1054. https://doi.org/10.3390/MA15031054.

Potthast, Antje, Thomas Rosenau, Richard Buchner, Thomas Röder, Gerald Ebner, Hartmut Bruglachner et al. 2002. The cellulose solvent system n,n-dimethylacetamide/lithium chloride revisited: the effect of water on physicochemical properties and chemical stability. Cellulose 9(1): 41–53. https://doi.org/10.1023/A:1015811712657/METRICS.

Priya, Ganesan, Balaraman Madhan, Uttamchand Narendrakumar, Rayadurgam Venkata Suresh Kumar and Inderchand Manjubala. 2021. *In vitro* and *in vivo* evaluation of carboxymethyl cellulose scaffolds for bone tissue engineering applications. ACS Omega 6(2): 1246–53. https://doi.org/10.1021/ACSOMEGA.0C04551/ASSET/IMAGES/MEDIUM/AO0C04551_M003.GIF.

Puls, Juergen, Steven A. Wilson and Dirk Hölter. 2011. Degradation of cellulose acetate-based materials: a review. Journal of Polymers and the Environment 19(1): 152–65. https://doi.org/10.1007/s10924-010-0258-0.

Raghavendra, Gownolla Malegowd, Tippabattini Jayaramudu, Kokkarachedu Varaprasad, Rotimi Sadiku, S. Sinha Ray and Konduru Mohana Raju. 2013. Cellulose–polymer–Ag nanocomposite fibers for antibacterial fabrics/skin scaffolds. Carbohydrate Polymers 93(2): 553–60. https://doi.org/10.1016/J.CARBPOL.2012.12.035.

Rahman, Md Saifur, Md Saif Hasan, Ashis Sutradhar Nitai, Sunghyun Nam, Aneek Krishna Karmakar, Md Shameem Ahsan et al. 2021. Recent developments of carboxymethyl cellulose. Polymers 13(8): 1345. https://doi.org/10.3390/POLYM13081345/S1.

Rahman, Mohammed Mizanur, Sanjida Afrin, Papia Haque, Md Minhajul Islam, Mohammed Shahidul Islam and Md Abdul Gafur. 2014. Preparation and characterization of jute cellulose crystals-reinforced poly(l-lactic acid) biocomposite for biomedical applications. International Journal of Chemical Engineering. https://doi.org/10.1155/2014/842147.

Rahmani Del Bakhshayesh, Azizeh, Nasim Annabi, Rovshan Khalilov, Abolfazl Akbarzadeh, Mohammad Samiei, Effat Alizadeh et al. 2018. Recent advances on biomedical applications of scaffolds in wound healing and dermal tissue engineering. Artificial Cells, Nanomedicine, and Biotechnology 46(4): 691–705. https://doi.org/10.1080/21691401.2017.1349778.

Rahmati, Maryam, David K. Mills, Aleksandra M. Urbanska, Mohammad Reza Saeb, Jayarama Reddy Venugopal, Seeram Ramakrishna et al. 2021. Electrospinning for tissue engineering applications. Progress in Materials Science 117(April): 100721. https://doi.org/10.1016/J.PMATSCI.2020.100721.

Ramli, Nor Amlizan and Tin Wui Wong. 2011. Sodium carboxymethylcellulose scaffolds and their physicochemical effects on partial thickness wound healing. International Journal of Pharmaceutics 403(1–2): 73–82. https://doi.org/10.1016/J.IJPHARM.2010.10.023.

Ran, Jiabing, Pei Jiang, Shinian Liu, Guanglin Sun, Pan Yan, Xinyu Shen et al. 2017. Constructing multi-component organic/inorganic composite bacterial cellulose-gelatin/hydroxyapatite double-network scaffold platform for stem cell-mediated bone tissue engineering. Materials Science & Engineering. C, Materials for Biological Applications 78(September): 130–40. https://doi.org/10.1016/J.MSEC.2017.04.062.

Rashtchian, Morvarid, Ahmad Hivechi, S. Hajir Bahrami, Peiman B. Milan and Sara Simorgh. 2020. Fabricating alginate/poly(caprolactone) nanofibers with enhanced bio-mechanical properties via cellulose nanocrystal incorporation. Carbohydrate Polymers 233(April): 115873. https://doi.org/10.1016/J.CARBPOL.2020.115873.

Raut, Mahendra P., Emmanuel Asare, Syed Mohammad Daniel Syed Mohamed, Elliot N. Amadi and Ipsita Roy. 2023. Bacterial cellulose-based blends and composites: versatile biomaterials for tissue engineering applications. International Journal of Molecular Sciences 24(2). https://doi.org/10.3390/IJMS24020986.

Rho, Jae Young, Richard B. Ashman and Charles H. Turner. 1993. Young's Modulus of trabecular and cortical bone material: ultrasonic and microtensile measurements. Journal of Biomechanics 26(2): 111–19. https://doi.org/10.1016/0021-9290(93)90042-D.

Rimdusit, Sarawut, Sorada Jingjid, Siriporn Damrongsakkul, Sunan Tiptipakorn and Tsutomu Takeichi. 2008. Biodegradability and property characterizations of methyl cellulose: effect of nanocompositing and chemical crosslinking. Carbohydrate Polymers 72(3): 444–55. https://doi.org/10.1016/j.carbpol.2007.09.007.

Sanchez-Salvador, Jose Luis, Ana Balea, M. Concepción Monte, Carlos Negro, Meaghan Miller, James Olson et al. 2020. Comparison of mechanical and chemical nanocellulose as additives to reinforce recycled cardboard. Scientific Reports 10(1): 1–14. https://doi.org/10.1038/s41598-020-60507-3.

Sannino, A., S. Pappadà, M. Madaghiele, A. Maffezzoli, L. Ambrosio and L. Nicolais. 2005. Crosslinking of cellulose derivatives and hyaluronic acid with water-soluble carbodiimide. Polymer 46(25): 11206–12. https://doi.org/10.1016/J.POLYMER.2005.10.048.

Sannino, Alessandro, Christian Demitri and Marta Madaghiele. 2009. Biodegradable cellulose-based hydrogels: design and applications. Materials 2(2): 353–73. https://doi.org/10.3390/MA2020353.

Saraiva, Sofia, Patrícia Pereira, C. T. Paula, R. C. Rebelo, Jorge F. J. Coelho, Arménio C. Serra et al. 2021. Development of electrospun mats based on hydrophobic hydroxypropyl cellulose derivatives. Materials Science and Engineering: C 131(December): 112498. https://doi.org/10.1016/J.MSEC.2021.112498.

Saxena, Varun, Abshar Hasan and Lalit M. Pandey. 2021. Antibacterial nano-biocomposite scaffolds of chitosan, carboxymethyl cellulose and Zn & Fe integrated hydroxyapatite (chitosan-CMC-FZO@HAp) for bone tissue engineering. Cellulose 28(14): 9207–26. https://doi.org/10.1007/S10570-021-04072-6/METRICS.

Schmidt, Christine E. and Jennie Baier Leach. 2003. Neural tissue engineering: strategies for repair and regeneration. Http://Dx.Doi.Org/10.1146/Annurev.Bioeng.5.011303.120731 5(November): 293–347. https://doi.org/10.1146/ANNUREV.BIOENG.5.011303.120731.

Schussler, Olivier, Pierre E. Falcoz, Juan C. Chachques, Marco Alifano and Yves Lecarpentier. 2021. Possible treatment of myocardial infarct based on tissue engineering using a cellularized solid collagen scaffold functionalized with Arg-Glyc-Asp (RGD) peptide. International Journal of Molecular Sciences 22(22): 12563. https://doi.org/10.3390/IJMS222212563.

Seddiqi, Hadi, Erfan Oliaei, Hengameh Honarkar, Jianfeng Jin, Lester C. Geonzon, Rommel G. Bacabac et al. 2021. Cellulose and its derivatives: towards biomedical applications. Cellulose. Vol. 28. Springer Netherlands. https://doi.org/10.1007/s10570-020-03674-w.

Septevani, Athanasia A., David A. C. Evans, Darren J. Martin and Pratheep K. Annamalai. 2018. Hybrid polyether-palm oil polyester polyol based rigid polyurethane foam reinforced with cellulose nanocrystal. Industrial Crops and Products 112(February): 378–88. https://doi.org/10.1016/J.INDCROP.2017.12.032.

Shahzadi, Iram, Muhammad Islam, Hamid Saeed, Ali Haider, Anum Shahzadi, Junaid Haider et al. 2022. Formation of biocompatible MgO/cellulose grafted hydrogel for efficient bactericidal and controlled release of doxorubicin. International Journal of Biological Macromolecules 220(November): 1277–86. https://doi.org/10.1016/J.IJBIOMAC.2022.08.142.

Shahzadi, Iram, Muhammad Islam, Hamid Saeed, Anum Shahzadi, Junaid Haider, Ali Haider et al. 2023. Facile synthesis of copolymerized cellulose grafted hydrogel doped calcium oxide nanocomposites with improved antioxidant activity for anti-arthritic and controlled release of doxorubicin for anti-cancer evaluation. International Journal of Biological Macromolecules 235(April): 123874. https://doi.org/10.1016/J.IJBIOMAC.2023.123874.

Shao, Wei, Hui Liu, Shuxia Wang, Jimin Wu, Min Huang, Huihua Min and Xiufeng Liu. 2016. Controlled release and antibacterial activity of tetracycline hydrochloride-loaded bacterial cellulose composite membranes. Carbohydrate Polymers 145(July): 114–20. https://doi.org/10.1016/J.CARBPOL.2016.02.065.

Shi, Qingfeng, Chengjun Zhou, Yiying Yue, Weihong Guo, Yiqiang Wu and Qinglin Wu. 2012. Mechanical properties and *in vitro* degradation of electrospun bio-nanocomposite mats from PLA and cellulose nanocrystals. Carbohydrate Polymers 90(1): 301–8. https://doi.org/10.1016/J.CARBPOL.2012.05.042.

Shuhui, Li, Jiajun Mao, Shuhui Li, Jianying Huang, Kai Meng, Guoqiang Chen et al. n.d. Recent Advances of multifunctional cellulose-based hydrogels. https://doi.org/10.1007/978-3-319-76573-0_5-1.

Silva, Izabela Gabriela Rodrigues da, Bruna Tássia Dos Santos Pantoja, Gustavo Henrique Doná Rodrigues Almeida, Ana Claudia Oliveira Carreira and Maria Angélica Miglino. 2022. Bacterial cellulose and ECM hydrogels: an innovative approach for cardiovascular regenerative medicine. International Journal of Molecular Sciences 23(7). https://doi.org/10.3390/IJMS23073955.

Singh, B. N., N. N. Panda, R. Mund and K. Pramanik. 2016. Carboxymethyl cellulose enables silk fibroin nanofibrous scaffold with enhanced biomimetic potential for bone tissue engineering application. Carbohydrate Polymers 151(October): 335–47. https://doi.org/10.1016/J.CARBPOL.2016.05.088.

Singh, Maninder, Alok R. Ray, Padma Vasudevan K. Verma and Sujoy K. Guha. 1979. Potential biosoluble carriers: biocompatibility and biodegradability of oxidized cellulose. Biomaterials, Medical Devices, and Artificial Organs 7(4): 495–512. https://doi.org/10.3109/10731197909118964.

Singh, Maninder, Alok R. Ray and Padma Vasudevan. 1982. Biodegradation studies on periodate oxidized cellulose. Biomaterials 3(1): 16–20. https://doi.org/10.1016/0142-9612(82)90055-2.

Siró, István and David Plackett. 2010. Microfibrillated cellulose and new nanocomposite materials: a review. Cellulose 17(3): 459–94. https://doi.org/10.1007/S10570-010-9405-Y.

Sperotto, Gabriela, Larissa Gabrieli Stasiak, João Pedro Maximino Gongora Godoi, Naiana Cristine Gabiatti and Samara Silva De Souza. 2021. A review of culture media for bacterial cellulose production: complex, chemically defined and minimal media modulations. Cellulose 28(5): 2649–73. https://doi.org/10.1007/S10570-021-03754-5.

Sreedharan, Mridula, Raji Vijayamma, Elena Liyaskina, Viktor V. Revin, Muhammad Wajid Ullah, Zhijun Shi et al. 2024. Nanocellulose-based hybrid scaffolds for skin and bone tissue engineering: a 10-year overview. Biomacromolecules 25(4): 2136–55. https://doi.org/10.1021/ACS.BIOMAC.3C00975.

Stanton, John, Ye Xue, Prabhdeep Pandher, Laura Malek, Tyler Brown, Xiao Hu and David Salas-de la Cruz. 2018. Impact of ionic liquid type on the structure, morphology and properties of silk-cellulose biocomposite materials. International Journal of Biological Macromolecules 108(March): 333–41. https://doi.org/10.1016/J.IJBIOMAC.2017.11.137.

Sturcova, A., D. C. Apperley, D. Kalaskar, S. J. Eichhorn and M. C. Jarvis. 2005. Relationship between the Micromorphology and Mechanical Properties of Cellulose. American Chemical Society.

Šturcová, Adriana, Geoffrey R. Davies and Stephen J. Eichhorn. 2005. Elastic modulus and stress-transfer properties of tunicate cellulose whiskers. Biomacromolecules 6(2): 1055–61. https://doi.org/10.1021/BM049291K.

Sunasee, Rajesh, Usha D. Hemraz and Karina Ckless. 2016. Cellulose nanocrystals: a versatile nanoplatform for emerging biomedical applications. Expert Opinion on Drug Delivery 13(9): 1243–56. https://doi.org/10.1080/17425247.2016.1182491.

Swingler, Sam, Abhishek Gupta, Hazel Gibson, Marek Kowalczuk, Wayne Heaselgrave and Iza Radecka. 2021. Recent advances and applications of bacterial cellulose in biomedicine. Polymers 13(3): 412. https://doi.org/10.3390/POLYM13030412.

Takeuchi, Yoshiko, Kaori Umemura, Kohei Tahara and Hirofumi Takeuchi. 2018. Formulation design of hydroxypropyl cellulose films for use as orally disintegrating dosage forms. Journal of Drug Delivery Science and Technology 46(August): 93–100. https://doi.org/10.1016/J.JDDST.2018.05.002.

Tamburaci, Sedef, Ceren Kimna and Funda Tihminlioglu. 2018. Novel phytochemical cissus quadrangularis extract–loaded chitosan/na-carboxymethyl cellulose–based scaffolds for bone regeneration. Journal of Bioactive and Compatible Polymers 33(6): 629–46. https://doi.org/10.1177/0883911518793913.

Tang, Jingyu, Luhan Bao, Xue Li, Lin Chen and Feng F. Hong. 2015. Potential of PVA-doped bacterial nano-cellulose tubular composites for artificial blood vessels. Journal of Materials Chemistry B 3(43): 8537–47. https://doi.org/10.1039/C5TB01144B.

Tang, Yufeng, Xiaoying Wang, Yan Li, Ming Lei, Yumin Du, John F. Kennedy et al. 2010. Production and characterisation of novel injectable chitosan/methylcellulose/salt blend hydrogels with potential application as tissue engineering scaffolds. Carbohydrate Polymers 82(3): 833–41. https://doi.org/10.1016/j.carbpol.2010.06.003.

Thirumala, Sreedhar, Jeffrey M. Gimble and Ram V. Devireddy. 2013. Methylcellulose based thermally reversible hydrogel system for tissue engineering applications. Cells 2(3): 460–75. https://doi.org/10.3390/CELLS2030460.

Thoorens, Gregory, Fabrice Krier, Bruno Leclercq, Brian Carlin and Brigitte Evrard. 2014. Microcrystalline cellulose, a direct compression binder in a quality by design environment—a review. International Journal of Pharmaceutics 473(1–2): 64–72. https://doi.org/10.1016/J.IJPHARM.2014.06.055.

Thunberg, Johannes, Theodoros Kalogeropoulos, Volodymyr Kuzmenko, Daniel Hägg, Sara Johannesson, Gunnar Westman et al. 2015. *In situ* synthesis of conductive polypyrrole on electrospun cellulose nanofibers: scaffold for neural tissue engineering. Cellulose 22(3): 1459–67. https://doi.org/10.1007/S10570-015-0591-5/FIGURES/3.

Tibolla, Heloisa, Franciele Maria Pelissari and Florencia Cecilia Menegalli. 2014. Cellulose nanofibers produced from banana peel by chemical and enzymatic treatment. LWT - Food Science and Technology 2 Part 2(59): 1311–18. https://doi.org/10.1016/J.LWT.2014.04.011.

Tohamy, Khairy M., Mostafa Mabrouk, Islam E. Soliman, Hanan H. Beherei and Mohamed A. Aboelnasr. 2018. Novel alginate/hydroxyethyl cellulose/hydroxyapatite composite scaffold for bone regeneration: *in vitro* cell viability and proliferation of human mesenchymal stem cells. International Journal of Biological Macromolecules 112(June): 448–60. https://doi.org/10.1016/J.IJBIOMAC.2018.01.181.

Torgbo, Selorm and Prakit Sukyai. 2018. Bacterial cellulose-based scaffold materials for bone tissue engineering. Applied Materials Today 11: 34–49. https://doi.org/10.1016/j.apmt.2018.01.004.

Torgbo, Selorm and Prakit Sukyai. 2020. Biodegradation and thermal stability of bacterial cellulose as biomaterial: the relevance in biomedical applications. Polymer Degradation and Stability 179(September): 109232. https://doi.org/10.1016/J.POLYMDEGRADSTAB.2020.109232.

Torres, Fernando G., Solene Commeaux and Omar P. Troncoso. 2012. Biocompatibility of bacterial cellulose based biomaterials. Journal of Functional Biomaterials 3(4): 864–78. https://doi.org/10.3390/JFB3040864.

Trache, Djalal, M. Hazwan Hussin, M. K. Mohamad Haafiz and Vijay Kumar Thakur. 2017. Recent progress in cellulose nanocrystals: sources and production. Nanoscale 9(5): 1763–86. https://doi.org/10.1039/C6NR09494E.

Tudoroiu, Elena Emilia, Cristina Elena Dinu-Pîrvu, Mădălina Georgiana Albu Kaya, Lăcrămioara Popa, Valentina Anuța, Răzvan Mihai Prisada et al. 2021. An overview of cellulose derivatives-based dressings for wound-healing management. Pharmaceuticals 14(12): 1215. https://doi.org/10.3390/PH14121215.

Tyagi, Neha. 2016. Production of cellulose from sugarcane molasses using gluconacetobacter intermedius snt-1: optimization & characterization. Journal of Cleaner Production 112: 71–80.

Ul-Islam, Mazhar, Taous Khan and Joong Kon Park. 2012. Nanoreinforced bacterial cellulose–montmorillonite composites for biomedical applications. Carbohydrate Polymers 89(4): 1189–97. https://doi.org/10.1016/J.CARBPOL.2012.03.093.

Ul-Islam, Mazhar, Waleed Ahmad Khattak, Muhammad Wajid Ullah, Shaukat Khan and Joong Kon Park. 2014. Synthesis of regenerated bacterial cellulose-zinc oxide nanocomposite films for biomedical applications. Cellulose 21(1): 433–47. https://doi.org/10.1007/S10570-013-0109-Y/FIGURES/9.

Umar Aslam Khan, Muhammad, Sajjad Haider, Adnan Haider, Saiful Izwan Abd Razak, Mohammed Rafiq Abdul Kadir, Saqlain A. Shah et al. 2021. Development of porous, antibacterial and biocompatible GO/n-HAp/bacterial cellulose/β-glucan biocomposite scaffold for bone tissue engineering. Arabian Journal of Chemistry 14(2): 102924. https://doi.org/10.1016/J.ARABJC.2020.102924.

Valle, Luís J. Del, Angélica Díaz and Jordi Puiggalí. 2017. Hydrogels for biomedical applications: cellulose, chitosan, and protein/peptide derivatives. Gels 3(3). https://doi.org/10.3390/GELS3030027.

Varghese, Rini Thresia, Reeba Mary Cherian, Cintil Jose Chirayil, Tijo Antony, Hanieh Kargarzadeh and Sabu Thomas. 2023. Nanocellulose as an avenue for drug delivery applications: a mini-review. Journal of Composites Science 7(6): 210. https://doi.org/10.3390/JCS7060210.

Viera, Rose G. P., Guimes Rodrigues Filho, Rosana M. N. de Assunção, Carla da Carla, Júlia G. Vieira and Grasielle S. de Oliveira. 2007. Synthesis and characterization of methylcellulose from sugar cane bagasse cellulose. Carbohydrate Polymers 67(2): 182–89. https://doi.org/10.1016/j.carbpol.2006.05.007.

Vuoriluoto, Maija, Hannes Orelma, Meri Lundahl, Maryam Borghei and Orlando J. Rojas. 2017. Filaments with affinity binding and wet strength can be achieved by spinning bifunctional cellulose nanofibrils. Biomacromolecules 18(6): 1803–13. https://doi.org/10.1021/ACS.BIOMAC.7B00256.

Wada, Masahida, Laurent Heux and Junji Sugiyama. 2004. Polymorphism of cellulose I family: reinvestigation of cellulose IVI. Biomacromolecules 5(4): 1385–91. https://doi.org/10.1021/BM0345357.

Wahid, Fazli, Xiang Jun Zhao, Xue Qing Zhao, Xiao Fang Ma, Na Xue, Xiao Zhi Liu et al. 2021. Fabrication of bacterial cellulose-based dressings for promoting infected wound healing. ACS Applied Materials & Interfaces 13(28): 32716–28. https://doi.org/10.1021/ACSAMI.1C06986.

Walther, Andreas, Jaakko V. I. Timonen, Isabel Díez, Antti Laukkanen and Olli Ikkala. 2011. Multifunctional high-performance biofibers based on wet-extrusion of renewable native cellulose nanofibrils. Advanced Materials (Deerfield Beach, Fla.) 23(26): 2924–28. https://doi.org/10.1002/ADMA.201100580.

Wang, Baoxiu, Xiangguo Lv, Shiyan Chen, Zhe Li, Jingjing Yao, Xufeng Peng et al. 2018. Use of heparinized bacterial cellulose based scaffold for improving angiogenesis in tissue regeneration. Carbohydrate Polymers 181(February): 948–56. https://doi.org/10.1016/J.CARBPOL.2017.11.055.

Wang, Hui Min David, Xiao Chun Li, Duu Jong Lee and Jo Shu Chang. 2017. Potential biomedical applications of marine algae. Bioresource Technology 244(November): 1407–15. https://doi.org/10.1016/J.BIORTECH.2017.05.198.

Wang, Jing, Javad Tavakoli and Youhong Tang. 2019. Bacterial cellulose production, properties and applications with different culture methods—a review. Carbohydrate Polymers 219(September): 63–76. https://doi.org/10.1016/J.CARBPOL.2019.05.008.

Wang, Xiaoju, Fang Cheng, Jun Liu, Jan Henrik Smått, David Gepperth, Mika Lastusaari et al. 2016. Biocomposites of copper-containing mesoporous bioactive glass and nanofibrillated cellulose: biocompatibility and angiogenic promotion in chronic wound healing application. Acta Biomaterialia 46(December): 286–98. https://doi.org/10.1016/J.ACTBIO.2016.09.021.

Wang, Yaping, Junmin Qian, Na Zhao, Ting Liu, Weijun Xu and Aili Suo. 2017. Novel hydroxyethyl chitosan/cellulose scaffolds with bubble-like porous structure for bone tissue engineering. Carbohydrate Polymers 167(July): 44–51. https://doi.org/10.1016/J.CARBPOL.2017.03.030.

Wang, Zhao Mei, Lin Li, Bi Sheng Zheng, Nodirali Normakhamatov and Si Yuan Guo. 2007. Preparation and anticoagulation activity of sodium cellulose sulfate. International Journal of Biological Macromolecules 41(4): 376–82. https://doi.org/10.1016/j.ijbiomac.2007.05.007.

Wardhono, Endarto Yudo, Nufus Kanani and Alfirano Alfirano. 2020. A simple process of isolation microcrystalline cellulose using ultrasonic irradiation. Journal of Dispersion Science and Technology 41(8): 1217–26. https://doi.org/10.1080/01932691.2019.1614947.

Wasilewska, Katarzyna and Katarzyna Winnicka. 2019. Ethylcellulose–a pharmaceutical excipient with multidirectional application in drug dosage forms development. Materials 12(20): 3386. https://doi.org/10.3390/MA12203386.

Wei, Pingdong, Lei Wang, Fang Xie and Jie Cai. 2022. Strong and tough cellulose–graphene oxide composite hydrogels by multi-modulus components strategy as photothermal antibacterial platform. Chemical Engineering Journal 431(March): 133964. https://doi.org/10.1016/J.CEJ.2021.133964.

Wolff, Michael W., Corina Siewert, Sylvia Lehmann, Sara Post Hansen, Rene Djurup, Rene Faber et al. 2010. Capturing of cell culture-derived modified vaccinia ankara virus by ion exchange and pseudo-affinity membrane adsorbers. 105(4): 761–69. https://doi.org/10.1002/bit.22595.

Wu, Dan, Per Isaksson, Stephen J. Ferguson and Cecilia Persson. 2018. Young's Modulus of trabecular bone at the tissue level: a review. Acta Biomaterialia 78(September): 1–12. https://doi.org/10.1016/J.ACTBIO.2018.08.001.

Wu, Jian, Yudong Zheng, Xiaoxiao Wen, Qinghua Lin, Xiaohua Chen and Zhigu Wu. 2014. Silver nanoparticle/bacterial cellulose gel membranes for antibacterial wound dressing: investigation *in vitro* and *in vivo*. Biomedical Materials (Bristol, England) 9(3). https://doi.org/10.1088/1748-6041/9/3/035005.

Wu, Qiang, Yujie Meng, Katherine Concha, Siqun Wang, Yanjun Li, Lingfei Ma et al. 2013. Influence of temperature and humidity on nano-mechanical properties of cellulose nanocrystal films made from switchgrass and cotton. Industrial Crops and Products 48(July): 28–35. https://doi.org/10.1016/J.INDCROP.2013.03.032.

Xi, Tingfei, Y. M. Chen, Yudong Zheng, Tingting Guo, Jiaquan Hou, Yizao Wan et al. 2009. *In vitro* cytotoxicity of bacterial cellulose scaffolds used for tissue-engineered bone. Http://Dx.Doi.Org/10.1177/0883911509102710 24(SUPPL.1): 137–45. https://doi.org/10.1177/0883911509102710.

Xia, Jian, Hao Zhang, Faquan Yu, Ying Pei and Xiaogang Luo. 2020. Superclear, porous cellulose membranes with chitosan-coated nanofibers for visualized cutaneous wound healing dressing. ACS Applied Materials and Interfaces 12(21): 24370–79. https://doi.org/10.1021/ACSAMI.0C05604/SUPPL_FILE/AM0C05604_SI_001.PDF.

Xing, Qi, Feng Zhao, Si Chen, James McNamara, Mark A. DeCoster and Yuri M. Lvov. 2010. Porous biocompatible three-dimensional scaffolds of cellulose microfiber/gelatin composites for cell culture. Acta Biomaterialia 6(6): 2132–39. https://doi.org/10.1016/J.ACTBIO.2009.12.036.

Xu, Dezhong, Yanan Cheng, Weijie Lin, Shibo Han, Shuai Wu, Ajoy Kanti Mondal et al. 2024. Di-aldehyde tunicate cellulose nanocrystal (D-TCNC) aerogels for drug delivery: effect of D-TCNC composition on aerogel structure and release properties. International Journal of Biological Macromolecules 256(January): 128345. https://doi.org/10.1016/J.IJBIOMAC.2023.128345.

Xu, Mengda and Jiangping Song. 2021. Targeted therapy in cardiovascular disease: a precision therapy era. Frontiers in Pharmacology 12(April): 623674. https://doi.org/10.3389/FPHAR.2021.623674/BIBTEX.

Xu, Xuezhu, Fei Liu, Long Jiang, J. Y. Zhu, Darrin Haagenson and Dennis P. Wiesenborn. 2013. Cellulose nanocrystals vs. cellulose nanofibrils: a comparative study on their microstructures and effects as polymer reinforcing agents. ACS Applied Materials and Interfaces 5(8): 2999–3009. https://doi.org/10.1021/AM302624T/ASSET/IMAGES/MEDIUM/AM-2012-02624T_0018.GIF.

Yamanaka, S., K. Watanabe, N. Kitamura, M. Iguchi, S. Mitsuhashi, Y. Nishi et al. 1989. The structure and mechanical properties of sheets prepared from bacterial cellulose. Journal of Materials Science 24(9): 3141–45. https://doi.org/10.1007/BF01139032/METRICS.

Yan, Bai Yi, Bai Lu Liu, Ling Yu Yi, Gui Ling Wang, Bai Zhong Xue and Ying Xie. 2020. Doxorubicin-loaded in situ gel combined with biocompatible hydroxyethyl cellulose hemostatic gauze for controlled release of drugs and prevention of breast cancer recurrence postsurgery. ACS Biomaterials Science and Engineering 6(10): 5959–68. https://doi.org/10.1021/ACSBIOMATERIALS.0C01187/SUPPL_FILE/AB0C01187_SI_001.PDF.

Yang, Jun, Chun Rui Han, Xue Ming Zhang, Feng Xu and Run Cang Sun. 2014. Cellulose nanocrystals mechanical reinforcement in composite hydrogels with multiple cross-links: correlations between dissipation properties and deformation mechanisms. Macromolecules 47(12): 4077–86. https://doi.org/10.1021/MA500729Q/SUPPL_FILE/MA500729Q_SI_001.PDF.

Yang, Mao, Wenjuan Zhen, Hui Chen and Zhihua Shan. 2016. Biomimetic design of oxidized bacterial cellulose-gelatin-hydroxyapatite nanocomposites. Journal of Bionic Engineering 13(4): 631–40. https://doi.org/10.1016/S1672-6529(16)60334-7.

Yang, Zhangqiang, Xiaojie Li, Junhui Si, Zhixiang Cui and Kaiping Peng. 2019. Morphological, mechanical and thermal properties of poly(lactic acid) (PLA)/cellulose nanofibrils (CNF) composites nanofiber for tissue engineering. Journal Wuhan University of Technology, Materials Science Edition 34(1): 207–15. https://doi.org/10.1007/S11595-019-2037-7/METRICS.

Ye, Shan, Lei Jiang, Chen Su, Zhongjie Zhu, Yanyi Wen and Wei Shao. 2019. Development of gelatin/bacterial cellulose composite sponges as potential natural wound dressings. International Journal of Biological Macromolecules 133(July): 148–55. https://doi.org/10.1016/J.IJBIOMAC.2019.04.095.

Ye, Shan, Lei Jiang, Jimin Wu, Chen Su, Chaobo Huang, Xiufeng Liu et al. 2018. Flexible amoxicillin-grafted bacterial cellulose sponges for wound dressing: *in vitro* and *in vivo* evaluation. ACS Applied Materials & Interfaces 10(6): 5862–70. https://doi.org/10.1021/ACSAMI.7B16680.

Yu, Hou Yong, Zong Yi Qin, Yan Nan Liu, Long Chen, Na Liu and Zhe Zhou. 2012. Simultaneous improvement of mechanical properties and thermal stability of bacterial polyester by cellulose nanocrystals. Carbohydrate Polymers 89(3): 971–78. https://doi.org/10.1016/J.CARBPOL.2012.04.053.

Yu, Rui, Hualei Zhang and Baolin Guo. 2021. Conductive biomaterials as bioactive wound dressing for wound healing and skin tissue engineering. Nano-Micro Letters 14(1). https://doi.org/10.1007/S40820-021-00751-Y.

Yu, Sujie, Jianzhong Sun, Qianqian Wang, Jian Wu and Jun Liu. 2023. Extraction of bioactive polysaccharide from Ulva prolifera biomass waste toward potential biomedical application. International Journal of Biological Macromolecules 235(April): 123852. https://doi.org/10.1016/J.IJBIOMAC.2023.123852.

Yuan, Qi, Jing Bian and Ming-Guo Ma. 2021. Advances in biomedical application of nanocellulose-based materials: a review. Current Medicinal Chemistry 28(40): 8275–95. https://doi.org/10.2174/0929867328666201130124501.

Zaman, Ahsan, Fei Huang, Man Jiang, Wei Wei and Zuowan Zhou. 2020. Preparation, properties, and applications of natural cellulosic aerogels: a review. Energy and Built Environment 1(1): 60–76. https://doi.org/10.1016/j.enbenv.2019.09.002.

Zanchetta, Enio, Eya Damergi, Bhavish Patel, Tobias Borgmeyer, Horst Pick, Adrian Pulgarin et al. 2021. Algal cellulose, production and potential use in plastics: challenges and opportunities. Algal Research 56(June): 102288. https://doi.org/10.1016/J.ALGAL.2021.102288.

Zeng, Aoqiong, Ruijin Yang, Yanjun Tong and Wei Zhao. 2023. Functional bacterial cellulose nanofibrils with silver nanoparticles and its antibacterial application. International Journal of Biological Macromolecules 235(April): 123739. https://doi.org/10.1016/J.IJBIOMAC.2023.123739.

Zennifer, Allen, Praseetha Senthilvelan, Swaminathan Sethuraman and Dhakshinamoorthy Sundaramurthi. 2021. Key advances of carboxymethyl cellulose in tissue engineering & 3D bioprinting applications. Carbohydrate Polymers 256(March): 117561. https://doi.org/10.1016/J.CARBPOL.2020.117561.

Zha, Fangwen, Wei Chen, Lu Hao, Chunsheng Wu, Meng Lu, Lifeng Zhang et al. 2020. Electrospun cellulose-based conductive polymer nanofibrous mats: composite scaffolds and their influence on cell behavior with electrical stimulation for nerve tissue engineering. Soft Matter 16(28): 6591–98. https://doi.org/10.1039/D0SM00593B.

Zhang, Hao, Jin Wu, Jun Zhang and Jiasong He. 2005. 1-allyl-3-methylimidazolium chloride room temperature ionic liquid: a new and powerful nonderivatizing solvent for cellulose. Macromolecules 38(20): 8272–77. https://doi.org/10.1021/MA0505676.

Zhang, Qilei, Dongqiang Lin and Shanjing Yao. 2015. Review on biomedical and bioengineering applications of cellulose sulfate. Carbohydrate Polymers 132: 311–22. https://doi.org/10.1016/j.carbpol.2015.06.041.

Zhang, Tiantian, Tao Zuo, Danning Hu and Chunyu Chang. 2017. Dual physically cross-linked nanocomposite hydrogels reinforced by tunicate cellulose nanocrystals with high toughness and good self-recoverability. ACS Applied Materials & Interfaces 9(28): 24230–37. https://doi.org/10.1021/ACSAMI.7B06219.

Zhang, Yanbo, Xiaochen Liu, Liangdan Zeng, Jin Zhang, Jianlin Zuo, Jun Zou et al. 2019. Polymer fiber scaffolds for bone and cartilage tissue engineering. Advanced Functional Materials 29(36): 1–20. https://doi.org/10.1002/adfm.201903279.

Zhang, Zhao Yu, Yi Sun, Yu Dong Zheng, Wei He, Ying Ying Yang, Ya Jie Xie et al. 2020. A biocompatible bacterial cellulose/tannic acid composite with antibacterial and anti-biofilm activities for biomedical applications. Materials Science and Engineering: C 106(January): 110249. https://doi.org/10.1016/J.MSEC.2019.110249.

Zhao, Xuan, Chuanshan Zhao, Yifei Jiang, Xingxiang Ji, Fangong Kong, Tong Lin et al. 2020. Flexible cellulose nanofiber/Bi2Te3 composite film for wearable thermoelectric devices. Journal of Power Sources 479(December): 229044. https://doi.org/10.1016/J.JPOWSOUR.2020.229044.

Zhijiang, Cai, Hou Chengwei and Yang Guang. 2012. Poly(3-hydroxubutyrate-co-4-hydroxubutyrate)/bacterial cellulose composite porous scaffold: preparation, characterization and biocompatibility evaluation. Carbohydrate Polymers 87(2): 1073–80. https://doi.org/10.1016/J.CARBPOL.2011.08.037.

Zimoch, Jakub, Dominika Zielinska, Katarzyna Michalak-Micka, Dominic Rütsche, Roland Böni, Thomas Biedermann et al. 2021. Bio-engineering a prevascularized human tri-layered skin substitute containing a hypodermis. Acta Biomaterialia 134(October): 215–27. https://doi.org/10.1016/J.ACTBIO.2021.07.033.

Zinge, Chinmay and Balasubramanian Kandasubramanian. 2020. Nanocellulose based biodegradable polymers. European Polymer Journal 133(June): 109758. https://doi.org/10.1016/J.EURPOLYMJ.2020.109758.

Zmejkoski, Danica, Dragica Spasojević, Irina Orlovska, Natalia Kozyrovska, Marina Soković, Jasmina Glamočlija et al. 2018. Bacterial cellulose-lignin composite hydrogel as a promising agent in chronic wound healing. International Journal of Biological Macromolecules 118(Pt A): 494–503. https://doi.org/10.1016/J.IJBIOMAC.2018.06.067.

Zou, Pengfei, Jiaxin Yao, Ya Nan Cui, Te Zhao, Junwei Che, Meiyan Yang et al. 2022. Advances in cellulose-based hydrogels for biomedical engineering: a review summary. Gels 8(6): 364. https://doi.org/10.3390/GELS8060364.

Zulkifli, Farah Hanani, Fathima Shahitha Jahir Hussain, W. S. W. Harun and Mashitah M. Yusoff. 2019. Highly porous of hydroxyethyl cellulose biocomposite scaffolds for tissue engineering. International Journal of Biological Macromolecules 122(February): 562–71. https://doi.org/10.1016/J.IJBIOMAC.2018.10.156.

CHAPTER 3

Exploring Chitosan

Its Role in Fabricating Bioactive Scaffolds

Jila Majidi,[1,2] *Zahra Taherian,*[1,3] *Mohadeseh Saraei,*[4] *Fatemeh Koeini,*[5] *Saghar Hoseinzadeh,*[6] *Alireza Ghasemi,*[7] *Sara Simorgh,*[1,2] *Sanaz Alizadeh,*[8,*] *Narendra Pal Singh Chauhan*[9,*] and *Mazaher Gholipourmalekabadi*[1,4,10,*]

1. Introduction

The chemical formula of chitosan is ($C_6H_{11}NO_4$), making it a linear polymer (Figure 1) (Abd El-Hack et al., 2020). Chitin is deacetylated to create this polymer, which is then randomly distributed into β-(1→4)-linked D-glucosamine and

[1] Cellular and Molecular Research Center, Iran University of Medical Sciences, Tehran, Iran.
[2] Department of Tissue Engineering & Regenerative Medicine, Faculty of Advanced Technologies in Medicine, Tehran, Iran.
[3] Department of Biology, Faculty of Science, University of Gilan, Rasht, Iran.
[4] Department of Medical Biotechnology, Faculty of Allied Medicine, Iran University of Medical Sciences, Tehran, Iran.
[5] Biomedical Engineering Department, Amirkabir University of Technology (Tehran Polytechnic), Tehran, Iran.
[6] Department of Medical Biotechnology, Faculty of Advanced Medical Sciences, Tabriz University of Medical Sciences, Tabriz, Iran.
[7] Student Research Committee, Alborz University of Medical Sciences, Karaj, Iran.
[8] Burn Research Center, Iran University of Medical Science, Tehran, Iran.
[9] Department of Chemistry, Faculty of Science, Bhupal Nobles' University, Udaipur-313002, Rajasthan, India.
[10] NanoBiotechnology and Regenerative Medicine Innovation Group, Noavarn Salamat ZHINO (PHC), Tehran, Iran.
* Corresponding authors: s.alizadeh@rsct.ir; alizadehsannaz@gmail.com, narendrapalsingh14@gmail.com, mazaher.gholipour@gmail.com; mazaher.gholipour@iums.ac.ir

DOI: 10.1201/9781003518792-3

N-acetyl-D-glucosamine units. Because chitosan produces amino sugars when it is hydrolyzed, it is non-toxic and biocompatible (Petroni et al., 2023; Qian et al., 2023; Ghormade et al., 2017; Hirano et al., 1989).

The special qualities of chitosan and its derivatives, including carboxymethyl chitosan, make them suitable for biomedical applications. One of chitosan's most important properties is the abundance of polar groups joined by hydrogen bonds. Although chitosan is insoluble in most organic solvents and water, it is still possible to modify it so that it is more soluble in water and performs better in medical devices. Aqueous acidic solutions can dissolve chitosan's primary amine groups when it is protonated in an acidic environment. Depending on the degree of deacetylation and processing parameters, chitosan's molecular weight can vary from 20 to 1200 kDa. The most often used technique for figuring out chitosan's molecular weight is viscometry analysis (Aranaz et al., 2021; Elkomy et al., 2022; Kluczka, 2024).

Chitosan use in drug delivery systems has advanced significantly over the past decade. As an example, chitosan has been used in a nasal device designed to deliver

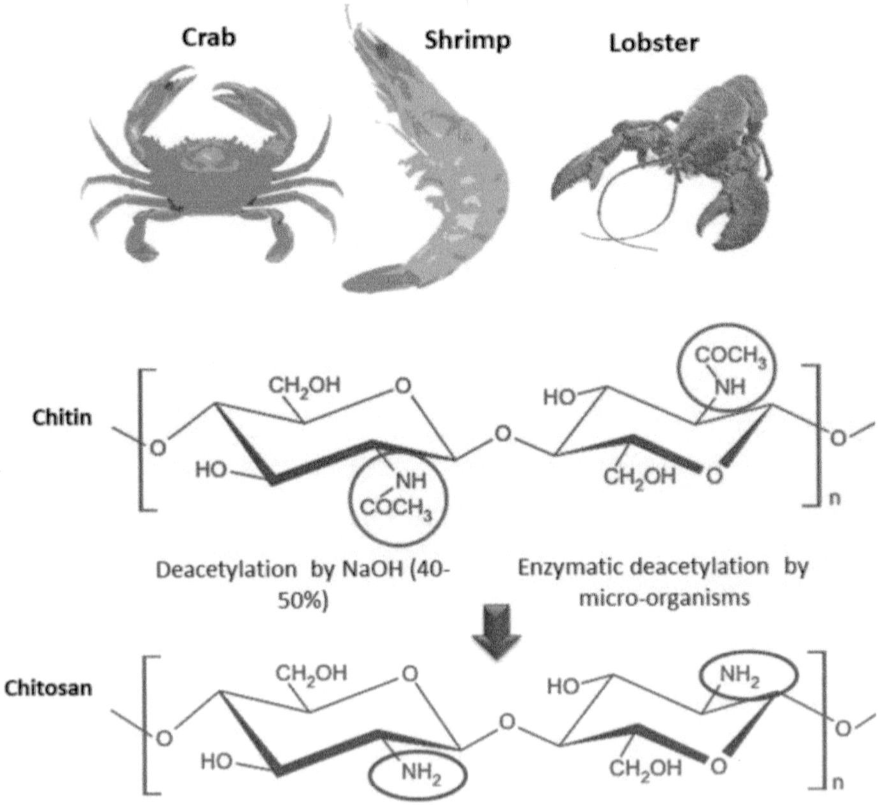

Figure 1. The chemical formula of chitosan; β-(1→4)-linked D-glucosamine (deacetylated unit) and N-acetyl-D-glucosamine (acetylated unit) (Abd El-Hack et al., 2020).

insulin to diabetics. Also, the antimicrobial activity of chitosan is also remarkable. During protonation, chitosan acquires a positive charge, which can neutralize negative charges on microbial cell membranes. As a result, chitosan is an excellent candidate for use in antimicrobial dressings and wound care products. Furthermore, chitosan's ability to form films, coupled with its biocompatibility, makes it a useful biomedical material. The biocompatibility and biodegradability of chitosan make it ideal for healing wounds quickly and effectively. It maintains a moist wound environment while forming a barrier against bacterial infections, accelerating the healing process. Moreover, chitosan is capable of combining with other bioactive agents to be used as a composite material to enhance therapeutic properties (Elkomy et al., 2022; Lv et al., 2023; Rodríguez-Vázquez et al., 2015a; Wang et al., 2021; Yan et al., 2021a).

Among the many applications of chitosan, tissue engineering has attracted significant interest. Because of its structural similarities to glycosaminoglycans, which are natural components of the extracellular matrix, chitosan makes an ideal scaffold material for tissue regeneration. Chitosan-based scaffolds are capable of supporting cell adhesion, proliferation, and differentiation, which are essential for successful tissue engineering. Moreover, chitosan can be chemically modified to alter its properties for specific tissue engineering applications, such as bone, cartilage, or skin regeneration (Aranaz et al., 2021; Tahrir et al., 2014; Oryan and Sahvieh, 2017; Rodríguez-Vázquez et al., 2015a).

Researchers have extensively studied chitosan's mucoadhesive properties as well as its ability to modulate drug release kinetics. It is possible to encapsulate a broad range of therapeutic molecules, including small drugs and large proteins, with chitosan, allowing for targeted and controlled delivery. For oral, nasal, and transdermal drug delivery systems, where controlled release and enhanced bioavailability are crucial, this capability is particularly advantageous. There is a wide range of potential uses for chitosan beyond those traditionally associated with biomedicine. Research has explored the use of chitosan-based nanoparticles for gene delivery, which can protect genetic material while facilitating its uptake by cells. The development of gene therapy and treatment for genetic disorders will now be possible (Elkomy et al., 2022; Liu et al., 2021; Meng et al., 2024).

Ultimately, chitosan is a versatile and promising biopolymer with many potential biomedical uses. Due to its unique properties, such as biocompatibility, biodegradability, antimicrobial activity, and film-forming capabilities, it makes an ideal candidate for drug delivery, wound healing, tissue engineering, and beyond. It is unquestionably likely that continued research and development in this field will lead to new and innovative applications for chitosan in the field of biomedicine.

2. Chitosan Structure and Derivatives

Chitosan is a highly significant biopolymer, mainly in the field of biomedical applications, due to its unique characteristics such as biocompatibility, biodegradability, and non-toxicity. It derives from chitin. Chitin is an organic compound composed of N-acetyl-D-glucosamine units. It is mostly present in the outer shells of crustaceans and insects and in the cell walls of fungi (Kean and Thanou, 2011; Dash et al., 2011; Pichyangkura and Chadchawan, 2015; Iber et al., 2021). Chitin is converted into chitosan through a process known as deacetylation, which involves the removal of acetyl groups from chitin. This process leads to the production of a polymer that is mainly made of β-(1→4)-linked D-glucosamine units. This chapter explores the structural properties of chitosan, the creation of its precursor, chitin, and the several techniques used to prepare chitosan (Kean and Thanou, 2011; Hussain et al., 2013).

The structure of chitosan is characterized by the presence of amino groups ($-NH_2$) along its polymer backbone, which has a substantial impact on its chemical properties and biological functions. The acetylation degree (DA) and deacetylation degree (DD) are crucial factors that determine the structural and functional characteristics of chitosan. The DA denotes the mole fraction of N-acetylated units, whereas the DD indicates the proportion of deacetylated units in the polymer (Younes and Rinaudo, 2015). The relationship between these factors is inverse; as the degree of acetylation falls, the degree of deacetylation increases, indicating a higher concentration of unbound amine groups. The interaction between DD and chitosan is vital since it directly affects the biological activity and solubility of chitosan. Chitosan with a higher degree of deacetylation (DD) exhibits enhanced biological activity and improved solubility in water. This is attributed to the increased presence of protonated amine groups, which facilitate interactions with water and biological tissues (Nilsen-Nygaard et al., 2015).

Furthermore, the functioning of chitosan is influenced by its molecular weight (MW) in addition to DD. Chitosan with a lower molecular weight generally shows improved bioactivity and solubility, making it very desired for a wide range of applications (Huang et al., 2004). Chitosan with a molecular weight below 30 kDa is often soluble in water without the need for acidification. However, chitosan with a larger molecular weight requires acidic conditions to dissolve. This is especially pertinent in biological and pharmacological contexts, where solubility plays a crucial role (Shuai et al., 2013; Nilsen-Nygaard et al., 2015).

3. Chemical Chitosan Properties

Chitosan, a (1,4)-linked 2-amino-2-deoxy-β-d-glucan, is a naturally occurring polymer derived from chitin through N-deacetylation. When the degree of deacetylation (DD) of chitin surpasses 50%, the resulting substance is recognized as chitosan, which ranks as the second most abundant natural polymer after cellulose. The degree of deacetylation, representing the percentage of glucosamine units in the chitosan molecule, is a critical parameter influencing the molecular structure and properties of chitosan. Complete deacetylation, yielding 100% glucosamine units, is unattainable; thus, chitosan exists as a copolymer comprising both N-acetylglucosamine and glucosamine units (Younes et al., 2015).

Chitosan is a structural component found in the exoskeletons of shellfish and insects, as well as the cell walls of bacteria and fungi. It is predominantly sourced from the shells of crustaceans such as crabs, lobsters, and shrimp (Omari et al., 2012). This biopolymer exhibits a range of physicochemical, biological, and technological properties, which will be explored in this section. The functional properties of chitosan, including its solubility, crystallinity, molecular weight, and viscosity, are significantly influenced by its degree of deacetylation (Rinaudo, 2008; Cheng et al., 2020).

Chitosan's inherent biocompatibility, biodegradability, and lack of toxicity and antigenicity have made it a promising candidate for a wide range of biomedical applications. Its cationic nature, a unique characteristic among abundant polysaccharides, contributes to its functionality in these applications. For instance, the positive charge of chitosan potentially allows for enhanced absorption of peptide and protein drugs through mucoadhesion to mucosal surfaces (e.g., gastric or nasal mucosa) or by facilitating the opening of tight junctions between epithelial cells. This property, combined with its ability to promote wound healing and exhibit bacteriostatic effects, makes chitosan a valuable component in wound dressings and drug delivery systems. Furthermore, chitosan's hydroxyl and amino groups allow it to interact with other negatively charged hydrocolloids or small molecules, forming complexes that can improve the mouth-feel of pharmaceutical formulations (Li and Huang, 2012).

The versatility of chitosan extends beyond biomedical applications. Its unique combination of biocompatibility, biodegradability, and reactive functional groups allows for its use in diverse forms, including hydrogels, micro/nanoparticles, beads, sponges, foams, membranes, films, and micro/nanofibers (Harugade et al., 2023). This broad applicability showcases its potential in various technological fields, underscoring its significance as a valuable biomaterial.

3.1 Solubility

Chitin, the most abundant organic material after cellulose, is primarily extracted from the shells of crustaceans and the exoskeletons of arthropods. Despite its prevalence, chitin's applications are limited due to its poor solubility. However, when subjected to deacetylation in an alkaline environment, chitin transforms into chitosan and becomes soluble in aqueous acid media (Roberts and Roberts, 1992).

Chitosan is a weak base with a pKa value of 6.2–7.0 and is insoluble in water and organic solvents (Tungtong et al., 2012). It readily dissolves in dilute aqueous acidic solutions ($pH < 6.5$), which can protonate a substantial fraction of glucosamine units into soluble $R\text{–}NH^+_3$ (Qin et al., 2006). Various acids, including acetic, lactic, formic, nitric, phosphoric, and hydrochloric acid, can be employed to solubilize chitosan (Cardile et al., 2008; Wang et al., 2004). However, it's worth noting that chitosan becomes insoluble at pH levels above 6.5 as the amine groups are deprotonated (Kumar et al., 2004). As limited aqueous solubility of chitosan under physiological conditions and in organic solvents reduces its potential in the pharmaceutical and biomedical applications, many attempts have been made to overcome this drawback by modifying polymer molecular weight, degree of acetylation, or using chemical modification method (Aranaz et al., 2021). There have been several reports on reducing the molecular weight of chitosan by enzymatic hydrolysates (Ikeda et al., 1993), lysozyme degradation (Nordtveit et al., 1996), shear-force treatment and acid-hydrolysis (Austin et al., 1981). These techniques have shown promise in producing chitosan with higher solubility compared to other methods. Another strategy focuses on controlling the deacetylation process to yield water-soluble chitosan, although this method may not be suitable for both small and large-scale production (Kurita et al., 1991). Chemical modifications have emerged as a powerful tool for improving chitosan's solubility. These include carboxymethylation, sulphonation, quaternization of amino groups, and grafting of hydrophilic groups onto chitosan (Fei Liu et al., 2001).

Numerous studies have demonstrated that these modifications can significantly enhance chitosan's solubility in water over a wide pH range. For instance, PEGylation of chitosan has been shown to increase its solubility in aqueous solutions and organic solvents while also imparting self-aggregation properties in aqueous solutions (Ouchi et al., 1998). Similarly, the conjugation of methoxy poly(ethylene glycol) (MPEG) to chitosan has improved its solubility in aqueous solutions across a broad pH range (4.0–11.0) and in various organic solvents such as DMF, DMSO, ethanol, and acetone (Jeong et al., 2008). Kumar and co-workers have used tyrosinase to graft chlorogenic acid onto chitosan, resulting in water solubility under basic conditions. The enzyme in this study was tyrosinase, which converts a wide range of phenolic substrates into electrophilic o-quinones. The o-quinones are freely diffusible and can undergo reaction with the nucleophilic amino groups of chitosan. When the amount of chlorogenic acid used in the modification reaction exceeded 30% relative to chitosan's amino groups, the modified chitosan exhibited solubility under both acidic and basic conditions (Kumar et al., 1999). Elsewhere, Sogias and colleagues tried to modulate the inter- or intramolecular forces in chitosan and its crystallinity to improve its aqueous solubility over a broader pH range. They found that re-acetylation of chitosan by anhydride acetic or fully N-deacetylated, resulting in its half-acetylated derivative, led to the disruption of chitosan crystallinity and consequently expanded its solubility window up to pH > 7.4. Also, the physical approach included the use of additives, urea, and guanidine hydrochloride, with abilities to disrupt intra- and inter macro-molecular hydrogen bonding and alter the solubility window of chitosan. When both physical and chemical approaches are used in combination, chitosan solubility can be achieved over a remarkably broad pH range (between 1 to 12) (Sogias et al., 2010).

3.2 Viscosity

The Mark–Houwink–Sakurada equation, $[\eta] = K \cdot M^a$, is crucial in determining the viscosity of polymers, including chitosan. In this equation, $[\eta]$ represents intrinsic viscosity, M denotes molecular weight, while a and K are experimentally determined constants known as the Mark-Houwink exponent and constant, respectively. These parameters are influenced by factors such as molecular weight distribution, polymer-solvent interaction, amino group content and distribution, pH, ionic strength of the solution, and the degree of deacetylation of chitosan (Halabalova et al., 2011).

The viscosity, non-Newtonian flow properties (including elastic properties and shear-thinning behavior), and flow activation energies of chitosan solutions increase as the degree of deacetylation rises, which is associated with chain flexibility (Wang and Xu, 1994). These characteristics also intensify with higher chitosan concentrations and lower ionic strength (Cho et al., 2006). The electric charge density on the chitosan molecule can be altered by changing pH, ionic strength, concentration, time, or the nature of the media. These modifications in electric charge density affect the molecule's conformation, subsequently influencing its viscoelastic properties. For example, in chitosan-organic acid solutions, apparent viscosities increase as pH decreases from 5.0 to 2.0. Conversely, in chitosan-HCl solutions, the opposite trend is observed. In chitosan-NaOH solutions, viscosities decrease as pH increases from 5.0 to 6.0, then rise again as pH further increases to 6.0–6.4 (Chen et al., 1994).

Apart from above-mentioned factors, temperature significantly impacts the viscosity of chitosan solutions. Pogodina et al. (Pogodina et al., 1986) reported that apparent viscosities of chitosan decreased with increasing temperature. Similarly, Chen et al. (Chen and Tsaih, 1998) investigated chitosans with identical degrees of deacetylation but varying molecular weights, revealing that intrinsic viscosities decrease linearly as temperature rises. Viscosity generally exhibits a proportional relationship with molecular weight, particularly for high molecular-weight materials. The molecular weight and viscosity of chitosan solutions can influence its biodegradation and drug release properties *in vivo*. Chitosan with larger molecular weight typically has longer molecular chains, resulting in greater relative movement resistance between molecules and consequently higher viscosity (Kofuji et al., 2005).

4. Biological Chitosan Properties

Chitosan is an important biomaterial due to its biodegradability, biocompatibility, anti-oxidant activity, and anti-microbial activity (Levengood and Zhang, 2014). It has been found to accelerate wound healing by modulating the function of inflammatory cells and promoting the formation of granulation tissue. The cationic nature of chitosan contributes to its interesting biological properties, such as promoting clotting and possessing antimicrobial properties (Dai et al., 2011). Additionally, its functionality in terms of possessing amine and hydroxyl groups can be utilized for various applications, including drug delivery and bone regeneration. Chitosan's hydrophilic nature also supports the adhesion and proliferation of cells. In addition, chitosan has also been implicated in promoting osteogenic differentiation of mesenchymal stem cells (Klokkevold et al., 1996; Yang et al., 2009).

4.1 Biodegradability

For tissue engineering, the ideal scaffold should break down at the same rate that new tissue grows. The scaffold won't be able to sustain the host tissue sufficiently if the rate of degradation outpaces the rate of regeneration, which will lead to ineffective regeneration. Under physiological conditions, chitosan undergoes both physical (such as swelling, cracking, and dissolution) and chemical (including depolymerization, oxidation, and non-enzymatic and enzymatic hydrolysis) degradation (Hu et al., 2012). The hydrolytic degradation of the glycosidic bonds between polysaccharide units occurs at a faster rate, making non-enzymatic hydrolytic mechanisms a minor part of chitosan degradation. Chitosan can be enzymatically degraded *in vivo* by lysozyme, a polycationic protein found in the extracellular matrix of human bone tissue (Deb et al., 2018). Lysozyme breaks the chitosan chain by cleaving the glycosidic bonds between polysaccharide units in the polymer. This results in a reduction of the molecular weight of the polymer until eventual solubility and removal of degradation products occur. The degradation products are nontoxic, mainly composed of glucosamine and saccharide, which can then be easily extracted from the body without interference with organs. The degradation rate by each mechanism is related to the degree of crystallinity, which is controlled by the DD, where higher DD results in a lower degradation rate due to closer chain packing and hydrogen bonding. The lysozyme concentration in the extracellular matrix of human tissues can increase up to 1000-fold the amount usually found in serum (0.95–2.45 μm) (Deb et al., 2018; Hu et al., 2012; Lazaridou et al., 2022). Therefore, it is important to determine the degradation rate of chitosan scaffolds at lower and higher concentrations of lysozyme to examine scaffold stability under physiological conditions (Hu et al., 2012).

4.2 Biocompatibility

Biomaterials must exhibit biocompatibility, meaning they should not cause adverse reactions when in contact with the body and should be able to interact harmoniously with the structures and cells of the human body without causing non-specific reactions. Clinical tests have shown no signs of inflammation or allergic reactions following the implantation, injection, topical application, or ingestion of chitosan in the human body (Lazaridou et al., 2022). This is attributed to the fact that chitosan is composed of GlcN and GlcNac, which are natural components of mammalian tissues (Ressler, 2022). Studies have supported the biocompatibility of the material and its direct correlation with the degree of deacetylation (DD) of the material used (Rizeq et al., 2019). Initial reports on toxicity did not show any adverse effects with chitosan films when subjected to standard "*in vivo*" toxicity tests to assess their safety. Furthermore, studies on the biocompatibility of chitosan films with different DD levels have shown that films with a DD between 69 and 74% induced a relatively acute inflammatory reaction due to rapid biodegradation, with almost complete resorption after 4 weeks of implantation. On the other hand, films with a higher DD (between 74 to 90%) resulted in a milder inflammatory reaction and slower degradation rate. This is consistent with the understanding that rapidly biodegradable biomaterials can trigger acute inflammation due to the significant production of low-molecular-weight compounds within a short period. It was concluded that films with $\geq$ 84% DD exhibited a reaction to softer tissue due to their slower degradation rate (Rodríguez-Vázquez et al., 2015b).

4.3 Hemostatic

Effective control of bleeding in trauma injuries is essential for reducing fatality rates. Hemorrhage from a major artery accounts for 50% of injuries sustained in combat settings and is responsible for 31% of mortality in civilian populations. Hemorrhage disrupts the flow of blood, which in turn affects the circulation of blood in small blood vessels and leads to a decrease in oxygen supply to the brain and body. Failure to promptly treat the patient may result in the development of hemorrhagic shock. Metabolic acidosis and hemorrhagic hypothermia can disrupt the coagulation system, making it more difficult for hemostasis to occur (Khan and g, 2019; Hu et al., 2018). Tactical combat casualty care (TCCC) has utilized combat gauze for eight years. Chitosan-based bleeding control dressings are now approved as alternatives to conventional equipment due to manufacturing advances. None of the bandages on the market can cure war injuries. Many animal experiments have demonstrated that chitosan bandages reduce bleeding even when blood coagulation is impeded. They are cost-effective. Chitosan's hemostatic efficacy depends on its molecular weight and DDA during purification. Higher deacetylation (DDA) increases erythrocyte and platelet aggregation, which begins hemostasis. Because platelets and red blood cells contain negative charges on their surfaces due to phosphatidylcholine, phosphatidylethanolamine, and sialic acid groups, respectively, chitosan's positive charge promotes blood coagulation (Figure 2) (Khan and Mujahid, 2019). Studies show that platelets and erythrocytes' negative charges cause electrostatic resistance, which prevents them from aggregating to start hemostasis (Gheorghiță et al., 2023;

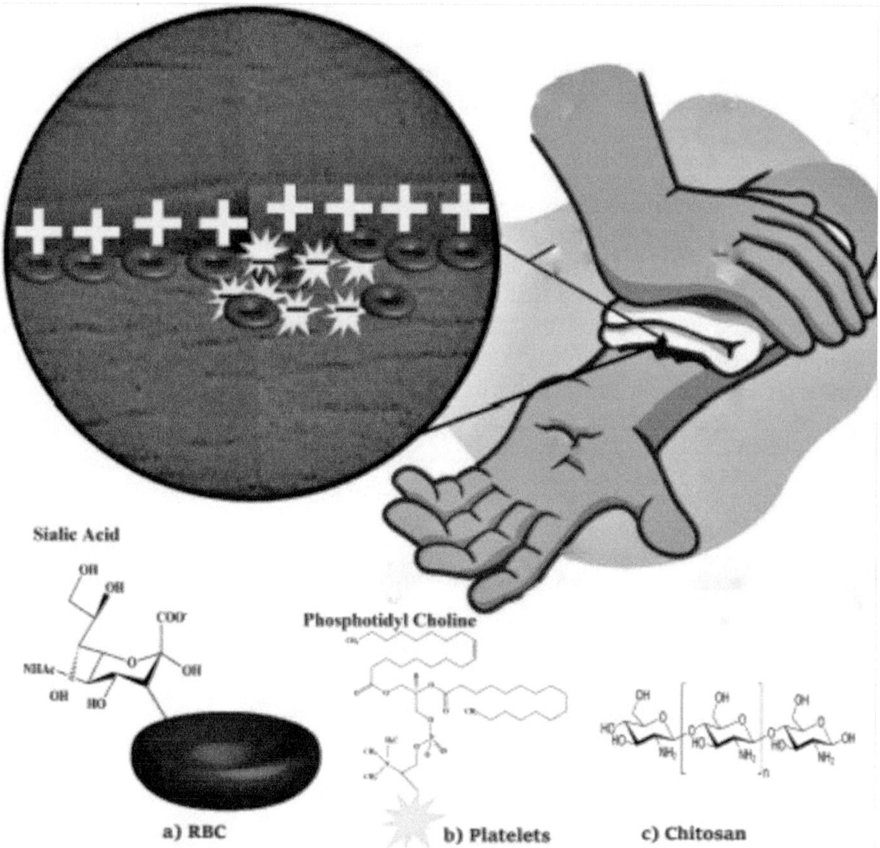

Figure 2. Hemostatic mechanism of chitosan-based material: The interaction between positively charged chitosan and negatively charged molecules on the surface of erythrocytes causes the aggregation of erythrocytes (Khan and Mujahid 2019).

Wang et al., 2019; Khan and Mujahid, 2019). Electrostatic interactions between chitosan (poly-N-acetyl glucosamine) amino groups and surface charges help erythrocytes aggregate. Platelet activation causes hemostasis (Pusateri et al., 2003).

Clinical studies have shown that bandages based on chitosan are useful in controlling wounds. These studies have assessed the effectiveness of Celox Gauze and chitosan-based bandages in prehospital care in conflict environments as well as civilian hospital care. All of the investigations looked closely at safety concerns related to chitosan-based dressings, and no issues were found in patients treated with these dressings (Celox particles and HemCon bandages) in civilian prehospital care settings or the military (Butler et al., 2014).

4.4 Antimicrobial Activity of Chitosan

In recent years, the emergence of antibiotic resistance and the rapid spread of drug-resistant strains have made treating microbial infections a serious challenge. It is indeed crucial to explore alternative solutions (O'Rourke et al., 2020). This has led to extensive studies on the antimicrobial potential of natural compounds, including biopolymers. They have found wide applications in the biomedicinally field due to their desirable properties, including biocompatibility, biodegradability, low toxicity, minimal immunogenicity, and cost-effectiveness and availability (Narayan, 2023). Chitosan, a unique bio-based polymer, is a linear hydrophilic polysaccharide derived from chitin. It has a wide-ranging antimicrobial effect against various organisms, such as bacteria, algae, yeasts, and fungi. Researchers have investigated chitosan in multiple forms (solutions, films, and composites) *in vivo* and *in vitro* experiments (Guarnieri et al., 2022). The antimicrobial activity of chitosan is also limited by low solubility, which can be attributed to its rigid crystalline structure. Modifying the free amino and hydroxyl groups, such as through quaternization and hydrophilic substitution, enhances solubility, resulting in derivatives with various biomedical applications, thereby demonstrating the potential of chitosan and its derivatives in combating microbial infections (Ma et al., 2008).

The antibacterial mechanism of chitosan and its derivatives remains a subject of ongoing research and is not fully understood; however, a few possible mechanisms are considered.

The most acceptable mechanism is the interaction between chitosan's polycationic structure and the microbial membrane's negative charge. As the number of positively charged amino groups in chitosan increases, its antimicrobial activity also increases due to the electrostatic reaction. Moreover, at higher chitosan concentrations, the chains accumulate and form a coating on bacteria, resulting in reduced binding and antimicrobial activity (Goy et al., 2009; Confederat et al., 2021).

Another mechanism is the passage of molecules with low molecular weight and small nanoparticles through the cell wall binding to DNA and preventing mRNA synthesis. The molecular size certainly affects this action, and the cell wall is impermeable to high-weight molecules and large nanoparticles. These particles can create a barrier on the wall to prevent the transfer of vital substances into the bacterial cell (Confederat et al., 2021; Yan et al., 2021b).

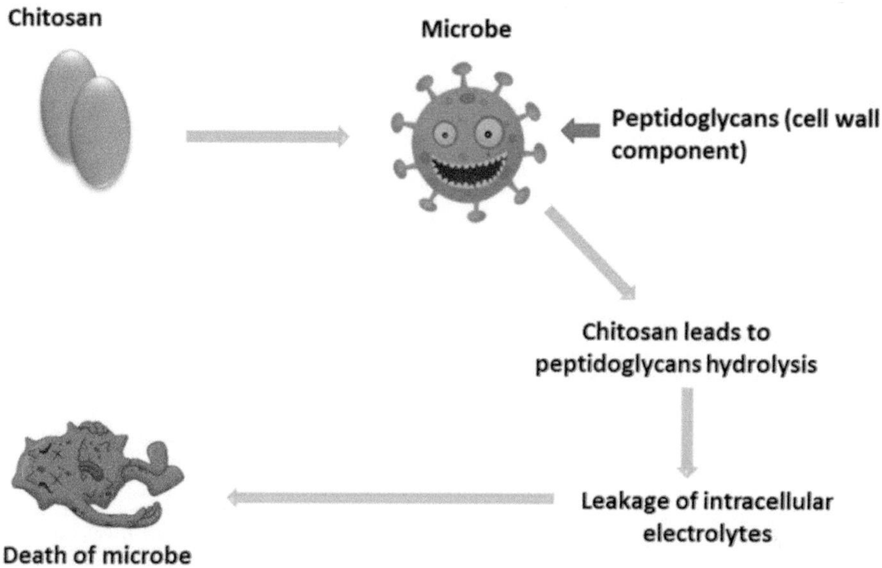

Figure 3. The proposed antimicrobial mechanism of chitosan (Abd El-Hack et al., 2020).

The third proposed mechanism is Chitosan's ability to chelate (bind) with metals. By chelation, amine groups with their positively charged uptake metals (such as copper, iron, or zinc). Metal ions are vital for bacterial growth and enzymatic processes. This starvation of bacterial nutrients leads to the inhibition of growth and cell death (Figure 3) (Abd El-Hack et al., 2020).

4.5 Antifungal Activity

Fungal infections are on the rise globally, especially among immunocompromised individuals with prolonged stays in intensive care units. Unfortunately, treatment options are limited, and outcomes are often unfavorable. The difficulties in managing fungal infections arise from the limited availability of treatment choices and the notable toxicity of current medications. Furthermore, the growing resistance to antifungal agents severely limits their effectiveness, making new therapeutic agents of great interest. Chitosan's antimicrobial effects are more pronounced against fungi than bacteria. This activity depends on molecular weight, degree of deacetylation, concentration, and pH. Optimal effectiveness is observed at 1% and 5% concentrations, particularly at lower pH levels (Confederat et al., 2021).

According to research, chitosan's antifungal activity involves inhibiting spore germination and mycelium growth. It also likely promotes cell wall degradation, disrupts cell membrane integrity, and inhibits ribosome biogenesis. Transcriptome analysis shows altered gene expression related to essential cellular processes such as the cell wall, membrane, and ribosome functions (Meng et al., 2020).

4.6 *Antiviral Activity*

Efforts to develop new drugs and vaccines have increased in recent years due to the lack of suitable antiviral medicines for emerging viral infections. The use of natural biopolymers as potential antiviral candidates holds promise for the development of bio-based antiviral drugs. The chitosan's remarkable properties and antiviral effectiveness make it an attractive platform for designing new derivatives and targeted delivery systems against viruses. Research into the antiviral properties of chitosan and its derivatives has been limited due to challenges and specific requirements associated with virus cultivation. Their precise mechanism of antiviral effect also remains incompletely understood, and researchers continue to explore this area (Jaber et al., 2022).

The antiviral property of chitosan and its derivatives is primarily based on ionic interactions between the positive charge of the amino groups and the negative charge of the virus surface. This interaction prevents the binding of the virus, entry of the viral genome, and its replication in the host cell, ultimately leading to the virus's demise (Issar and Arora, 2022).

In conclusion, the need for further research to fully understand the antimicrobial mechanism of chitosan and its derivatives is evident, highlighting the ongoing nature of the study and the potential for future discoveries. In addition, advances in the field of nanotechnology and the development of CS-based nanocomposites, chemical modifications of chitosan with antimicrobial peptides, or combining it with other agents like antibiotics or natural extracts can significantly enhance antimicrobial treatments resulting in synergistic effects, presenting new strategies for developing more effective antimicrobial products. Also, evaluating the long-term safety and effectiveness of these in clinical settings and their potential use with other therapeutic agents requires more research in the future.

4.7 *Antioxidant Activity*

Chitosan is known for its multifaceted biological activities, including antimicrobial, antifungal, and antioxidant capabilities. While the antibacterial properties of chitosan have been thoroughly investigated, its antioxidant properties have received less attention but are becoming increasingly intriguing. Antioxidants play a vital role in counteracting the harmful effects of free radicals and reactive oxygen species (ROS), which can result in oxidative stress, cellular harm, aging, and the development of different diseases. Chitosan and its derivatives have a high potential for use in food preservation, medicine, and cosmetics due to their antioxidant capability (Kołodziejska et al., 2021; Mohan and Pandey, 2016; Cheung et al., 2015; Wang, Yan et al., 2024b).

The antioxidant action of chitosan is mostly due to the amino and hydroxyl groups present in its glucosamine monomer units. These functional groups can transfer electrons or hydrogen atoms to free radicals, effectively neutralizing them and reducing oxidative damage. The antioxidant activity in chitosan can vary greatly depending on parameters such as its molecular weight, degree of deacetylation, and the precise environmental circumstances in which it is used (Islam et al., 2017a; Leong et al., 2021; Wang, Yan et al., 2024b).

The molecular weight (MW) of chitosan is a crucial factor in determining its antioxidant capacity. Studies have demonstrated that Chito oligosaccharides (COS), which are low molecular weight chitosan, have considerably more antioxidant action than high molecular weight chitosan. The heightened efficacy of low molecular weight chitosan can be attributed to their reduced viscosity and delicate solubility, which enables improved interaction with free radicals. In addition, chitosan with low molecular weight has a higher quantity of unbound amino and hydroxyl groups that can be used for scavenging radicals. On the other hand, chitosan with a high molecular weight tends to create hydrogen bonds inside its structure. This restricts the accessibility of these functional groups for antioxidant activity (Wang, Yan et al., 2024b; Ngo and Kim, 2014; Shariatinia, 2019).

The capacity of chitosan to eliminate free radicals is commonly evaluated using many tests, such as the DPPH (2,2-diphenyl-1-picrylhydrazyl) radical scavenging assay, the ABTS (2,2'-azino-bis (3-ethylbenzothiazoline-6-sulfonic acid)) radical scavenging assay, and the reducing power assay. Low molecular weight chitosan has consistently shown better efficacy in neutralizing free radicals compared to their high molecular weight counterparts in various studies. For example, studies have shown that chitosan with a molecular weight below 5 kDa has a DPPH scavenging activity of more than 80%, which is considerably higher than that of chitosan with a molecular weight above 100 kDa (Mathew and Abraham, 2006; Shahidi and Abuzaytoun, 2005a; Wang, Yan et al., 2024a).

The level of deacetylation (DD) is a significant component that affects the antioxidant activity of chitosan. The degree of deacetylation quantifies the proportion of glucosamine units in chitosan relative to N-acetylglucosamine units. Increasing the degree of deacetylation typically leads to a greater number of unbound amino groups, which can augment the antioxidant activity. Nevertheless, the correlation between deacetylation and antioxidant activity is not simple, as an extremely high level of deacetylation might result in decreased solubility and modified molecular structure, which may ultimately affect the effectiveness of antioxidants (Wang, Yan et al., 2024a; Shariatinia, 2019; Kittur et al., 2002).

Chemical modification can greatly boost the antioxidant activity of chitosan, in addition to its inherent features. An established method involves the attachment of antioxidant molecules onto the chitosan structure. Chitosan has undergone grafting with phenolic chemicals like gallic acid, ferulic acid, and epigallocatechin, which are renowned for their strong antioxidant qualities. The chitosan derivatives have shown significantly enhanced radical scavenging activity in comparison to unmodified chitosan. Chitosan that was modified with gallic acid demonstrated a DPPH scavenging activity of 89.5% in a particular research investigation. Additionally, it showed enhanced reducing power and metal chelation activity. The enhanced antioxidant activity in these derivatives is a result of the combined impacts of chitosan's intrinsic characteristics and the potent radical scavenging capabilities of the grafted compounds (Elkomy, Yan et al., 2024b; Yen et al., 2008; QYein et al., 2002).

Environmental variables, such as pH and temperature, also affect the antioxidant activity of chitosan. The solubility and structure of chitosan are influenced by pH, which in turn might impact its interaction with free radicals. Chitosan undergoes protonation and increases its solubility at low pH, hence improving its capacity to remove free radicals. Nevertheless, chitosan's antioxidant ability can be diminished due to the breakdown caused by severe pH circumstances. Temperature is a crucial component that can significantly impact chitosan. Elevated temperatures can expedite the deterioration of chitosan and diminish its antioxidant activity. On the other hand, moderate temperatures can increase the ability of low molecular-weight chitosan to dissolve, thereby enhancing its capacity to remove free radicals (Islam et al., 2017b; Crini and Badot, 2008).

Chitosan not only directly scavenges radicals, but also demonstrates antioxidant capabilities by chelating metals. Transition metals, including iron and copper, can catalyze the formation of extremely reactive hydroxyl radicals via Fenton reactions. Chitosan can attach to these metal ions, decreasing their availability to engage in these processes, therefore minimizing oxidative damage. The metal chelation capability of chitosan is crucial in food preservation since it effectively inhibits lipid oxidation and prolongs the shelf life of perishable items (Shahidi and Abuzaytoun, 2005b; Vinsova and Vavrikova, 2011).

Chitosan's antioxidant action has a wide range of applications that show great potential. Chitosan is employed as a natural preservative in the food industry to inhibit oxidative deterioration and prolong the shelf life of food goods. Due to its biocompatibility and non-toxic properties, it is a desirable substitute for synthetic antioxidants, which might potentially have negative health effects. The pharmaceutical and cosmetic sectors are investigating the potential of chitosan and its derivatives in preventing diseases and aging that are caused by oxidative stress. Chitosan-based films and coatings are currently being created to enable the controlled release of antioxidant compounds. This development presents novel approaches for drug delivery and skincare (Wang, Yan et al., 2024b; Wang, Yuan et al., 2024).

Ultimately, although the antioxidant activity of chitosan has received less research attention compared to its antimicrobial capabilities, it holds substantial promise, especially when taking into account the multiple elements that can augment its efficacy. The improved ability of chitosan to scavenge radicals is attributed to the presence of low molecular weight chitosan, derivatives that are grafted with antioxidant molecules, and environmental conditions that promote solubility and availability of functional groups. Through continuous research and development, the antioxidant capabilities of chitosan may have wider use in several domains such as food preservation and biomedicine. This could provide natural and effective answers to difficulties connected to oxidative stress.

4.8 Anti-Inflammatory Properties

Inflammation plays a pivotal role in the innate immune response, serving as a critical mechanism to counter various stimuli, including physical trauma, microbial infiltration, ultraviolet radiation exposure, and immunological reactions (Ngo et al., 2015). In the context of biomaterial implantation, the immune response presents one of the most significant challenges, necessitating a comprehensive understanding of the interactions between these materials and immune cells (Vasconcelos et al., 2015).

The introduction of medical devices and tissue-engineered constructs into the body may elicit a foreign body reaction (FBR) localized around the implant site, potentially leading to systemic inflammation. Consequently, there is a pressing need to develop biomaterials capable of mitigating FBRs while simultaneously promoting tissue healing and regeneration. To effectively modulate the immune response, ideal biomaterials should possess the capacity to stimulate the production of anti-inflammatory cytokines, such as interleukin-10 (IL-10), which are known to attenuate inflammation and facilitate the healing process (Yi et al., 2020). Conversely, these biomaterials should also demonstrate the ability to regulate the levels of pro-inflammatory mediators. This includes modulating the expression of pro-inflammatory cytokines, including tumor necrosis factor (TNF)-α, interleukin (IL)-1, interleukin-6 (IL-6), and interleukin-8 (IL-8), as well as interferons (IFNs) (Ma et al., 2016; Andrade del Olmo et al., 2022), Additionally, the biomaterials should exhibit properties that inhibit the formation of oxidative stress species, such as nitric oxide (NO) and reactive oxygen species (ROS) (Hardy et al., 2018).

Early investigations into chitosan's immunological properties revealed its capacity to enhance the functionality of inflammatory cells, particularly macrophages, by inducing cytokine production and upregulating the expression of activation markers (Moran et al., 2018b). The intrinsic immunostimulatory effects of chitosan have been attributed to the specific distribution pattern of d-glucosamine (GlcN) and β-1,4-linked N-acetyl-d-glucosamine (GlcNAc) along its oligomeric chain (Porporatto et al., 2003). Moreover, the degree of deacetylation and molecular weight of chitosan have been identified as key determinants of its anti-inflammatory properties. Research has demonstrated that high molecular weight chitosan promotes the production of various pro-inflammatory mediators in macrophages, including IL-1, TNF-α, granulocyte-macrophage colony-stimulating factor (GM-CSF), NO, and IL-6. Conversely, low molecular weight chitosan has been shown to enhance the release of TNF-α and IL-1β in macrophages. Notably, the internalization of these low molecular weight chitosans appears to be mediated by a lectin-like receptor with mannose specificity, suggesting a distinct mechanism of action compared to their high molecular weight counterparts (Niu et al., 2021; Ribeiro et al., 2021). Given the immunological functions of chitosan, recent research has focused on exploring chitosan and its composites as biomedical materials with specific functions such as the ability to modify the host immune response. A notable example of this approach is the work of Friedman et al. (Friedman et al., 2013) who synthesized chitosan-alginate nanoparticles (NPs) for the treatment of dermatological conditions with both infectious and inflammatory components. Their investigation centered on the potential of these nanoparticles to modulate the inflammatory cytokines and chemokines induced by Propionibacterium acnes (P. acnes), a bacterium commonly associated with acne. The researchers examined the effects of chitosan-alginate NPs on cytokine production in human monocytes and keratinocytes, two cell types crucial in the skin's immune response. Their findings demonstrated that the chitosan-alginate NPs successfully inhibited the production of key inflammatory cytokines, specifically IL-12p40 and IL-6, in a dose-dependent manner (Friedman et al., 2013).

Research by Ayaz et al. (Ayaz et al., 2021) has demonstrated that the anti-inflammatory properties of chitosan/gelatin composite cryogels can be influenced by the ratio of chitosan to gelatin during their decay. An examination of pro-inflammatory cytokine production, specifically TNFα and IL6, revealed that the suppression of inflammatory responses in mammalian macrophages exposed to various cryogel formulations in powder form is largely dependent on chitosan activity. Cryogels with higher chitosan content exhibited anti-inflammatory effects, as evidenced by TNFα production levels in LPS-activated macrophages. Conversely, increasing the gelatin content neutralized this effect, resulting in a more biocompatible and immunologically inert material. In another study, to assess the potential foreign body reaction triggered by device implantation in damaged bone tissue, which involves the activation of inflammatory cascades, researchers evaluated the impact of bioactivated chitosan-based scaffolds on inflammatory responses under physiological conditions. This study utilized human mesenchymal stem cells (hMSCs) and an *in vitro* model of osteoporosis, created through a co-culture of osteoblasts and macrophages. The findings indicated that bioactivated scaffolds containing osteoinductive signals can: (1) inhibit the synthesis of inflammatory mediators such as IL-1β; (2) decrease oxidative stress metabolites; and (3) promote the generation of anti-inflammatory markers like IL-10 in hMSCs (Fasolino et al., 2019).

The anti-inflammatory properties of O-carboxymethylated chitosan (O-CMC), derived from chitosan, were investigated *in vivo* for their effects on allergic reactions and allergic asthma. The synthesized polymers were assessed for anti-inflammatory capabilities by analyzing acute edema in rat hind paws induced by carrageenan. The results demonstrated that various O-CMC formulations were capable of neutralizing the inflammatory response in both phases of carrageenan-induced hind paw edema. The first phase is characterized by the release of histamine, kinins, and 5-HT, while the second phase is associated with the release of various inflammatory mediators, including leukotrienes, proteases, and arachidonate metabolites such as prostaglandins, particularly those of the E series (Adnan et al., 2020).

5. Preparation of Chitin and Chitosan

The process of extracting chitin from crab shells consists of three main stages: demineralization, deproteinization, and deacetylation. These operations are predominantly carried out by chemical or biological approaches, each of which has unique benefits and difficulties (Kim, 2010).

5.1 Chemical Methods

Chemical methods play a crucial role in fabricating bioactive scaffolds using chitosan, enhancing its properties and functionalities for biomedical applications. Various techniques have been developed to modify chitosan's structure at the molecular level, allowing for improved biocompatibility, mechanical strength, and bioactivity.

5.1.1 Demineralization

Demineralization is the initial stage in the chemical process of extracting chitin, in which calcium carbonate and other mineral constituents are eliminated from crab shells. Hydrochloric acid (HCl) is frequently employed for this objective, as it reacts with the calcium carbonate in the shells, resulting in the formation of water-soluble calcium chloride that can be effortlessly rinsed away. The efficiency of demineralization is influenced by the concentration of HCl and the length of the reaction, which are crucial parameters. Usually, hydrochloric acid (HCl) concentrations of up to 10% are used, and the reaction is allowed to proceed for a period of 2 to 3 hours. Nevertheless, acetic acid and other alternative acids have been investigated as substitutes for HCl, mostly due to the environmental issues linked to its use. Although acetic acid is not as effective at removing calcium, it does produce a more eco-friendly by-product called calcium acetate, which has great solubility in water (Jung et al., 2005; Younes and Rinaudo, 2015).

5.1.2 Deproteinization

Following the process of demineralization, deproteinization is performed to eliminate the proteins and other organic substances from the shells, resulting in the extraction of chitin. The chemical utilized in this stage is sodium hydroxide (NaOH), which is commonly employed with concentrations varying from 1% to 10%. The reaction generally takes place at high temperatures (ranging from 65°C to 100°C) for several hours. The selection of the base and reaction conditions is of utmost importance, as they have a significant impact not only on the effectiveness of protein removal but also on the preservation of the chitin structure. Although NaOH is commonly utilized, there is ongoing research into potassium hydroxide (KOH) as a more environmentally friendly option. This is because the resulting potassium salts from KOH can be utilized as fertilizers, unlike the sodium salts created by NaOH, which are more detrimental to the environment (Ahmed and Ikram, 2017; Kaczmarek et al., 2019; Kou et al., 2021).

5.1.3 Deacetylation

Another process of preparing chitosan is the deacetylation of chitin, which entails eliminating acetyl groups to generate the unbound amine groups that are distinctive of chitosan. This procedure is commonly performed by utilizing a substantial concentration of NaOH at increased temperatures. The extent of deacetylation attained during this stage is a crucial determinant of the ultimate characteristics of chitosan. Elevated temperatures and extended reaction durations often yield an increased DD, resulting in chitosan with improved biological activity and solubility. Deacetylation can be performed using heterogeneous or homogeneous procedures, with the former being more prevalent in industrial environments due to its straightforwardness and efficiency (Kou et al., 2021; de Queiroz Antonino et al., 2017; Wang and Zhuang, 2022).

5.2 Biological Method

Unlike chemical approaches, biological methods for chitosan manufacture utilize enzymes or microbial fermentation to accomplish deproteinization and deacetylation. These approaches are typically regarded as more ecologically sustainable since they circumvent the employment of harsh chemicals and generate less perilous by-products. Nevertheless, they frequently necessitate extended processing durations and can lead to reduced yields and less uniform product quality (Kou et al., 2021; Thadathil and Velappan, 2014).

5.2.1 Enzymatic Deproteinization

Enzymatic deproteinization employs proteolytic enzymes to hydrolyze the proteins present in crustacean shells, resulting in the isolation of chitin. This technique is extremely precise and may be carried out using less harsh conditions compared to chemical deproteinization, hence minimizing the likelihood of chitin degradation. Nevertheless, the expenses associated with enzymes and the requirement for meticulous regulation of reaction conditions are notable disadvantages (Ruiz and Corrales, 2017; Younes et al., 2014).

5.2.2 Fermentation

Fermentation-based techniques utilize lactic acid bacteria to generate organic acids capable of demineralizing and deproteinizing the shells. This technique has a slower rate compared to chemical approaches and necessitates meticulous regulation of microbial activity. However, it provides the benefit of yielding chitosan that is devoid of any chemical residues. Fermentation is a highly desirable method for producing chitosan of exceptional purity, specifically for use in biomedical applications (Arbia et al., 2013; Kim, 2010).

The synthesis of chitosan from chitin entails a sequence of chemical or biological procedures, each presenting distinct difficulties and advantages. Although chemical procedures are well-established and very effective, they give rise to environmental problems because of the utilization of harsh chemicals. Conversely, biological processes, while being more sustainable, are less efficient and necessitate greater time and resources. The selection of the technique relies on the intended use of chitosan, where chemical methods are favoured for large-scale industrial manufacturing, while biological methods are more appropriate for generating chitosan with high purity for delicate applications.

6. Significance of Chitosan as Scaffolds in Tissue Engineering

Biocompatibility and safety play a crucial role in the effectiveness of wound dressings in medicine. A material like Chitosan, renowned for its safety and biocompatibility, is excellent for this purpose. Infection is one of the most challenging problems in wound management. Microbial invasion is one of the major causes of infection. The inherent antimicrobial properties of chitosan are particularly helpful in this context, as they help to prevent wound infections by neutralizing microbes (Radulescu et al., 2015; Shahid-ul-Islam and Butola, 2019; Lin et al., 2023).

As an effective modulator of the inflammatory response, chitosan also plays a crucial role in the promotion of wound healing as a potent anti-inflammatory agent. To promote wound healing, chitosan-based dressings need to be used to reduce inflammation factors in the wounds. Besides minimizing tissue damage, this anti-inflammatory property also promotes the healing of the tissue. Furthermore, it has been demonstrated that chitosan can accelerate the coagulation process, which is an essential step in wound healing. Because chitosan enhances coagulation, blood clotting is promoted more rapidly, so wounds are sealed more quickly and blood loss is reduced. For acute wounds, where quick clotting is essential to prevent excessive bleeding, this property is particularly beneficial (Ahmed et al., 2019; Chaby et al., 2007).

The intrinsic properties of chitosan make it not only an excellent material for the production of bioactive agents but also for enhancing the immune system and stimulating growth. The presence of these agents can significantly improve the healing process of wound dressings that are based on chitosan by enhancing the healing process. These substances contain bioactive chemicals that can accelerate the healing process by stimulating cellular activity required for tissue regeneration and repair. When these bioactive compounds are combined with chitosan, wound care is improved in a synergistic way (Li et al., 2021; Moran et al., 2018a; Wang et al., 2021).

Using chitosan as a scaffold material for tissue engineering is another demonstration of the versatility and effectiveness of the material. Supporting tissues such as chitosan scaffolds mimic the natural extracellular matrix, which facilitates cell attachments, proliferation, and differentiation. When it comes to the regeneration of different tissues, including skin, bone, and cartilage, structure is crucial. It is possible to chemically adjust chitosan to better meet the needs of various tissue engineering processes by adjusting its characteristics to these specifications (Man et al., 2016; Oryan et al., 2016, 2017).

Porous chitosan scaffolds are particularly useful because they allow nutrients and oxygen to diffuse, which are vital for cell survival and function. Furthermore, porosity aids in the removal of waste products, which promotes the growth of healthy tissues. Additionally, the mechanical properties of chitosan can be tailored to support specific types of tissues being regenerated. Growth factors can be incorporated into chitosan scaffolds to enhance their effectiveness in tissue engineering. To regenerate tissue, these growth factors promote migration, proliferation, and differentiation of cells. Growth factors are released from chitosan scaffolds in a controlled and localized manner, which optimizes the healing process (Shariatinia and Jalali, 2018; Man et al., 2016; Szymańska et al., 2022).

Consequently, chitosan is an essential component of tissue engineering scaffolds and wound dressings. The material's biocompatibility, antimicrobial activity, ability to modulate inflammation, and compatibility with bioactive agents make it ideal for wound care and tissue regeneration. Developing chitosan-based technologies will improve patient outcomes and advance medical science with innovative solutions in tissue engineering.

6.1 Chitosan-based Scaffolds in Tissue Engineering

Chitosan-based scaffolds can be fabricated utilizing a variety of approaches. The initial techniques employed to fabricate porous chitosan scaffolds include solvent casting and particle leaching. The process entails dissolving chitosan in a solvent, subsequently pouring it into a mold containing a porogen, and finally immersing it in water to extract the porogen. This technique enables accurate control over the size and porosity of pores, but it can lead to scaffolds that have limited mechanical strength (Silva et al., 2007; Freyman et al., 2001). Freeze-drying is a common technique that involves freezing a solution of chitosan using liquid nitrogen and subsequently sublimating it under vacuum conditions. This leads to the formation of a permeable framework with significant cavities and interconnections, which are crucial for the field of tissue engineering. Nevertheless, this approach can result in scaffolds that have insufficient mechanical properties (Diop et al., 2015). Electrospinning is a versatile technique employed to create nanofibrous chitosan scaffolds that closely resemble the extracellular matrix (ECM). This procedure entails subjecting a chitosan solution to a high-voltage electric field, which leads to the formation of exceedingly thin fibers that accumulate on a grounded substrate. These scaffolds possess a significant ratio of surface area to volume, which promotes the attachment and growth of cells. Fiber diameter and scaffold porosity can be tailored by adjusting solution properties, applied voltage, and collecting distance (Agarwal et al., 2008; Schiffman and Schauer, 2008). 3D printing, or additive manufacturing, is a sophisticated technique that enables precise manipulation of chitosan scaffolds. This method entails the application of chitosan-based ink in thin layers, which then undergo crosslinking to form a robust and reliable structure. 3D printing can produce scaffolds with complex forms and diverse levels of porosity, specifically designed to fulfill the unique mechanical and biological requirements of various tissues (Murphy and Atala, 2014).

7. Functionalization and Enhancement of Chitosan-Based Scaffolds

Chitosan, a polymer produced from chitin found in nature, is extensively utilized in tissue engineering because of its ability to interact well with living organisms, break down gradually over time, and have beneficial biological characteristics. Nevertheless, it is important to note that this material has several drawbacks, including low mechanical strength and fast degradation. Therefore, it is necessary to implement functionalization and improvement techniques to optimize its performance as a scaffold material (Pandey et al., 2017). The addition of biomolecules such as growth factors, peptides, and proteins is essential for increasing the biological activity of scaffolds made from chitosan. Growth factors such as basic fibroblast growth factor (bFGF) and vascular endothelial growth factor (VEGF) stimulate the formation of new blood vessels (angiogenesis) and the repair of tissues. On the other hand, peptides like the RGD sequence imitate components of the extracellular matrix, thereby improving cell communication and cellular reactions. Incorporating proteins such as collagen and fibronectin into chitosan scaffolds might enhance their resemblance to the natural extracellular matrix (ECM), facilitating cell adhesion and dissemination (Patel and Mikos, 2004; Coathup et al., 2012; Ho et al., 2009).

Crosslinking is an essential technique for altering the mechanical characteristics and degradation rates of chitosan scaffolds. Chemical crosslinking techniques, such as glutaraldehyde, genipin, or carbodiimides, create strong and stable covalent connections between chitosan chains, hence improving the mechanical strength and stability of the scaffold. Physical crosslinking techniques, such as ionic crosslinking using multivalent ions or thermal treatment, provide a gentler option and can enhance mechanical characteristics without sacrificing biocompatibility (Berger et al., 2004; Augustine et al., 2021).

The integration of nanoparticles into chitosan scaffolds has emerged as a viable method to improve their functionality. Nanoparticles have the potential to be utilized in drug delivery systems, allowing for the precise and regulated release of medicinal substances, hence augmenting the scaffold's bioactivity. Chitosan scaffolds can be enhanced with antimicrobial nanoparticles, like silver or zinc oxide nanoparticles, to hinder bacterial infections. This integration lowers the chances of infection and enhances the overall effectiveness of tissue regeneration (Rinaudo, 2006; Mikušová and Mikuš, 2021).

At some point, the improvement of chitosan-based scaffolds in tissue engineering relies on the functionalization and augmentation achieved through the incorporation of biomolecules, crosslinking techniques, and nanoparticles.

7.1 Chitosan/Composites Scaffolds

In tissue engineering, the construction of scaffolds frequently necessitates the incorporation of several materials to get the intended mechanical characteristics, biocompatibility, and functioning. Composite scaffolds, which integrate chitosan with other materials, are especially significant due to their ability to create a customized environment for tissue development. When combined with other polymers, ceramics, or bioactive compounds, chitosan improves the mechanical strength, stability, and biological functioning of the scaffold. This synergy enhances the performance of the scaffold in particular applications, such as bone regeneration or wound healing, where scaffolds made of a single material may not meet the requirements (Signorini et al., 2023).

7.1.1 Chitosan-Based Hydrogels

Chitosan hydrogels can be created by either chemical or physical crosslinking techniques. Chemical crosslinking produces a strong and inflexible network, whereas physical crosslinking relies on ionic, hydrogen, or hydrophobic contacts. The selection of crosslinking technique has a substantial influence on the mechanical characteristics, porosity, and degradation rate of the hydrogel, all of which are vital for tissue engineering. Composite hydrogels can be improved by integrating nanoparticles, polymers, or bioactive compounds, such as silver nanoparticles for antibacterial effects or growth factors for promoting cell growth and specialization. Responsive hydrogels that can react to changes in the environment have been created to control drug-delivery systems (Berger et al., 2004; Pita-López et al., 2021).

Chitosan hydrogels are extensively used in drug delivery systems, wound healing, cartilage and soft tissue engineering, and regenerative medicine. They stimulate tissue regeneration, preserve a moist environment, assist cell migration, and aid in the proliferation and specialization of chondrocytes in the process of repairing cartilage (Jayakumar, Prabaharan et al., 2011; Boateng et al., 2008).

Multiple research projects have proven the effectiveness of chitosan-based hydrogels in tissue engineering. Research conducted *in vivo* has demonstrated that the combination of chitosan hydrogels and growth factors significantly enhances tissue regeneration in comparison to scaffolds without composites. The results highlight the potential of chitosan hydrogels in regenerative medicine and therapeutic settings (Thein-Han and Misra, 2009; Pita-López et al., 2021).

7.1.2 Chitosan-Based Sponges

Chitosan sponges are created through the processes of freeze-drying or porogen leaching, which have an impact on the porosity and mechanical strength of the scaffold. Freeze-drying forms a porous structure, whereas porogen leaching eliminates a substance that forms pores, enabling precise control over the size and distribution of the pores. These strategies have a direct impact on the sponge's capacity to facilitate cell infiltration and nutrient diffusion, which are crucial for tissue engineering purposes (Dragan and Dinu, 2020; Ahmed and Ikram, 2016).

Composite sponges can be improved by combining chitosan with bioactive ceramics or growth factors and polymers, which enhances their ability to support certain cellular activities and promote bone development. Surface modification strategies, such as applying a layer of bioactive chemicals, augment the scaffold's capacity to facilitate targeted biological functions. Chitosan sponges are highly efficient in bone tissue engineering, hemostatic sponges, nerve and cartilage regeneration, and *in vivo* investigations. They exhibit excellent performance in preclinical bone and cartilage regeneration models, leading to improved tissue integration and accelerated healing times compared to non-composite sponges. Chitosan sponges, when mixed with other bioactive materials, have demonstrated successful performance in preclinical models of bone and cartilage regeneration in living organisms. This suggests that they have the potential to be used in clinical settings in the future (Ahmed and Ikram, 2016; Xia et al., 2020).

7.1.3 Chitosan/Nano-fibrous Scaffolds

Nanofibers are attractive biological and pharmacological transporters. Nanofibers' unique properties include (I) high surface area to volume ratio, (II) high porosity surface with nanometer dimensions, (III) suitable mechanical properties (swelling, stress, strain, thickness, etc.) with adjusted synthesis process, (IV) analog to extracellular model, and (V) easy large-scale production, successfully used in medical applications (Wani et al., 2017; de Farias et al., 2019). Fiber sizes and architectures were altered by varying concentrations, viscosities, types of polymer solutions, and preparation techniques. More than other nanomaterials, nanofibers can better encapsulate drugs due to their size and structure (Ibrahim et al., 2023; de Farias et al., 2019). First treatment bursts of single-layer nanofibers discharged drugs. This did not occur in multi-layer and core-shell nanofiber release systems. Pharmaceuticals are loaded into core-shell nanofibers. Hydrophobic nanofibers slowed drug release in multi-layer and core-shell carriers. Multi-drug formulations dual-sensitive systems with temperature and pH-sensitive polymers, and drug/nanofibers-magnetic nanoparticle matrices for hyperthermia treatment and drug delivery can improve anticancer drug/nanofiber systems. Making these carriers kill more cancer cells (Wani et al., 2017).

Chitosan can be manufactured into nanofibers, nanoparticles, microparticles, nanobeads, hydrogels, membranes, scaffolds, sponges, and others. Electrospinning using acetic acid (90%), trifluoroacetic acid, and different solvents/cosolvents produces chitosan nanofibers (Shikhi-Abadi and Irani, 2021). These chemicals have a broad surface area, high porosity, low toxicity, compatibility with living organisms, natural degradation, renewable nature, low immunological reaction, and better clinical functioning. The electrospinnability of pure chitosan is limited by its high viscosity, rheology, swelling, brittleness, stiffness, polycationic charge, low solubility, quick degradation, and extensive inter/intrachain hydrogen bonding. To address these difficulties, chitosan solutions were mixed with synthetic polymers, metal nanoparticles, and nano clays before electrospinning (Jayakumar et al., 2010).

7.1.4 Chitosan/Nanoceramics Scaffolds

Ceramics have been widely used in combination with chitosan to promote the bioactivity of chitosan-based scaffolds. Hydroxyapatite (HAP) is commonly utilized in TE as one of the most frequently used (Fourie et al., 2022). Both hydroxyapatite (HAP) and chitosan possess unique advantages that make them excellent biomaterials. The purpose of integrating HAP into chitosan scaffolds is to address the shortcomings linked with each material. For instance, chitosan is introduced to counteract the fragility of HAP, whereas HAP serves to enhance the bioactivity and mechanical characteristics of chitosan (Baharlouei and Rahman, 2022; Scalera et al., 2023). Nano-sized hydroxyapatite (nHAP) is commonly utilized and provides added benefits when incorporated into biopolymer matrices like chitosan. The use of nHAP introduces nontopographic features that mimic the nanostructure of bone and enhance the scaffold's mechanical strength (Aguilar et al., 2019). However, one drawback of incorporating nanoparticles into chitosan is the reduction in scaffold porosity, which directly impacts mechanical properties. This is attributed to the distribution of nanoparticles within the polymer matrix and the formation of intermolecular hydrogen bonds between the amine groups of chitosan and the hydroxyl group of the nanoparticles (Scalera et al., 2023).

β-Tricalcium phosphate (β-TCP) is a calcium phosphate ceramic that is sometimes combined with chitosan, although not as frequently as HAP. The use of chitosan and β-TCP composite materials has been observed in the creation of scaffolds and injectable hydrogels. One challenge of mixing β-TCP with chitosan is that the degradation rate of β-TCP is slower than that of chitosan, leading to difficulties in controlling the degradation rate of the composite scaffolds (Ressler et al., 2023). However, the incorporation of β-TCP into chitosan scaffolds lowers the chitosan degradation rate because it shields the polymer from lysozyme's enzymatic breakdown. As a result, the biodegradation rate is slowed down, prolonging the scaffold's structural integrity and encouraging greater bone formation. Additionally, it has been discovered that adding β-TCP improves chitosan's mechanical qualities while decreasing its antibacterial effectiveness against *Escherichia coli* (Knabl et al., 2019). The combination of calcium phosphates with chitosan may impact the antimicrobial properties of chitosan, potentially negating an advantage that could be beneficial in bone tissue engineering and limiting the application of these composite materials (Piaia et al., 2021).

Bioactive ceramic materials, such as bioactive glasses, are integrated into polymer-based scaffolds to enhance properties like reducing the brittleness of bioactive glass and addressing the poor mechanical features of chitosan (Daskalakis et al., 2022). *In vivo*, bioactive glasses form a bond with surrounding bone tissue by creating a calcium phosphate layer that is later altered by bone cells, making it a superior option when compared to HAP. In addition to promoting osteointegration, the byproducts released from bioactive glasses may play a role in stimulating gene expression that regulates osteogenesis (Sergi et al., 2020). Nano-sized bioactive glass particles are preferred over their micro-sized counterparts due to their more favorable mechanical properties and increased bioactivity in chitosan membranes (Erol-Taygun et al., 2019). Nano-titania, also known as titanium dioxide (TiO_2), has been a focus of biomedical research due to its distinct physiochemical characteristics. While these materials are not as commonly utilized as calcium phosphate-based composites due to their potential toxicity and non-biodegradable nature, combining TiO_2 with a natural polymer like chitosan can help mitigate toxic effects, reduce adverse reactions, and enhance biodegradability (Anaya-Esparza et al., 2020). The mechanical properties of chitosan matrices are enhanced by the addition of TiO_2, likely due to their uniform dispersion within the chitosan matrix and the interaction between the amine group on chitosan and the Lewis acidic sites of TiO_2 (Madkour et al., 2021). Zirconia, a crystalline dioxide of zirconium, is another widely used material, particularly in the fields of dentistry, orthopedics, and bone implant materials. Zirconia is favored for medical devices due to its favorable mechanical properties and biocompatibility (Jayakumar, Ramachandran et al., 2011).

While chitosan combined with ceramic and bio-glass materials fulfills the fundamental criteria for TE, there are still specific limitations that need to be addressed, involving the degradation rate of scaffolds and enhancing the structural and mechanical properties. Therefore, further research is necessary to optimize the utilization and combination of these materials (Aranaz et al., 2021).

8. Limitations and Future Prospective

Chitosan, a naturally derived polymer, has garnered significant attention in the field of tissue engineering due to its biodegradability, antimicrobial activity, low toxicity, and structural similarity to glycosaminoglycans found in the extracellular matrix (ECM). These properties make chitosan an attractive candidate for applications in regenerative medicine, particularly in hard and soft tissue engineering (Croisier and Jérôme, 2013). However, despite its promise, chitosan does face several limitations that need to be addressed to fully exploit its potential.

One of the major challenges associated with chitosan is its low solubility, which arises from strong inter- and intra-molecular hydrogen bonding within its rigid crystalline structure. This limits its solubility in neutral or basic environments, restricting its use in certain biomedical applications. Moreover, pure chitosan lacks sufficient mechanical strength and does not support proper cell adhesion, which is crucial for tissue regeneration (Logithkumar et al., 2016). To overcome these limitations and enhance chitosan's performance, researchers have focused on various modifications. The integration of nanotechnology has led to the development of nanoparticles, nanofibers, and nanocomposites, improving the overall performance of chitosan in tissue regeneration (Wan et al., 2008). Furthermore, combining chitosan with other biocompatible and biodegradable polymers has resulted in the creation of films, nano/microparticles, and hydrogels, which serve as effective drug-delivery systems for applications such as tissue regeneration and wound treatment (Yan et al., 2012). Despite these advancements, challenges remain. In bone tissue engineering (BTE), for instance, further improvements are needed to enhance the mechanical strength of chitosan-based scaffolds to better mimic the load-bearing properties of natural bone. Similarly, in neural tissue engineering (NTE), achieving precise tissue mimicry remains a significant hurdle. The development of scaffolds that can effectively guide neural cell growth and differentiation is still an ongoing area of research.

Another promising approach to improving chitosan's functionality in tissue engineering is 3D bioprinting. This modern fabrication process allows for the precise placement of biological materials, cells, and biomolecules onto scaffolds, enabling the creation of customized structures with tailored biological and mechanical properties (Del Valle et al., 2017). While 3D bioprinting has shown great potential in replicating natural tissue architectures, there are still challenges to overcome, such as the need for better control over pore size, mechanical strength, and the long-term stability of printed constructs. Additionally, the use of cytotoxic solvents and the creation of abnormal-sized pores during traditional fabrication methods like freeze-drying and solvent casting continue to limit the broader application of chitosan-based materials (Taghizadeh et al., 2022). In the case of electrospinning, pure chitosan's high viscosity, stiffness, and polycationic nature further complicate its processing into nanofibers (Qasim et al., 2018). To address these issues, researchers utilized alkali treatment (Homayoni et al., 2009) or blended chitosan with synthetic polymers such as polyethylene oxide (PEO), polycaprolactone (PCL), and polyvinyl alcohol (PVA) to improve its electrospinnability and overall performance (Pakravan et al., 2011; Prasad et al., 2015).

Looking forward, the adaptability and biocompatibility of chitosan-based composites continue to make them valuable assets in tissue engineering. Ongoing research aims to fine-tune their properties to meet the specific demands of different tissue types, such as controlled biodegradability, optimal mechanical strength, and favorable surface chemistries for cell attachment and proliferation. With continued advancements in fabrication techniques and material modifications, chitosan-based biomaterials hold the potential to significantly reshape the future landscape of regenerative medicine. In conclusion, while chitosan and its composites offer numerous advantages in tissue engineering, overcoming the existing limitations through innovative material designs and fabrication methods will be crucial for their widespread clinical application.

References

Abd El-Hack, Mohamed E., Mohamed T. El-Saadony, Manal E. Shafi, Nidal M. Zabermawi, Muhammad Arif, Gaber Elsaber Batiha et al. 2020. Antimicrobial and antioxidant properties of chitosan and its derivatives and their applications: A review. International Journal of Biological Macromolecules 164: 2726–44.

Adnan, Sherjeel, Nazar Muhammad Ranjha, Muhammad Hanif and Sajid Asghar. 2020. O-Carboxymethylated chitosan; A promising tool with *in-vivo* anti-inflammatory and analgesic properties in albino rats. International Journal of Biological Macromolecules 156: 531–36.

Agarwal, Seema, Joachim H. Wendorff and Andreas Greiner. 2008. Use of electrospinning technique for biomedical applications. Polymer 49: 5603–21.

Aguilar, Alicia, Naimah Zein, Ezeddine Harmouch, Brahim Hafdi, Fabien Bornert, Damien Offner et al. 2019. Application of chitosan in bone and dental engineering. Molecules 24: 3009.

Ahmed, Fatma, Faiza M. Soliman, Mohamed A. Adly, Hamdy A. M. Soliman, Mansour El-Matbouli and Mona Saleh. 2019. Recent progress in biomedical applications of chitosan and its nanocomposites in aquaculture: A review. Research in Veterinary Science 126: 68–82.

Ahmed, Shakeel and Saiqa Ikram. 2016. Chitosan based scaffolds and their applications in wound healing. Achievements in the Life Sciences 10: 27–37.

Ahmed, Shakeel and Saiqa Ikram. 2017. Chitosan: Derivatives, Composites and Applications (John Wiley & Sons).

Anaya-Esparza, Luis Miguel, José Martín Ruvalcaba-Gómez, Claudia Ivette Maytorena-Verdugo, Napoleón González-Silva, Rafael Romero-Toledo, Selene Aguilera-Aguirre et al. 2020. Chitosan-TiO_2: A versatile hybrid composite. Materials 13: 811.

Andrade del Olmo, Jon, Leyre Pérez-Álvarez, Virginia Sáez-Martínez, Sandra Benito-Cid, Leire Ruiz-Rubio, Raúl Pérez-González et al. 2022. Wound healing and antibacterial chitosan-genipin hydrogels with controlled drug delivery for synergistic anti-inflammatory activity. International Journal of Biological Macromolecules 203: 679–94.

Aranaz, Inmaculada, Andrés R. Alcántara, Maria Concepción Civera, Concepción Arias, Begoña Elorza, Angeles Heras Caballero et al. 2021. Chitosan: An overview of its properties and applications. Polymers 13: 3256.

Aranaz, Inmaculada, Andrés R. Alcántara, Maria Concepción Civera, Concepción Arias, Begoña Elorza, Angeles Heras Caballero et al. 2021. Chitosan: An overview of its properties and applications. Polymers 13.

Aranaz, Inmaculada, Andrés R. Alcántara, Maria Concepción Civera, Concepción Arias, Begoña Elorza, Angeles Heras Caballero et al. 2021. Chitosan: An overview of its properties and applications. Polymers 13: 3256–56.

Arbia, Wassila, Leila Arbia, Lydia Adour and Abdeltif Amrane. 2013. Chitin extraction from crustacean shells using biological methods—a review. Food Technology and Biotechnology 51: 12–25.

Augustine, Robin, Hashim Alhussain, Alap Ali Zahid, Syed Raza Ur Rehman, Rashid Ahmed and Anwarul Hasan. 2021. Crosslinking strategies to develop hydrogels for biomedical applications. Nano Hydrogels: Physico-Chemical Properties and Recent Advances in Structural Designing: 21–57.

Austin, P. R., C. J. Brine, J. E. Castle and J. P. Zikakis. 1981. Chitin: New facets of research. Science (New York, N.Y.) 212: 749–53.

Ayaz, Furkan, Didem Demir and Nimet Bölgen. 2021. Differential anti-inflammatory properties of chitosan-based cryogel scaffolds depending on chitosan/gelatin ratio. Artificial Cells, Nanomedicine, and Biotechnology 49: 682–90.

Baharlouei, Parnian and Azizur Rahman. 2022. Chitin and chitosan: prospective biomedical applications in drug delivery, cancer treatment, and wound healing. Marine Drugs 20: 460.

Berger, J., M. Reist, J. M. Mayer, O. Felt, N. A. Peppas and R. Gurny. 2004. Structure and interactions in covalently and ionically crosslinked chitosan hydrogels for biomedical applications. Eur. J. Pharm. Biopharm. 57: 19–34.

Boateng, Joshua S., Kerr H. Matthews, Howard N. E. Stevens and Gillian M. Eccleston. 2008. Wound healing dressings and drug delivery systems: a review. J. Pharm. Sci. 97: 2892–923.

Butler, Frank K., Russ S. Kotwal, Michael A. Dubick and Jeffrey A. Bailey. 2014. Management of external hemorrhage in tactical combat casualty care: chitosan-based hemostatic gauze dressings. Change 13: 05.

Cardile, V., G. Frasca, L. Rizza, F. Bonina, C. Puglia, A. Barge et al. 2008. Improved adhesion to mucosal cells of water-soluble chitosan tetraalkylammonium salts. International Journal of Pharmaceutics 362: 88–92.

Chaby, Guillaume, Patricia Senet, Michel Vaneau, Philippe Martel, Jean Claude Guillaume, Sylvie Meaume et al. 2007. Dressings for acute and chronic wounds: A systematic review. Archives of Dermatology 143: 1297–304.

Chen, R. H., W. C. Lin and J. H. Lin. 1994. Effects of pH, ionic strength, and type of anion on the rheological properties of chitosan solutions. Acta Polymerica 45: 41–46.

Chen, Rong Huei and Min Larng Tsaih. 1998. Effect of temperature on the intrinsic viscosity and conformation of chitosans in dilute HCl solution. International Journal of Biological Macromolecules 23: 135–41.

Cheng, Jiaqi, Huaping Zhu, Jianlian Huang, Jianxin Zhao, Bowen Yan, Shenyan Ma et al. 2020. The physicochemical properties of chitosan prepared by microwave heating. Food Science & Nutrition 8: 1987–94.

Cheung, Randy Chi Fai, Tzi Bun Ng, Jack Ho Wong and Wai Yee Chan. 2015. Chitosan: An update on potential biomedical and pharmaceutical applications. Mar. Drugs 13: 5156–86.

Cho, Jaepyoung, Marie-Claude Heuzey, André Bégin and Pierre J. Carreau. 2006. Viscoelastic properties of chitosan solutions: Effect of concentration and ionic strength. Journal of Food Engineering 74: 500–15.

Coathup, Melanie J., Karin A. Hing, Sorousheh Samizadeh, Oliver Chan, Yvette S. Fang, Charlie Campion et al. 2012. Effect of increased strut porosity of calcium phosphate bone graft substitute biomaterials on osteoinduction. Journal of Biomedical Materials Research Part A 100: 1550–55.

Confederat, Luminita Georgeta, Cristina Gabriela Tuchilus, Maria Dragan, Mousa Sha'at and Oana Maria Dragostin. 2021. Preparation and antimicrobial activity of chitosan and its derivatives: A concise review. Molecules 26: 3694.

Crini, Grégorio and Pierre-Marie Badot. 2008. Application of chitosan, a natural aminopolysaccharide, for dye removal from aqueous solutions by adsorption processes using batch studies: A review of recent literature. Progress in Polymer Science 33: 399–447.

Croisier, Florence and Christine Jérôme. 2013. Chitosan-based biomaterials for tissue engineering. European Polymer Journal 49: 780–92.

Dai, Tianhong, Masamitsu Tanaka, Ying-Ying Huang and Michael R. Hamblin. 2011. Chitosan preparations for wounds and burns: antimicrobial and wound-healing effects. Expert Review of Anti-infective Therapy 9: 857–79.

Dash, Mamoni, Federica Chiellini, Raphael M. Ottenbrite and Emo Chiellini. 2011. Chitosan—A versatile semi-synthetic polymer in biomedical applications. Progress in Polymer Science 36: 981–1014.

Daskalakis, Evangelos, Boyang Huang, Cian Vyas, Anil Ahmet Acar, Ali Fallah, Glen Cooper et al. 2022. Novel 3D bioglass scaffolds for bone tissue regeneration. Polymers 14: 445.

de Farias, Bruna Silva, Tito Roberto Sant'Anna Cadaval Junior and Luiz Antonio de Almeida Pinto. 2019. Chitosan-functionalized nanofibers: A comprehensive review on challenges and prospects for food applications. International Journal of Biological Macromolecules 123: 210–20.

de Queiroz Antonino, Rayane Santa Cruz Martins, Bianca Rosa Paschoal Lia Fook, Vítor Alexandre de Oliveira Lima, Raid Ícaro de Farias Rached, Eunice Paloma Nascimento Lima et al. 2017. Preparation and characterization of chitosan obtained from shells of shrimp (Litopenaeus vannamei Boone). Marine Drugs 15: 141.

Deb, Payel, Ashish B. Deoghare, Animesh Borah, Emon Barua and Sumit Das Lala. 2018. Scaffold development using biomaterials: a review. Materials Today: Proceedings 5: 12909–19.

Del Valle, Luís J., Angélica Díaz and Jordi Puiggalí. 2017. Hydrogels for biomedical applications: cellulose, chitosan, and protein/peptide derivatives. Gels 3: 27.

Diop, M., N. Auberval, A. Viciglio, A. Langlois, W. Bietiger, C. Mura et al. 2015. Design, characterisation, and bioefficiency of insulin–chitosan nanoparticles after stabilisation by freeze-drying or cross-linking. International Journal of Pharmaceutics 491: 402–08.

Dragan, Ecaterina Stela and Maria Valentina Dinu. 2020. Advances in porous chitosan-based composite hydrogels: Synthesis and applications. Reactive and Functional Polymers 146: 104372.

Elkomy, Mohammed H., Adel A. Ali and Hussein M. Eid. 2022. Chitosan on the surface of nanoparticles for enhanced drug delivery: A comprehensive review. Journal of Controlled Release : Official Journal of the Controlled Release Society 351: 923–40.

Erol-Taygun, Melek, Irem Unalan, Maizlinda Izwana Binti Idris, João F. Mano and Aldo R. Boccaccini. 2019. Bioactive glass-polymer nanocomposites for bone tissue regeneration applications: a review. Advanced Engineering Materials 21: 1900287.

Fasolino, Ines, Maria Grazia Raucci, Alessandra Soriente, Christian Demitri, Marta Madaghiele, Alessandro Sannino et al. 2019. Osteoinductive and anti-inflammatory properties of chitosan-based scaffolds for bone regeneration. Materials Science and Engineering: C 105: 110046.

Fei Liu, Xiao, Yun Lin Guan, Dong Zhi Yang, Zhi Li and Kang De Yao. 2001. Antibacterial action of chitosan and carboxymethylated chitosan. Journal of Applied Polymer Science 79: 1324–35.

Fourie, Jaundrie, Francois Taute, Louis du Preez and Deon De Beer. 2022. Chitosan composite biomaterials for bone tissue engineering—a review. Regenerative Engineering and Translational Medicine: 1–21.

Freyman, T. M., I. V. Yannas and L. J. Gibson. 2001. Cellular materials as porous scaffolds for tissue engineering. Progress in Materials Science 46: 273–82.

Friedman, Adam J., Jenny Phan, David O. Schairer, Jackson Champer, Min Qin, Aslan Pirouz et al. 2013. Antimicrobial and anti-inflammatory activity of chitosan–alginate nanoparticles: a targeted therapy for cutaneous pathogens. Journal of Investigative Dermatology 133: 1231–39.

Gheorghiță, Daniela, Horațiu Moldovan, Alina Robu, Ana-Iulia Bița, Elena Grosu, Aurora Antoniac et al. 2023. Chitosan-based biomaterials for hemostatic applications: a review of recent advances. International Journal of Molecular Sciences 24: 10540.

Ghormade, V., E. K. Pathan and M. V. Deshpande. 2017. Can fungi compete with marine sources for chitosan production? International Journal of Biological Macromolecules 104: 1415–21.

Goy, Rejane C., Douglas de Britto and Odilio BG Assis. 2009. A review of the antimicrobial activity of chitosan. Polímeros 19: 241–47.

Guarnieri, Anna, Micaela Triunfo, Carmen Scieuzo, Dolores Ianniciello, Elena Tafi, Thomas Hahn et al. 2022. Antimicrobial properties of chitosan from different developmental stages of the bioconverter insect Hermetia illucens. Scientific Reports 12: 8084.

Halabalova, Vera, Lubomir Simek and Pavel Mokrejs. 2011. Intrinsic viscosity and conformational parameters of chitosan chains. Rasayan J. Chem 4: 223.

Hardy, Alexandre, Cendrine Seguin, Anaïs Brion, Philippe Lavalle, Pierre Schaaf, Sylvie Fournel et al. 2018. β-Cyclodextrin-Functionalized Chitosan/Alginate Compact Polyelectrolyte Complexes (CoPECs) as Functional Biomaterials with Anti-Inflammatory Properties. ACS Applied Materials and Interfaces 10: 29347–56.

Harugade, Anuja, Atul P. Sherje and Anil Pethe. 2023. Chitosan: A review on properties, biological activities and recent progress in biomedical applications. Reactive and Functional Polymers 191: 105634–34.

Hirano, Shigehiro, Hisaya Tsuchida and Norio Nagao. 1989. N-acetylation in chitosan and the rate of its enzymic hydrolysis. Biomaterials 10: 574–76.

Ho, Yi-Cheng, Fwu-Long Mi, Hsing-Wen Sung and Pi-Li Kuo. 2009. Heparin-functionalized chitosan–alginate scaffolds for controlled release of growth factor. International Journal of Pharmaceutics 376: 69–75.

Homayoni, Homa, Seyed Abdolkarim Hosseini Ravandi and Masoumeh Valizadeh. 2009. Electrospinning of chitosan nanofibers: Processing optimization. Carbohydrate Polymers 77: 656–61.

Hu, Junli, Yaping Hou, Hyejin Park, Bogyu Choi, Siying Hou, Amy Chung et al. 2012. Visible light crosslinkable chitosan hydrogels for tissue engineering. Acta Biomaterialia 8: 1730–38.

Hu, Zhang, Dong-Ying Zhang, Si-Tong Lu, Pu-Wang Li and Si-Dong Li. 2018. Chitosan-based composite materials for prospective hemostatic applications. Marine Drugs 16: 273.

Huang, Min, Eugene Khor and Lee-Yong Lim. 2004. Uptake and cytotoxicity of chitosan molecules and nanoparticles: effects of molecular weight and degree of deacetylation. Pharmaceutical Research 21: 344–53.

Hussain, Md Rabiul, Murshid Iman and Tarun K. Maji. 2013. Determination of degree of deacetylation of chitosan and their effect on the release behavior of essential oil from chitosan and chitosan-gelatin complex microcapsules. Int. J. Adv. Eng. Appl. 6: 4–12.

Iber, Benedict, Nor Kasan, Donald Torsabo and Jenifer Omuwa. 2021. A review of various sources of chitin and chitosan in nature. Journal of Renewable Materials 10: 42–49.

Ibrahim, Marwan A., Mona H. Alhalafi, El-Amir M. Emam, Hassan Ibrahim and Rehab M. Mosaad. 2023. A review of chitosan and chitosan nanofiber: Preparation, characterization, and its potential applications. Polymers 15: 2820.

Ikeda, Ikuo, Michihiro Sugano, Katsuko Yoshida, Eiji Sasaki, Yasushi Iwamoto and Kouta Hatano. 1993. Effects of chitosan hydrolysates on lipid absorption and on serum and liver lipid concentration in rats. Journal of Agricultural and Food Chemistry 41: 431–35.

Islam, S., M. A. Rahman Bhuiyan and M. N. Islam. 2017a. Chitin and chitosan: Structure, properties and applications in biomedical engineering. J. Polym. Environ 25: 854–66.

Islam, S., M. A. Rahman Bhuiyan and M. N. Islam. 2017b. Chitin and chitosan: structure, properties and applications in biomedical engineering. Journal of Polymers and the Environment 25: 854–66.

Issar, Upasana and Richa Arora. 2022. Antiviral potency of chitosan, its derivatives, and nanocomposites. *In*: Chitosan-Based Nanocomposite Materials: Fabrication, Characterization and Biomedical Applications (Springer).

Jaber, Nisrein, Mayyas Al-Remawi, Faisal Al-Akayleh, Najah Al-Muhtaseb, Ibrahim SI Al-Adham and Phillip J. Collier. 2022. A review of the antiviral activity of Chitosan, including patented applications and its potential use against COVID-19. Journal of Applied Microbiology 132: 41–58.

Jayakumar, R., M. Prabaharan, S. V. Nair and H. Tamura. 2010. Novel chitin and chitosan nanofibers in biomedical applications. Biotechnology Advances 28: 142–50.

Jayakumar, R., Roshni Ramachandran, V. V. Divyarani, K. P. Chennazhi, H. Tamura and S. V. Nair. 2011. Fabrication of chitin–chitosan/nano TiO_2-composite scaffolds for tissue engineering applications. International Journal of Biological Macromolecules 48: 336–44.

Jayakumar, R., M. Prabaharan, P. T. Sudheesh Kumar, S. V. Nair and H. Tamura. 2011. Biomaterials based on chitin and chitosan in wound dressing applications. Biotechnol. Adv. 29: 322–37.

Jeong, Young Il, Don Gon Kim, Mi Kyeong Jang and Jaw Woon Nah. 2008. Preparation and spectroscopic characterization of methoxy poly(ethylene glycol)-grafted water-soluble chitosan. Carbohydrate Research 343: 282–89.

Jung, Woo-Jin, Gyung-Hyun Jo, Ju-Hee Kuk, Kil-Yong Kim and Ro-Dong Park. 2005. Demineralization of crab shells by chemical and biological treatments. Biotechnology and Bioprocess Engineering 10: 67–72.

Kaczmarek, Michal Benedykt, Katarzyna Struszczyk-Swita, Xingkang Li, Miroslawa Szczęsna-Antczak and Maurycy Daroch. 2019. Enzymatic modifications of chitin, chitosan, and chitooligosaccharides. Frontiers in Bioengineering and Biotechnology 7: 243.

Kean, Thomas and Maya Thanou. 2011. Chitin and chitosan: sources, production and medical applications.

Khan, Munezza A. and Mohammad Mujahid. 2019. A review on recent advances in chitosan based composite for hemostatic dressings. International Journal of Biological Macromolecules 124: 138–47.

Kim, Se-Kwon. 2010. Chitin, Chitosan, Oligosaccharides and their Derivatives: Biological Activities and Applications (CRC Press).

Kittur, Farooqahmed S., K. V. Harish Prashanth, Kadimi Udaya Sankar and Rudrapatnam N. Tharanathan. 2002. Characterization of chitin, chitosan and their carboxymethyl derivatives by differential scanning calorimetry. Carbohydrate Polymers 49: 185–93.

Klokkevold, Perry R., Lorese Vandemark, E. Barrie Kenney and George W. Bernard. 1996. Osteogenesis enhanced by chitosan (poly-N-acetyl glucosaminoglycan) *in vitro*. Journal of Periodontology 67: 1170–75.

Kluczka, Joanna. 2024. Chitosan: structural and chemical modification, properties, and application. International Journal of Molecular Sciences 25.

Knabl, Ludwig, Bettina Kuppelwieser, Astrid Mayr, Wilfried Posch, Michaela Lackner, Débora Coraça-Huber et al. 2019. High percentage of microbial colonization of osteosynthesis material in clinically unremarkable patients. Microbiologyopen 8: e00658.

Kofuji, Kyoko, Chun-Jun Qian, Masumi Nishimura, Ikumi Sugiyama, Yoshifumi Murata and Susumu Kawashima. 2005. Relationship between physicochemical characteristics and functional properties of chitosan. European Polymer Journal 41: 2784–91.

Kołodziejska, M., K. Jankowska, M. Klak and M. Wszoła. 2021. Chitosan as an underrated polymer in modern tissue engineering. Nanomaterials (Basel) 11.

Kou, Shijie Gabriel, Linda M. Peters and Michael R. Mucalo. 2021. Chitosan: A review of sources and preparation methods. International Journal of Biological Macromolecules 169: 85–94.

Kumar, Guneet, Paul J. Smith and Gregory F. Payne. 1999. Enzymatic grafting of a natural product onto chitosan to confer water solubility under basic conditions. Biotechnology and Bioengineering 63: 154–65.

Kumar, M., N. V. Ravi, R. A. A. Muzzarelli, C. Muzzarelli, H. Sashiwa and A. J. Domb. 2004. Chitosan chemistry and pharmaceutical perspectives. Chemical Reviews 104: 6017–84.

Kurita, Keisuke, Mami Kamiya and Shin Ichiro Nishimura. 1991. Solubilization of a rigid polysaccharide: Controlled partial N-Acetylation of chitosan to develop solubility. Carbohydrate Polymers 16: 83–92.

Lazaridou, Maria, Dimitrios N. Bikiaris and Dimitrios A. Lamprou. 2022. 3D bioprinted chitosan-based hydrogel scaffolds in tissue engineering and localised drug delivery. Pharmaceutics 14: 1978.

Leong, Yoong Kit, Fan-Chiang Yang and Jo-Shu Chang. 2021. Extraction of polysaccharides from edible mushrooms: Emerging technologies and recent advances. Carbohydr. Polym. 251: 117006.

Levengood, Sheeny K. Lan and Miqin Zhang. 2014. Chitosan-based scaffolds for bone tissue engineering. Journal of Materials Chemistry B 2: 3161–84.

Li, Ji and Qingrong Huang. 2012. Rheological properties of chitosan–tripolyphosphate complexes: From suspensions to microgels. Carbohydrate Polymers 87: 1670–77.

Li, Xiaomin, Ronge Xing, Chaojie Xu, Song Liu, Yukun Qin, Kecheng Li et al. 2021. Immunostimulatory effect of chitosan and quaternary chitosan: A review of potential vaccine adjuvants. Carbohydrate Polymers 264.

Lin, Xuezheng, Xiaochen Gong, Qicheng Ruan, Wei Xu, Chunjing Zhang and Kai Zhao. 2023. Antimicrobial application of chitosan derivatives and their nanocomposites. Current Medicinal Chemistry 30: 1736–55.

Liu, Ying, Jia Li Chen, Pei Fang Li and Ning Ning. 2021. The effect of chitosan in wound healing: a systematic review. Advances in Skin & Wound Care 34: 262–66.

Logithkumar, R., A. Keshavnarayan, S. Dhivya, A. Chawla, S. Saravanan and N. Selvamurugan. 2016. A review of chitosan and its derivatives in bone tissue engineering. Carbohydrate Polymers 151: 172–88.

Lv, Shenghua, Shanshan Zhang, Jingjing Zuo, Shan Liang, Juhui Yang, Jialin Wang et al. 2023. Progress in preparation and properties of chitosan-based hydrogels. International Journal of Biological Macromolecules 242.

Ma, Guiping, Dongzhi Yang, Yingshan Zhou, Ming Xiao, John F. Kennedy and Jun Nie. 2008. Preparation and characterization of water-soluble N-alkylated chitosan. Carbohydrate Polymers 74: 121–26.

Ma, Li, Chuan an Shen, Lei Gao, Da wei Li, Yu ru Shang, Kai Yin et al. 2016. Anti-inflammatory activity of chitosan nanoparticles carrying NF-κB/p65 antisense oligonucleotide in RAW264.7 macropghage stimulated by lipopolysaccharide. Colloids and Surfaces B: Biointerfaces 142: 297–306.

Madkour, Metwally, Khaled D. Khalil and Fakhreia A. Al-Sagheer. 2021. Heterogeneous hybrid nanocomposite based on chitosan/magnesia hybrid films: Ecofriendly and recyclable solid catalysts for organic reactions. Polymers 13: 3583.

Man, Zhentao, Xiaoqing Hu, Zhenlong Liu, Hongjie Huang, Qingyang Meng, Xin Zhang et al. 2016. Transplantation of allogenic chondrocytes with chitosan hydrogel-demineralized bone matrix hybrid scaffold to repair rabbit cartilage injury. Biomaterials 108: 157–67.

Mathew, Sindhu and T. Emilia Abraham. 2006. Studies on the antioxidant activities of cinnamon (Cinnamomum verum) bark extracts, through various *in vitro* models. Food Chemistry 94: 520–28.

Meng, Di, Betchem Garba, Yun Ren, Man Yao, Xiaoshuang Xia, Mingyan Li, and Yun Wang. 2020. Antifungal activity of chitosan against Aspergillus ochraceus and its possible mechanisms of action. International Journal of Biological Macromolecules 158: 1063–70.

Meng, Hongqi, Ying Zhao, Hang Cai, Di You, Ying Wang, Siyu Wu et al. 2024. Hydrogels containing chitosan-modified gold nanoparticles show significant efficacy in healing diabetic wounds infected with antibiotic-resistant bacteria. International Journal of Nanomedicine 19: 1539–56.

Mikušová, Veronika and Peter Mikuš. 2021. Advances in chitosan-based nanoparticles for drug delivery. International Journal of Molecular Sciences 22: 9652.

Mohan, Sandya Madan and Bhawana Pandey. 2016. Antimicrobial activity of Oxalis corniculata Linn. International Journal of Science and Research 5(7): 575–578.

Moran, Hannah B. T., Joanna L. Turley, Mats Andersson and Ed C. Lavelle. 2018a. Immunomodulatory properties of chitosan polymers. Biomaterials 184: 1–9.

Moran, Hannah B. T., Joanna L. Turley, Mats Andersson and Ed C. Lavelle. 2018b. Immunomodulatory properties of chitosan polymers. Biomaterials, 184: 1–9.

Murphy, Sean V. and Anthony Atala. 2014. 3D bioprinting of tissues and organs. Nature Biotechnology 32: 773–85.

Narayan, Shoba. 2023. A systematic review of different classes of biopolymers and their use as antimicrobial agents. Russian Journal of Bioorganic Chemistry 49: 262–87.

Ngo, D. H. and S. K. Kim. 2014. Antioxidant effects of chitin, chitosan, and their derivatives. Adv Food Nutr Res 73: 15–31.

Ngo, Dai-Hung, Thanh-Sang Vo, Dai-Nghiep Ngo, Kyong-Hwa Kang, Jae-Young Je, Hoang Nguyen-Duc Pham et al. 2015. Biological effects of chitosan and its derivatives. Food Hydrocolloids 51: 200–16.

Nilsen-Nygaard, Julie, Sabina P. Strand, Kjell M. Vårum, Kurt I. Draget and Catherine T. Nordgård. 2015. Chitosan: Gels and interfacial properties. Polymers 7: 552–79.

Niu, Wei, Yuelin Dong, Ziwei Fu, Jiajie Lv, Ligui Wang, Zhenhai Zhang et al. 2021. Effects of molecular weight of chitosan on anti-inflammatory activity and modulation of intestinal microflora in an ulcerative colitis model. International Journal of Biological Macromolecules 193: 1927–36.

Nordtveit, Ragnhild J., Kjell M. Vårum and Olav Smidsrød. 1996. Degradation of partially N-acetylated chitosans with hen egg white and human lysozyme. Carbohydrate Polymers 29: 163–67.

O'Rourke, Aubrie, Sinem Beyhan, Yongwook Choi, Pavel Morales, Agnes P. Chan, Josh L. Espinoza et al. 2020. Mechanism-of-action classification of antibiotics by global transcriptome profiling. Antimicrobial Agents and Chemotherapy 64: 10.1128/aac. 01207-19.

Omari, Khaled W., Jessica E. Besaw and Francesca M. Kerton. 2012. Hydrolysis of chitosan to yield levulinic acid and 5-hydroxymethylfurfural in water under microwave irradiation. Green Chemistry 14: 1480–87.

Oryan, Ahmad, Soodeh Alidadi, Amin Bigham-Sadegh and Ali Moshiri. 2016. Comparative study on the role of gelatin, chitosan and their combination as tissue engineered scaffolds on healing and regeneration of critical sized bone defects: an *in vivo* study. Journal of Materials Science: Materials in Medicine 27.

Oryan, Ahmad, Soodeh Alidadi, Amin Bigham-Sadegh and Ali Moshiri. 2017. Effectiveness of tissue engineered based platelet gel embedded chitosan scaffold on experimentally induced critical sized segmental bone defect model in rat. Injury 48: 1466–74.

Oryan, Ahmad and Sonia Sahvieh. 2017. Effectiveness of chitosan scaffold in skin, bone and cartilage healing. International Journal of Biological Macromolecules 104: 1003–11.

Ouchi, Tatsuro, Hidetoshi Nishizawa and Yuichi Ohya. 1998. Aggregation phenomenon of PEG-grafted chitosan in aqueous solution. Polymer 39: 5171–75.

Pakravan, Mehdi, Marie Claude Heuzey and Abdellah Ajji. 2011. A fundamental study of chitosan/PEO electrospinning. Polymer 52: 4813–24.

Pandey, Anamika R., Upasana S. Singh, Munira Momin and Chintan Bhavsar. 2017. Chitosan: Application in tissue engineering and skin grafting. Journal of Polymer Research 24: 1–22.

Patel, Z. S. and A. G. Mikos. 2004. Angiogenesis with biomaterial-based drug- and cell-delivery systems. J Biomater Sci Polym Ed 15: 701–26.

Petroni, Simona, Irene Tagliaro, Carlo Antonini, Massimiliano D'Arienzo, Sara Fernanda Orsini, João F. Mano et al. 2023. Chitosan-based biomaterials: insights into chemistry, properties, devices, and their biomedical applications. Marine Drugs 21.

Piaia, Lya, Simone S. Silva, Joana M. Gomes, Albina R. Franco, Emanuel M. Fernandes, Flávia C. M. Lobo et al. 2021. Chitosan/β-TCP composites scaffolds coated with silk fibroin: A bone tissue engineering approach. Biomedical Materials 17: 015003.

Pichyangkura, Rath and Supachitra Chadchawan. 2015. Biostimulant activity of chitosan in horticulture. Scientia Horticulturae 196: 49–65.

Pita-López, María Luisa, Gabriela Fletes-Vargas, Hugo Espinosa-Andrews and Rogelio Rodríguez-Rodríguez. 2021. Physically cross-linked chitosan-based hydrogels for tissue engineering applications: A state-of-the-art review. European Polymer Journal 145: 110176.

Pogodina, N. V., G. M. Pavlov, S. V. Bushin, A. B. Mel'Nikov, Ye B. Lysenko, L. A. Nud'Ga et al. 1986. Conformational characteristics of chitosan molecules as demonstrated by diffusion-sedimentation analysis and viscometry. Polymer Science USSR 28: 251–59.

Porporatto, Carina, Ismael D. Bianco, Clelia M. Riera and Silvia G. Correa. 2003. Chitosan induces different L-arginine metabolic pathways in resting and inflammatory macrophages. Biochemical and Biophysical Research Communications 304: 266–72.

Prasad, Tilak, E. A. Shabeena, D. Vinod, T. V. Kumary and P. R. Anil Kumar. 2015. Characterization and in vitro evaluation of electrospun chitosan/polycaprolactone blend fibrous mat for skin tissue engineering. Journal of Materials Science: Materials in Medicine 26: 1–13.

Pusateri, Anthony E., Simon J. McCarthy, Kenton W. Gregory, Richard A. Harris, Luis Cardenas, Albert T. McManus et al. 2003. Effect of a chitosan-based hemostatic dressing on blood loss and survival in a model of severe venous hemorrhage and hepatic injury in swine. Journal of Trauma and Acute Care Surgery 54: 177–82.

Qasim, Saad B., Muhammad S. Zafar, Shariq Najeeb, Zohaib Khurshid, Altaf H. Shah, Shehriar Husain et al. 2018. Electrospinning of chitosan-based solutions for tissue engineering and regenerative medicine. International Journal of Molecular Sciences 19: 407–07.

Qian, Lu, Kai Zhang, Xin Guo and Miao Yu. 2023. What happens when chitin becomes chitosan? A single-molecule study. RSC Adv. 13: 2294–300.

Qin, C., Y. Du, L. Xiao, Z. Li and X. Gao. 2002. Enzymic preparation of water-soluble chitosan and their antitumor activity. Int J Biol Macromol 31: 111–7.

Qin, Caiqin, Huirong Li, Qi Xiao, Yi Liu, Juncheng Zhu and Yumin Du. 2006. Water-solubility of chitosan and its antimicrobial activity. Carbohydrate Polymers 63: 367–74.

Radulescu, Marius, Denisa Ficai, Ovidiu Oprea, Anton Ficai, Ecaterina Andronescu, and Alina Holban. 2015. Antimicrobial chitosan based formulations with impact on different biomedical applications. Current Pharmaceutical Biotechnology 16: 128–36.

Ressler, Antonia. 2022. Chitosan-based biomaterials for bone tissue engineering applications: a short review. Polymers 14: 3430.

Ressler, Antonia, Roope Ohlsbom, Andreja Žužić, Arjen Gebraad, Erkka J. Frankberg, Toni-Karri Pakarinen et al. 2023. Chitosan/collagen/Mg, Se, Sr, Zn-substituted calcium phosphate scaffolds for bone tissue engineering applications: A growth factor free approach. European Polymer Journal 194: 112129.

Ribeiro, José Carlos Viana, Tereza Cristina Marques Forte, Samia Jéssica Silva Tavares, Fábia Karine Andrade, Rodrigo Silveira Vieira and Vilma Lima. 2021. The effects of the molecular weight of chitosan on the tissue inflammatory response. Journal of Biomedical Materials Research Part A 109: 2556–69.

Rinaudo, Marguerite. 2006. Chitin and chitosan: Properties and applications. Prog. Polym. Sci. 31: 603–32.

Rinaudo, Marguerite. 2008. Main properties and current applications of some polysaccharides as biomaterials. Polymer International 57: 397–430.

Rizeq, Balsam R., Nadin N. Younes, Kashif Rasool and Gheyath K. Nasrallah. 2019. Synthesis, bioapplications, and toxicity evaluation of chitosan-based nanoparticles. International Journal of Molecular Sciences 20: 5776.

Roberts, George A. F. and George A. F. Roberts. 1992. Solubility and solution behaviour of chitin and chitosan. Chitin Chemistry: 274–329.

Rodríguez-Vázquez, Martin, Brenda Vega-Ruiz, Rodrigo Ramos-Zúñiga, Daniel Alexander Saldaña-Koppel and Luis Fernando Quiñones-Olvera. 2015a. Chitosan and its potential use as a scaffold for tissue engineering in regenerative medicine. BioMed Research International 2015.

Rodríguez-Vázquez, Martin, Brenda Vega-Ruiz, Rodrigo Ramos-Zúñiga, Daniel Alexander Saldaña-Koppel and Luis Fernando Quiñones-Olvera. 2015b. Chitosan and its potential use as a scaffold for tissue engineering in regenerative medicine. BioMed Research International 2015: 821279.

Ruiz, Gustavo Adolfo Muñoz and Hector Fabio Zuluaga Corrales. 2017. Chitosan, chitosan derivatives and their biomedical applications. Biological Activities and Application of Marine Polysaccharides 87.

Scalera, F., S. I. A. Pereira, A. Bucciarelli, D. M. Tobaldi, A. Quarta, F. Gervaso et al. 2023. Chitosan-hydroxyapatite composites made from sustainable sources: A morphology and antibacterial study. Materials Today Sustainability 21: 100334.

Schiffman, Jessica D. and Caroline L. Schauer. 2008. A review: electrospinning of biopolymer nanofibers and their applications. Polymer Reviews 48: 317–52.

Sergi, Rachele, Devis Bellucci and Valeria Cannillo. 2020. A review of bioactive glass/natural polymer composites: State of the art. Materials 13: 5560.

Shahid-ul-Islam and B. S. Butola. 2019. Recent advances in chitosan polysaccharide and its derivatives in antimicrobial modification of textile materials. International Journal of Biological Macromolecules 121: 905–12.

Shahidi, F. and R. Abuzaytoun. 2005a. Chitin, chitosan, and co-products: chemistry, production, applications, and health effects. Adv Food Nutr Res 49: 93–135.

Shahidi, Fereidoon and Reem Abuzaytoun. 2005b. Chitin, chitosan, and co-products: Chemistry, production, applications, and health effects. In, Adv Food Nutr Res (Elsevier).

Shariatinia, Z. 2019. Pharmaceutical applications of chitosan. Adv Colloid Interface Sci 263: 131–94.

Shariatinia, Zahra and Azin Mazloom Jalali. 2018. Chitosan-based hydrogels: Preparation, properties and applications. International Journal of Biological Macromolecules 115: 194–220.

Shikhi-Abadi, Parvaneh Ghaderi and Mohammad Irani. 2021. A review on the applications of electrospun chitosan nanofibers for the cancer treatment. International Journal of Biological Macromolecules 183: 790–810.

Shuai, Hung-Hsun, Chung-Yao Yang, I. Hans, Chen Harn, Roger L. York, Tzu-Chun Liao et al. 2013. Using surfaces to modulate the morphology and structure of attached cells–a case of cancer cells on chitosan membranes. Chemical Science 4: 3058–67.

Signorini, L., G. Marenzi, A. Facente, B. Marrelli, R. M. Marano, A. Valletta et al. 2023. Critical overview on pure chitosan-based scaffolds for bone tissue engineering: clinical insights in dentistry. Int J Med Sci 20: 1527–34.

Silva, G. A., P. Ducheyne and R. L. Reis. 2007. Materials in particulate form for tissue engineering. 1. Basic concepts. J. Tissue Eng. Regen. Med. 1: 4–24.

Sogias, Ioannis A., Vitaliy V. Khutoryanskiy and Adrian C. Williams. 2010. Exploring the factors affecting the solubility of chitosan in water. Macromolecular Chemistry and Physics 211: 426–33.

Szymańska, Emilia, Michał Wojasiński, Robert Czarnomysy, Renata Dębowska, Iwona Łopianiak, Kamil Adasiewicz et al. 2022. Chitosan-enriched solution blow spun poly(ethylene oxide) nanofibers with poly(dimethylsiloxane) hydrophobic outer layer for skin healing and regeneration. International Journal of Molecular Sciences 23.

Taghizadeh, Mohsen, Ali Taghizadeh, Mohsen Khodadadi Yazdi, Payam Zarrintaj, Florian J. Stadler, Joshua D. Ramsey et al. 2022. Chitosan-based inks for 3D printing and bioprinting. Green Chemistry 24: 62–101.

Tahrir, G. Farzaneh, Fariba Ganji and Taha Ahooyi. 2014. Injectable thermosensitive chitosan/glycerophosphate-based hydrogels for tissue engineering and drug delivery applications: a review. Recent Patents on Drug Delivery & Formulation 9: 107–20.

Thadathil, Nidheesh and Suresh Puthanveetil Velappan. 2014. Recent developments in chitosanase research and its biotechnological applications: A review Food Chemistry 150: 392–99.

Thein-Han, W. W. and R. D. K. Misra. 2009. Biomimetic chitosan–nanohydroxyapatite composite scaffolds for bone tissue engineering. Acta Biomaterialia 5: 1182–97.

Tungtong, Siriwat, Siriporn Okonogi, Sombat Chowwanapoonpohn, Weerachai Phutdhawong and Songwut Yotsawimonwat. 2012. Solubility, viscosity and rheological properties of water-soluble chitosan derivatives. Maejo International Journal of Science and Technology 6: 315.

Vasconcelos, Daniela P., Madalena Costa, Isabel F. Amaral, Mário A. Barbosa, Artur P. Águas and Judite N. Barbosa. 2015. Modulation of the inflammatory response to chitosan through M2 macrophage polarization using pro-resolution mediators. Biomaterials 37: 116–23.

Vinsova, Jarmila and Eva Vavrikova. 2011. Chitosan derivatives with antimicrobial, antitumour and antioxidant activities—a review. Current Pharmaceutical Design 17: 3596–607.

Wan, Ying, Hua Wu, Xiaoying Cao and Siqin Dalai. 2008. Compressive mechanical properties and biodegradability of porous poly(caprolactone)/chitosan scaffolds. Polymer Degradation and Stability 93: 1736–41.

Wang, Chih-Hsin, Juin-Hong Cherng, Chuan-Chieh Liu, Tong-Jing Fang, Zhi-Jie Hong, Shu-Jen Chang et al. 2021. Procoagulant and antimicrobial effects of chitosan in wound healing. International Journal of Molecular Sciences 22.

Wang, Jianlong and Shuting Zhuang. 2022. Chitosan-based materials: Preparation, modification and application. Journal of Cleaner Production 355: 131825.

Wang, Jingjing, Yuning Yuan, Yu Liu, Xiang Li and Shengjun Wu. 2024. Application of chitosan in fruit preservation: A review. Food Chemistry: X 23: 101589.

Wang, Tao, Mahir Turhan and Sundaram Gunasekaran. 2004. Selected properties of pH-sensitive, biodegradable chitosan–poly(vinyl alcohol) hydrogel. Polymer International 53: 911–18.

Wang, Wei and Deshi Xu. 1994. Viscosity and flow properties of concentrated solutions of chitosan with different degrees of deacetylation. International Journal of Biological Macromolecules 16: 149–52.

Wang, Yi-Wen, Chuan-Chieh Liu, Juin-Hong Cherng, Chien-Seng Lin, Shu-Jen Chang, Zhi-Jie Hong et al. 2019. Biological effects of chitosan-based dressing on hemostasis mechanism. Polymers 11: 1906.

Wang, Zhihua, Yongbin Yan, Zhengmao Zhang, Changchun Li, Lanfei Mei, Ruyi Hou et al. 2024a. Effect of chitosan and its water-soluble derivatives on antioxidant activity. Polymers 16: 867.

Wang, Zhihua, Yongbin Yan, Zhengmao Zhang, Changchun Li, Lanfei Mei, Ruyi Hou et al. 2024b. Effect of chitosan and its water-soluble derivatives on antioxidant activity. Polymers (Basel) 16.

Wani, Saima, Hasham S. Sofi, Shafquat Majeed and Faheem A. Sheikh. 2017. Recent trends in chitosan nanofibers: from tissue-engineering to environmental importance: a review. Mater Sci Res India 14: 89–99.

Xia, Guixue, Dongqing Zhai, Yue Sun, Lin Hou, Xiaofan Guo, Lixia Wang et al. 2020. Preparation of a novel asymmetric wettable chitosan-based sponge and its role in promoting chronic wound healing. Carbohydrate Polymers 227: 115296.

Yan, Dazhong, Yanzhen Li, Yinling Liu, Na Li, Xue Zhang and Chen Yan. 2021a. Antimicrobial properties of chitosan and chitosan derivatives in the treatment of enteric infections. Molecules (Basel, Switzerland) 26.

Yan, Dazhong, Yanzhen Li, Yinling Liu, Na Li, Xue Zhang and Chen Yan. 2021b. Antimicrobial properties of chitosan and chitosan derivatives in the treatment of enteric infections. Molecules 26: 7136.

Yan, Zhang, Li Ruixin, Fan Yubo, Liu Hao, Guo Yong, Wang Liang et al. 2012. Microarchitectural and mechanical characterization of chitosan/hydroxyapatite/demineralized bone matrix composite scaffold. Journal of Porous Materials 19: 251–59.

Yang, Xiaochuan, Xuening Chen and Hongjun Wang. 2009. Acceleration of osteogenic differentiation of preosteoblastic cells by chitosan containing nanofibrous scaffolds. Biomacromolecules 10: 2772–78.

Yen, Ming-Tsung, Joan-Hwa Yang and Jeng-Leun Mau. 2008. Antioxidant properties of chitosan from crab shells. Carbohydrate Polymers 74: 840–44.

Yi, Peng, Xiongfeng Xu, Bo Qiu and Huajie Li. 2020. Impact of chitosan membrane culture on the expression of pro-and anti-inflammatory cytokines in mesenchymal stem cells. Experimental and Therapeutic Medicine 20: 3695–702.

Younes, Islem, Sawssen Hajji, Véronique Frachet, Marguerite Rinaudo, Kemel Jellouli and Moncef Nasri. 2014. Chitin extraction from shrimp shell using enzymatic treatment. Antitumor, antioxidant and antimicrobial activities of chitosan. International Journal of Biological Macromolecules 69: 489–98.

Younes, Islem and Marguerite Rinaudo. 2015. Chitin and chitosan preparation from marine sources. Structure, properties and applications. Marine Drugs 13: 1133–74.

Younes, Islem, Marguerite Rinaudo, David Harding and Hitoshi Sashiwa. 2015. Chitin and chitosan preparation from marine sources: structure, properties and applications. Marine Drugs 13: 1133–74.

CHAPTER 4

Collagen Composites for Tissue Engineering

Pravina Rathore

1. Introduction

The wonder protein found abundantly in cell's extracellular matrix, finding its explicit application for rejuvenating lost/damaged tissues for a better human life is "Collagen".

The word collagen is derived from the Greek word *kolla*, meaning "glue", with suffix *-gen*, denoting "producing".

It is the prime component of connective tissues as cartilage, bones, tendons, ligaments, and skin, making up from 25% to 35% of the whole-body protein content. Also it is also amply present in the dentin in teeth, the corneas of eyes, blood vessels, the gut and the inter-vertebral disc. Along with these they present in muscle tissue, where it is a major component of the endomysium, constituting 1–2% of muscle tissue. In terms of weight of the skeletal muscle tissue, it accounts for 6% of the total weight (Owczarzy, 2020). It has been reported in fossils from the Mesozoic and Paleozoic. Collagen plays a vital function in providing support and strength structurally to various tissues of skin, bones, tendons, ligaments and other connective tissues. Due to its unique properties and functions along with the ability to change owing to various physicochemical or enzymatic processes to form a matrix, it finds its applications in medical, cosmetic and other industries (Shoulders and Raines, 2009; Ricard-Blum, 2011; Sorushanova et al., 2019; Gustavo et al., 2021; Barbara, 2024).

Department of Botany, Bhupal Nobles' University, Udaipur - 313002, Rajasthan, India.
Email: pravinarathore@gmail.com

DOI: 10.1201/9781003518792-4

Tissue Engineering

The science to develop or help in developing the lost or damaged tissues/organs is Tissue engineering. This can be done *in-vitro* or *in-vivo*. The aim of tissue engineering is to build up biological alternatives that replace, sustain and improve tissue function or growth (Ikada, 2006). Tissue engineering is an interdisciplinary science revolving around biology, chemistry and engineering to get a fruitful output. The basic principle is to recreate a microenvironment and provide ingredients to form and develop new tissues. The ingredients used must imitate the native forms. Also they need to be nonhazardous and biodegradable. Tissue engineering use biocomponents, scaffolds or cells to rejuvenate new tissues. Scaffolds are 2/3-D structures, natural or synthetic, providing a surface for cell adherence and growth. Cells possessing dividing capacity can also be part of tissue engineering. Biocomponents are molecules of living origin and includes simple to complex ones. These biomolecules used in tissue engineering must be biocompatible, flexible, strong in mechanical properties and biodegradable. One such biomolecule to have surfaced strongly in tissue engineering is Collagen (Prafulla et al., 2020; Chapekar, 2000; Han et al., 2020).

2. Molecular Structure

Collagen is highly heterogeneous, due to its varied amino acid composition, structure and presence in various specified tissues. However all are based on same structural frame. A basic structural unit of collagen is a left-handed α-helix molecule called the *tropocollagen*, The tropocollagen is composed of three polypeptide, spirally wound chains. 2 of these are same (α-1) and one (α-2) is different. These chains are rich in 3 amino acids viz., proline, glycine and hydroxyproline. The main collagen amino acid is Glycine which is present at every third amino acid position in a single polypeptide chain (approx. 35%). Proline is another prominent amino acid, accounting for 12% of all amino acids. The most common amino acid sequence of collagen are glycine-proline-X and glycine-X-hydroxyproline, where X is any amino acid other than glycine, proline or hydroxyproline. The repeatable amino acids are important for the collagen structure. The repetitive presence of Glycine permits the formation of intra and inter hydrogen bonds in its molecules (Brodsky, 1997). These are responsible for the strength of the entire helical structure. The amino acid composition and their amount in polypeptide chains differ between different types of collagen. The unique feature of collagen molecule is that, amounts of basic and acidic amino acids present are nearly at equimolar concentration (Figure 1).

The three left-handed helices (topocollagen unit) are twisted together into a right-handed triple helix or “super helix”, that is approximately 300 nm long and 1.5 nm in diameter. A cooperative quaternary structure is stabilized by many hydrogen bonds. The tropocollagen unit aggregate to form many tropocollagens which then bind together to form a fibril, and many of these then form a fibre. The aggregation of tropocollagen molecules occur by the electrostatic and hydrophobic side bonds (Shoulders and Raines, 2009; Sorushanova et al., 2019; Gustavo et al., 2021).

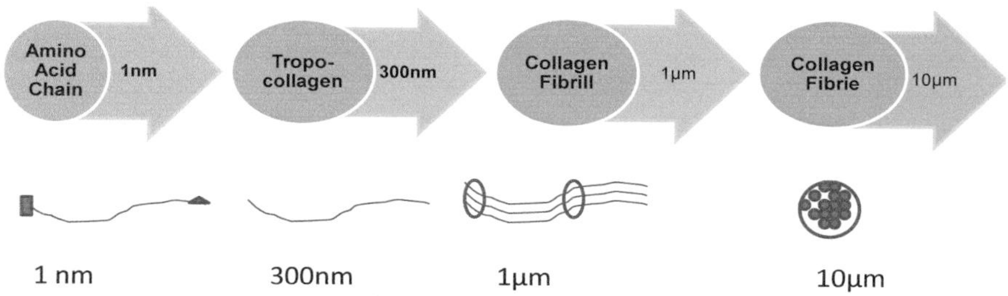

Figure 1. Diameteric succession from amino acids to collagen fibre.

3. Types

The collagen is a natural polymers constituting 1/3 of the total mass of proteins in the human body (Shoulders and Raines, 2009). The studies till date reveal 28 types of collagen based on their structural, compositional and distribution in humans. The collagen proteins are divided two main groups based on the structure: fibrillar collagens and non-fibrillar collagens. Based on subunit composition, numbering in Roman numerals (I–XXIX) is carried out to describe the types of collagen. Additionally, Arabic numerals are used to indicate the polypeptide chains composition (α1–α6).
Of these five most common types are given in Table 1.

The availability of such diversity in Collagen is attributed to differences in the expression of genes encoding enzymes or various post-translational modifications. The biosynthesis of collagen starts with the transcription unit formed from various genes, its translation, modification of translational unit, i.e., the initial polypeptide. These post translational changes include enzymatic glycosylation and hydroxylation of Lysine, cleavage of procollagen and natural crosslinking to form the quaternary structure. All these activities result in varied forms of collagen.

Table 1. Types of collagen and their availability in various human tissues.

Type I	Skin, tendon, vasculature, organs, bone (main component of the organic part of bone)
Type II	Cartilage (main collagenous component of cartilage)
Type III	Reticulate (main component of reticular fibers), commonly found alongside type
Type IV	Forms basal lamina, the epithelium-secreted layer of the basement membrane
Type V	Cell surfaces, hair, and placenta

Types of Collagen as distributed in different forms is depicted in Figure 2.

Collagen is present naturally in many organisms and has been extracted from them for commercial uses since last century. It is found in all mammals and the abundant sources are reported from various animals as chicken, pig, rabbit and deer especially their skin and bones from which its obtained in the form of broths (Silvipriya et al., 2015; León-López, 2019). Plants like Citrus fruits, leafy vegetables, papaya, guava, sesame seeds and almonds are rich in vegan collagen (Brodsky and

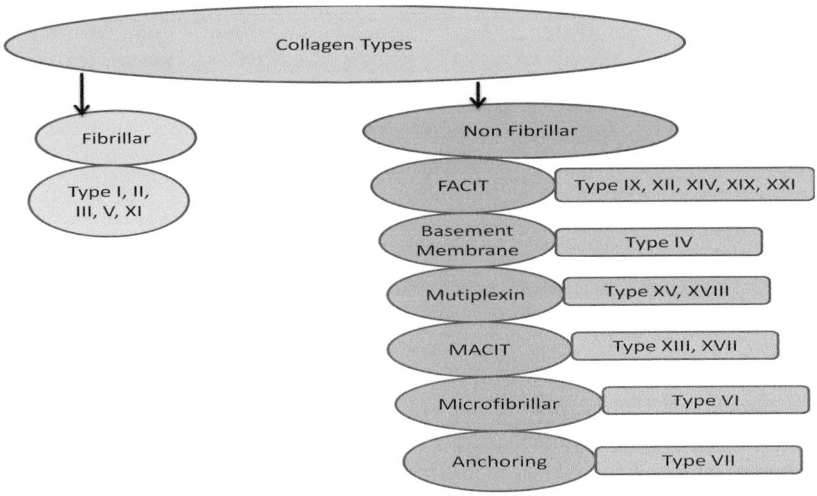

Figure 2. Various forms of collagen.

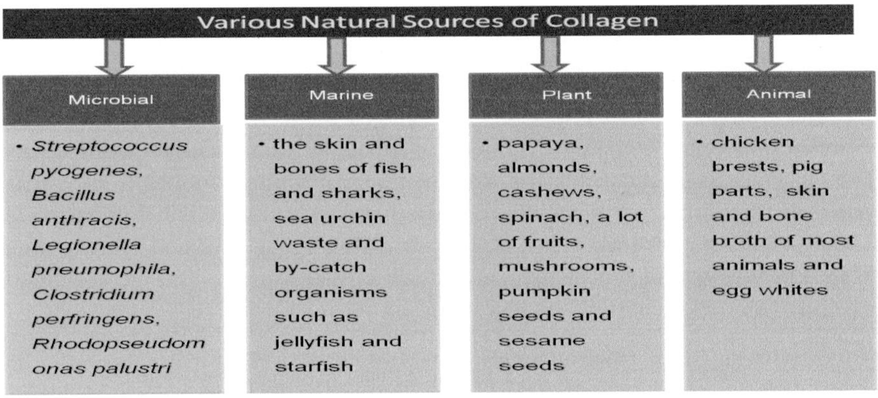

Figure 3. Various natural sources of collagen.

Kaplan, 2013; Deane et al., 2020). Similarly, fish and other marine organism are considered best collagen sources (Yamada et al., 2013; Geahchan et al., 2022; Furtado et al., 2022). Collagen like protein has been reported from bacteria as *Streptococcus pyogenes, Bacillus anthracis, Legionella pneumophila, Clostridium perfringens, Solibacter usitatus, Streptococcus pyogenes*, *Rhodopseudomonas palustri* – which are considered as microbial collagen sources (Duarte et al., 2014; Yu et al., 2014). Different natural sources for collagen is shown in Figure 3.

4. Properties

The collagen molecule weighs approximately 300,000 Da, Its capacity to bind with water is enormous allowing it to form gels. The same capacity allows it to be good constituent in texturing and thickening. The ability to form emulsion and foam gives it an useful end in many medical applications. The characteristic surface related properties like stabilization, adhesion, cohesion, film-forming properties, and a protective function for the colloid also makes it highly utilizable ingredient. It is considered as a fine surfactant. It has prime biocompatibility and has limited immunogenicity. It can enter lipid free surfaces. However, all properties of collagen are dependent on the age, as aging leads to higher cross-linking of collagen polymer and thus reduction in its mechanical strength. Properties of collagen is depicted in Figure 4.

Collagen is generally insoluble in water and oils owing to its cross-linked structure. When denatured, its cross links are damaged and its capacity to bind with water is increased. However the pH of the vicinity, water content and degree of cross linking determines the denaturation of Collagen protein due to temperature. On denaturation, the triple helix unwind and chains separate. This loose mass of chains on cooling has high water absorbing capacity and is called gelatin. These properties leads to the use of collagen and gelatin in pharmaceutical, medical and bioengineering industries.

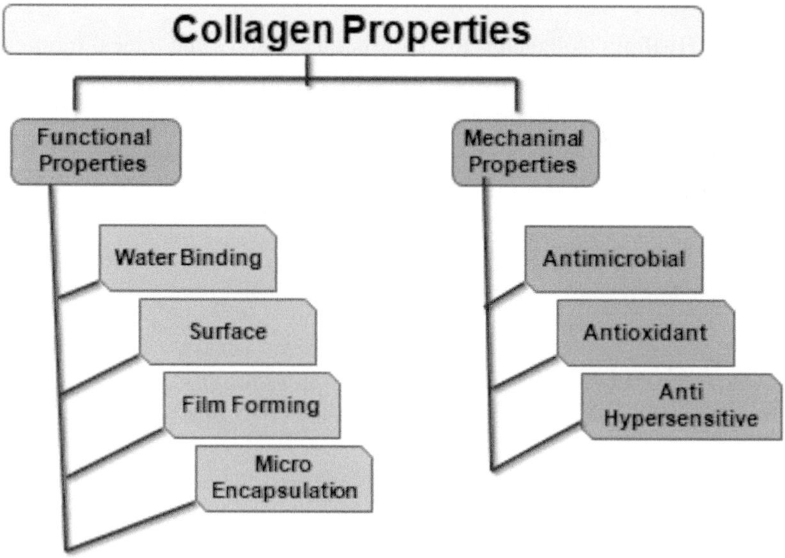

Figure 4. Different types of properties of collagen.

5. Applications of Collagen

Owing to its biocompatibility, hydration power, biodegradability, aggregating, fibrous and film forming capacities, collagen is effectively used in food, medical, pharmaceutical and biotechnological fields. The various applications of collagen includes:

A. Food Industry:

a. **Gelatin Production:** Collagen is hydrolyzed to produce gelatin, which is used in food products like desserts, marshmallows, and gummy candies.

b. **Functional Foods:** Collagen peptides are used as protein supplements in various food and beverage products due to their bioavailability and health benefits.

c. **Casings** for sausages (Hashim et al., 2015; Liu et al., 2015).

B. Cosmetics and Personal Care:

a. **Skin Care Products:** Collagen is a popular ingredient in moisturizers, creams, and serums aimed at improving skin elasticity and reducing signs of aging.

b. **Hair Care:** Collagen is included in shampoos and conditioners to strengthen hair strands and improve overall hair health (Al-Nimry et al., 2021; Sionkowska et al., 2020; Jadach et al., 2024).

C. Medical and Pharmaceutical Uses:

a. **Wound Healing:** Collagen is used in wound dressings and topical formulations to promote healing and tissue regeneration.

b. **Surgical Sutures:** Collagen-based sutures are biocompatible and can be absorbed by the body over time.

c. **Bone Grafting:** Collagen matrices are used as scaffolds for bone regeneration and repair.

d. **Dermal Fillers:** Collagen injections are used in cosmetic procedures to reduce wrinkles and restore volume to the skin (Sangeetha et al., 2019; Zhou et al., 2023; Movaniya et al., 2021; Justin et al., 2023; Sokolakis et al., 2022).

D. Nutritional Supplements:

a. **Joint Health:** Collagen supplements are taken orally to support joint health and reduce symptoms of osteoarthritis.

b. **Skin Health:** Oral collagen supplements are believed to promote skin elasticity and hydration from within (Martínez-Puig et al., 2023; Sionkowska et al., 2020; Al-Atif et al., 2022).

E. Biomedical Engineering:

a. **Tissue Engineering:** Collagen scaffolds are used as a biomaterial for tissue engineering applications, such as growing new skin or repairing damaged tissues.

b. **Drug Delivery:** Collagen matrices can be engineered to deliver drugs or therapeutic agents to specific tissues or organs (Makwana et al., 2022; Soliman et al., 2022; Gál et al., 2020).

F. Research and Development:

a. **Cell Culture:** Collagen-coated surfaces are used in cell culture applications to provide a natural environment for cell growth and differentiation.

b. **Biochemical Studies:** Collagen is studied for its role in diseases like arthritis and for its interactions with other biomolecules (Caliari and Burdick, 2016; Lin et al., 2023; Chua and Lim, 2023; Gopalakrishnan et al., 2022).

G. Musical Instruments: Animal Collagen glues are still used in making musical instruments such as fine violins and guitars, which may have to be reopened for repairs as softening again upon reheating (Colin, 2018; Meyer, 2019; Olden et al., 2020). Different commercial available collagen scaffolds are shown in Figure 5.

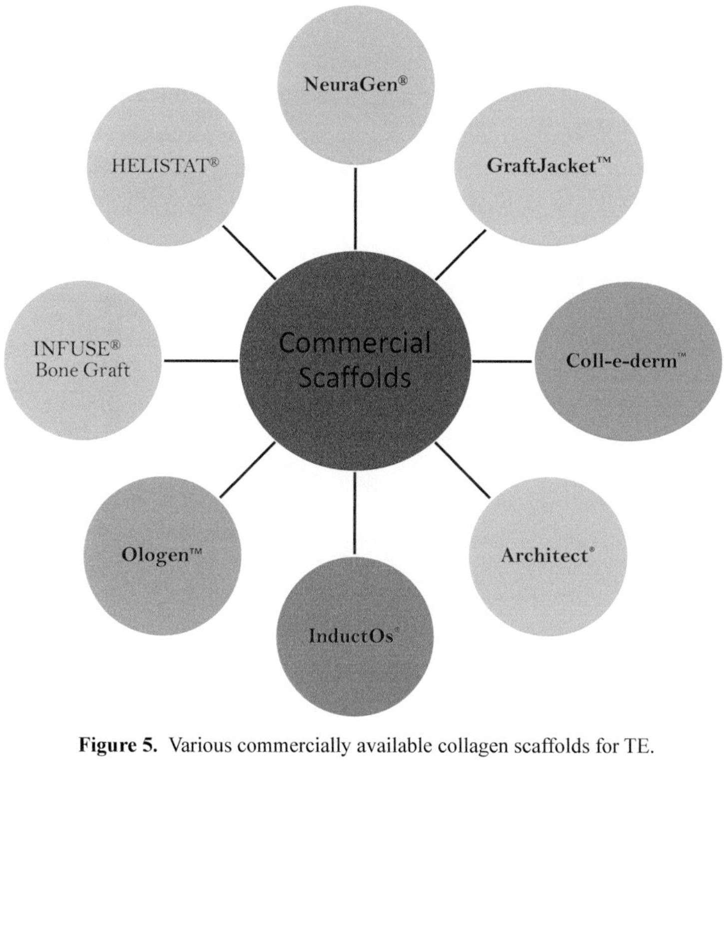

Figure 5. Various commercially available collagen scaffolds for TE.

6. Biosynthesis of Collagen

A. Natural Process

i. 44 genes transcribed in nucleus- each has "*COL*" prefix

ii. N-Terminal Signal Sequence recognized by RER for translation to form **pre-pro-peptide**

iii. The **pre-pro-peptide** directly enters the ER lumen for Post-transcriptional changes to form **preprocollagen.** The formation of pre-pro-collagen formation occurs in 3 steps-

  a. Removal of N-terminal Signal Sequence=prepeptide formed

  b. Hydroxylation of lysines and pralines by 'prolyl hydroxylase' and 'lysyl hydroxylase' enzymes respectively. This facilitates cross-linking of peptides

  c. Glycosylation (Glucose/Galactose) of hydroxyl group of lysines

iv. The resulting such 3 prepeptides twists in helical form to give rise to **Procollagen,** which is packed into a vesicle to be transferred to golgi apparatus for further editing.

v. In Golgi Apparatus, oligosaccharide/s unit/s is further added to this molecule to form the mature Procollagen, which is packaged into a secretory vesicle to be move out to extracellular space.

vi. In the extracellular space, membrane bound enzymes called Collagen Peptidases removes the ends to form **troppocollagen** – the unit of collagen fibril.

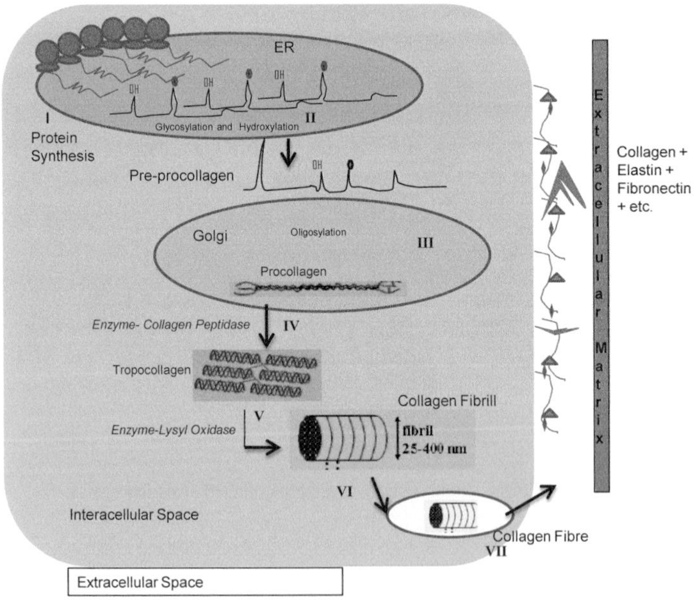

Figure 6. Biosynthesis of collagen in cell-A schematic presentation.

vii. A copper dependent enzyme, *Lysyl Oxidase,* present in the exracellular matrix, act on lysines and hydroxylysines producing aldehyde groups. The aldehyde group of trpocollagen forms covalent bonds forming a Collagen Fibril- the unit of Collagen (Gelse et al., 2003; Shoulders and Raines, 2009). Atypical synthesis of collagen is depicted in Figure 6.

B. Artificial Process

Owing to its polymeric properties and biological origin, collagen has become a favoured molecule in bioengineering. However, natural collagen faces problems like purity and consistency, biocompatibility and being of animal origin faces the ethical issues. To overcome these problems synthetic collagen/collagen mimic products came into existence (Hu et al., 2022). This synthesized collagen could prove useful biomaterial for varied bioengineering tasks (Luo et al., 2013; Valot et al., 2021).

Collagen can be produced by utilizing 2 techniques:

i. Chemical Synthesis

ii. Collagen/Collagen mimic proteins RDT (Recombinant DNA technology)

The chemical synthesis is found to produce collagen fragments which are short of vital characteristics of collagen, i.e., lacked triple-helical conformation, irregular resistance to enzymatic degradation and abrupt proline and lysine modifications (Fertala, 2020). The collagen construction requires substantial folding of protein chains not achievable specifically by chemical methods. Due to its application in biomedical fields, collagen requires to be biocompatible and consistent, therefore the RDT methods were worked-out.

Collagen being a massive molecule to be active, has to be produced in high quantity. Artificial production of collagen through recombinant technology (Schnieke et al., 1987) is challenged by not only its large structure but also with complex folding and modifications that are dependent on specified enzyme systems and modifiers (Fertala, 2020). Mammalian cells were first to be used for expression of recombinant DNA, which later shifts to smaller unicellular microbes like *E. coli* and Yeast (Bulleid et al., 2000).

For RDT, constructing and expressing Genes coding for collagen (Table 2) were used which could produce both homo and hetromeric fibrils (Stein et al., 2009). Later, instead of full length collagen, short fragments to be modified as required were produced to meet out applications of the collagen (Sutherland et al., 2018). The advent of bacterial tri-helical peptides, similar to collagen is found to be more stable at higher temperature and biocompatible. Their large scale production is easy. Thus the bacterial collagen is finding more relevance in biomedical application (Yu et al., 2014; An et al., 2014).

Table 2. Collagen mimic protein gene/s found in various bacteria.

Stable triple-helix like collagen from Bacteria and their respective Genes		
S. No.	Bacterial Genus	Genes
1	*Streptococcus pyogenes*	SclA/Scl1 gene
2	*Bacillus anthracis*	BclA gene
3	*Legionella pneumophila*	Lcl gene
4	*Clostridium perfringens*	
5	*Solibacter usitatus*	
6	*Streptococcus pyogenes*	SclB/Scl2 gene
7	*Rhodopseudomonas palustri*	

7. Collagen Composites

Collagen has proved as one of the best natural biomaterial for various uses in medical field. Collagen being easily available, inert, acceptable to biological systems and of fibrillar nature makes it a highly valued biomaterial. It can be processed into sheets, dispersion solutions, injectable solutions of viscous nature, tubes, sponges, foams or nano-fibrous powders that renders it as a choice for use in biomaterial (Sionkowska and Kozłowska, 2010). Collagen forms various composites with bioceramics like glass, HP, etc., Carbons, polymers and other gives rise to structures to be used in biomedical fields. The various techniques utilized to study the composites formed from collagen includes FR-IR (Fourier Transform Infrared Spectroscopy), SEM (Scanning Electron Microscopy) and XRD (X-ray Diffraction) (John et al., 2001).

A. Collagen Composites with HA {Hydroxyapatite [Ca10(PO4)6(OH)2]}:

Collagen in its self is a fragile component with negligible mechanical strength. Adding a product as HA (Hydroxyapatite) to collagen improves the mechanical strength, stability and surface area for cellular attachment required in tissue engineering (Geiger et al., 2003; Yunus et al., 2015). It has been noted by various researchers that Composites of HA with collagen are influenced by concentration of collagen and HA, method adopted and the crosslinking method used for the composites (Calabrese et al., 2016). These composites are further improved if instead of micro-sized HA, nano-sized HA particles are used, which increase the surface area for cell adhesion. The HA directly forms chemical bond with bone tissue therefore Collagen-HA composites are highly successful for bone tissue engineering. Research has proved that Collagen-HA composites enhance bioactivity on rough surfaces as bones. Also the high ability of Collagen-HA composites to bind bones and the osteoconductivity makes it's a choice material for bone tissue engineering (Lickorish et al., 2004). To prepare these scaffolds Direct Blending, SBF immersion and co-precipitation methods have been used.

B. Collagen Composites with Bioactive Glass (BG):

Bioactive glass is ceramics which are crystal-glass nanocomposites which have one or more crystalline phases with controlled size and morphology in a glass matrix and are made through controlled crystallization of glasses. BG shows biocompatibility and easily bonds with hard or soft tissues rendering it to be useful biomaterial. The nano-particles of BG have shown to improve bone and teeth mineralization. BG is also known to be antibacterial as it releases ions like P, Ca and Si which causes pH and osmotic pressure changes at implantation sites. Since this feature is unrelated to bacterial resistance, its efficacy in TE is further enhanced. The commercial product Bioglass 45S5® is already verified for angiogenesis (Leu et al., 2008; Wu et al., 2013). The collagen-BG composite is expected to and have proved efficient in TE owing to osteoconductive, osteoinductive, angigenic, biocompatible and microbial resistant nature of the composites. Also, nanosized particle improves the mechanical strength and bone regenerativity (Gorustovich et al., 2010; Izquierdo-barba et al., 2012).

Another silicate, Wollasonite is also found equally appropriate in TE to form composites with collagen. Its also similar to BG as releases Si and Ca ions for microbial safety and osteogenesis. Studies shows that it improves mechanical strength of collagen scaffolds (Li and Chang, 2013; Wang et al., 2013).

C. Collagen Composites with Carbon Based Components:

i. CNTs (Carbon Nano Tubes)

Carbon, a non-metallic inorganic compound, special in physical characteristics due to its atomic configuration framework which differs in its various forms (Wang et al., 2020). The nano range allotrops of C includes carbon black (CB), carbon fiber, graphene carbon nanotubes (CNTs), and fullerene. Graphene is a 2-D framework of C-nanoparticles. When rolled it becomes CNT-Carbon Nano Tubes (De Volder et al., 2013). Owing to their unique properties, CNTs are finding great opportunities in the field of TE. They have exceptional mechanical, electrical, thermal and optical potential to be harnessed in various fields. The length, dimension and functional groups of CNTs are available at desired format to meet the needs of TE. The CNTs occur in 2 forms-

a. SWCNT (Single Walled CNT) - Formed by rolling of Graphene sheet. Has a diameter of 0.4–5.0 Nanometers. They have great tensile strength and electrical capabilities.

b. MWCNT (Multi Walled CNT) – Formed of multiple SWCNT nested inside each other to form a 1–20 nm diameter tubular structure. They are highly resistant to heat and chemically inert. They make surfaces corrosion free when layered above them, making them harder and durable. They have tremendous tensile strength. These strong attributes make them a choice material in TE (Hada et al., 2023).

CNTs are reported to adhere to cell surfaces strongly hence are biocompatible also. Their extraordinary tensile strength, durability, inertness, thermal stability, biocompatibility and various conformations are utilizable appropriately in TE. Their scaffolds with collagen are found encouraging in bone repair and regeneration. Collagen composites with CNT fabricated with Freeze drying method have given notable results in TE of bones (Sheng et al., 2015; Türk et al., 2018).

ii. GO (Graphene Oxide)

Graphene, pure carbon product of nano range has honey comb like lattice structure (Bahrami et al., 2021). Graphene forms a monolayer. The chemical oxidation of graphite/graphene gives GO (Graphene Oxide). The oxygen functional group, when reduced gives rGO (reduced Graphene Oxide). Both GO and rGO are utilized in electronics, photonics and biomedicenes (López-Diaz et al., 2020). GO has large surface area, great dispersal ability and hydrophilicity. Along with this, it shows biocompatibility. Its collagen scaffolds are non-cytotoxic so easily used in TE. The oxygen group in Go provides it aqueous dispersibilty that makes it good material for biosensing, drug delivery vector, bioimaging and TE (López-Diaz et al., 2020). GO is reported to improve bio-interaction with functional groups like carboxylic, phenol hydroxyl and epoxide present on tissue surfaces, resulting in enhanced adherence and prolifertation of tissues. Collagen-Go scaffolds are reported to accelerate wound healing and promoting osteogenic differentiation (Bahrami et al., 2021).

D. Collagen Composites with Polymers:

Tissue engineering is highly dependent on scaffold and its properties. Numerous synthetic components which are biodegradable, immunological reaction-less and flexible to be processed are used and investigated for TE (BaoLin and Ma 2014).

I. Natural Polymers

Collagen with naturally occurring polymers that are part of the extra cellular matrix, like glycosaminoglycans or silk fibroin have given encouraging results in TE. Glycosaminoglycans (GAGs) are negatively charged polymer of disaccharides occurring in extracellular matrix of all mammalian tissues. These molecules adorned with the capabilities of biocompatibility, low immunogenicity, biodegradability and high resorbability due to which they become ideal for biomedical applications (Mathew-Steiner et al., 2021). The four main GAGs used are heparin/heparan sulfate, chondroitin sulfate/dermatan sulfate, keratan sulfate, and hyaluronic acid. These GAGs are important part of cellular matrix and responsible for binding with elicitors of growth like cytokines and GF (Growth Factors). Studies suggest that Collagen-GAGs (CG) scaffolds amplify the proliferation of cells as well as their migration. The CGs have shown stable matrix formation for improved osteo-differentiation and regeneration. Due to these properties CG is finding increased use in skin, cartilage, tendon and nerve tissue engineering. Nanoparticulate mineralzed CG Composites were found to be highly effective in healing of cranial defects (Liv et al., 2015).

Another naturally occurring polymer rich in serine and glycine is the silk fibrion (SF), a protein with remarkable mechanical properties, hydrophobicity, biocompatibility, biodegradability and bioresorbability suitable for TE applications.

It represents the protein part of the silk thread. It can be easily processed into fibre, mats, meshes, hydrogel or sponge using different fabrication techniques like spin coating, electrospinning, freeze drying, physical, and chemical cross-linking or 3-D printing. SF with collagen has already been used and found significant in osteogenesis and neural recoveries. They are found to be advantageous for cell proliferation, differentiation, migration and adhesion in many studies investigated for bone, cartilage, ligament, tendon, skin, wound healing and tympanic membrane regeneration (Sun et al., 2021).

II. Artificial Polymers

Biodegradable polymers synthesized chemically, like polycaprolactone (PCL), polyglycolic acid (PGA), polylactic acid (PLA), polyvinyl alcohol (PVA), polyethylene terephthalate (PET), etc. have also been used to form diverse scaffolds with collagen for TE (Ferreira et al., 2012).

PGA is linear aliphatic polyester polymer which is a biodegradable thermoplastic. It is known to be forming strong fibers. Studies suggest that scaffold of PGA and collagen are influenced by weight ratio of the two. PGA is found to loosen its mechanical integrity in 2–4 weeks and developed into synthetic absorbable suture. Also it has been fabricated into numerous nonwoven fabrics for TE (Hiraoka et al., 2023). The Collagen-PGA sponges were found to enhance the initial attachment, proliferation rate and osteogenic differentiation of the bone marrow-MSCs (BM-MSC) (Toosi et al., 2016).

PLA is a biodegradable thermoplastic polymer of lactic acid. It is a biocompatible chemically synthesized polymer which is mechanically stable, versatile and processed easily. These properties make it a choice material for tissue engineering. However, for its slow degradation, limited bioactivity and low hydrophilicity, it is combined with other polymers which can overcome these shortcomings for TE. Collagen-PLA (COL-PLA) scaffolds are one such composite which have been used in engineering of various tissues. Studies found that COL-PLA scaffolds were highly successful in bone TE due to their high biocompatibility, osteoconductivity and similarity to bone mineral phase leading to quick recovery (Cárdenas-Aguazaco et al., 2023). Woven COL-PLA scaffold showed high potential in tendon repair (Xie et al., 2022). Electrospun COL-PLA scaffold also proved excellent in nerve tissue repair (Aseer et al., 2024).

PVA polyvinyl alcohol is an odorless, colorless synthetic polymer forming biocompatible hydrogels which find extensive use in medical applications. It has low protein adhesion and low toxicity. Lan et al. (2020) found that hydrogels prepared from Collagen and PVA were highly effective in cartilage tissue engineering and regeneration. Similarly, Leng et al. (2020) prepared a composite film from Collagen and PVA, activated with curcumin nanoparticles for wound dressing and found it capable in wound management and rapid healing of the wound. Marin et al. (2018) also recommended Collagen-PVA sponges for wound healing and skin grafts. Peng et al. (2011) tested hemolysis, anaphylaxis, pyrogen and acute systemic toxicity for *in vivo* implants and found hydrogel prepared from collagen-PVA by cyclic freeze-drying promising. Bai et al. (2020) found tannic acid to improve the mechanical strength and bioactivity of Collagen-PVA hydrogels.

PET - polyethylene terephthalate is thermoplastic polymer resin used in fibres, containers and thermoforming. It has been used in tissue Engineering to prepare scaffolds along with collagen.

PCL is a semicrystalline polymer known for its high solubility and ability to form blends, low melting point and biocompatibility, thus fulfill characteristics of a good biomaterial. It is a hydrophobic polymer that degrades very slowly (up to 1 year) and hence aptly combined with collagen for TE (Mateos-Timoneda et al., 2009). Wang et al. (2020) reported rapid bone regeneration and promontory effect on osteogenesis from scaffold prepared from Collagen-Hyroxoapatite and PCL. Weisgerber et al. (2016) found high increase in elastic modulus when PCL was added to composites of collagen and glycosaminoglycan. Different Collagen composites are tabulated in Table 3.

Table 3. Various collagen composites components and their application/s in TE.

S. No.	Contributor to Scaffold with Collagen	Used for tissue engineering	Reference
1	Hydroxyapatite	Bone	Geiger et al., 2003; Yunus et al., 2015
2	Bioglass	Teeth	Denise et al., 2023
3	Carbon Nano Tubes	Bone	Wang et al., 2020
4	Graphine Oxide	Bone and Wound Healing	Bahrami et al., 2021
5	Glycosaminoglycans	Cranial Tissue	Kalbitze et al., 2015
6	Silk Fibrion	Bone, cartilage, ligament, tendon, skin and tympanic membrane regeneration	Weizhen et al., 2021
7	Polycaprolactone (PCL)	Bone, Skin	Mateos-Timoneda et al., 2009
			Wang et al., 2020
8	Polyglycolic acid (PGA)	Bone and Cartilage	Tripathi et al., 2023
			Toosi et al., 2016
9	Wollasonite	Bone	Li et al., 2013
			Wang et al., 2013
10	PVA (Polyvinyl alcohol)	Cartilage	Wei wei et al. (2020)
		Skin Graft	Qing Qing et al. (2020)
		Wound Healing	Peng et al. (2018)
			Marin et al. (2018)

8. Fabrication Techniques

The repair, replacement or regeneration of damaged or lost tissue using scaffolds, biomaterial or bioactive factors which mimic the original is the basic approach of tissue engineering. The biocompatibility, biodegradability, cell surface attachment, mineral transport, porosity, mechanical strength and the structure of scaffold should meet the basic requirement of particular tissues. The structure of scaffold with higher surface area, interconnectivity and porosity are detrimental aspects of tissue engineering. Other than contributing substitutes, the method of fabrication also influences the scaffolds. The conventional fabrication techniques involves electrospinning, freeze drying, gas foaming while the laser sintering and 3-D printing are the recent developed technique for fabricating tissue scaffolds (Shahrzad et al., 2022). These techniques have been used for preparing collagen based scaffolds. Here, 3 techniques that are frequently used for scaffold preparation are discussed.

A. Electrospinning

As the name suggests, electrospinning is the electrohydrodynamic process generating nano fibers or particle from a liquid phase (polymer solution) under electric current. A reservoir (typically a syringe), a pump, a high voltage power source and a collector are the various parts of a electrospinner. The polymer solution is filled in the reservoir. To prepare collagen scaffolds, various composites of collagen with natural or artificial polymer at requisite concentrations are prepared to match that of the desired native tissue. Through electrospinning, different shapes, porosity and structure matrices or fibers can be constructed from collagen and components. Also, spinning parameters like solvent-solute concentration, needle-ground distance or the applied strength of current can be used to vary the scaffold properties so that versatile architecture scaffolds can be formed. Along with this, mechanical strength and resistance of scaffolds can be fine tuned to requirement by cross-linking for *in vivo/in vitro* studies. A variety of cross linkers have been used in production of collagen composites (Figure 7).

It is observed that collagen electrospun scaffold degrade fast than natural collagen and has lower mechanical properties, thus limiting the use of collagen scaffolds for *in vitro* culture or implantation. Hence, crosslinking electrospun collagen using polymers or cabodimides, etc. can highly improve the scaffold properties and their applications. When 2/more components are used to prepare scaffolds, co-spinning, coaxial electrospinning, layering or coating techniques are used to further enhance the scaffold properties for tissue engineering (Blackstone et al., 2021). Jelena et al. (2012) prepared synthetic human elastin for dermal tissue engineering by electrospun collagen composite scaffold. Meng et al. (2012) reported accelerated cell adhesion and growth of NIH3T3 cells with electrospun Collagen-PHVB [poly(3-hydroxybutyrate-co-3-hydroxyvalerate)] nanofibre scaffold. Wu et al. (2018) found electrospun collagen-PVP scaffold encouraging for eye corneal tissue engineering. Blackstone et al. (2021) in a review revealed that electrospun collagen based scaffolds were aptly similar to native tissues engineered for *in vitro* and *in vivo* studies.

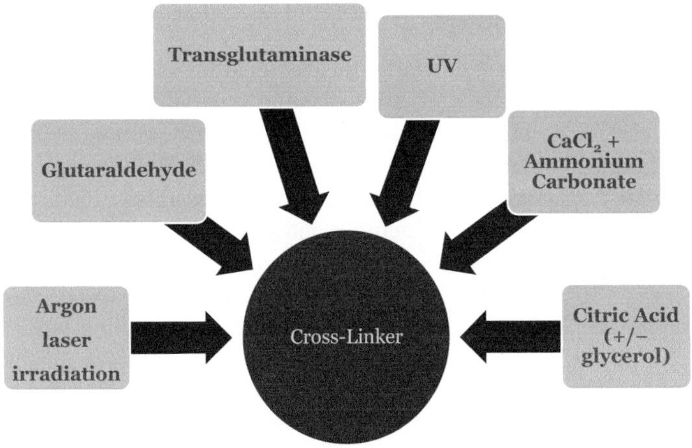

Figure 7. Various cross-linkers used in electrospinning of collagen scaffolds.

B. Freeze-Drying

The method of evaporating ice is literal meaning of freeze drying. In this process residual solvent is removed to produce dry powder. It is highly recommended process for fabrication, when components susceptible to temperature are involved. Scaffolds prepared using this technique imitate structure, mechanical and bioactivite properties of the resident tissue hence collagen based scaffolds for implants and replacement of tissues have been prepared by freeze drying technique (Mogoşanu et al., 2016). Yoon et al. (2005) enhanced osteoblastic cell responses to freeze dried Collagen-Fluorapatite composites. Lowe et al. (2016) found collagen scaffolds prepared from freeze drying aligned for regeneration and growth of rat dermal fibroblasts and axons emerging from chick dorsal root ganglia explants. Cai and Yang (2011) showed better cytocompatibility with 3T3 fibroblast cells on bacterial cellulose (BC)/collagen composites prepared by freeze drying.

C. 3D Bioprinting

Printing computer assisted scaffolds in 3-Dimensions using biomaterials in which material is deposited, joined or solidified layer by layer is 3D bioprinting. This automated technique permits wide architectural complex designing as per needs of tissue engineering effectively (K Prem and Naidu, 2024). 3D bioprinting is considered superior over other techniques as it can vary pore size, pore interconnectivity in a desired architectural frame of scaffolds as per need. Also, cost of fabrication is lowered using 3D bioprinting. Integrating living cells and growth factors during fabrication are the other noteworthy advantages of this technique (Chuang et al., 2023). Collagen and chitosan composites prepared by 3D printing exhibited vigorous cell growth on the surface of scaffolds (Suo et al., 2021). Zengh et al. (2024) studied the 3D printing of collagen composite materials enabling improved personalized customization of scaffolds.

9. Conclusions and Future Perspectives

Collagen molecule though forms an integral part of extra cellular matrix and governs lots of physio-chemical transformations, using it solo in today's technically advanced era for tissue engineering seems impractical. It lacks the requisite mechanical and bioactive properties to meet the needs of current tissue engineering. Wound healing, tissue regeneration or replacements are the major fields where Collagen composites are finding their huge applications. The future for collagen based scaffolds is promising and new arenas needs to be unfolded for its applications. Reducing cost of production, easy accessibility and betterment in application is being looked forward for collagen based products. Improving biocompatibility and bioactivity also needs attention. Also Regenerating whole lost organ using collagen composites is a major thrust area for the future research. Combing TE with Gene Therapy is also looked upon as the application of collagen based scaffolds. The collagen based composites are serving mankind in their best known way currently and further research in these is expected to uplift humanity to another level.

References

Al-Atif, H. 2022. Collagen supplements for aging and wrinkles: a paradigm shift in the fields of dermatology and cosmetics. Dermatology Practical and Conceptual 12(1): e2022018. https://doi.org/10.5826/dpc.1201a18.

Al-Nimry, S., A. A. Dayah, I. Hasan and R. Daghmash. 2021. Cosmetic, biomedical and pharmaceutical applications of fish gelatin/hydrolysates. Marine Drugs 19(3): 145. https://doi.org/10.3390/md19030145.

An, B., D. L. Kaplan and B. Brodsky. 2014. Engineered recombinant bacterial collagen as an alternative collagen-based biomaterial for tissue engineering. Frontiers in Chemistry 2: 40. https://doi.org/10.3389/fchem.2014.00040.

Aseer, M., M. Taleb, Z. Arabpour, Q. A. Bhatti and H. Ghanbari. 2024. Potential of collagen/PLA-based nanofibrous scaffold to support PC12 cells and neural repair. Journal of Contemporary Medical Sciences 10(1). https://doi.org/10.22317/jcms.v10i1.1493.

Bahrami, S., N. Baheiraei and M. Shahrezaee. 2021. Biomimetic reduced graphene oxide coated collagen scaffold for *in situ* bone regeneration. Sci Rep 11: 16783. https://doi.org/10.1038/s41598-021-96271-1.

Bai, Z., T. Wang, X. Zheng, Y. Huang, Y. Chen and W. Dan. 2020. High strength and bioactivity polyvinyl alcohol/collagen composite hydrogel with tannic acid as cross-linker. Polymer Engineering and Science 61. 10.1002/pen.25574. https://doi.org/10.1002/pen.25574.

BaoLin, G. and P. X. Ma. 2014. Synthetic biodegradable functional polymers for tissue engineering: a brief review. Science China. Chemistry 57(4): 490–500. https://doi.org/10.1007/s11426-014-5086-y.

Barbara, K., K. Agnieszka and K. Joanna. 2020. Biologically inspired collagen/apatite composite biomaterials for potential use in bone tissue regeneration—a review. Materials 13(1748): 1–17. doi:10.3390/ma13071748.

Blackstone, B. N., S. C. Gallentine and H. M. Powell. 2021. Collagen-based electrospun materials for tissue engineering: a systematic review in Bioengineering 8(3): 39. https://doi.org/10.3390/bioengineering8030039.

Brodsky, B. and J. A. Ramshaw. 1997. The collagen triple-helix structure. Matrix biology. Journal of the International Society for Matrix Biology 15(8-9): 545–554. https://doi.org/10.1016/s0945-053x(97)90030-5.

Brodsky, B. and D. L. Kaplan. 2013. Shining light on collagen: expressing collagen in plants. Tissue Eng Part A 19(13-14): 1499–501. doi: 10.1089/ten.TEA.2013.0137.

Bulleid, N. J., D. C. John and K. E. Kadler. 2000. Recombinant expression systems for the production of collagen. Biochemical Society Transactions 28(4): 350–353.

Cai, Z. and G. Yang. 2011. Bacterial cellulose/collagen composite: Characterization and first evaluation of cytocompatibility. J Appl Polym Sci. 120(5): 2938–294412. https://doi.org/10.1002/app.33318.

Calabrese, G., R. Giuffrida, C. Fabbi, E. Figallo, D. Lo Furno, R. Gulino et al. 2016. Collagen-hydroxyapatite scaffolds induce human adipose derived stem cells osteogenic differentiation *in vitro*. PloS One 11(3): e0151181. https://doi.org/10.1371/journal.pone.0151181.

Caliari, S. and J. Burdick. 2016. A practical guide to hydrogels for cell culture. Nat Methods 13: 405–414. https://doi.org/10.1038/nmeth.3839.

Cárdenas-Aguazaco, W., B. Camacho, E. Y. Gómez-Pachón, A. L. Lara-Bertrand and I. Silva-Cote. 2023. Electrospun scaffolds of polylactic acid, collagen, and amorphous calcium phosphate for bone repair. Pharmaceutics 15(11): 2529. https://doi.org/10.3390/pharmaceutics15112529.

Chapekar, M. S. 2000. Tissue engineering: Challenges and opportunities. J. Biomed. Mater. Res. 53: 617–620. https://doi.org/10.1002/1097-4636(2000)53:6<617::AID-JBM1>3.0.CO;2-C.

Chua, P. and W. K. Lim. 2023. The strategic uses of collagen in adherent cell cultures. Cell Biology International 47(2): 367–373. https://doi.org/10.1002/cbin.11966.

Chuang, G., W. Jiacheng, Z. Yiming and L. Hong. 2023. Construction of 3D bioprinting of HAP/collagen scaffold in gelation bath for bone tissue engineering. Regenerative Biomaterials 10: rbad067. https://doi.org/10.1093/rb/rbad067.

Colin G. Scanes, 2018. Chapter 2 - Animal Attributes Exploited by Humans (Nonfood Uses of Animals). pp. 13–40. Editor(s): Colin, G., S. Scanes and R. Toukhsati. Animals and Human Society. Academic Press, ISBN 9780128052471. https://doi.org/10.1016/B978-0-12-805247-1.00002-2.

De Volder, M. F., S. H. Tawfick, R. H. Baughman and A. J. Hart. 2013. Carbon nanotubes: present and future commercial applications. Science (New York) 339(6119): 535–539. https://doi.org/10.1126/science.1222453.

Deane, C. S., J. J. Bass, H. Crossland, B. E. Phillips and P. J. Atherton. 2020. Animal, plant, collagen and blended dietary proteins: effects on musculoskeletal outcomes. Nutrients 12(9): 2670. https://doi.org/10.3390/nu12092670.

Ferreira, A. M., P. Gentile, V. Chiono and G. Ciardelli. 2012. Collagen for bone tissue regeneration. Acta Biomaterialia 8(9): 3191–3200. https://doi.org/10.1016/j.actbio.2012.06.014.

Fertala, A. 2020. Three decades of research on recombinant collagens: reinventing the wheel or developing new biomedical products? Bioengineering 7(4): 155. https://doi.org/10.3390/bioengineering7040155.

Furtado, M., L. Chen, Z. Chen, A. Chen and W. Cui. 2022. Development of fish collagen in tissue regeneration and drug delivery. Engineered Regeneration 3(3): 217–231. ISSN 2666-1381. https://doi.org/10.1016/j.engreg.2022.05.002.

Gál, R., P. Mokrejš, P. Mrázek, J. Pavlačková, D. Janáčová and J. Orsavová. 2020. Chicken heads as a promising by-product for preparation of food gelatins. Molecules (Basel, Switzerland) 25(3): 494. https://doi.org/10.3390/molecules25030494.

Gao, C., S. Peng, P. Feng, and C. Shuai. 2017. Bone biomaterials and interactions with stem cells. Bone Research 5: 17059. https://doi.org/10.1038/boneres.2017.59.

Geahchan, S., P. Baharlouei and A. Rahman. 2022. Marine collagen: a promising biomaterial for wound healing, skin anti-aging, and bone regeneration. Marine Drugs 20(1): 61. https://doi.org/10.3390/md20010061.

Geiger, M., R. H. Li and W. Friess. 2003. Collagen sponges for bone regeneration with rhBMP-2. Advanced Drug Delivery Reviews 55(12): 1613–1629. https://doi.org/10.1016/j.addr.2003.08.010.

Gelse, K., E. Pöschl and T. Aigner. 2003. Collagens—structure, function, and biosynthesis. Advanced Drug Delivery Reviews 55(12): 1531–1546. https://doi.org/10.1016/j.addr.2003.08.002.

Gopalakrishnan, S., A. Gupta, D. Matuwana, J. J. Amante, J. Kearney Cathal and M. Rotello Vincent. 2022. *In vitro* cell culture models for ultrasound treatments using collagen-based scaffolds. MethodsX 9: 101909, ISSN 2215-0161. https://doi.org/10.1016/j.mex.2022.101909.

Gorustovich, A. A., J. A. Roether and A. R. Boccaccini. 2010. Effect of bioactive glasses on angiogenesis: a review of *in vitro* and *in vivo* evidences. Tissue Engineering-Part B Reviews 16(2): 199–207. https://doi.org/10.1089/ten.TEB.2009.0416.

Gustavo, A. R., B. Sara, M. Miguelangel, B. José and V. Rick 2021. Collagen Type I biomaterials as scaffolds for bone tissue engineering. Polymers 13(4): 599. https://doi.org/10.3390/polym13040599.

Hada, V., K. Chaturvedi, A. Singhwane, N. Siraj, A. Gupta, N. Sathish et al. 2023. Nanoantibiotic effect of carbon-based nanocomposites: epicentric on graphene, carbon nanotubes and fullerene composites: a review. 3 Biotech 13(5): 147. https://doi.org/10.1007/s13205-023-03552-9.

Han, F., J. Wang, L. Ding, Y. Hu, W. Li, Z. Yuan et al. 2020. Tissue engineering and regenerative medicine: achievements, future, and sustainability in Asia. Frontiers in Bioengineering and Biotechnology 8: 83. https://doi.org/10.3389/fbioe.2020.00083.

Hashim, P., M. Ridzwan, J. Bakar and D. Hashim. 2015. Collagen in food and beverage industries. International Food Research Journal 22: 1–8.

Hiraoka, Y., Y. Kimura, H. Ueda and Y. Tabata. 2003. Fabrication and biocompatibility of collagen sponge reinforced with poly(glycolic acid) fiber. Tissue Engineering 9(6): 1101–1112. https://doi.org/10.1089/10763270360728017.

Hu, J., J. Li, J. Jiang, L. Wang, J. Roth, K. N. McGuinness et al. 2022. Design of synthetic collagens that assemble into supramolecular banded fibers as a functional biomaterial testbed. Nature Communications 13(1): 6761. https://doi.org/10.1038/s41467-022-34127-6.

Ikada, Yoshito. 2006. Challenges in tissue engineering. Journal of the Royal Society Interface 3.10: 589–601. https://doi.org/10.1098/rsif.2006.0124.

Izquierdo-barba, I., D. Arco, Y. Sakamoto, O. Terasaki, A. Lopez-Noriega and Vallet-Regí, M. 2008. High-performance mesoporous bioceramics mimicking bone mineralization. Chemistry of Materials 20. DOI:10.1021/cm800172x.

Jadach, B., Z. Mielcarek and T. Osmałek. 2024. Use of collagen in cosmetic products. Current Issues in Molecular Biology 46(3): 2043–2070. https://doi.org/10.3390/cimb46030132.

Jelena, R., G. W. Steven, L. Zhe, K. M. M. Peter, J. Y. Cara, W. Yiwei et al. 2012. Electrospun synthetic human elastin:collagen composite scaffolds for dermal tissue engineering. Acta Biomaterialia 8(10): 3714–3722. https://doi.org/10.1016/j.actbio.2012.06.032.

John, A., L. Hong, Y. Ikada and Y. Tabata. 2001. A trial to prepare biodegradable collagen–hydroxyapatite composites for bone repair. Journal of Biomaterials Science, Polymer Edition 12(6): 689–705. https://doi.org/10.1163/156856201316883485.

Justin, A. W., F. Cammarata, A. A. Guy, S. R. Estevez, S. Burgess, H. Davaapil et al. 2023. Densified collagen tubular grafts for human tissue replacement and disease modelling applications. Biomaterials Advances 145: 213245. https://doi.org/10.1016/j.bioadv.2022.213245.

K. Prem A. and D. J. Naidu. 2024. A comprehensive review of 3D printing techniques for biomaterial-based scaffold fabrication in bone tissue engineering. Annals of 3D Printed Medicine 13: 100141. ISSN 2666-9641. https://doi.org/10.1016/j.stlm.2023.100141.

Lan, W., M. Xu, X. Zhang, L. Zhao, D. Huang, X. Wei et al. 2020. Biomimetic polyvinyl alcohol/type II collagen hydrogels for cartilage tissue engineering. Journal of Biomaterials Science. Polymer Edition 31(9): 1179–1198. https://doi.org/10.1080/09205063.2020.1747184.

Leng, Q., Y. Li, X. Pang, B. Wang, Z. Wu, Y. Lu et al. 2020. Curcumin nanoparticles incorporated in PVA/collagen composite films promote wound healing. Drug Delivery 27(1): 1676–1685. https://doi.org/10.1080/10717544.2020.1853280.

León-López, A., A. Morales-Peñaloza, V. M. Martínez-Juárez, A. Vargas-Torres, D. I. Zeugolis and G. Aguirre-Álvarez. 2019. Hydrolyzed collagen-sources and applications. Molecules (Basel, Switzerland) 24(22): 4031. https://doi.org/10.3390/molecules24224031.

Leu, A. and J. K. Leach. 2008. Proangiogenic potential of a collagen/bioactive glass substrate. Pharmaceutical Research 25(5): 1222–1229. https://doi.org/10.1007/s11095-007-9508-9.

Li, H. and J. Chang. 2013. Stimulation of proangiogenesis by calcium silicate bioactive ceramic. Acta Biomaterialia 9(2): 5379–5389. https://doi.org/10.1016/j.actbio.2012.10.019.

Lickorish, D., J. A. Ramshaw, J. A. Werkmeister, V. Glattauer and C. R. Howlett. 2004. Collagen-hydroxyapatite composite prepared by biomimetic process. Journal of Biomedical Materials Research. Part A 68(1): 19–27. https://doi.org/10.1002/jbm.a.20031.

Lin, C. R., S. H. L. Tsai, K. Y. Huang, P. A. Tsai, H. Chou and S. H. Chang. 2023. Analgesic efficacy of collagen peptide in knee osteoarthritis: a meta-analysis of randomized controlled trials. Journal of Orthopaedic Surgery and Research 18(1): 694. https://doi.org/10.1186/s13018-023-04182-w.

Liu, D., M. Nikoo, G. Boran, P. Zhou and J. M. Regenstein. 2015. Collagen and gelatin. Annual Review of Food Science and Technology 6: 527–557. https://doi.org/10.1146/annurev-food-031414-111800.

Liv, K., F. Katja, M. Stephanie, S. Matthias and P. Tilo. 2015. Glycosaminoglycan functionalization of mechanically and topologically defined collagen I matrices J. Mater. Chem. B. 3: 8902–8910. DOI https://doi.org/10.1039/C5TB01737H.

López-Diaz, D., M. D. Merchán and M. M. Velázquez. 2020. The behavior of graphene oxide trapped at the air water interface. Advances in Colloid and Interface Science 286: 102312. https://doi.org/10.1016/j.cis.2020.102312.

Lowe, C .J., I. M. Reucroft, M. C. Grota and D. I. Shreiber. 2016. Production of highly aligned collagen scaffolds by freeze-drying of self-assembled, fibrillar collagen gels. ACS Biomater Sci Eng. 2(4): 643–651. doi: 10.1021/acsbiomaterials.6b00036.

Luciano, B. 2023. Chapter 27 - Periodontal drug delivery: advances and challenges. pp. 751–782. *In*: Amit Kumar Nayak, Md Saquib Hasnain, Bibek Laha and Sabyasachi Maiti (eds.) . Advanced and Modern Approaches for Drug Delivery. Academic Press. ISBN 9780323916684. https://doi.org/10.1016/B978-0-323-91668-4.00019-8.

Luo, T. and K. L. Kiick. 2013. Collagen-like peptides and peptide-polymer conjugates in the design of assembled materials. European Polymer Journal 49(10): 2998–3009. https://doi.org/10.1016/j.eurpolymj.2013.05.013.

Makwana, P., S. I. Vadiya, S. Khetani and N. Mehta. 2022. Use of collagen sheets as graft material for myringoplasty. Indian Journal of Otolaryngology and Head and Neck Surgery 74(3): 4314–4318. https://doi.org/10.1007/s12070-021-02977-z.

Marin, Ș., M. Georgiana, A. Kaya, M. V. Ghica, C. D. Lăcrămioara, I. Denisa et al. 2018. Collagen-polyvinyl alcohol-indomethacin biohybrid matrices as wound dressings. Pharmaceutics 10(4): 224. https://doi.org/10.3390/pharmaceutics10040224.

Marques, C. F., G. S. Diogo, S. Pina, J. M. Oliveira, T. H. Silva and R. L. Reis. 2019. Collagen-based bioinks for hard tissue engineering applications: a comprehensive review. Journal of Materials Science. Materials in Medicine 30(3): 32. https://doi.org/10.1007/s10856-019-6234-x.

Martin, R. F. 2009. Skin surgery and minor procedures. Foreword. The Surgical clinics of North America 89(3): xiii–xv. https://doi.org/10.1016/j.suc.2009.04.001.

Martínez-Puig, D., E. Costa-Larrión, N. Rubio-Rodríguez and P. Gálvez-Martín. 2023. Collagen supplementation for joint health: the link between composition and scientific knowledge. Nutrients 15(6): 1332. https://doi.org/10.3390/nu15061332.

Mateos-Timoneda, M. A. 2009. Chapter 8 - Polymers for bone repair. pp. 231–251. *In*: Josep, A. Planell, Serena M. Best, Damien Lacroix and Antonio Merolli (eds.). Bone Repair Biomaterials. Woodhead Publishing (Series in Biomaterials). ISBN 9781845693855. https://doi.org/10.1533/9781845696610.2.231.

Mathew-Steiner, S. S., S. Roy and C. K. Sen. 2021. Collagen in wound healing. Bioengineering 8: 63. https://doi.org/10.3390/bioengineering8050063.

Meng, W., S. Y. Kim, J. Yuan, J. C. Kim, O. H. Kwon, Kawazoe, Naoki K. et al. 2007. Electrospun PHBV/collagen composite nanofibrous scaffolds for tissue engineering. Journal of Biomaterials Science, Polymer Edition 18(1): 81–94. https://doi.org/10.1163/156856207779146114.

Meyer, M. 2019. Processing of collagen based biomaterials and the resulting materials properties. Biomedical Engineering (online) 18(1): 24. https://doi.org/10.1186/s12938-019-0647-0.

Mogoşanu, G., A. Grumezescu, L. Mogoantă, L. Bejenaru and C. Bejenaru. 2016. Applications of nanobiopolymers for soft tissue engineering. pp. 83–109. *In*: Alexandru Mihai Grumezescu (ed.). Nanobiomaterials in Soft Tissue Engineering. William Andrew Publishing. ISBN 9780323428651 https://doi.org/10.1016/B978-0-323-42865-1.00004-0.

Movaniya, P. N., T. R. Makwana, N. N. Desai, K. G. Makwana and H. B. Patel. 2021. Efficacy of collagen membrane graft in intraoral surgery - an evaluative study. Annals of Maxillofacial Surgery 11(1): 42–48. https://doi.org/10.4103/ams.ams_192_20.

Olden, J., J. Vitule, J. Cucherousset and M. Kennard. 2020. There's more to fish than just food: exploring the diverse ways that fish contribute to human society. Fisheries 45(1): 453–464. https://doi.org/10.1002/fsh.10443.

Owczarzy Aleksandra, Kurasiński, Robert, Kuli Karolina, Rogóż Wojciech, Szkudlarek Agnieszka and Maciążek-Jurczyk Małgorzata. 2020. Collagen - structure, properties and application. Engineering of Biomaterials 156: 17–23. doi:10.34821/eng.biomat.156.2020.17-23.

Peng, Z., Z. Peng and Y. Shen. 2011. Study on biological safety of polyvinyl alcohol/collagen hydrogel as a tissue substitute (II). Journal of Macromolecular Science- Part A 48(8): 632–636. https://doi.org/10.1080/15226514.2011.586281.

Prafulla K. Chandra, Shay Soker and Anthony Atala. 2020. Chapter 1 - Tissue engineering: current status and future perspectives. pp. 1–35. Editor(s): Robert Lanza, Robert Langer, Joseph P. Vacanti and Anthony Atala, Principles of Tissue Engineering (Fifth Edition), Academic Press, ISBN 9780128184226, https://doi.org/10.1016/B978-0-12-818422-6.00004-6.

Ricard-Blum, S. 2011. The collagen family. Cold Spring Harbor Perspectives in Biology 3(1): a004978. https://doi.org/10.1101/cshperspect.a004978.

Sangeetha, K., A. V. Jisha Kumari, E. Radha and P. N. Sudha. 2019. Chapter 3-Pharmaceutical applications of collagen. *In*: Amit Kumar Nayak, Md Saquib Hasnai and Dilipkumar Pal (eds.). Natural Polymers for Pharmaceutical Applications: Volume 3: Animal-Derived Polymers (1st ed.). Apple Academic Press. https://doi.org/10.1201/9780429328350.

Schnieke, A., M. Dziadek, J. Bateman, T. Mascara, K. Harbers, R. Gelinas et al. 1987. Introduction of the human pro alpha 1(I) collagen gene into pro alpha 1(I)-deficient Mov-13 mouse cells leads to formation of functional mouse-human hybrid type I collagen. Proceedings of the National Academy of Sciences of the United States of America 84(3): 764–768. https://doi.org/10.1073/pnas.84.3.764.

Shahrzad, R., Mahshid Ma., M. Sanaz and K. Zahra. 2022. Chapter 7 - Polymer nanocomposites for biomedical applications. pp. 175–215. *In*: Ahmed Barhoum, Jaison Jeevanandam and Michael K. Danquah (eds.). Micro and Nano Technologies, Fundamentals of Bionanomaterials, Elsevier. ISBN 9780128241479. https://doi.org/10.1016/B978-0-12-824147-9.00007-8.

Sheng, C., L. Hejun, L. Kezhi, L. Jinhua and Z. Leilei. 2015. A dense and strong bonding collagen film for carbon/carbon composites. Applied Surface Science 347(30): 307–314. https://doi.org/10.1016/j.apsusc.2015.04.081.

Shoulders, M. D. and R. T. Raines. 2009. Collagen structure and stability. Annual Review of Biochemistry 78: 929–958. https://doi.org/10.1146/annurev.biochem.77.032207.120833.

Silvipriya, K. S., K. K. Kumar, A. R. Bhat, B. D. Kumar, A. John and P. Lakshmanan. 2015. Collagen: Animal Sources and Biomedical Application by Journal of Applied Pharmaceutical Science 5(03): 123–127. DOI: 10.7324/JAPS.2015.50322.

Sionkowska, A. and J. Kozłowska. 2010. Characterization of collagen/hydroxyapatite composite sponges as a potential bone substitute. International Journal of Biological Macromolecules 47(4): 483–487. https://doi.org/10.1016/j.ijbiomac.2010.07.002.

Sionkowska, A., A. Katarzyna, M. Katarzyna and G. Magdalena. 2020. Collagen based materials in cosmetic applications: a review. Materials 13(19): 4217. https://doi.org/10.3390/ma13194217.

Sokolakis, I., N. Pyrgidis and G. Hatzichristodoulou. 2022. The use of collagen fleece (TachoSil) as grafting material in the surgical treatment of Peyronie's disease. A comprehensive narrative review. Int J Impot Res 34: 260–268. https://doi.org/10.1038/s41443-020-00401-8.

Soliman, A. M., S. L. Teoh and S. Das. 2022. Fish gelatin: current nutritional, medicinal, tissue repair applications, and as a carrier of drug delivery. Current Pharmaceutical Design 28(12): 1019–1030. https://doi.org/10.2174/1381612828666220128103725.

Sorushanova, A., L. M. Delgado, Z. Wu, N. Shologu, A. Kshirsagar, R. Raghunath et al. 2019. The collagen suprafamily: from biosynthesis to advanced biomaterial development. Advanced Materials (Deerfield Beach, Fla.) 31(1): e1801651. https://doi.org/10.1002/adma.201801651.

Stein, H., M. Wilensky, Y. Tsafrir, M. Rosenthal, R. Amir, T. Avraham et al. 2009. Production of bioactive, post-translationally modified, heterotrimeric, human recombinant type-I collagen in transgenic tobacco. Biomacromolecules 10(9): 2640–2645. https://doi.org/10.1021/bm900571b.

Sun, W., D. A. Gregory, M. A. Tomeh and X. Zhao. 2021. Silk fibroin as a functional biomaterial for tissue engineering. International Journal of Molecular Sciences 22(3): 1499. https://doi.org/10.3390/ijms22031499.

Suo, H., J. Zhang, M. Xu and L. Wang. 2021. Low-temperature 3D printing of collagen and chitosan composite for tissue engineering. Materials Science and Engineering: C 123: 111963. http://doi.org/10.1016/j.msec.2021.111963.

Sutherland, T. D., M. G. Huson and T. D. Rapson. 2018. Rational design of new materials using recombinant structural proteins: Current state and future challenges. Journal of Structural Biology 201(1): 76–83. https://doi.org/10.1016/j.jsb.2017.10.012.

Toosi, S., H. Naderi-Meshkin, F. Kalalinia, M. T. Peivandi, H. HosseinKhani, A. R. Bahrami et al. 2016. PGA-incorporated collagen: Toward a biodegradable composite scaffold for bone-tissue engineering. Journal of Biomedical Materials Research. Part A 104(8): 2020–2028. https://doi.org/10.1002/jbm.a.35736.

Türk, S., I. Altınsoy, G. Çelebi Efe, M. Ipek, M. Özacar and C. Bindal. 2018. 3D porous collagen/functionalized multiwalled carbon nanotube/chitosan/hydroxyapatite composite scaffolds for bone tissue engineering. Materials Science and Engineering. C. Materials for Biological Applications 92: 757–768. https://doi.org/10.1016/j.msec.2018.07.020.

Valot, L., M. Maumus, L. Brunel, J. Martinez, M. Amblard, D. Noël et al. 2021. A collagen-mimetic organic-inorganic hydrogel for cartilage engineering. Gels (Basel, Switzerland) 7(2): 73. https://doi.org/10.3390/gels7020073.

Wang, X., F. Tang, Q. Cao, X. Qi, M. Pearson, M. Li et al. 2020. Comparative study of three carbon additives: carbon nanotubes, graphene, and fullerene-C60, for synthesizing enhanced polymer nanocomposites. Nanomaterials 10(5): 838. https://doi.org/10.3390/nano10050838.

Wang, C., K. Lin, J. Chang and J. Sun. 2013. Osteogenesis and angiogenesis induced by porous β-CaSiO(3)/PDLGA composite scaffold via activation of AMPK/ERK1/2 and PI3K/Akt pathways. Biomaterials 34(1): 64–77. https://doi.org/10.1016/j.biomaterials.2012.09.021.

Weisgerber, D. W., K. Erning, C. L. Flanagan, S. J. Hollister and B. A. C. Harley. 2016. Evaluation of multi-scale mineralized collagen-polycaprolactone composites for bone tissue engineering. Journal of the Mechanical Behavior of Biomedical Materials 61: 318–327. https://doi.org/10.1016/j.jmbbm.2016.03.032.

Wu, Z., B. Kong, R. Liu, W. Sun and S. Mi. 2018. Engineering of corneal tissue through an aligned PVA/collagen composite nanofibrous electrospun scaffold. Nanomaterials (Basel) 24;8(2): 124. doi: 10.3390/nano8020124.

Wu, C. and J. Chang. 2013. A review of bioactive silicate ceramics. Biomedical Materials (Bristol, England) 8(3): 032001. https://doi.org/10.1088/1748-6041/8/3/032001.

Xie, Y., F. Zhang, O. Akkus and M. W. King. 2022. A collagen/PLA hybrid scaffold supports tendon-derived cell growth for tendon repair and regeneration. Journal of Biomedical Materials Research. Part B, Applied Biomaterials 110(12): 2624–2635. https://doi.org/10.1002/jbm.b.35116.

Yamada, S., H. Nagaoka, M. Terajima, N. Tsuda, Y. Hayashi and M. Yamauchi. 2013. Effects of fish collagen peptides on collagen post-translational modifications and mineralization in an osteoblastic cell culture system. Dental Materials Journal 32(1): 88–95. https://doi.org/10.4012/dmj.2012-220.

Yamazaki, C. M., Y. Kadoya, K. Hozumi, H. Okano-Kosugi, S. Asada, K. Kitagawa et al. 2010. A collagen-mimetic triple helical supramolecule that evokes integrin-dependent cell responses. Biomaterials 31(7): 1925–1934. https://doi.org/10.1016/j.biomaterials.2009.10.014.

Yoon, B. H., H. W. Kim, S. H. Lee, C. J. Bae, Y. H. Koh, Y. M. Kong et al. 2005. Stability and cellular responses to fluorapatite-collagen composites. Biomaterials 26(16): 2957–2963. https://doi.org/10.1016/j.biomaterials.2004.07.062.

Yu, Z., B. An, J. A. Ramshaw and B. Brodsky. 2014. Bacterial collagen-like proteins that form triple-helical structures. Journal of Structural Biology 186(3): 451–461. https://doi.org/10.1016/j.jsb.2014.01.003.

Zhang, D., X. Wu, J. Chen and K. Lin. 2017. The development of collagen based composite scaffolds for bone regeneration. Bioactive Materials 3(1): 129–138. https://doi.org/10.1016/j.bioactmat.2017.08.004.

Zheng, L., N. Tseomashko, A. Voronova, A. Vasil'kov, X. Hu and X. Wang. 2024. Recent advances of collagen composite biomaterials for biomedical engineering: antibacterial functionalization and 3D-printed architecturalization. Collagen and Leather 6: 22. https://doi.org/10.1186/s42825-024-00164-8.

Zhou, N., Y.-D. Liu, Y. Zhang, T.-W. Gu and L.-H. Peng. 2023. Pharmacological functions, synthesis, and delivery progress for collagen as biodrug and biomaterial. Pharmaceutics 15(5): 1443. https://doi.org/10.3390/pharmaceutics15051443.

CHAPTER 5

Gelatin as a Biomaterial for Tissue Engineering

Aayushi Randhawa,[1,2] *Keya Ganguly*[1,3] and *Ki-Taek Lim*[1,2,3,*]

1. Introduction

The significant demand for tissue and organ replacements through innovative biological alternatives are generated through tissue-specific cells and by using natural or artificial polymeric substrates due to scarcity of tissues and organ donors (Kang, Tabata et al., 1999). Since collagen is the primary fibrous protein component of cartilage, bones, and skin, gelatin is a protein derivative derived from collagen through partial hydrolysis. Factors such as types and sources have an intrinsic influence on the characteristics of the gelatin (Gómez-Guillén, Giménez et al., 2011). The rheological characteristics of gelatin play a major role in determining its suitability for specific applications (Stainsby, 1987). The primary characteristics that best characterize the entire commercial standard of gelatin are its thermal stability and gel strength, aside from fundamental physical-chemical features, including solubility, structural parameters, transparency, odour, colour, and taste. The thermal stability and gel strength are predominantly influenced by the molecular characteristics of gelatin, particularly concerning two primary elements: (i) molecular weight distribution resulting from manufacturing conditions and (ii) species-determined amino acid content (Gómez-Guillén, Turnay et al., 2002; Gómez-Guillén, Giménez et al., 2011). The amount of amino acids, mainly proline, and hexaproline, are crucial for gelation. The durability of the triple helix structure in renatured gelatin correlates

[1] Department of Biosystems Engineering, Kangwon National University, Chuncheon-24341, Republic of Korea.
[2] Interdisciplinary Program in Smart Agriculture, Kangwon National University, Chuncheon-24341, Republic of Korea.
[3] Institute of Forest Science, Kangwon National University, Chuncheon-24341, Republic of Korea.
* Corresponding author: ktlim@kangwon.ac.kr

DOI: 10.1201/9781003518792-5

to the overall amount of pyrrolidine amino acids present since the nucleation region, considering the Pro+Hyp dense regions of the molecules, seems most likely to be associated with the development of the nucleation zone (Ledward, 1986). Generally, the gelatin contains a potent emulsifying ability and prevent soy milk, milk proteins from aggregating into gastric acid once they reach the stomach, and considered as advantageous for the breakdown of food. Type A and type B are the two primary varieties of gelatin that can be created depending on how collagen is pretreated. Acid treated gelatin, known as type A has an isoelectric point between pH 6 and 9. It is more frequently used to treat the least covalently crosslinked collagen present in pig skin. The more complicated collagen present in bovine skins treated with alkaline method with an isoeletric point of pH 5 is termed as type B gelatin (Lv, Huang et al., 2019). Collagen denaturation results in the production of low-intrinsic-viscosity gelatin solution, known as thermo-responsive gelation, which gels at a temperature lower than 37°C. The gradual and reversible process of gelatin gelation referred to as coil to helix transition, is ascribed to the significant amount of peptide bonds present in the macromolecular framework of the gelatin. Furthermore, the proline and 4-hydroxyproline content of the parent collagen influences the gelation temperature (Nikkhah, Akbari et al., 2016).

2. Characteristics of Gelation

Gelatin is composed of eighteen different types of complicated amino acids. Of these, proline, glycine, and hydroxyproline make up around 57% of the compound. The remaining 43% of the material is composed of other notable amino acids, including glutamic acid, arginine, alanine, and aspartic acid. In addition to having a variety of double and single unfolded hydrophilic chains, gelatin is made up of 50.5% carbon, 25.2% oxygen, 17% nitrogen, and 6.8% hydrogen. Additionally, chemically the gelatin is comprised of various polypeptide chains such as the a-chain consists of one polymer, the b-chain contains two covalently crosslinked chains, and the c-chain made of three covalently crosslinked chains (molar masses of about 90×10^3, 180×10^3, and 300×10^3 D) (Karim and Bhat, 2008; Manikandan, Thirupathi Kumara Raja et al., 2018). As an amphoteric protein, gelatin has an isoionic point that ranges from pH 5.0 to pH 9.0, contingent upon the source material and method of synthesis. In hot water, gelatin dissolves entirely, while it swells in cold water. The required dry gelatin structure must be released at a temperature higher than 60°C. When compared to other biopolymers, gelatin has particular set of characteristics. Initially, gelatin macromolecules have functional groups that are both basic and acidic (Figure 1a&b) (Bagal-Kestwal, Pan et al., 2019).

Additionally, at low temperatures, it possesses the ability to create a particular triple-stranded spiral shape that is missing in the majority of synthetic polymers. Several factors, including the presence of covalent connections, molecular weight, and concentration, influence how quickly gelatin forms its helical structure. Gelatin dissolves in various solutions of polyhydric alcohols like propylene glycol, glycerol, etc., but is insoluble in fewer polar organic solvents, including acetone, benzene, dimethylformamide, and primary alcohols. Additionally, it dissolves in strong polar organic solvents, notably formamide, trifluoroethanol, and acetic acid (Vold and

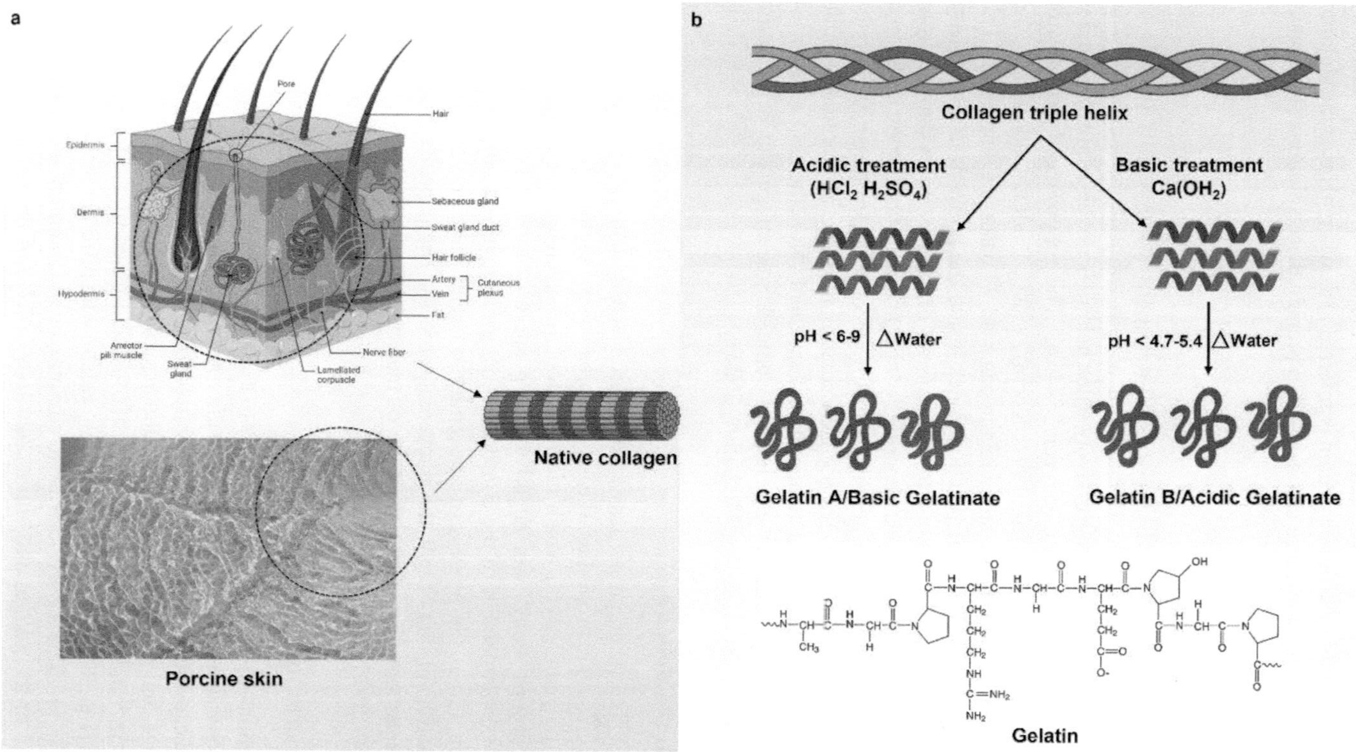

Figure 1. The preparation of gelatin. (a) Basic structure of collagen found in porcine skin. (b) Preparation and conversion of collagen triple helix structure into the gelatin polymer (Bagal-Kestwal, Pan et al., 2019). Created with Biorender.com.

Vold, 1983). When heated in an environment with a relative humidity of more than 60%, gelatin loses its capacity to dissolve and swell. Still, it possesses an extended shelf life in the absence of moisture or dry conditions. At low temperatures, liquid gelatin exhibits remarkable stability. H^+ and OH^- are the only non-colloidal ions present in gelatin solution. Isoionic gelatin is referred to as when pH is at the isoionic point. An ion exchange procedure can be used to prepare these solutions (Finch and Jobling, 1977).

2.1 Gelation Ability

Gelatin is slightly soluble in cold water, with a hypothetical solubility of 34% in water. The solubility of gelatin is enhanced as the temperature increases. The formation of a gel is initiated progressively when gelatin is mixed with water, and it starts to dissolve at a temperature between 35–45°C. The gel produced with gelatin is often thermoreversible (Keenan, 1994; Koepff, 1997). Undoubtedly, this is the most intriguing feature of the gelatin. Gelatin can hold 99 parts in a single part. The establishment and association of junction zones inside the network is a necessary step in the gel structure. In contrast to the majority of hydrocolloids derived from polysaccharides, gelatin forms gels, regardless of pH and does not require additional reactive agents. The solution of gelatin gradually becomes more dense as it cools, leading to a sol-to-gel transition. The sol-gel transition is known as thermoreversible gel because it is reversible and temperature-dependent. Nonetheless, a series of cooling and heating procedures can lead to some degree of gelatin degradation (Bagal-Kestwal, Pan et al., 2019; Mikhailov, 2023).

2.2 Thickness Capacity

The two most significant properties of gelatin are viscosity and gel strength. Gelatin possesses hygroscopic properties. It easily absorbs water and holds it in its structure. Gelatin is a polyampholyte, which means it contains uncharged hydrophobic and hydrophilic groups along with negative and positive charges. The selective and non-selective entanglement of the conformationally disorganized polymeric chains in the solvent allows the process of thickening to occur. The concentration should reach a certain level that begins the thickening process. The capacity of gelatin to thicken depends significantly on the type of material, temperature, and pH of the system (Smith, 1921; Howe, 2000).

2.3 Film-Forming Characteristics

When a solution of gelatin is applied thinly to a surface and then dried, it emerges as a film. This characteristic has been used in microencapsulation and the production of soft and hard capsules. Gelatin films can be used to create biosensors and immobilize cells, enzymes, and proteins. The gelatin film also plays an important role in various tissue regeneration, such as the development of gelatin-based films for cornea tissue regeneration (Liu, Ren et al., 2013). Blending gelatin with different biopolymers, such as chitosan, to produce a polyelectrolytic complex by the interaction of carboxylate ions ($-COO^-$) of the gelatin and ammonium ions ($-NH_3^+$) of the chitosan promotes the environment optimum for cell proliferation and further promote the organization of skin tissue. The gelatin films play a vital role in the biomedical field owing to its cost-effectiveness and mechanical properties (Yin, Yao et al., 1999).

3. 3D Printing Methods of Gelatin-Based Inks

Micro molding, the primary method of producing 3D objects made of gelatin, promotes a variety of cell functions. However, the ability to introduce conventional 3D design complexity, resolution, and precision into the fabrication process has made 3D printing a dominant technology (Hölzl, Lin et al., 2016; Alhaskawi, Zhou et al., 2024). A critical factor in the 3D printing of gelatin, especially for the extrusion printing technique, is the viscosity of the gel. The inherent temperature sensitivity of pure gelatin and its low viscosity at room temperature or above hamper the printing process (Catros, Guillotin et al., 2011). The gelatin-based ink can be modified due to the variable chemical and mechanical properties associated with various tissues and to simulate the tissue extracellular matrix (ECM) in nature. Optimization of concentrations, functionalization with different polymers, and the introduction of second phases in gelatin are a few of these modifications. These variables may also affect the properties of the gelatin inks, modifying their suitability for particular tissues (Waidi, Kariim et al., 2024).

3.1 Vat Photopolymerization

Vat photopolymerization refers to the process of additive manufacturing that produces the object of interest in a layer-by-layer manner using photo-induced polymerization of liquid photopolymers. Early in the 1980s, stereolithography (SLA) was invented as the first vat photopolymerization technique (Kodama, 1981). Later, in order to meet the demands of printing speed and resolution, a number of new vat photopolymerization techniques were generated that utilized the fundamental principles of SLA. These techniques include digital light processing (DLP), two-photon polymerization (TPP), continuous liquid interface production (CLIP), and computed axial lithography (CAL) (Lu, Mapili et al., 2006; Tumbleston, Shirvanyants et al., 2015; Kelly, Bhattacharya et al., 2019). All vat photopolymerization techniques, with the exception of CAL, share a common printing setup and procedure. With the exception of material condition and the light source, the printer used for vat photopolymerization is comparable to that of SLA. In accordance with the printer arrangement, the light source orientation and the build platform movement can be adjusted. In order to cure the photopolymer resin and create a thin layer of structured solid through the shifting of the light source, light is emitted throughout the printing process. A fresh layer of fluid photopolymer flows in the printing area to create the subsequent layers, while the building platform slides by one layer thickness after the previous layer is printed (Figure 2a). This procedure continues in a layer-by-layer fashion until the intended 3D objects are achieved (Zhou, Fu et al., 2020).

Due to limited mechanical properties and its rapid degradation rate of gelatin hydrogels, which restricts its application in the biological field. The functionalization of methacrylate groups with gelatin improves the mechanical strength and

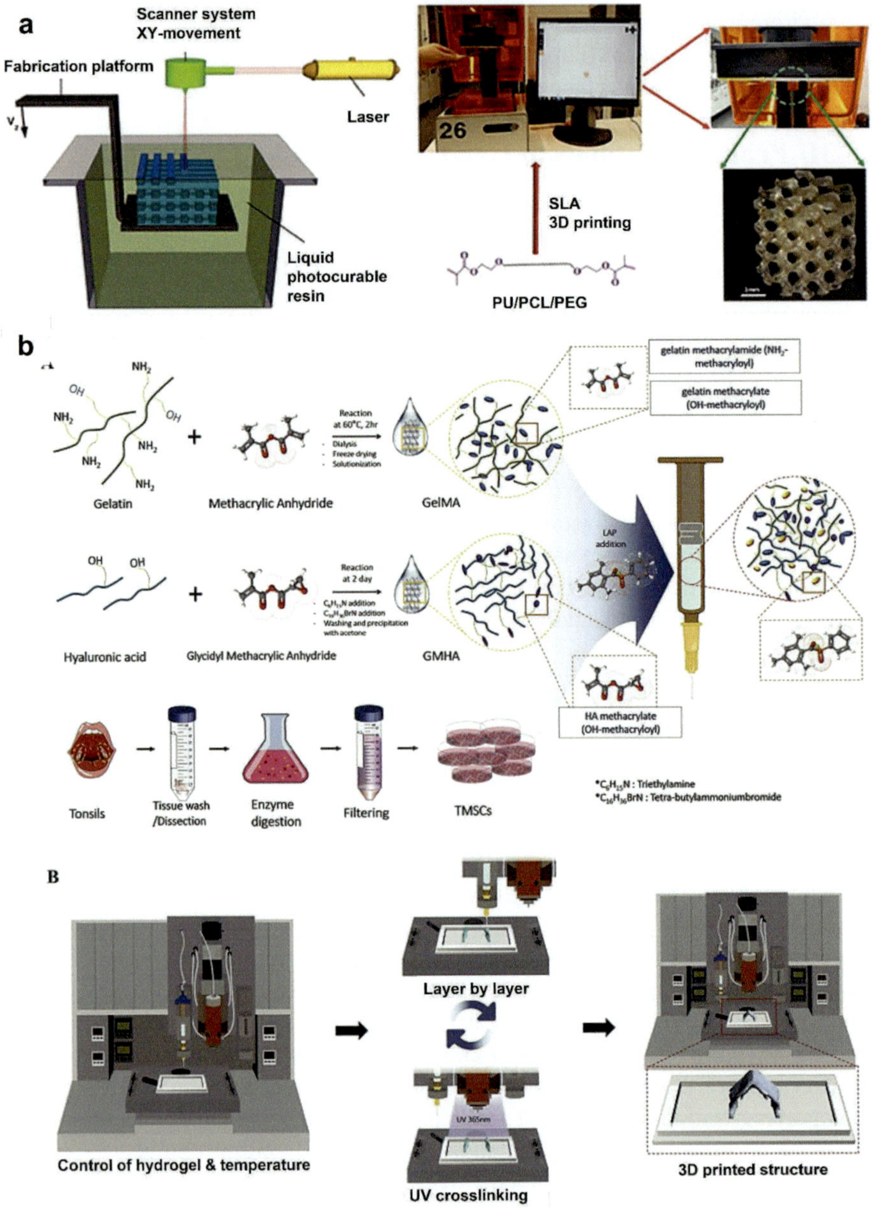

Figure 2. Different types of 3D printing techniques to develop gelatin-based hydrogel scaffolds. (a) An SLA printer operated on photopolymerization principles (Randhawa, Dutta et al., 2023). (b) Material extrusion-based printing of hyaluronic acid incorporated hydrogel for tissue engineering applications (Song, Li et al., 2022).

photocrosslinking abilities, making it suitable for photopolymerization-based 3D printing methods (Lee, Park et al., 2020; Randhawa, Dutta et al., 2023). Thus 3D printed GelMA hydrogels are well suited for an extensive spectrum of tissue engineering applications, and these UV-crosslinked hydrogels have also been used to create various structures such as pre-vascularized dental pulp, soft hydrogels, cartilage-like constructs, etc. (Song, Li et al., 2022; Randhawa, Dutta et al., 2023).

3.2 Extrusion-Based Printing

For the purpose of producing therapeutically appropriate structures, the extrusion-based printing technique is suggested to be the most favorable choice (Malda, Visser et al., 2013; Herzog, Franke et al., 2024). During the printing process, with the help of a cartridge and nozzle, the polymer ink is deposited on the printing platform in a layer-by-layer fashion (Groll, Burdick et al., 2018). In the field of tissue engineering, two types of inks are involved, generally biomaterial ink, referred to as polymer ink without the incorporation of cells, and bioink, allowing the cells and other biological molecules to be mixed with the polymer ink. The extrusion printers can be operated mechanically or pneumatically. The rheological properties, printing rate, and optimum flow rate are some of the criteria that determine the printing precision in this technique. This printing method requires cheap, easy-to-use equipment and little maintenance. High production volume, outstanding precision (down to micrometer level), and remarkable design versatility are the other characteristics that make this technology unique (Schwab, Levato et al., 2020). Many computer-aided design (CAD) models are available for free or at a cost. These models are straightforward, quick to use, and comprise medical photos of the tissue of interest. A schematic depiction of an extrusion-based printer is shown in Figure 2b.

Since gelatin and its derived substances possess good rheological properties, robust biocompatibility, low cytotoxicity, great accessibility, and flexible/easy handling, they are frequently utilized as an excellent material for extrusion-assisted hydrogel printing. Nevertheless, gelatin's intrinsic weak mechanical properties and temperature sensitivity might make it challenging to maintain acceptable form of precision printing accuracy, especially at physiological conditions and mechanical attributes resembling the target tissue. The latter issue must be resolved by adding a different molecule to gelatin, imparting enzymatic and chemical crosslinking capabilities (Figure 3a&b). Since raw gelatin only undergoes temperature-based physical crosslinking, such alterations facilitate post-printing stabilities. Furthermore, the yield plays an important role in choosing the optimal polymeric inks for extrusion printing. For instance, efficient gel flow via the nozzle requires a static yield stress (αstat). Conversely, the regulated gel flow is determined by the dynamic yield stress (αdyn). To enable the uninterrupted filament extrusion from the needle, and to provide the shear thinning characteristics required for extrusion printing, the ink must have stiff and slightly deformable properties (Smith, Basu et al., 2018). Gelatin is most commonly printed at temperatures well lower than its melting point because it gels and becomes thick at these temperatures, making it easy to print (Waidi, Kariim et al., 2024).

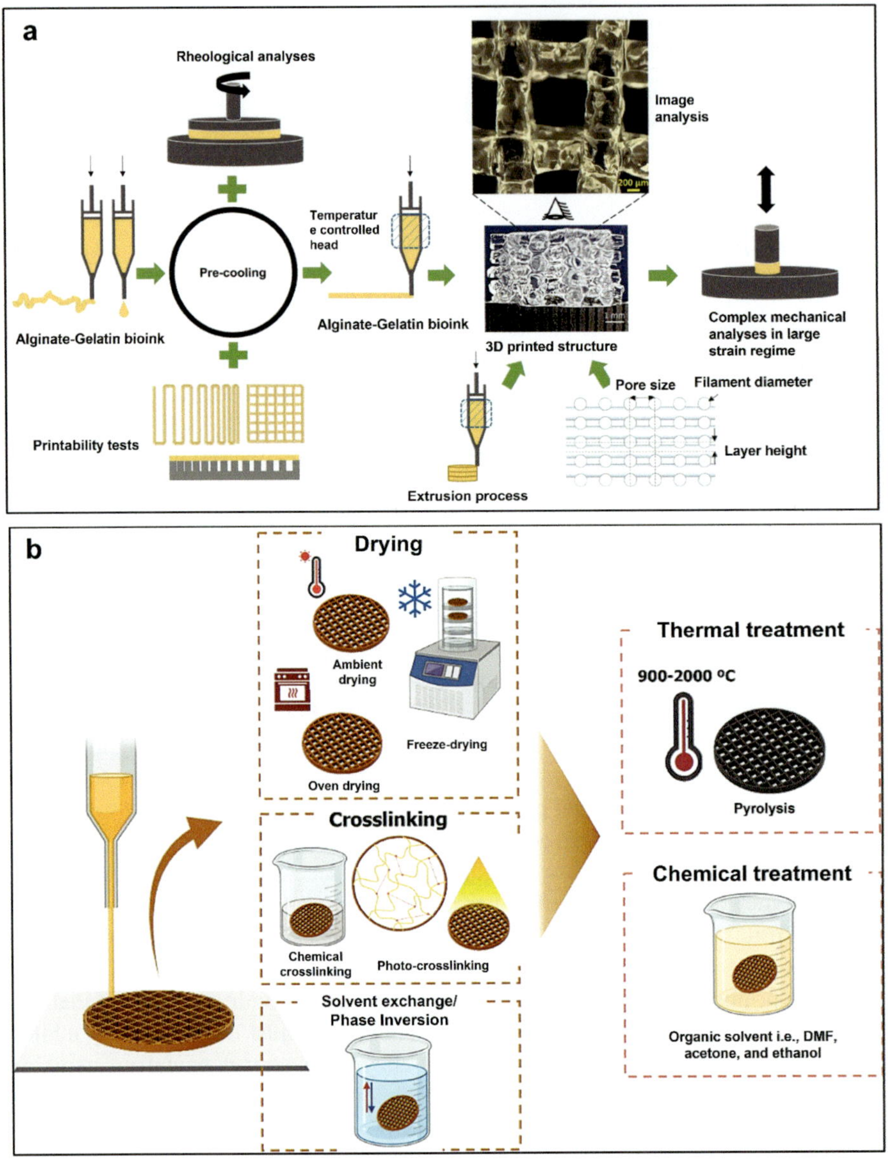

Figure 3. Diagrammatic illustration of the necessary steps taken to enhance the printing process. (a) A pre-cooling step with rheological investigations and printability testings allowed a gelatin-alginate-based bioink to 3D print various structures smoothly with varied meso-structures (Burge, Dawson-Hughes et al., 2007). (b) Different post-printing treatments for 3D printed structures (Hutmacher, 2000).

4. Biomedical Applications of Gelatin-Based Biomaterials

Gelatin serves as a biopolymer that is completely absorbable, biodegradable, and biocompatible (Karim and Bhat, 2008). Gelatin has been the subject of intense interest for a variety of applications in tissue engineering due to these qualities. The subsequent part will discuss research on the application of gelatin-based composites for tissue engineering in the areas of the bone, eyes, and the cardiovascular system.

4.1 Bone Tissue Engineering

Among the most prevalent crippling injuries with a significant treatment cost—nearing tens of billions of dollars in the United States alone—is musculoskeletal diseases (Burge, Dawson-Hughes et al., 2007). Biomimetic structures have been used in place of traditional ones for bone regeneration, restoration, and treatment since the invention of tissue engineering (Hutmacher, 2000). The formation of bones is a multifaceted and ever-changing process that begins with the infiltration of osteoprogenitor cells (Bose, Roy et al., 2012). Proliferation and differentiation of the cells result in the release of proteins and the eventual formation of bone (Khan, Yaszemski et al., 2008; Bose, Roy et al., 2012). The advanced level of frameworks, which stimulate bone development and increase vascularization driven through multiple growth-promoting agents, have been developed as a result of adapting these procedures (Yamada, Tabata et al., 1997; Khan, Yaszemski et al., 2008; Garg, Singh et al., 2012). Several gelatin-based hydrogels are used for bone regeneration applications. In a study by Ganguly et al., a soft hydrogel based on GelMA incorporated with cellulose nanocrystals was developed to study the osteogenic differentiation of hydrogel under a pulsatile pressure stimulation of 5–20 kPa. The stimulation was given to cells cultured on GelMA-CNC every 20 minutes for 7 days. The *in vitro* results revealed an increase in the osteogenic properties compared to the control group. Furthermore, the study also revealed that a pressure stimuli on a GelMA-CNC based soft hydrogel improves the reactive oxygen scavenging properties of the cells (Figure 4a) (Ganguly, Dutta et al., 2023).

It is interesting to note that a range of biological molecules that contribute to the development of bone can be designed for regulated release into gelatin-based structures (Yamada, Tabata et al., 1997; Hong, Tabata et al., 1998). Numerous groups have used this approach to promote bone renewal and development. For instance, it was demonstrated that regulating the release of basic fibroblastic growth factor (bFGF) through acidic gelatin improved bone formation in rabbit skull models with damaged locations (Yamada, Tabata et al., 1997). BMP-2-laden gelatin hydrogels were used in a different study that produced similar outcomes (Hong, Tabata et al., 1998). For possible uses in bone tissue engineering, the mechanism of vascularization can also be encouraged by VEGF, FGF, or BMP-2 (Tabata, Hijikata et al., 1994; Tabata and Ikada, 1999; Young, Wong et al., 2005; Kempen, Lu et al., 2009). The purpose of these gelatin-based composites is to disperse growth hormones sequentially to promote bone development and angiogenesis at the damaged site (Patel, Young et al., 2008; Kempen, Lu et al., 2009). In addition to these uses, gelatin-based hydrogels have also been employed for antibacterial purposes by releasing antimicrobial

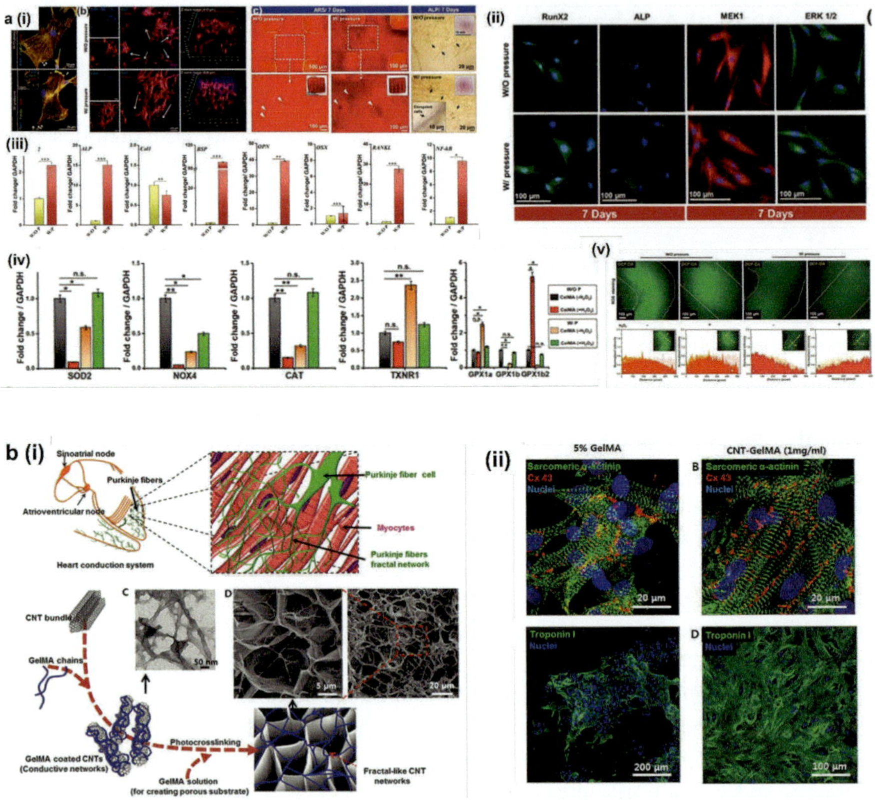

Figure 4. Application of gelatin-based materials in various tissue engineering applications. (a) *In vitro* osteogenic properties of GelMA-CNC soft hydrogel. (i) The cell cytoskeleton arrangement and mineralization study of the mesenchymal stem cells cultured on GelMA-CNC hydrogel, (ii) Fluorescence photographs of the osteogenic protein markers Runx2, ALP, and signaling pathway protein markers such as MEK1, and ERK1/2, (iii) Real-time qPCR investigation of the various osteogenic genes including, *Runx2, ALP, COL1, BSP, OPN, OSX, RANKL, NF-κβ*. (iv–v) Estimation of the ROS scavenging potential of the GelMA-CNC hydrogel by gene expression and DCF-DA staining of the cultured cells (Ganguly, Dutta et al., 2023). (b) CNT-incorporated gelatin structures in cardiac tissue engineering application (i) Schematic depiction of the separated cardiac conduction network displaying the Purkinje fibers found in the heart's inner ventricle walls and the process of creating CNT networks resembling fractals and incorporated in GelMA hydrogel. (ii) Morphological evaluation of the cardiac cells cultured on CNT-GelMA hydrogel. Immunostaining images of sarcomeric R-actin (green), troponin 1 (green), and nuclei (blue) of the cells cultured on hydrogels (Whitcher, Srinivasan et al., 2001).

products and antibiotics following implantation and fracture (Kuijpers, Engbers et al., 1998; Silvio and Bonfield, 1999; Yaffe, Herman et al., 2003).

4.2 Cardiovascular Tissue Engineering

Thirty percent of deaths worldwide are caused by cardiovascular diseases (CVDs), which are the primary cause of mortality (Organization, 2011). The primary obstacles to treating heart damage brought on by CVDs are the scarcity of donors and the incapacity of existing treatments to stimulate myocardial regeneration (Kempen, Lu et al., 2009) adequately. The myocardium comprises heavily vascularized tissue made up of fibroblasts and myocytes arranged in an intricate structure of extracellular matrix (ECM) derived from collagen. The area of biological engineering has garnered much interest recently for its ability to create cardiac tissue constructions that closely resembles the natural organization of myocardial tissue. The most popular types of constructs utilized in cardiac tissue engineering are hydrogels with porous, fibrous structures, which are similar to numerous other tissue-engineered constructs (Rane and Christman, 2011).

In cardiac tissue engineering procedures, gelatin has been employed as an appropriate scaffolding biological substance due to its appealing characteristics as a biopolymer. For example, embryonic rat cardiac muscle cells have been seeded into commercially accessible gelatin foams such as Gelfoam®, a water-insoluble gelatin sponge, to generate viable cardiac grafts (Li, Jia et al., 1999). In either the *in vitro* as well as *in vivo* (implantation) tests, it was shown that the cells could proliferate and beat on their own amid the three-dimensional (3D) milieu of the gelatin matrix. In order to develop scaffolds with mechanical qualities and a tunable rate of degradation that resemble the structure of natural cardiac tissue and regulate cellular organization, gelatin has also been combined with other synthetic materials like polylactic acid (PLA), polycaprolactone (PCL), and poly(glycerol-sebacate) (PGS) (Li, Yau et al., 2000; Ozawa, Mickle et al., 2004; Nerurkar, Elliott et al., 2007; Ifkovits, Devlin et al., 2009; Kharaziha, Nikkhah et al., 2013; Tamayol, Akbari et al., 2013). For instance, in a new study, the behavioral features such as contraction, differentiation, and alignment of cardiac fibroblasts and cardiomyocytes cultured on gelatin/PGS scaffold were precisely regulated by tuning the mechanical properties of the scaffold (Kharaziha, Nikkhah et al., 2013). The appropriate functioning of the tissue in designed grafts depends on the electromechanical synchronization of cardiac cells. Gelatin serves as an insulating biomaterial, and electroconductive materials such as carbon nanotubes (CNTs) incorporation into a GelMA hydrogel result in improved mechanical and electrical characteristics of the tissue matrix. Additionally, the rhythmic activity of the cardiomyocytes also improved (Figure 4b) (Shin, Jung et al., 2013).

4.3 Ophthalmology

Ophthalmic illnesses and disorders affect millions of people globally. phthalmic illnesses and disorders affect millions of people globally. The leading causes of blindness among these include corneal abnormalities, including keratoconus and corneal cystrophies (Whitcher, Srinivasan et al., 2001). The fragile, transparent, dome-like outermost layer of the eye, made up of three distinct layers of endothelium, stroma, and epithelium, is called the cornea. The cornea is crucial to vision because it helps focus light, filters some ultraviolet (UV) rays from the surroundings, and shields the inner workings of the eye from dangerous particles outside (Mimura, Yokoo et al., 2012). For many disorders linked to ophthalmology, the replacement of tissue through donating an eye is an extremely widely approved treatment. However, the achievement rate of cornea transplants have been hampered by a shortage of suitable donors, difficulties storing the tissue, and a significant risk of rejections (Whitcher, Srinivasan et al., 2001; Mimura, Yokoo et al., 2012). The gelatin biomaterials are mechanically stable and transparent, promote adhesion of cells and proliferation, and have been demonstrated to be appealing candidates for ocular tissue restoration. For transplantation uses, gelatin in a variety of porous films have been utilized as an appropriate transporter for corneal endothelium and fibroblast precursor cells (JUMBLATT, MAURICE et al., 1980; Mimura, Amano et al., 2008). Because of their porosity microstructure, these scaffolds aid in cell attachment and migration, facilitating the transport of nutrients and oxygen to the cells. Gelatin sheets with aligned fibers have been developed and implanted with corneal cells in order to more closely resemble the highly ordered architecture of the native tissue of the cornea (Tonsomboon and Oyen, 2013). With directional loading circumstances, the highly ordered nanofibers within the scaffolds produced distinct tensile properties (Tonsomboon and Oyen, 2013), and they also improved cellular orientation (Zhang, He et al., 2011). Further investigation has shown that the optical and mechanical attributes of the construct can be modified by mixing gelatin with other materials such as hyaluronic acid (HA), collagen I, and alginate and adjusting the proportions of each constituent (Zhang, He et al., 2011; Liu, Ren et al., 2013; Tonsomboon and Oyen, 2013). Furthermore, it was demonstrated that adding alginate increased the tensile elastic modulus by a factor of five (Tonsomboon and Oyen, 2013).

5. Conclusion and Future Perspectives

Several natural biopolymers have been extensively studied for various applications, such as tissue engineering, food processing, and the development of biosensors. Gelatin is a naturally occurring biomolecule that plays a significant role in the biomedical and tissue engineering field. Gelatin can be employed in the tissue engineering field in various forms, such as, 3D printed structures, including hydrogels, scaffolds, films, and foams. The mechanical characteristics can be tuned by incorporating different polymers. Despite the remarkable progress made by gelatin, most of the concerns should be addressed to improve the clinical significance of gelatin in tissue engineering applications. For instance, obtaining a particular kind of porosity in a hydrogel for an infiltration of a specific cell type to avoid the infiltration of other biological molecules are still challenging. In addition to the *in vitro* studies, *in vivo* studies utilizing suitable animal models are required to predict the efficacy of the developed structures. The development of new gelatin-based materials can greatly improve the current limitations.

Acknowledgements

This study was supported by the "Basic Science Research Program" through the "National Research Foundation of Korea" funded by the "Ministry of Education" (NRF2018R1A6A1A03025582, NRF2019R1D1A3A03103828, and NRF2022R1I1A3063302). This work was also supported by Innovative Human Resource Development for Local Intellectualization program through the Institute of Information & Communications Technology Planning & Evaluation (IITP) grant funded by the Korea government (MSIT) (IITP-2024-RS-2023-00260267).

References

Alhaskawi, A., H. Zhou, Y. Dong, X. Zou, S. H. A. Ezzi, V. G. Kota et al. 2024. Advancements in 3D-printed artificial tendon. Journal of Biomedical Materials Research Part B: Applied Biomaterials 112(2): e35364.

Bagal-Kestwal, D. R., M. Pan and B. H. Chiang. 2019. Properties and applications of gelatin, pectin, and carrageenan gels. Biomonomers for Green Polymeric Composite Materials: 117–140.

Bose, S., M. Roy and A. Bandyopadhyay. 2012. Recent advances in bone tissue engineering scaffolds. Trends in Biotechnology 30(10): 546–554.

Burge, R., B. Dawson-Hughes, D. H. Solomon, J. B. Wong, A. King and A. Tosteson. 2007. Incidence and economic burden of osteoporosis-related fractures in the United States, 2005–2025. Journal of Bone and Mineral Research 22(3): 465–475.

Catros, S., B. Guillotin, M. Bačáková, J.-C. Fricain and F. Guillemot. 2011. Effect of laser energy, substrate film thickness and bioink viscosity on viability of endothelial cells printed by Laser-Assisted Bioprinting. Applied Surface Science 257(12): 5142–5147.

Finch, C. and A. Jobling. 1977. The science and technology of gelatin. The Science and Technology of Gelatin: 258–260.

Ganguly, K., S. D. Dutta, A. Randhawa, D. K. Patel, T. V. Patil and K. T. Lim. 2023. Transcriptomic changes toward osteogenic differentiation of mesenchymal stem cells on 3D-printed GelMA/CNC hydrogel under pulsatile pressure environment. Advanced Healthcare Materials 12(11): 2202163.

Garg, T., O. Singh, S. Arora and R. Murthy. 2012. Scaffold: a novel carrier for cell and drug delivery. Critical Reviews™ in Therapeutic Drug Carrier Systems 29(1).

Gómez-Guillén, M. C., J. Turnay, M. Fernández-Dıaz, N. Ulmo, M. A. Lizarbe and P. Montero. 2002. Structural and physical properties of gelatin extracted from different marine species: a comparative study. Food Hydrocolloids 16(1): 25–34.

Gómez-Guillén, M. C., B. Giménez, M. E. López-Caballero and M. P. Montero. 2011. Functional and bioactive properties of collagen and gelatin from alternative sources: A review. Food Hydrocolloids 25(8): 1813–1827.

Groll, J., J. A. Burdick, D.-W. Cho, B. Derby, M. Gelinsky, S. C. Heilshorn et al. 2018. A definition of bioinks and their distinction from biomaterial inks. Biofabrication 11(1): 013001.

Herzog, J., L. Franke, Y. Lai, P. Gomez Rossi, J. Sachtleben and D. Weuster-Botz. 2024. 3D bioprinting of microorganisms: principles and applications. Bioprocess and Biosystems Engineering 47(4): 443–461.

Hölzl, K., S. Lin, L. Tytgat, S. Van Vlierberghe, L. Gu and A. Ovsianikov. 2016. Bioink properties before, during and after 3D bioprinting. Biofabrication 8(3): 032002.

Hong, L., Y. Tabata, M. Yamamoto, S. Miyamoto, K. Yamada, N. Hashimoto et al. 1998. Comparison of bone regeneration in a rabbit skull defect by recombinant human BMP-2 incorporated in biodegradable hydrogel and in solution. Journal of Biomaterials Science, Polymer Edition 9(9): 1001–1014.

Howe, A. M. 2000. Some aspects of colloids in photography. Current Opinion in Colloid & Interface Science 5(5-6): 288–300.

Hutmacher, D. W. 2000. Scaffolds in tissue engineering bone and cartilage. Biomaterials 21(24): 2529–2543.

Ifkovits, J. L., J. J. Devlin, G. Eng, T. P. Martens, G. Vunjak-Novakovic and J. A. Burdick. 2009. Biodegradable fibrous scaffolds with tunable properties formed from photo-cross-linkable poly (glycerol sebacate). ACS Applied Materials & Interfaces 1(9): 1878–1886.

JUMBLATT, M. M., D. M. MAURICE and B. D. SCHWARTZ. 1980. A gelatin membrane substrate for the transplantation of tissue cultured cells. Transplantation 29(6): 498–499.

Kang, H.-W., Y. Tabata and Y. Ikada. 1999. Fabrication of porous gelatin scaffolds for tissue engineering. Biomaterials 20(14): 1339–1344.

Karim, A. A. and R. Bhat. 2008. Gelatin alternatives for the food industry: recent developments, challenges and prospects. Trends in Food Science & Technology 19(12): 644–656.
Keenan, T. R. 1994. 16. GELATIN. Handbook of Biodegradable Polymers 7: 307.
Kelly, B. E., I. Bhattacharya, H. Heidari, M. Shusteff, C. M. Spadaccini and H. K. Taylor. 2019. Volumetric additive manufacturing via tomographic reconstruction. Science 363(6431): 1075–1079.
Kempen, D. H., L. Lu, A. Heijink, T. E. Hefferan, L. B. Creemers, A. Maran et al. 2009. Effect of local sequential VEGF and BMP-2 delivery on ectopic and orthotopic bone regeneration. Biomaterials 30(14): 2816–2825.
Khan, Y., M. J. Yaszemski, A. G. Mikos and C. T. Laurencin. 2008. Tissue engineering of bone: material and matrix considerations. Jbjs 90(Supplement_1): 36–42.
Kharaziha, M., M. Nikkhah, S.-R. Shin, N. Annabi, N. Masoumi, A. K. Gaharwar et al. 2013. PGS: Gelatin nanofibrous scaffolds with tunable mechanical and structural properties for engineering cardiac tissues. Biomaterials 34(27): 6355–6366.
Kodama, H. 1981. Automatic method for fabricating a three-dimensional plastic model with photo-hardening polymer. Review of Scientific Instruments 52(11): 1770–1773.
Koepff, P. 1997. Photographic gelatin and major quality aspects–a general overview. The Imaging Science Journal 45(3-4): 246–248.
Kuijpers, A. J., G. H. Engbers, P. B. van Wachem, J. Krijgsveld, S. A. Zaat, J. Dankert et al. 1998. Controlled delivery of antibacterial proteins from biodegradable matrices. Journal of Controlled Release 53(1-3): 235–247.
Ledward, D. 1986. Gelation of gelatin. Functional-Properties of Food Macromolecules: 171–201.
Lee, J. S., H. S. Park, H. Jung, H. Lee, H. Hong, Y. J. Lee et al. 2020. 3D-printable photocurable bioink for cartilage regeneration of tonsil-derived mesenchymal stem cells. Additive Manufacturing 33: 101136.
Li, R.-K., Z.-Q. Jia, R. D. Weisel, D. A. Mickle, A. Choi and T. M. Yau. 1999. Survival and function of bioengineered cardiac grafts. Circulation 100(suppl_2): II-63–Ii-69.
Li, R.-K., T. M. Yau, R. D. Weisel, D. A. Mickle, T. Sakai, A. Choi et al. 2000. Construction of a bioengineered cardiac graft. The Journal of Thoracic and Cardiovascular Surgery 119(2): 368–375.
Liu, Y., L. Ren and Y. Wang. 2013. Crosslinked collagen–gelatin–hyaluronic acid biomimetic film for cornea tissue engineering applications. Materials Science and Engineering: C 33(1): 196–201.
Lu, Y., G. Mapili, G. Suhali, S. Chen and K. Roy. 2006. A digital micro-mirror device-based system for the microfabrication of complex, spatially patterned tissue engineering scaffolds. Journal of Biomedical Materials Research Part A: An Official Journal of The Society for Biomaterials, The Japanese Society for Biomaterials, and The Australian Society for Biomaterials and the Korean Society for Biomaterials 77(2): 396–405.
Lv, L.-C., Q.-Y. Huang, W. Ding, X.-H. Xiao, H.-Y. Zhang and L.-X. Xiong. 2019. Fish gelatin: The novel potential applications. Journal of Functional Foods 63: 103581.
Malda, J., J. Visser, F. P. Melchels, T. Jüngst, W. E. Hennink, W. J. Dhert et al. 2013. 25th anniversary article: engineering hydrogels for biofabrication. Advanced Materials 25(36): 5011–5028.
Manikandan, A., S. Thirupathi Kumara Raja, T. Thiruselvi and A. Gnanamani. 2018. Engineered fish scale gelatin: An alternative and suitable biomaterial for tissue engineering. Journal of Bioactive and Compatible Polymers 33(3): 332–346.
Mikhailov, O. V. 2023. Gelatin as it is: history and modernity. International Journal of Molecular Sciences 24(4): 3583.
Mimura, T., S. Amano, S. Yokoo, S. Uchida, S. Yamagami, T. Usui et al. 2008. Tissue engineering of corneal stroma with rabbit fibroblast precursors and gelatin hydrogels. Molecular Vision 14: 1819.
Mimura, T., S. Yokoo and S. Yamagami. 2012. Tissue engineering of corneal endothelium. Journal of Functional Biomaterials 3(4): 726–744.
Nerurkar, N. L., D. M. Elliott and R. L. Mauck. 2007. Mechanics of oriented electrospun nanofibrous scaffolds for annulus fibrosus tissue engineering. Journal of Orthopaedic Research 25(8): 1018–1028.

Nikkhah, M., M. Akbari, A. Paul, A. Memic, A. Dolatshahi-Pirouz and A. Khademhosseini. 2016. Gelatin-based biomaterials for tissue engineering and stem cell bioengineering. Biomaterials from Nature for Advanced Devices and Therapies: 37–62.

Organization, W. H. 2011. Global Status Report on Noncommunicable Diseases 2010, World Health Organization.

Ozawa, T., D. A. Mickle, R. D. Weisel, K. Matsubayashi, T. Fujii, P. W. Fedak et al. 2004. Tissue-engineered grafts matured in the right ventricular outflow tract. Cell Transplantation 13(2): 169–177.

Patel, Z. S., S. Young, Y. Tabata, J. A. Jansen, M. E. Wong and A. G. Mikos. 2008. Dual delivery of an angiogenic and an osteogenic growth factor for bone regeneration in a critical size defect model. Bone 43(5): 931–940.

Randhawa, A., S. D. Dutta, K. Ganguly, D. K. Patel, T. V. Patil and K. T. Lim. 2023. Recent advances in 3D printing of photocurable polymers: types, mechanism, and tissue engineering application. Macromolecular Bioscience 23(1): 2200278.

Rane, A. A. and K. L. Christman. 2011. Biomaterials for the treatment of myocardial infarction: a 5-year update. Journal of the American College of Cardiology 58(25): 2615–2629.

Schwab, A., R. Levato, M. D'Este, S. Piluso, D. Eglin and J. Malda. 2020. Printability and shape fidelity of bioinks in 3D bioprinting. Chemical Reviews 120(19): 11028–11055.

Shin, S. R., S. M. Jung, M. Zalabany, K. Kim, P. Zorlutuna, S. b. Kim et al. 2013. Carbon-nanotube-embedded hydrogel sheets for engineering cardiac constructs and bioactuators. ACS Nano 7(3): 2369–2380.

Silvio, L. D. and W. Bonfield. 1999. Biodegradable drug delivery system for the treatment of bone infection and repair. Journal of Materials Science: Materials in Medicine 10(10): 653–658.

Smith, C. 1921. Osmosis and swelling of gelatin. Journal of the American Chemical Society 43(6): 1350–1366.

Smith, P. T., A. Basu, A. Saha and A. Nelson. 2018. Chemical modification and printability of shear-thinning hydrogel inks for direct-write 3D printing. Polymer 152: 42–50.

Song, P., M. Li, B. Zhang, X. Gui, Y. Han, L. Wang et al. 2022. DLP fabricating of precision GelMA/HAp porous composite scaffold for bone tissue engineering application. Composites Part B: Engineering 244: 110163.

Stainsby, G. 1987. Gelatin gels. Advances in Meat Research (USA) 4.

Tabata, Y., S. Hijikata and Y. Ikada. 1994. Enhanced vascularization and tissue granulation by basic fibroblast growth factor impregnated in gelatin hydrogels. Journal of Controlled Release 31(2): 189–199.

Tabata, Y. and Y. Ikada. 1999. Vascularization effect of basic fibroblast growth factor released from gelatin hydrogels with different biodegradabilities. Biomaterials 20(22): 2169–2175.

Tamayol, A., M. Akbari, N. Annabi, A. Paul, A. Khademhosseini and D. Juncker. 2013. Fiber-based tissue engineering: Progress, challenges, and opportunities. Biotechnology Advances 31(5): 669–687.

Tonsomboon, K. and M. L. Oyen. 2013. Composite electrospun gelatin fiber-alginate gel scaffolds for mechanically robust tissue engineered cornea. Journal of the Mechanical Behavior of Biomedical Materials 21: 185–194.

Tumbleston, J. R., D. Shirvanyants, N. Ermoshkin, R. Janusziewicz, A. R. Johnson, D. Kelly et al. 2015. Continuous liquid interface production of 3D objects. Science 347(6228): 1349–1352.

Vold, R. D. and M. J. Vold. 1983. Colloid and interface chemistry. (No Title).

Waidi, Y. O., I. Kariim and S. Datta. 2024. Bioprinting of gelatin-based materials for orthopedic application. Frontiers in Bioengineering and Biotechnology 12: 1357460.

Whitcher, J. P., M. Srinivasan and M. P. Upadhyay. 2001. Corneal blindness: a global perspective. Bulletin of the World Health Organization 79(3): 214–221.

Yaffe, A., A. Herman, H. Bahar and I. Binderman. 2003. Combined local application of tetracycline and bisphosphonate reduces alveolar bone resorption in rats. Journal of Periodontology 74(7): 1038–1042.

Yamada, K., Y. Tabata, K. Yamamoto, S. Miyamoto, I. Nagata, H. Kikuchi et al. 1997. Potential efficacy of basic fibroblast growth factor incorporated in biodegradable hydrogels for skull bone regeneration. Journal of Neurosurgery 86(5): 871–875.

Yin, Y. J., K. D. Yao, G. X. Cheng and J. B. Ma. 1999. Properties of polyelectrolyte complex films of chitosan and gelatin. Polymer International 48(6): 429–432.

Young, S., M. Wong, Y. Tabata and A. G. Mikos. 2005. Gelatin as a delivery vehicle for the controlled release of bioactive molecules. Journal of Controlled Release 109(1-3): 256–274.

Zhang, F., C. He, L. Cao, W. Feng, H. Wang, X. Mo et al. 2011. Fabrication of gelatin–hyaluronic acid hybrid scaffolds with tunable porous structures for soft tissue engineering. International Journal of Biological Macromolecules 48(3): 474–481.

Zhou, L. Y., J. Fu and Y. He. 2020. A review of 3D printing technologies for soft polymer materials. Advanced Functional Materials 30(28): 2000187.

CHAPTER 6

Hyaluronic Acid-Based Hydrogels for Bone Tissue Engineering

Aysu Aydınoğlu

1. Introduction

1.1 Bone Tissue Engineering: An Overview

Bone tissue engineering is an interdisciplinary field that combines information in materials science, biology, and engineering to conceptualize new and innovative approaches for repairing or rebuilding irreparably damaged or diseased bone tissues (Orban et al., 2002). The approach to bone tissue engineering is that it attempts to design biologically viable bone, rationally imparting the attributes of natural bone, facilitating the healing or regenerative process and is easily integrated within host tissue (Henkel et al., 2013).

Bone is a highly dynamic and complex integral tissue that plays a critical role in the support of the body structure, protection of vital organs, body movement facilitation, and acting as a reserve for vital minerals (Henkel et al., 2013). However, defects in bone can occur in response to a range of factors, including trauma, diseases like osteoporosis and osteonecrosis, or congenital defects (Einhorn, 1996). Traditional methods like autografts, allografts, and synthetic materials have been used to address bone defects, but each presents inherent limitations. Although generally considered as the gold standard, autografts are hampered by donor site morbidity and availability (Baron, 2008). Currently, allografts entail risks of immune rejection or disease transmission from the donor. In fact, it is more common that

Department of Metallurgy and Materials Science Engineering, Faculty of Chemistry and Metallurgy, Yildiz Technical University, Turkey.
Email: aysukarakas@gmail.com

DOI: 10.1201/9781003518792-6

neither synthetic materials nor combinations thereof exactly duplicate the combined biological and mechanical properties expected from a natural bone's performance (Pape et al., 2010).

Consequently, bone tissue engineering has emerged as an attractive and promising approach to address the shortcomings and limitations of traditional bone reconstructive techniques. In this context, the triad of elements includes scaffolds, cells, and signaling molecules (Yang et al., 2020). Scaffolds are defined as three-dimensional structures where the cells attach and grow, finally differentiating in engineered bone tissues. Thus, the scaffold in bone tissue engineering must be biodegradable, biocompatible, and should have mechanical properties like those in natural bone. They can be fabricated from a variety of materials, including natural polymers like collagen and hyaluronic acid, synthetic polymers, ceramics, and composites (Orban et al., 2002).

Cells are a vital component for the regeneration of bone, especially mesenchymal stem cells (MSCs) that are able to differentiate into osteoblasts, which are responsible for the regeneration of bone. Such stem cell types may either be derived from the bone marrow, adipose tissue, or induced pluripotent stem cells. The fate and functions generally of cells within the scaffold environment underlies the potential to control the differentiation of bone tissue engineering (Guarino et al., 2014). Furthermore, signaling biomolecules like growth factors are often incorporated into the scaffold to promote cellular migration, proliferation, and osteogenic differentiation (Hosseinkhani et al., 2014). Signalling molecules regulate 'behaviors' both at the cell and tissue levels in tissue turnover by reception of signals by the molecules involved. Again, the first position amongst potent osteoinductive growth factors are bone morphogenetic proteins, which also belong in the queue of signalling molecules in bone tissue engineering. Such signalling molecules can be embedded either into the scaffold material or directly delivered into the site of defect (Zhu et al., 2020).

Biopolymer-based composite advances have contributed strongly to bone tissue engineering. Among them, hyaluronic acid-based hydrogels have stunningly excellent biocompatibilities with tunable properties and capabilities to engender cell proliferation and cell differentiation (Beauvais et al., 2016). HA is a naturally occurring glycosaminoglycan, a key component of the supporter and developing extracellular matrix, with the highest levels of tissue hydration and tissue repair. As a scaffold material, HA-based hydrogels can be used to mimic the natural bone environment and offer a perfect framework for bone regeneration (Reddy et al., 2021; Xu et al., 2012).

In conclusion, bone tissue engineering creates a gateway of expectation for new regenerative treatments, compared to the classic ones from the previous century, through concerted actions of the scaffold, cells, and signaling molecules incorporation. Continuing advancement in the development of advanced biopolymer-based composites with the HA-based hydrogels further drives innovation in this field, thereby continuing to bring us closer to effective and sustainable bone regeneration therapies.

1.2 Hyaluronic Acid: An Ideal Biomaterial for Bone Tissue Engineering

Hyaluronic acid is a naturally occurring glycosaminoglycan that is abundantly located within the extracellular matrix of connective tissues. Its various unique characteristics placed it at the centre of the critical elements for tissue engineering, especially regenerating the bone or other hard tissues. HA's biocompatibility (Volpi et al., 2009), biodegradability (Allison et al., 2006), and tendency to formation in hydrogels (Xing et al., 2020) make it an excellent material to develop scaffolds supporting tissue regeneration and reparation.

One of the main contributions to the importance of HA in tissue engineering is its inherent biocompatibility (Mori et al., 2004). Since HA is a naturally occurring human body component, it is well-accepted and normally does not induce immunologic operations. This aspect is especially crucial for tissue engineering applications, whereby HA scaffolds will not cause an integration problem with the host tissue, predisposing the HA to rejection or an adverse reaction. Moreover, biocompatibility for HA, to the extent that chemical modifications could be precipitated without harming its compatibility, will allow the development of further enhancements of its properties as tailored for the required application for specific scaffolds (Mero et al., 2014).

Another key advantage in tissue engineering is that HA has the ability to form hydrogels. The three-dimensional water-swollen networks can mimic the natural extracellular matrix which imparts an environment of cell attachment, proliferation, and differentiation favoring to them (Bae et al., 2014). HA-based hydrogels can be easily formulated and tuned to achieve desired mechanical and biological properties for versatile application in numerous tissue engineering formats. In the specific example of bone tissue engineering, the HA hydrogel would create a desirable environment that would support the growth and differentiation of osteoblasts—the bone-forming cells—therefore promoting bone regeneration (Liu et al., 2013).

One of the most important factors with HA—biodegradability—has a great value in the application to tissue engineering: HA is able to degrade with the help of natural enzyme processes that take place in the living organism; therefore, scaffold degradation matches the process of tissue regeneration. Through this degradation process, there can be gradual replacement of the scaffold by newly formed tissues, which eventually leads to the restoration of native tissue structure and function. The degraded HA-based hydrogels must ideally control degraded material support of the scaffold temporally until new tissue has formed without leaving any material that may evoke chronic complications (Yeom et al., 2014).

Besides its properties as a structural molecule, HA also plays major roles in cell signaling and tissue repair. It is able to bind to many different cell surface receptors, among which are CD44 and RHAMM, and by this means influence cell behavior, with effects on cell migration, proliferation, and differentiation continued to be important processes in tissue repair. HA can also stabilize growth factors through binding or interacting with them to influence activity. By modulating this meaning, HA-based scaffolds may be designed for delivering growth factors in a controlled manner, which may further help in tissue regeneration (Fakhari et al., 2013).

In the field of bone tissue engineering, HA is one of the important factors for osteogenesis. According to studies, HA can enhance the osteogenic differentiation of MSCs, thus it becomes very important in bone formation. Hydrogels based on HA can provide the conducive environment in which the MSCs can easily differentiate into osteoblasts to facilitate the new bone tissue formation (Rahar et al., 2011). For instance, HA-based scaffolds have been developed for use in the healing of bone defects to enhance bone cell growth and integration into host tissue (Avila et al., 2008). For example, in another study, HA hydrogels are explored as depots for growth factors with aims to potentiate bone regeneration by controllably releasing such signaling molecules directly at the site of injury (Dicker et al., 2014).

It can be summarized, hence, that apart from biocompatibility, the formation capabilities of hydrogels, biodegradability, and cell signaling, hyaluronic acid is important tissue engineering due to its important component. In turn, it sustains growth and differentiation of bone-forming cells, supporting the further tissue regeneration and integration into host tissue. As research in the area continues to advance, HA is expected to play a more crucial role in the ways to develop effective and innovative tissue engineering therapies.

1.3 Chapter Objectives

This chapter is an overview of the design and application of hyaluronic acid-based hydrogels in bone tissue engineering. On account of its biocompatibility and biodegradability, besides its role in the extracellular matrix, HA stands as an ideal material for tissue engineering and, particularly, for bone regeneration.

The chapter opened with a look at these intrinsic properties of HA that set the base for its application in tissue engineering. We further proceed to the synthesis and characterization of HA-based hydrogels while looking into some cross-linking methods applied to these hydrogels for mechanical properties, degradation rates, and even for bioactive molecule encapsulation. All such modifications enrich the osteoinductive or osteoconductive potential of such hydrogels in all bone tissue engineering applications.

This chapter further looks at the newer developments in this area, with special interest in composite scaffold formation by combination of HA hydrogels with other biocompatible materials. On the matter, we show the incorporation of growth factors and nanoparticles to increase the functionality of HA hydrogels, and we deal with innovative techniques of fabrication, such as 3D printing, for its ability to create constructs strongly resembling the natural architecture of bone.

The last chapter goes more theoretically into the updated clinical application of these HA-based hydrogels of bone tissue engineering for the view of successes achieved in the area of bone defect repair. We will also address the problems arising from this approach, including host tissue incorporation, scale-up of HA hydrogels manufacturing processes, and mechanical strength. These are key factors to consider in the translation of laboratory research to a clinical environment.

It finally outlines future perspectives and current efforts in research that are aimed at overcoming the present limitations of HA-based hydrogels. This chapter is thus oriented to indicate the potential of these materials for the revolutionizing of strategies connected with bone repair and regeneration and therefore retains the need for further research and development in this direction.

Therefore, the primary goal has been to give insight on how this would be helpful to researchers, clinicians, and students in the creation of a new generation of biomaterials in bone tissue engineering. The attainment of such benefits underlines the crucial role that HA-based hydrogels play in developing effective strategies for bone repair and regeneration within the scope of regenerative medicine and tissue engineering.

2. Hyaluronic Acid: Properties and Function

2.1 Chemical Structure and Composition

Hyaluronic acid, also known as hyaluronan, is a homogenous, unbranched glycosaminoglycan that consists of repeated disaccharide units of β1-3-D-N-acetyl glucosamine and β1-4-D-glucuronic acid (Weigel et al., 1997). As shown in Figure 1, the linear polymer is one of the most abundant components of the extracellular matrix of mammalian cells and tissues and is involved in various physiological activities corresponding to cell activities (Itano et al., 2009; Weigel, 2015). The size of HA in vivo varies according to the tissue type, reflecting its multifunctionality in different biological contexts. The size of HA is the major point affecting its biological activity and mechanical property in tissue engineering (Soranzo et al., 2004).

HA is naturally synthesized by a class of integral plasma membrane proteins known as HAS chitin synthases: HAS1, HAS2, and HAS3 in vertebrates (Borzacchiello et al., 2015). These glycosyltransferases mediate the polymerization of HA through the creation of two distinct types of glycosidic bonds forming its repeating disaccharide structure. Each HA synthase isoform makes a different contribution to the synthesis of HA; that is, there are tissue and cell-specific expression patterns (Haider et al., 2017). While HAS1 is responsible for baseline HA synthesis, HAS2 has been found to participate during embryogenesis, specifically in morphogenesis, cell migration, proliferation, and angiogenesis. Finally, HAS3 expression in carcinogenesis has been implicated in many cases. Interestingly, HAS1 and HAS2 synthesize HA of molecular weight about 2×10^5—2×10^6 Daltons, whereas HAS3 synthesizes HA with a molecular weight of about 10^5 Daltons (Conduah et al., 2009; Henderson et al., 2015).

Expression of the HA synthase genes by these many types of growth factors, cytokines, and signaling molecules is regulated. The genes, located on chromosomes 8, 16, and 19, are induced to transcribe in response to a wide variety of stimuli and give rise to HA synthase proteins in the endoplasmic reticulum that then traffic to the plasma membrane via the secretory pathway. This spatiotemporal regulation ensures appropriate modulation of HA synthesis during different developmental stages and in response to pathological conditions. Tissue differential distribution of HA varies at individual developmental stages and in pathological conditions, under the influence of spatiotemporally regulated transcription of the three synthases. There is abundant HA in fetal tissues, but it is partially replaced during development with collagen

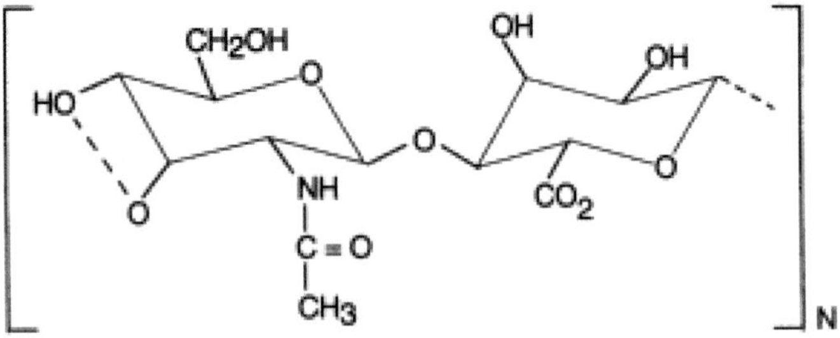

Figure 1. Hyaluronic acid's chemical structure [29].

fibers and proteoglycans, enabling mature tissues to do more rigorous mechanical tasks (Avila et al., 2008).

The inherent ability of HA to bind and retain water molecules in its molecule is an intrinsic property of its chemical structure, forming hydrogels with high water content. Such a characteristic feature is then fundamental to the construction of a more hydrated ECM by virtue of cellular phenomena involving migration, proliferation, and differentiation. In such a milieu, the HA-based hydrogels reproduce a medium somewhat similar in nature to the natural ECM, promoting growth and differentiation of osteoblastic lineage cells and other cells involved in bone tissue engineering that pertain to bones (Grieco et al., 2022).

Probably, one of the most important features of HA's chemical structure is its potential for chemical modification, which turns out to result in a number of derivatives for special applications (Pang et al., 2018). Table 1 shows the common hyaluronic acid modifications and their applications.

HA hydrogel crosslinking can enhance HA molecules' mechanical properties and stability so they can be used in load-bearing applications in bone tissue engineering. A covalent cross-linking between chains of HA is realized with chemical agents such as divinyl sulfone, glutaraldehyde, or carbodiimides (Casale et al., 2019). Henderson et al. (Henderson et al., 2015) showed that such hydrogels have increased resistance to enzymatic degradation. For example, Hylan G-F 20 is a crosslinked derivative of HA and is used as a viscous supplement, which is quite beneficial for the management of osteoarthritis by providing better lubrication and shock absorption (Conduah et al., 2009).

Modification of HA through esterification represents another route to produce HA derivatives. In HYAFF® derivatives, for example, which are obtained by the reaction of HA with anhydrides and in which the hydroxyl groups of HA have been modified with the formation of esters, the degradation rate becomes more controlled, and its mechanical properties improved, thus making this material useful for several tissue engineering applications (Avila et al., 2008; Grieco et al., 2022). On the other hand, sulfation introduces sulfate groups into HA that have increased affinity to bind growth factors and cytokines at the cell surface, transmitting signals to modulate cell activity and promoting tissue regeneration. Sulfated HA represents a very useful tool whenever high bioactivity is required (Pang et al., 2018).

Table 1. Common hyaluronic acid modifications and their applications.

Modification	Description	Applications
Crosslinking	Forming covalent bonds between HA chains using agents like divinyl sulfone	Enhancing mechanical properties and stability for load-bearing applications in bone tissue engineering
Esterification	Modifying hydroxyl groups to form esters, such as HYAFF® derivatives	Controlled degradation rates and improved mechanical properties for various tissue engineering applications
Sulfation	Introducing sulfate groups to enhance binding affinity for growth factors and cytokines	Promoting cell signaling and tissue regeneration, particularly valuable in applications requiring enhanced bioactivity
Conjugation	Attaching peptides, proteins, or drugs to HA	Improving cell adhesion and proliferation (e.g., RGD peptide conjugation) or enabling controlled drug release at the site of tissue regeneration

Its utility is further extended by conjugation of HA with other molecules, such as peptides, proteins, or drugs. The RGD peptide-conjugated HA enhances cell adhesion and proliferation. In the case of HA conjugated with drugs, it allows for controlled release of therapeutic agents at the site of tissue regeneration. These functionalizations enhance the biological activity and application scope of HA in tissue engineering (Allison et al., 2006).

In contrast to all the other glycosaminoglycans, which are synthesized as proteoglycans on core proteins - among them chondroitin/dermatan sulfates, heparin/heparan sulfate, and keratin sulfate - in the Golgi apparatus, HA is synthesized at the inner surface of the plasma membrane as a free linear polymer from cytosolic substrates: UDP-glucuronate and UDP-N-acetyl glucosamine. Thus, HA represents a unique synthesis pathway that makes it special, compared to other GAGs, in features that underlie its special properties (Casale et al., 2019; Volpi et al., 2009).

The United States Pharmacopeia has set the standard regarding the required quality and purity levels of HA to be used for medical and pharmaceutical purposes. It requires HA to be highly pure with very low impurities such as proteins, nucleic acids, and endotoxins in order to minimize any kind of immune response. One another important parameter of HA, controlled by USP standards, is the molecular weight. It impacts its viscosity, hydration capacity, and biological activity. USP standards specify ranges that are suitable for various applications, ensuring that the HA used is appropriate for an intended use (Falcone et al., 2006).

Another critical aspect regulated by USP standards is sterility. The HA products need to be sterile in order to prevent infections. Methods, such as autoclaving or filtration, are verified so that HA properties are not modified due to the process of sterilization. The endotoxin levels unpacked in units of EU (endotoxin units) per milliliter are rigidly controlled, since it may cause fever with some serious adverse reactions. Furthermore, USP standards regulate residual solvents used in the synthesis of HA to ensure that the end product is free from hazardous residues (Suwan et al., 2012; Sze et al., 2016).

USP standards stipulate the physicochemical properties of HA, which are very important to possess consistent performance for medical applications. These properties become important in ensuring that the HA products will perform the functions intended and provide a reliable scaffold for tissue engineering (Fallacara et al., 2018).

Its chemical structure and composition, with all its derivatives, make hyaluronic acid an amazingly flexible platform for advanced scaffolds in tissue engineering. By adherence to the USP standards, HA products guarantee safety and efficacy for clinical application. This will secure an important role in furthering regenerative medicine and tissue engineering. Differential gene expression and regulation of HA synthases are emblematic of a sophisticated biological control over HA synthesis to ensure appropriate function in such a wide range of tissues at different stages in development (Dicker et al., 2014).

Moreover, the possibility of chemically modifying HA and preparing specific derivatives makes HA very useful and efficient in tissue engineering, especially for bone tissue regeneration.

2.2 Biological Functions and Mechanisms

HA plays a regulatory role in the immune system and inflammation. HA mediates immune response through its interactions with the immune cell effectors, such as macrophages and dendritic cells, in the microenvironment, especially during the time of tissue injury or inflammation (Dicker et al., 2014). For example, low molecular weight HA fragments generated during tissue injury can act as DMAMPs through interaction with the ligand-binding sites of immune cellular toll-like receptors to provoke an inflammatory response (Petrey et al., 2014). The high molecular weight HA gives anti-inflammatory effects by stabilizing independent natural killer extracellular matrix and mediating the resolution of inflammation. Key among the signatures that drive changes in immune cell behavior is the formation of HA cables by the high molecular weight HA. These form in states of inflammation and extend from the surface to promote binding of leukocytes, altering their behavior. Among these alterations are ones that permit the carriage of leukocytes and progenitors that are likely to participate in inflammation and tissue repair (Petrey et al., 2014; Ruppert et al., 2014).

Also, another major function of HA relates to the growth of new blood vessels: angiogenesis, a process important for tissue regeneration and repair. HA supports angiogenesis by associating with the endothelial cells to promote the proliferation and migration of these cells, which would constitute new blood vessels. This has been seen to occur through the binding of HA to receptors, including CD44 and RHAMM, within the endothelial cells, thereby activating signaling pathways that will promote the formation of new blood vessels (Alameddine et al., 2014; Savani et al., 2001; Slevin et al., 2007). Regrowth of new bone tissue through adequate blood supply—this is an important condition for angiogenesis in bone tissue engineering, the one that enhances the survival and integration of new tissue (Pardue et al., 2008; Sivaraj et al., 2016).

Different from several other GAGs, including chondroitin/ dermatan sulphates, heparin/heparan sulphate and keratin sulphate, that are first assembled with core proteins early in the Golgi apparatus, HA synthesis as a simple linear polymer at the inner surface of the plasma membrane from cytosolic substrates; UDP-glucuronate and UDP-N-acetyl glucosamine. Elaborating the unique synthetic pathway of HA that makes the difference between other GAGs exist with reciprocal contribution serenades to its uniqueness (Caon et al., 2021).

In brief, effectiveness of hyaluronic acid underpins tissue engineering and regenerative medicine because of interfacing in a fluid process with biologic roles and mechanism, among others. It interfaces fluidly with tissue hydration, cell signaling, immune response, angiogenesis, and ECM stabilization (Xu et al., 2012). These properties, being biocompatible, amenable to various chemical modifications HA, are one of the unequivocal components in developing advanced scaffolds for bone tissue engineering and other regenerative applications. As research further continues to reveal the systematic interactions and mechanisms of HA, only then its potential to renovate tissue engineering and regenerative medicine will continue another step further up.

2.3 Role in Bone Regeneration

Hyaluronic acid is one of the major factors in the complex process of bone regeneration, showing unique biological properties that support both renovation and new formation of bone tissue. Due to its intrinsic biocompatibility, excellent hydrophilicity, and ability to interact with various cellular and molecular components, it presents itself as an ideal candidate for applications in tissue engineering procedures relating to bones, primarily when creating hydrogels (Liu et al., 2018).

One major effect HA has in bone regeneration, however, is that it forms hydrogels that provide 3D scaffolds for the growth of new tissues. An HA-based hydrogel system in a hydrated form emulates very well to the natural component of extracellular matrixes of bone tissue, reflecting optimal conditions for cell adhesion, proliferation, and differentiation—processes that are all integral to successful tissue regeneration. It is this viscoelastic nature of HA hydrogels that is critical in providing mechanical support for newly forming bone (Borzacchiello et al., 2015; Xing et al., 2020).

HA is an active player in osteogenesis, as observed through the promotion of proliferation and differentiation of osteogenic cells. HA interacts with mesenchymal stem cells—a precursor to bone-forming osteoblasts—through specific receptors such as CD44. This event enhances the attachment and subsequent proliferation of MSCs on the HA scaffold, hence providing a conducive environment for their differentiation into osteoblasts (Florea et al., 2020; Hwang et al., 2023). Besides, it can be chemically modified through the incorporation of osteoinductive agents like bone morphogenetic proteins to further stimulate the differentiation of MSCs into osteoblasts, making them even more osteogenic in HA-based hydrogels (Duan et al., 2013; Yeom et al., 2014).

In other words, adequate vascularization or new formation of blood vessels is necessary for bone regeneration so that newly formed bone tissue would have normal access to required nutrition and that waste products are removed (Alqahtani, 2023). Self HA plays a critical role in promoting angiogenesis to let newly formed bone tissues thrive; it interplays with endothelial cells on their proliferation and migration that is induced by receptors CD44 and RHAMM. This interplay initiates multiple signals that result in the neo-formation of vessels within the HA-based scaffold for the newly regenerating bone tissue to ensure its adequate oxygenation and nourishing for survival and integration (Al-Fotawi et al., 2023; Carvalho et al., 2023).

An inflammatory response is a critical phase of tissue repair which needs to be balanced for optimal healing. Excessive inflammation has an impaired bone regeneration capability. HA, especially high molecular weight HA, has anti-inflammatory properties in terms of cell and matrix stabilization and resolution of inflammation. Such immunomodulatory action sets ground for a conducive environment for bone healing and regeneration (Lee et al., 2002; Ruppert et al., 2014).

HA-based hydrogels are designed for controlled degradation, and they naturally break down over some time. This degradation is tailored to match the formation of new tissue; therefore, the scaffold provides temporary mechanical support and, with time, is replaced by newly formed bone tissue in an ideal manner. Even more importantly, the degradation products of HA are non-toxic and biocompatible and are naturally metabolized by the body, eliminating resultant late-term unfavourable effects (Xing et al., 2020).

However, researchers have been trying many ways to increase the role of HA in bone regeneration through the combination of HA with other biocompatibles for preparing composite scaffolds. For example, hyaluronan mixed with ceramics like hydroxyapatite or tricalcium phosphate can enhance theirmechanical strength and osteoconductivity (Allison et al., 2006). Furthermore, combination of HA with peptides, growth factors, or nanoparticles have been used to enhance the bioactivity of HA while giving targeted delivery for pharmacological agents, which can significantly enhance the bone regeneration (Ballini et al., 2009).

In this context, HA-based hydrogels have already demonstrated great potential for bone regeneration inseau several application areas related to defects, fractures, and osteoarthritis. Needless to say, such work never appears to end, while researchers pursue innovative ways of optimizing the HA scaffolds for ever-better clinical results. The strategies currently under development include state-of-the-art fabrication techniques like 3D printing that can engineer complicated patient-specific scaffold architectures very closely resembling a natural bone structure.

3. Hyaluronic Acid-Based Hydrogels: Synthesis and Preparation

Hyaluronic acid-based hydrogels are one of the most promising biomaterials in bone tissue engineering, given their biocompatibility and biodegradability with tunable features. Synthesis and fabrication are two steps during the development of such hydrogels that identify their overall efficacy as scaffolds for bone regeneration. The methods chosen will master key characteristics—mechanical strength, degradation kinetics, and biological activity—all critical for guiding cell behavior and tissue formation.

3.1 Hydrogel Synthesis Methods

In general, cross-linking methods to synthesize hydrogels can be basically divided into two major categories: physical and chemical crosslinking. More recent approaches, however, include enzymatic crosslinking (Sayed, 2023). Each of the methods has unique advantages and is chosen according to the final properties desired to be presented by the hydrogel: mechanical strength, biocompatibility, or degradation rate. Next, the different cross-linking approaches and their applications in HA-based hydrogels synthesis will be described and detailed.

3.1.1 Physical Crosslinking

Physical crosslinking takes place through non-covalent interactions, which include hydrogen bonding, ionic interactions, and hydrophobic forces that bind or hold the hydrogel networks, making physical crosslinking not as a requirement to added chemical crosslinkers that would lower the biocompatibility of the hydrogel (Kamata et al., 2015; Tan et al., 2010). Physical methods of crosslinking are typically far more reversible, usually in response to the environment in terms of things like pH, temperature, or ionic strength.

One of the common techniques of crosslinking HA is that it is crosslinked via ionic interactions. The characteristics feature of such HA is that it is crosslinked with ionic interactions involving negatively charged carboxyl groups with multivalent cations, such as calcium or aluminum ions (França et al., 2021). One study was forming HA hydrogels using calcium ions. The researchers proved that ionic crosslinking of HA with calcium ions allows hydrogels to be achieved with adjustable mechanical and swelling properties by changing the concentration of calcium ions. This allows for great flexibility in altering the hydrogel properties according to very specific requirements for effective tissue engineering use (Gan et al., 2023).

Another of these is hydrogen bonding, a physical crosslinking method where HA molecules form hydrogen bonds between their hydroxyl and carboxyl groups under some specific conditions with the result of producing a gel network (França et al., 2021). For example, Liu et al. reported the formation mechanism of HA hydrogel induced by hydrogen bonding with the addition of polyvinyl alcohol (Liu et al., 2018). The hydrogen bonds formed between HA and PVA concluded a hydrogel of improved strength and stability. These hydrogels presented good biocompatibility with enhanced visible proliferation of chondrocytes and promising potential in cartilage tissue engineering.

Thus, hydrophobic interactions have also been utilized to drive the creation of physically crosslinked hydrogels. Hydrophobic groups can be introduced into HA chains, for example, by the introduction of alkyl chains, forming hydrophobic domains within the network (Khunmanee et al., 2017). For example, Jiang et al. developed octadecylamine-modified HA hydrogels, in which hydrophobic interactions of the alkyl chains form the physically crosslinked network (Jiang et al., 2009). These hydrogels have good mechanical properties and controlled drug release behavior, respectively, which showed promise for use in drug delivery applications for tissue engineering.

Another interesting approach to physical crosslinking is the use of temperature-responsive polymers. For instance, poly is a polymer that exhibits a lower critical solution temperature of around 32°C. Below this temperature, the polymer is water-soluble, but above it, the polymer becomes hydrophobic and precipitates out of the solution (Cook et al., 2021; Geever et al., 2006). Grafting PNIPAAm, a thermally responsive polymer, onto hyaluronic acid can result in the formation of temperature-sensitive hydrogels (Santos et al., 2010; Zhao et al., 2010). Figure 2 shows the procedure for synthesis of a PNIPAAm grafted HA copolymer. These hydrogels can transition into a gel state at physiological body temperature, enabling minimally invasive, injectable delivery. Studies have demonstrated that such HA-based, thermally responsive hydrogels support the growth and differentiation of mesenchymal stem cells, highlighting their promising potential for bone tissue engineering applications (Seliktar, 2012; Seol et al., 2013).

The hydrogels for responsiveness application are usually prepared by physical crosslinking methods, giving them environmental sensitivity. For example, pH sensitivity can be introduced with HA hydrogels that have been crosslinked by HMHB. These pH stimuli can be applied by introducing pK a-b values of acidic or basic type on the backbone of HA to get hydrogels that swell or shrink depending on the pH of the environment. Through it, it becomes very useful in making delivery systems for controlled drug release, releasing its payload in response to an acidic environment due to a wound or a tumor site (Hendi et al., 2020; Li et al., 2021).

Although physical crosslinking methods are simple and biocompatible, they usually provide the hydrogel with relatively low mechanical strength when compared to that of the chemically crosslinked counterpart. Hybrid crosslinking can be applied to the received dilemma, i.e., a combination of physical and mild chemical crosslinking for the enhancement of mechanical properties together with biocompatibility (Basara et al., 2019; Khunmanee et al., 2017). Studies have been performed with dual crosslinking systems that employ a combination of ionic and hydrogen bonding to form hydrogels where the mechanical strength and stability are improved (Khunmanee et al., 2017; Lin et al., 2015). This renders the development of physically cross-linked HA-based hydrogels a highly promising manner with intrinsic advantages like biocompatibility, tunable properties, and more, for applications in bone tissue engineering. However, its low mechanical strength issues in its application for weight-bearing purposes still remain debatable (Hwang et al., 2023). Further studies in hybrid crosslinking strategies and including reinforcement agents will greatly help in paving the way forward to overcome the challenges in load-bearing bone regeneration.

Physical crosslinking methods are often employed in preparing HA-based hydrogels due to their ability to form networks without potentially toxic chemical agents. These methods are particularly advantageous in biomedical applications with critical biocompatibility and non-toxicity. Table 2 below outlines the various physical crosslinking methods used for HA-based hydrogels, highlighting their advantages and disadvantages.

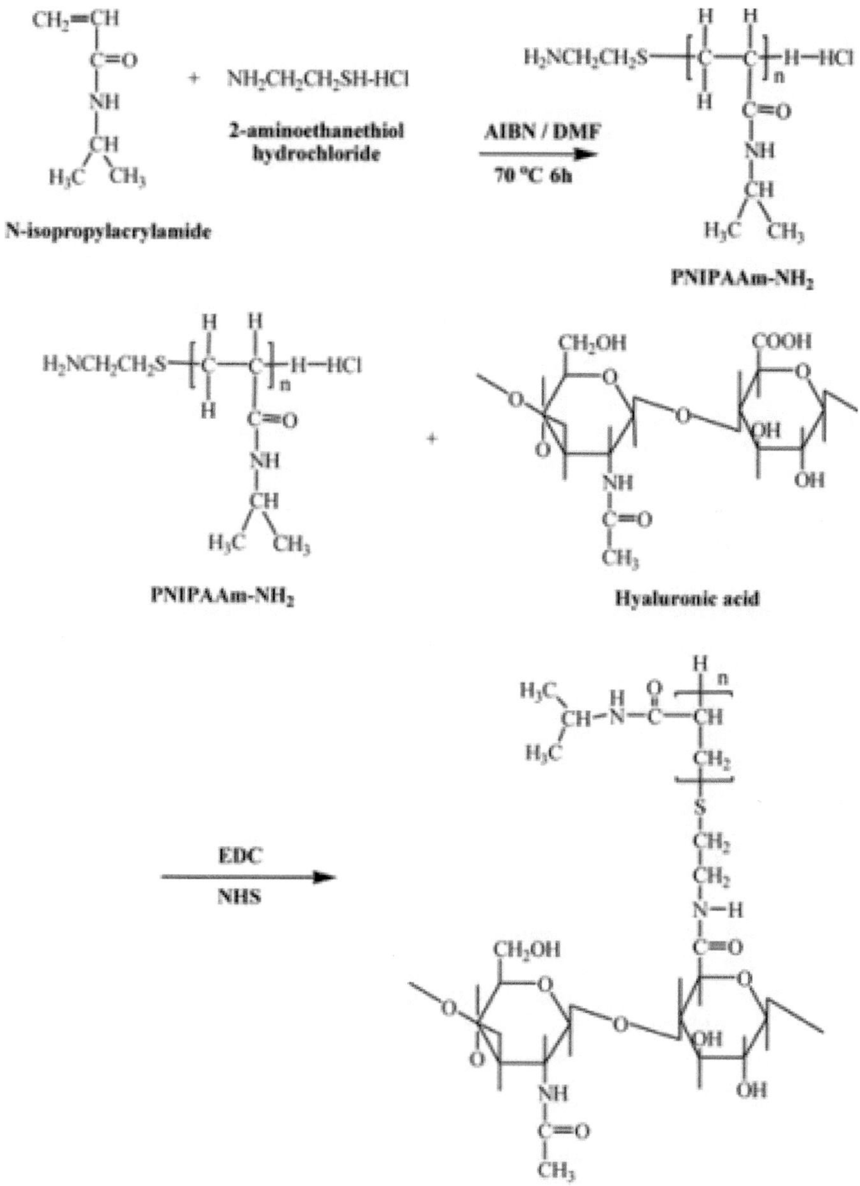

Figure 2. Synthetic scheme of amine-terminated telechelic PNIPAAm and PNIPAAm-grafted HA copolymer [72].

Table 2. Advantages and disadvantages of physical crosslinking methods for HA-based hydrogels.

Physical Crosslinking Method	**Advantages**	**Disadvantages**
Ionic Crosslinking	Does not require chemical crosslinkers, occurs under biocompatible conditions.	Ionic bonds may be weak, limiting the mechanical strength of hydrogels.
Hydrogen Bonding	Does not involve chemical crosslinkers, ensuring biocompatibility.	Hydrogen bonds can easily break, affecting the long-term stability of hydrogels.
Hydrophobic Interactions	Facilitates spontaneous crosslinking through hydrophobic interactions.	Hydrophobic regions may have restricted solubility in water.
Physical Aggregation	Aggregation processes are often quick and reversible under changing environmental conditions.	Achieving precise control over the aggregation process can be difficult.
Thermal Gelation	Gelation can be easily managed by adjusting temperature.	Thermal gelation works only within specific temperature ranges, limiting its application.

3.1.2 Chemical Crosslinking

This chemical crosslinking process makes a very powerful procedure to obtain stable HA-based hydrogels presenting crosslinked networks when looking into HA-based hydrogels in load-bearing bone tissue engineering. The cross-linking is achieved through covalent bonding between HA molecules, providing good control over the properties of the hydrogel, such as stiffness, degradation, and swelling behavior (Sturabotti et al., 2022). However, the direct use of chemical crosslinkers raises potential cytotoxicity and biocompatibility issues, further emphasizing the need for a very detailed process of evaluation and validation before possible clinical translation (Pérez-Luna et al., 2018).

The chemical structure of HA highlighting the three most commonly used sites of covalent modification, the carboxylic groups, hydroxyl group, and $-NHCOCH_3$ group as it is shown in Figure 1 (Walimbe et al., 2017). Various chemical agents and techniques have been used with varying advantages, each for covalent cross-linking of HA to realize specific tissue engineering applications. Many reagents can condense carboxyl and amino groups to form amide bonds. The hydroxyl and carboxyl groups present in the HA structure can be modified and then crosslinked using small bi-functional or multi-functional crosslinking agents such as di- or poly-epoxides, divinyl sulfone, or glutaraldehyde (Khunmanee et al., 2017). For crosslinking hydrogels, the most commonly used agents are carbodiimides, carbonyl diimidazole, and others (Figure 3). A key feature of the mechanism of action of these reagents is the initial activation of the carboxyl group. These reagents facilitate the formation of amide or ester bonds through a two-step process. First, the reagent forms a reactive adduct with the carboxyl group. Subsequently, nucleophilic attack at the activated species eliminates the activating moiety, resulting in the formation of

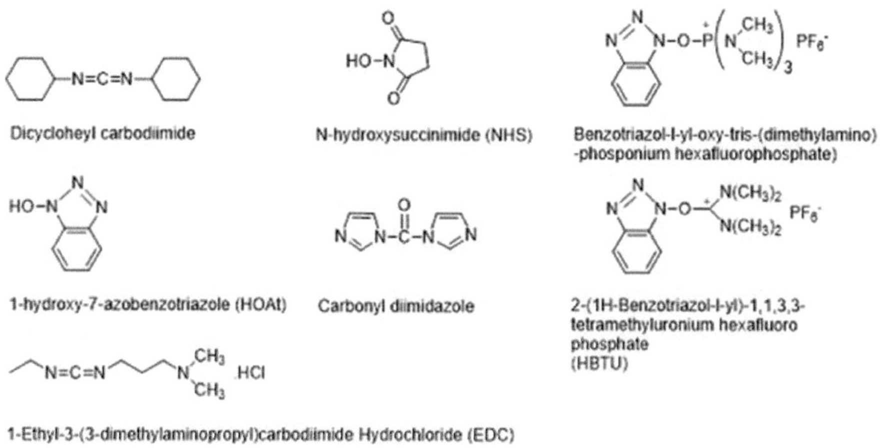

Figure 3. Chemical agents for the activation of hyaluronic acid (HA)-COOH [64].

a bond that does not involve the incorporation of the crosslinking agent. The amino group or alcohol group serves as the nucleophile in these reactions.

The chemical modification of –OH groups can be divided into four reaction types: ether formation, ester formation, hemiacetal formation, and oxidation. Reagents used in these reactions are summarized in Figure 4. Hydrogels can be crosslinked using divinyl sulfone or ethylene sulfide under alkaline conditions, which minimizes degradation of hyaluronic acid compared to high temperature methods. In contrast, glutaraldehyde crosslinking, although widely employed, requires acidic conditions, is unstable, and is cytotoxic, necessitating extensive purification of the final product. Conversely, Toole et al. demonstrated that divinyl sulfone crosslinking of hyaluronic acid produces hydrogels with significantly optimized mechanical properties and a controllable degradation rate. This approach enables the fine-tuning of hydrogel characteristics by adjusting the concentration and type of crosslinking agent (Khunmanee et al., 2017; Toole, 2004).

However, some of the mentioned bifunctional crosslinkers present the potential for cytotoxicity and should be optimized and evaluated carefully for biocompatibility. Azide-alkyne cycloaddition has been an area of intense cross-linking of HA because of the high specificity of these reactions being highly efficient and happening at mild conditions. It offers control of the network structure and properties in the formation

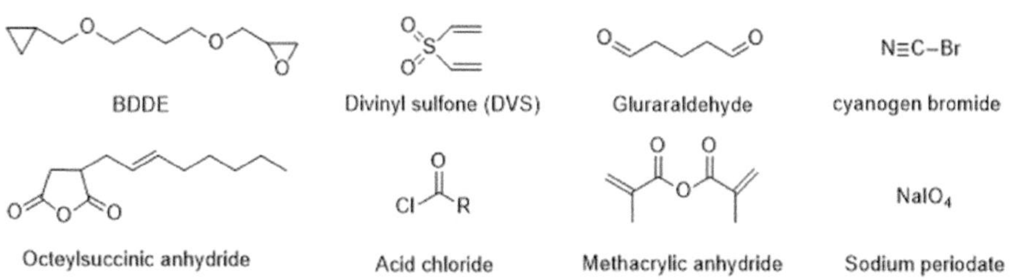

Figure 4. Crosslinking agents for modification of –OH groups [64].

of such gel networks (Sturabotti et al., 2022). An example of this includes hydrogels developed by Burdick and Prestwich (Burdick et al., 2011), where click chemistry was used to develop HA for hydrogels in such a way that their mechanical properties could be tailor-made along with their biodegradability for several tissue engineering applications. This is more striking as the click chemistry crosslinking strategies of biomaterials display attractive features due to their rigid ins and biocompatibility into biomedical applications. Photo-crosslinking allows control of the crosslinking process both in space and time, thus creating hydrogels with complex geometries and gradients in properties, as is important for advanced tissue engineering applications. Photo-reactive groups, such as methacrylates or acrylates, are attached to HA, which form covalent bonds upon exposure to ultraviolet or visible light (Luo et al., 2023; Buckley et al., 2023). The first methacrylated hyaluronic acid hydrogel was cross-linked using UV light, demonstrating this technique as a powerful tool for fabricating scaffolds with intricate architectures (Park et al., 2003).

Another versatile route to the chemical crosslinking of HA is the Michael addition reaction: a carbanion or another nucleophile performs nucleophilic addition to an α,β-unsaturated carbonyl compound—for example, either of the vinyl sulfone or acrylate groups—functionalized onto the HA chains (Censi et al., 2010; Jin et al., 2010; Li et al., 2017). Serban et al. (Serban et al., 2008) managed to obtain products of thiol-modified HA and the corresponding vinyl sulphone functionalized polyethylene glycol that formed hydrogels after a Michael type addition process, that had their mechanical properties and biodegradation rate tuned.

Table 3. Comparison of different chemical crosslinking methods for HA, summarizing their advantages, disadvantages, and key applications in bone tissue engineering.

Chemical Crosslinking Method	Advantages	Disadvantages	Key Applications
Glutaraldehyde Crosslinking	Provides high mechanical strength and stability.	Potential for residual toxicity; requires thorough washing.	Suitable for load-bearing bone scaffolds.
Carbodiimide Crosslinking	Produces minimal toxic byproducts, increasing biocompatibility.	Requires precise control over reaction conditions.	Ideal for creating hydrogels for bone regeneration.
Photo-Crosslinking	Enables precise spatial control of crosslinking using light.	Some photoinitiators may have cytotoxic effects.	Useful in creating complex structures for bone tissue engineering in 3D applications
Enzymatic Crosslinking	Utilizes natural enzymes, ensuring high biocompatibility.	Crosslinking may be slower compared to other methods.	Used for creating hydrogels that closely mimic natural bone matrix.
Click Chemistry	Provides highly specific and efficient crosslinking.	HA needs to be functionalized, adding complexity.	Promotes bone cell differentiation in engineered tissues.

To address the cytotoxicity concerns associated with certain chemical cross-linkers, the focus has shifted towards developing biocompatible alternatives that produce non-toxic byproducts during the cross-linking process. Cross-linked HA using polyethylene glycol diglycidyl ether and found excellent biocompatibility and controlled degradation of the resulting hydrogels. Such strategies are critical in the development of hydrogels; this would support tissue regeneration without causing adverse effects (Sturabotti et al., 2022).

These enable the fine-tuning of HA-based hydrogel properties that are used for the tissue engineering of bone. Table 3 compares different chemical crosslinking methods for HA, highlighting their advantages, disadvantages, and key applications in bone tissue engineering. Although classic crosslinkers will deliver strong mechanical properties, their existing concern over cytotoxicity needs careful optimization and testing. Advanced techniques, including the use of click chemistry, photo-crosslinking, and biocompatible crosslinkers, cross beyond such limitations and further open the door to research characterized by more secure and improved HA-based biomaterials in regenerative medicine.

3.1.3 Enzymatic Crosslinking

Enzymatic crosslinking is another bioinspired method for forming HA-based hydrogels. This approach takes advantage of the mild reaction conditions and specificity of enzymes to produce highly biocompatible materials. Enzymatic crosslinking is particularly useful in applications where preserving the bioactivity of incorporated cells and bioactive molecules is paramount (Gan et al., 2023; Jin et al., 2010; Raia et al., 2017).

Transglutaminase is an enzyme abundant in most tissues, and it mediates covalent bond formation between glutamine and lysine residues. By modifying HA with these amino acids, transglutaminase can be used to develop crosslinked hydrogels with tunable properties (Duarte et al., 2020). For example, Tan et al. (Tan et al., 2011) effectively demonstrated this approach, creating HA hydrogels that exhibit excellent biocompatibility and support the growth and proliferation of chondrocytes, making them promising for cartilage tissue engineering applications.

Another common enzyme used extensively in biotechnology is horseradish peroxidase (Bae et al., 2015). It catalyzes the oxidation of phenolic compounds in the presence of hydrogen peroxide to yield covalent crosslinks. Therefore, this mechanism can be exploited in crosslinking HA modified with phenolic groups (Azevedo et al., 2003). For example, Shin et al. fabricated HA-tyramine hydrogels crosslinked with HRP, and they exhibited high mechanical properties with tunable degradation rates. The hydrogels were biocompatible and promoted the growth of human adipose derived stem cells, which is good evidence for soft tissue engineering of the skin (Shin et al., 2015).

Another crosslinking enzyme is tyrosinase, as it catalyzes the oxidation of tyrosine residues resulting in di-tyrosine bonds. This approach can be used with HA, where tyrosine residues are first introduced to the polymer, followed by their subsequent oxidation with tyrosinase to obtain a crosslinked hydrogel (Choi et al., 2022; Kim et al., 2018). Specifically, Lee et al. (Lee et al., 2009) were the first to report HA-tyramine conjugates crosslinked by tyrosinase. Then, the hydrogel obtained exhibited suitable mechanical properties and biocompatibility and was able to support the growth of fibroblasts and, therefore, was suggested to be used in wound healing applications.

Laccase is an oxidative enzyme that can be used to catalyze the formation of dimeric phenolic bonds (Shraddha et al., 2011), providing another option for the enzymatic crosslinking of HA (Hwang et al., 2023; Mohit et al., 2020). When HA is conjugated with phenolic compounds, laccase can be employed to catalyze the crosslinking reaction, resulting in the formation of a stable hydrogel network. This method is highly biocompatible, making it suitable for encapsulating cells or other bioactive materials. Hsiao et al. demonstrated the effectiveness of laccase in crosslinking HA hydrogels, which were able to encapsulate human mesenchymal stem cells while maintaining cell viability, suggesting their potential for cell delivery and tissue engineering applications (Han et al., 2020; Khunmanee et al., 2017).

Enzyme-triggered click chemistry is a method currently regarded as the next generation, which combines both specificity and efficiency. This technique relies on the enzymatic catalysis of bio-orthogonal reactions whereby covalent bonds between HA molecules can be produced (Avila et al., 2008; Gottschalk et al., 2022). More recently, King et al. reported a successful application of enzyme-mediated click chemistry for the formation of HA hydrogels, the network structure of which was modulated, decoupled from hydrogel mechanical properties. These biocompatible

Table 4. Summary of different enzymes used for HA crosslinking, including their mechanisms, advantages, limitations, and key applications in tissue engineering.

Enzyme	Mechanism	Advantages	Limitations	Key Applications
Transglutaminase	Catalyzes the formation of covalent bonds between lysine and glutamine residues on HA.	High specificity with minimal side reactions.	Requires specific amino acid sequences.	Used for scaffolds in soft tissue regeneration.
Tyrosinase	Oxidizes tyrosine residues on HA, leading to dityrosine crosslinks.	Works under physiological conditions.	Only targets tyrosine residues.	Employed in wound healing and skin regeneration.
Laccase	Catalyzes the oxidation of phenolic groups, leading to crosslinking.	Produces water as a byproduct, ensuring high biocompatibility.	Requires optimal conditions.	Suitable for crosslinking in bone scaffolds.
Horseradish Peroxidase	Mediates the crosslinking of phenolic groups in the presence of hydrogen peroxide.	Gelation time can be controlled by adjusting peroxide concentration.	High concentrations of peroxide may be toxic.	Used in creating hydrogels for cardiac tissue repair.
Sortase A	Facilitates the ligation of proteins or peptides containing specific recognition motifs.	High specificity for target sequences.	Requires Peptide Engineering: Needs specific peptide sequences for crosslinking.	Applied in designing hydrogels with tailored properties for various tissue types.

hydrogels effectively fostered the differentiation of stem cells to both osteogenic and chondrogenic lineages and hold great promise for more advanced tissue engineering applications (King et al., 2010).

Enzymatically cross-linked HA-based hydrogels can be used as a highly effective and versatile method for the development of structures with enhanced biocompatibility and customized properties, and therefore this way of cross-linking has great potential to be used in a number of medical applications but with special importance to tissue engineering and regenerative medicine. Different enzymes can be used to achieve this crosslinking, each with its unique mechanism, advantages, and limitations. Table 4 provides a summary of these enzymes, focusing on their roles in HA crosslinking and their key applications in tissue engineering. Indeed, there is so much advancement in understanding the enzyme kinetics and the field of bio-orthogonal chemistry that improvements can be expected in the future regarding various modes of design with much progress likely in the direction of designing more sophisticated and functional HA-based biomaterials.

3.2 Hydrogel Preparation Techniques

Effective development toward bone tissue engineering requires the capacity to appreciate a level of precision in control of architecture and properties of hyaluronic acid-based hydrogels. Making such hydrogels is enabled by recent technological advancements in preparing hydrogels for such precise definitions. Several such methods have been developed by now, each consisting of various gains and abilities in tailoring hydrogels for a certain application in the appropriate style. Examples of such methods are the sol-gel process, electrospinning, and 3D printing, where each concerns scaffold design and each application.

3.2.1 Sol-Gel Process

The sol-gel process is one of the most versatile and widely used techniques for fabricating hyaluronic acid-based hydrogels (Khunmanee et al., 2017; Lee et al., 2019). This method allows for precise control over the porosity and mechanical properties of the hydrogels, which are crucial factors in bone tissue engineering. The process involves the transformation of a sol, a colloidal solution, into a solid gel network through a series of carefully controlled hydrolysis and condensation reactions (Grieco et al., 2022).

At the first step of the sol-gel process, HA is usually dissolved in a compatible solvent, generally water or an aqueous buffer solution, to form a homogeneous colloidal suspension. The sol may also contain silicates or metal alkoxides as precursors for further gelation. These precursors undergo a series of hydrolysis and condensation reactions to form a three-dimensional gel network (Bokov et al., 2021; D'Amora et al., 2019). By varying parameters such as, pH, temperature, and ionic strength of the solution, the gelation step can be triggered and controlled in sol–gel synthesis. At the gelation stage, or during it, colloidal particles in a sol dispersed in liquid agglomerate or cross-link, forming a continuous network that entraps the solvent within the pores (Aguilar, 2018; Deshmukh et al., 2020; Kurniawansyah et al., 2023; Ruel-Gariepy et al., 2004). The composition and concentration of the initial sol can be modified to change the kinetics of gelation and the final properties of the hydrogel. The possibility to control the parameters of gelation enables a possibility to tune, with a high accuracy, the microstructure of the hydrogel and, consequently, to tune the mechanical properties of the hydrogel for the envisaged applications (Caccavo et al., 2018; Lamberti et al., 2016). For instance, Kakizawa et al. investigated the impact of varying gelation conditions on the properties of HA-based hydrogels. They were able to demonstrate that changing the pH and temperature during gelation made it possible to tailor the hydrogel's microstructure and mechanical properties (Borzacchiello et al., 2015).

After gelation, the hydrogel can proceed to age: condensation reactions continue to take place, and the network structure strengthens and becomes more stable. This aging step contributes importantly to the long-term stability and mechanical integrity of these materials. This is then followed by additional drying, done to remove the solvent in excess, that will result in a mechanically rigid porous structure. Drying, in turn, determines the porosity and mechanical characteristics of the final hydrogel. As, for instance, it allows one to obtain quite large porous superlight aerogels, the supercritical drying in CO_2. In these terms, it was shown that HA-based aerogels prepared in such a way have a high porosity and large surface area, which makes them quite effective scaffolds in relation to the bone tissue engineering problem. Such properties favour such important processes in tissue regeneration as cell adhesion, proliferation, and differentiation (Zheng et al., 2020). Figure 5 illustrates the general scheme of the sol-gel synthesis approach used to prepare the HA-based nanocomposite hydrogels.

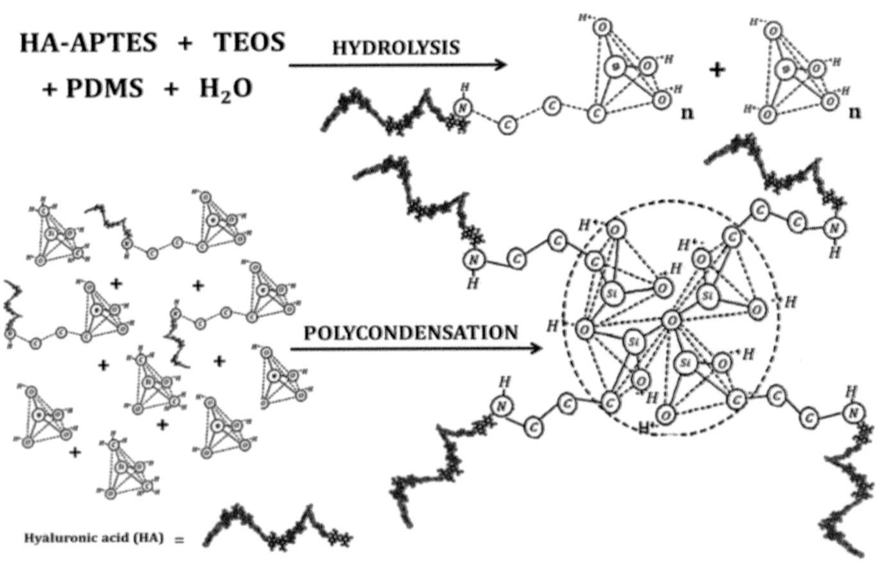

Figure 5. Schematic illustration the sol-gel synthesis of hyaluronate. APTES: (3-Aminopropyl) triethoxysilane; TEOS: Tetraethyl Orthosilicate; PDMS: Polydimethylsiloxane [115].

Silica-incorporated hyaluronic acid hydrogels prepared via the sol-gel process exhibited significant improvements in both mechanical strength and bioactivity. The enhanced properties of these hybrid hydrogels make them highly promising for applications in bone tissue engineering (Hwang et al., 2019). The sol-gel process has been demonstrated to provide a high level of control over the microstructure and properties of HA-based hydrogels. This flexibility makes the technique very promising for the production of scaffolds with tailored porosity and adequate mechanical strength for bone tissue engineering. By adjusting the processing parameters, it is possible to develop hydrogels with characteristics that align with the specific needs of the bone regeneration process, enabling the creation of better, more effective, and biocompatible bone substitutes (Lin et al., 2021).

3.2.2 Electrospinning

A dominant advance of the technique is through the work of electrospinning, developed into a robust, flexible approach to fabricating nanofibrous scaffolds from HA-based solutions. It is essentially a novel technique for the direct construction of extracellular matrix-mimicking fibrous matrices with similar complexities to nature. This method uses a strong electric field that results in the jet of a polymer solution that forms a network of ultra-fine fibres with diameters ranging from nanometers to micrometers (Jun et al., 2018).

Basically, the electrospinning process involves dissolving HA in a suitable solvent. Quite often, other polymers will be added to the solution to improve the spinning capability of the solution and also to modulate the properties of fibers ensuing from it. Viscosity and conductivity are the two most critical parameters of polymer solutions that affect the electrospinning process substantially and hence the characteristics of final nanofibers also. Uniformity in jet formation as well as the deposition of fiber is determined by suitable viscosity and nominal electrical conductivity (Snetkov et al., 2019). In this case, Kim et al. demonstrated the ability to electrospin nanofibers by blending hyaluronic acid with polycaprolactone. This combination leverages the biocompatibility of HA and the favorable mechanical properties of PCL, creating a material that can support cellular adhesion and proliferation, key factors for tissue engineering applications targeting bone defects (Kim et al., 2008).

The electrospinning setup consists of a polymer solution loaded into a syringe fitted with a needle, which acts as a spinneret. After that, a high voltage is applied across the spinneret and a grounded collector to create a strong electric field. From here, with increasing voltage, electrostatic forces acting on the polymer solution overcome its surface tension, thus causing a jet of fluid to erupt from the tip of the spinneret. This jet elongates quickly and disperses while passing through the electrical field, and subsequently, the solvent evaporation is accompanied by polymer solidification into sub-micron fibers (Reneker et al., 2008; Rutledge, 2008; Rutledge et al., 2007). Bhattarai et al. (Bhattarai et al., 2018) demonstrated proof that electrospinning could be used to obtain HA nanofibers suitable for tissue engineering applications. These were further optimized to achieve uniform and continuous fibers with the properties needed through, for example, process parameters such as applied voltage, flow rate, and distance between spinneret and collector.

The electrospun fibers are collected on the grounded collector, usually in the form of a nonwoven mat. Morphology and characteristics of the electrospun fibers, like the fiber diameter, alignment, and porosity, can also be controlled efficiently by the variation of a number of parameters within the electrospinning process itself. These parameters include applied voltage, solution flow rate, type of collector, and distance from spinneret to collector. Such a degree of control allows tuning the architecture and properties of the produced scaffold for a variety of tissue engineering applications (Dogan et al., 2009). For example, Li et al. (Li et al., 2004) formed aligned HA nanofibers by collecting using a rotating drum collector, which induced cells to tend to grow in this direction—a very desirable property where guided cell growth is required, such as in nerve regeneration and bone tissue engineering.

For bone tissue engineering, there are associated advantages of electrospinning. The nanofibrous scaffolds produced by this technique display an extremely large surface area-to-volume ratio, which may promote cellular attachment, spreading, and proliferation. Apart from that, control over fiber diameter and alignment allows one to manufacture scaffolds mimicking the natural ECM structure and mechanical properties. Such a biomimetic approach allows advanced cell-material interactions and may guide tissue regeneration, making electrospinning a very promising technique toward the development of advanced bone grafts or other tissue engineering constructs (Liu et al., 2020).

3.2.3 3D Printing

In the context of bone tissue engineering, additive manufacturing approaches have revolutionized the fabrication of patient-specific scaffolds from HA-based hydrogels. This methodology offers the opportunity for unprecedented control over architecture, pore size, and total geometry to form constructs that mimic the structure of natural bone.

First of all, the procedure requires the creation of a three-dimensional blueprint of the scaffold design in a computer-aided design program. This digital model may then be pre-designed to exactly replicate the geometry and dimensions of the bone defect site; this would ensure that the scaffold fits well in the patient. In this way, the personalized generation accommodates even those bone defects that are very complex or located in difficult anatomic locations. Once the digital model is created, loading of the HA-based hydrogel is either loaded into a 3D printer fitted with an extrusion nozzle or a specialized bioprinter with the capability to accurately deposit hydrogel layer-by-layer. The printing parameters such as extruded rate, layer thickness, printing speed, among others, are precariously optimized to ensure the deposited hydrogel forms an accurate structure of the scaffold. Schuurman et al. [ref] demonstrated high accuracy that bioprinting could achieve in manufacturing scaffolds based on HA with controlled deposition of several bioinks. By means of this strategy, scaffolds with gradient properties or scaffolds with heterogeneous cell distribution can be fabricated, thus mimicking the native tissue's complexity (Petta et al., 2020).

The cross-linking of the hydrogel after printing is a very important step: it stabilizes the scaffold and gives the desired mechanical properties. Cross-linking can be induced both physically and chemically or enzymatically, and different methods give advantages in the final properties of the scaffold. In the case of photo-crosslinkable hydrogels, the crosslinking is initiated by UV or visible light, permitting the preparation of very complex geometries with a high degree of accuracy. The visibility of the visible-light crosslinking in methacrylated HA hydrogels was shown to have excellent mechanical and biochemical properties by Kesti et al. (Kesti et al., 2015).

For instance, using CAD methodology, Murphy and Atala designed scaffolds with complex pore architectures in an attempt to enhance nutrient diffusion and cell migration—two of the major factors governing successful bone regeneration (Bishop et al., 2017). Similarly, Noh et al. developed HA-based hydrogels for bone regeneration applications, as illustrated in Figure 6. Their results demonstrated the potential of the HA-g-pHEA-Gelatin hydrogel as a bioink material for tissue engineering, including injectable and 3D bioprinting uses (Noh et al., 2019).

Besides, 3D printing for bone tissue engineering provides another advantage in the sense that it allows the scaffold architecture to be very accurately controlled. 3D printing enables the achievement of highly interconnected pore networks within the scaffolds, thus promoting cell seeding, nutrient diffusion, and waste removal—a factor highly important in cellular survival and the growth of tissues. Further, 3D printing can contribute topographical cues and biochemical signals in the scaffold to bring guiding forces for proper tissue regeneration.

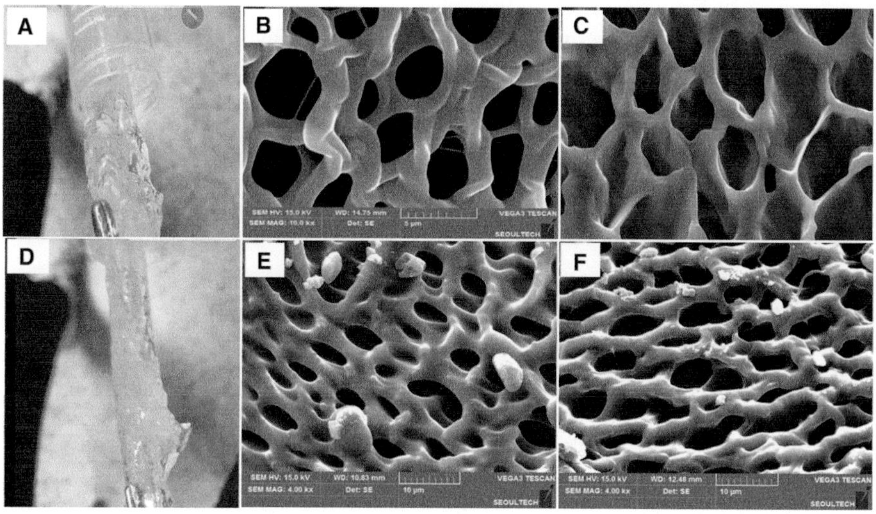

Figure 6. Digital (a, d) and surface (b, e) and cross-section (c, f) of scanning electron microscopy (b, c, e and f) morphologies of HA-g-pHEA-Gelatin hydrogel (a, b and c: 6% methacrylic anhydride and D, E and F: 8% methacrylic anhydride). The gel composition is 0.25 g HA, 3 mL HEA, 0.25 g Gelatin-MA, and the scale bars of (b and c) and (e and f) are 5 μm and 10 μm, respectively [129].

Table 5. Comparison of different fabrication techniques for HA hydrogels in bone tissue engineering.

Fabrication Technique	Control Over Porosity	Mechanical Properties	Biocompatibility	Processing Considerations
Sol-Gel Processing	Porosity can be influenced by the sol-gel transition and drying process.	Can create both soft and rigid gels depending on the drying process.	Typically biocompatible; residual solvents need to be removed.	Involves complex and time-consuming steps, including drying and aging.
Electrospinning	Produces nanofibrous structures with highly tunable porosity.	Fibrous structure enhances mechanical strength and elasticity.	Can incorporate bioactive molecules and cells, maintaining functionality.	Requires specialized equipment and careful parameter optimization.
3D Bioprinting	Enables precise design of porosity and architecture through CAD modeling.	Mechanical properties can be tailored by adjusting the material composition and printing parameters.	Allows for the incorporation of living cells and growth factors directly into the scaffold.	Requires advanced equipment and expertise in CAD and biomaterial handling.

It has been revealed that 3-D printing is a disruptive technology in the field of bone tissue engineering due to the capability it offers in fabricating scaffolds with unprecedented control over their design and fabrication. The ability of this technology to form complicated architecture with tailored properties offers high promise in the effectiveness of bone regeneration therapies.

In bone tissue engineering, the fabrication of HA hydrogels requires careful consideration of factors such as porosity, mechanical properties, biocompatibility, and ease of processing. Different fabrication techniques offer unique advantages and challenges. Table 5 shows the key aspects of three prominent techniques including, Sol-Gel Processing, Electrospinning, and 3D Bioprinting.

4. Hyaluronic Acid Hydrogels: Modification and Functionalization

Although HA hydrogels are inherently biocompatible and biodegradable, they very often must be modified and functionalized in some form to fulfill the potential of tissue engineering and regenerative medicine. Modifications of HA hydrogels can be through entrapping bioactive agents or through chemical modifications in various ways to show specific biological effects. The changes can increase cell-matrix adhesion and—in most cases—lead to controlled tissue regeneration or most often controlled drug delivery. This section explores the diverse approaches and strategies that have been applied to date in the tailoring of HA hydrogels toward improved functionality and therapeutic efficacy.

4.1 Chemical Modifications for Enhanced Properties

Hyaluronic acid-based hydrogels inherently possess excellent biocompatibility and biodegradability, yet they are often chemically modified to tailor their mechanical strength, stability, and biological response for specific tissue engineering and regenerative medicine applications. The modification process introduces new functional groups or facilitates linking of HA molecules with other agents, leading to the final properties of the hydrogel.

One of these most used types of modification strategies is methacrylation, which allows the incorporation of methacrylate groups into the HA backbone. Conjugation allows this modification to be written in with HA backbones to enable photo-crosslinking; upon irradiation, the methacrylated HA will undergo phase interconnection to form a stable hydrogel network. HA hydrogels that are methacrylated-modified show superior mechanical properties and are thus well suited for those applications where an increase, stiffness, or resistance is required against degradation (Jiang et al., 2009). For instance, Kim et al. (Kim et al., 2011) has reported on the potential of MeHA hydrogels in cartilage tissue engineering, showing significantly improved mechanical properties compared with HA-based hydrogels.

Sulfation involves the attachment of sulfate groups into the HA molecule to mimic the structures of other glycosaminoglycans such as heparin. The hydrogel, through the immobilization of these sulfate groups, has a better binding capacity to growth factors and cytokines, which are two large groups of signaling molecules critically deregulated in tissue regeneration. Sulfated HA hydrogels may facilitate cell signaling upon tissue regeneration by inducing the immobilization of these signaling molecules (Feng et al., 2017). Wang et al. (Yao et al., 2019) have demonstrated the effectiveness of sulfated HA hydrogels to promote osteoblast proliferation and differentiation and therefore they may find applications in bone tissue engineering.

Thiolation is the process through which HA introduces thiol groups into its backbone, to yield under oxidative conditions disulfide bonds. This strategy modifies hydrogels that are redox-responsive, through which they have the ability to degrade due to different environmental changes in the surrounding oxidative environment. Considering this special point of thiolated HA hydrogels, their applications in controlled drug release become quite tempting. The redox state control at tissue can further control the release rate dynamics of the loaded therapeutic agents from these hydrogels, providing local synthesis for a particular responsive drug delivery system in a spatiotemporally responsive manner (Grosso et al., 2021).

HA can be acetylated and deacetylated through the addition and removal of acetyl groups, respectively, to change the hydrophilicity of the hydrogel and its intermolecular interactions with cells and other molecules. Acetylated HA is not only more hydrophobic, but there is also better interaction between the polymer and hydrophobic drugs. This is useful in drug loading techniques and makes it possible to program controlled release profiles. As an example, acetylated HA hydrogels are very promising for the delivery of anti-inflammatory drugs for osteoarthritis treatment (Trombino et al., 2019).

4.2 Incorporation of Bioactive Molecules

Incorporation of bioactive molecules into HA hydrogels offers a powerful approach to enhance the biological activity of HA-derived hydrogels with an increasing therapeutic potential. This inclusion could range from embedding such molecules to their covalent linkage within the hydrogel matrix and would result in sophisticated bioactive materials with the potential to govern cellular behavior and support tissue regeneration.

Growth factors are potent regulators of cell proliferations, differentiation, and tissue regeneration; thus their presence in an HA hydrogel would confer localized, sustained delivery and high therapeutic efficacy toward that purpose. For instance, bone morphogenetic proteins, with well-documented osteoinductive effects, have been incorporated into HA hydrogels to produce bone. As demonstrated by Patel et al. in a critical-sized bone defect model, BMP-2-loaded hyaluronic acid hydrogels were able to significantly promote bone regeneration (Patel et al., 2019).

Biological activity in HA hydrogels can also modify the use of bioactive peptides. These are short amino acid sequences that are well known to have biological activities themselves. For instance, the RGD peptide is known to promote cell attachment by the interaction of integrin receptors of the cell membrane. Grafting of RGD peptides into HA hydrogels is to enhance cell attachment and proliferation. They may be used in many functions in tissue engineering. Burdick et al. showed that the HA hydrogels functionalized with RGD promoted the attachment and growth of osteoblasts and chondrocytes and are at the top of promising things in the aspect of bone and cartilage tissue engineering (Burdick et al., 2011).

Consequently, HA hydrogels could become a very desirable system for drug delivery systems given their biocompatibility, biodegradability, and tunable properties. Drugs can be physically entrapped in the hydrogel matrix by chemical conjugation with the HA molecules—therefore, by enabling release profiles 'with control'. Controlled release is effectively achieved by the method, thus minimizing systemic side effects and ensuring that the drug is specifically delivered to the affected tissue. For example, loading dexamethasone to HA hydrogels for sustained drug release intended for the treatment of inflammatory conditions, providing site-specific and extended anti-inflammatory effects-which may be practically used (Hickey et al., 2002).

Incorporation of enzymes in HA hydrogels brings in bio-catalytic functions: for the said case, one can only have specific biochemical reactions *in situ*. In this regard, this strategy has the potential for application where localized enzymatic activity will be achieved—for example, in wound healing. In another illustration, the incorporation of collagenase, an enzyme that degrades collagen, will swell HA hydrogels, and this degeneration of damaged collagen in wounds speeds up tissue repair (Chang et al., 2013).

Table 6 serves as a concise reference for understanding how different bioactive molecules contribute to the efficacy of HA hydrogels in bone tissue engineering, offering insights into their application in advancing bone healing and regeneration. Through the incorporation of bioactive molecules, HA hydrogels are converted from being unable to interfere support scaffolds into working biomaterials that

Table 6. Examples of bioactive molecules incorporated into HA hydrogels and their therapeutic targets.

Bioactive Molecule	Therapeutic Target
Bone Morphogenetic Proteins (BMPs)	Osteogenesis: Promotes bone formation and enhances bone healing.
Vascular Endothelial Growth Factor (VEGF)	Angiogenesis: Stimulates blood vessel formation to support bone tissue growth.
Transforming Growth Factor-beta (TGF-β)	Chondrogenesis and Osteogenesis: Induces cartilage and bone formation.
Fibroblast Growth Factor (FGF)	Cell Proliferation and Differentiation: Enhances the proliferation and differentiation of osteoblasts.
Platelet-Derived Growth Factor (PDGF)	Bone Healing: Promotes the recruitment and proliferation of osteogenic cells.
Insulin-Like Growth Factor (IGF-1)	Bone Matrix Formation: Stimulates bone matrix production and mineralization.
Bone Morphogenetic Proteins (BMPs)	Osteogenesis: Promotes bone formation and enhances bone healing.

can architect cellular behavior and stimulate tissue regeneration. Such materials harbor a high potential in many bioengineering areas, including tissue engineering, regenerative medicine, and various drug delivery.

4.3 HA-Based Composites

Although a great number of beneficial properties are attributed to HA hydrogels, the combination with other biomaterials in composite materials is a very potent approach for further improving mechanical and biological performance. This method uses the particular characteristics of all components to compose synergistic biomaterials for a specific tissue engineering use or application in regenerative medicine.

Immobilization of ceramics such as hydroxyapatite or tricalcium phosphate within HA hydrogel provided mechanical reinforcement and osteoconductivity. Such ceramics are able to mimic the mineral part of natural bone and thus provide a good environment for attachment, proliferation, and differentiation of bone cells. For example, HA/HA and HA/TCP composite hydrogels developed by Kim et al. showed improved support of bone cell growth and *in vivo* promotion of bone regeneration (Kim et al., 2014).

Combinations of HA and synthetic polymers, like polyethylene glycol (Yuan et al., 2019) or polycaprolactone (Hamlet et al., 2017), in the hydrogel provide control over mechanical properties and degradation rates. Such polymers would increase stability, strength, and elasticity of the formed hydrogel so that it is tailored for applications involving very specific mechanical requirements. Combined HA-PEG hydrogels have recently been reviewed as versatile platforms and vehicles either for cell encapsulation or drug delivery due to their tunable properties and biocompatibilities.

Blending HA with other natural biopolymers, such as collagen, gelatin, or chitosan, exploits their intrinsic biocompatibility and bioactivity. Quite often, these kinds of natural polymers bear cell-binding domains and hence are capable of mimicking the extracellular matrix, enhancing cell adhesion, proliferation, and differentiation. For example, composite hydrogels of HA-collagen have found application in the successful generation of biomimetic scaffolds for tissue engineering of cartilage and bone (Xu et al., 2012). Levett et al. have composites that support the growth and differentiation of chondrocytes and osteoblasts when demonstrating enablement for cartilage and bone regeneration (Levett et al., 2014).

Such beneficial properties of HA make the materials, in conjunction with other biomaterials, very useful in building a diverse library of composite hydrogels with tailored properties. Because of this versatility, HA-based composite hydrogels become prime candidates for an extremely wide range of biomedical applications related to tissue engineering and drug delivery and wound healing.

5. Characterization of Hyaluronic Acid Hydrogels

Characterization of hyaluronic acid hydrogels is important in understanding their physicochemical properties and relating them to performance in various biomedical applications. This section shall explore techniques characterizing structure, mechanical, and degradation properties of HA hydrogels and elaborate on what makes them good tissue engineering scaffolds for use in regenerative medicine.

5.1 Structural Characterization

Understanding the physicochemical structure of HA hydrogels is of a great interest to understand and predict their behavior in biological environments. Structural scale characterization can give information on morphology, composition, and molecular interactions quite usefully.

The surface morphology and microstructure of HA hydrogels can be analyzed, since the resolution is made possible by the visualization of SEM. The technique involves the use of a focused electron beam to scan the surface of a sample and obtain its high-resolution images, which clearly show the porous structure, fiber alignment, and general morphology of the hydrogel. For instance, Yang et al. used SEM imaging of the microstructure of HA-based hydrogels, showing evidence of their porous architecture with proper interconnected fiber networks, which is a critical property for cell infiltration and the diffusion of nutrients in tissue engineering (Qiao et al., 2020).

TEM provides even higher resolving power than the SEM and so allows for the inside of HA hydrogels to be visualized at a nanoscale. One can obtain images showing the internal features at high resolution in the hydrogel by passing an electron beam through an ultra-thin specimen. Zhao et al. used TEM to visualize the nanoscale fibrillar structure within HA hydrogels, which allowed unraveling their mechanical properties and molecular interactions with cells (Zhao et al., 2021).

FTIR spectroscopy is a powerful tool for high-precision characterization of the chemical composition and functional groups within HA hydrogels. This technique generates a unique spectral fingerprint, measuring the absorption of infrared light at different wavelengths, which provides insights into the molecular bonds and interactions present in the hydrogel. FTIR analysis can be used to verify the successful incorporation of functional groups onto the HA backbone, as demonstrated in studies by Burdick and Prestwich (Burdick et al., 2011).

NMR spectroscopy provides detailed information on molecular structure and dynamics of HA hydrogels. Characterizing the magnetic properties of atomic nuclei within the hydrogel, NMR can reveal chemical environment and interactions of HA molecules. For example, the application of 1H NMR or 13C NMR has been used to describe the degree of modification and cross-linking in HA hydrogels. These allowed determination of restricting information or chemical stability and functionality (Burdick et al., 2011).

The differences in information provided by the different techniques of microscopy and spectroscopy give a good complementary overview of structural features of HA hydrogels in correlation to their properties, in different biomedical applications.

5.2 Mechanical Properties

Mechanical properties are one of the most important parameters in ensuring the performance of hydrogels for various biomedical applications, particularly in tissue engineering. In tissue engineering, the hydrogels should bear not only a certain magnitude of physiological loads but also provide an appropriate mechanical environment similar to that of their native tissues. An in-depth understanding and control of such properties will tailor ready hydrogels for specific applications.

The testing for compression is the basic method of investigating the behavior of HA hydrogels under these compressive forces. This procedure gives information relating to factors, including the stiffness and elasticity, of the hydrogel and its ability to hold compression without any fracture or visible deformation. Generally, it will imply that during a test for compression with a cylindrical or cuboidal sample in this hydrogel, it would be subjected to a controlled compressive force, and deformation at that point would be measured. The compressive modulus, often reported as Young's modulus, represents the resistance of the hydrogel to its own deformation under compression and can be worked out from the stress-strain curve. For instance, Kim et al. investigated the mechanical characteristics of HAp-HA composite hydrogels for cartilage regeneration. Using compression testing, the results demonstrated that the compressive modulus of these hydrogels could be tunable due to the differences in HA concentration and crosslinking density, resulting in a material that bore the same kind of mechanical features as indigenous cartilage. This tunability is significant for providing an analogous mechanical environment compared with the target tissue in support of cell growth and differentiation (Kim et al., 2021).

Tensile testing determines the ability of the hydrogel to bear tensile applied forces, indicating its strength, flexibility, and ductility. Tensile testing is the process of material deformation by imposition and measurement of relative deformation when the force applied is controlled, at least until the point of failure, usually fracture or rupture. Important parameters about the tested material that could be derived from tensile testing include tensile strength, generally the maximum amount of the stress the hydrogel can bear without breaking, and elongation at break, which characterizes its ductility. Lee et al. also reported tensile testing results of HA-tyrosine hydrogels for soft tissue engineering. In comparison with unmodified HA hydrogels, tensile strength and elongation at break were significantly increased through the incorporation of tyrosine into the HA network. This latter modification improved these mechanical properties due to additional crosslinking and strengthening hydrogen-bonding interactions within the hydrogel network, upon chemical modifications, which suggests the potential of chemical alterations that might affect mechanical behavior of HA hydrogels (Lee et al., 2018).

Table 7. Summary of mechanical properties of different HA hydrogels for bone regeneration applications.

Type of HA Hydrogel	Compressive Modulus (kPa)	Tensile Strength (kPa)	Elongation at Break (%)	Storage Modulus (kPa)	Loss Modulus (kPa)	Potential Applications
Ionic Crosslinked HA Hydrogel	20-100	10-40	10-50	15-60	5-20	Early-Stage Bone Healing: Suitable for initial stabilization and supporting bone regeneration.
Chemically Crosslinked HA Hydrogel	200-500	50-200	5-30	150-400	50-150	Bone Scaffolds: High mechanical strength for load-bearing bone regeneration.
Hybrid HA-Collagen Hydrogel	100-300	30-100	10-40	100-300	30-100	Osteochondral Tissue Repair: Balances stiffness and flexibility for complex bone-cartilage interfaces.
Photocrosslinked HA Hydrogel	250-600	100-250	5-20	200-500	80-200	Customized Bone Implants: High strength and precision for 3D-printed bone structures.
Nanocomposite HA Hydrogel	300-800	150-300	5-15	300-700	100-250	Load-Bearing Bone Applications: Enhanced stiffness and mechanical performance for load-bearing implants.

Characterizing the viscoelastic properties of HA hydrogels is essential to understanding their behavior. These measurements reveal the hydrogel's storage modulus, representing the elastic component, and the loss modulus, representing the viscous component. These moduli are frequency-dependent, meaning their values vary with the applied oscillatory strain. Elisseeff and colleagues conducted pioneering research on the rheological behavior of methacrylated HA hydrogels, demonstrating that these materials have tunable viscoelastic properties. The storage and loss moduli can be modulated by the degree of methacrylation and crosslinking density. This tunability makes methacrylated HA hydrogels versatile biomaterials, as their mechanical properties can be tailored to match the target tissue requirements in various biomedical applications (Khunmanee et al., 2017).

Such mechanical testing strategies can be combined to give a basic understanding of the mechanical behavior of HA hydrogels for a given biomedical application. This information is of paramount importance in designing and developing HA-based biomaterials with tailored mechanical properties that will result in regeneration of native tissues in our body.

The mechanical properties of HA hydrogels are crucial for their application in bone regeneration, where they must provide sufficient support, stability, and strength to facilitate bone repair and regeneration. Table 7 summarizes the mechanical properties of various HA hydrogels specifically tailored for bone tissue engineering applications. These hydrogels are designed to meet the specific demands of bone regeneration, such as supporting load-bearing areas, providing initial stabilization, and integrating with surrounding bone tissue. The information provided helps in selecting the most appropriate HA hydrogel type for different stages and requirements of bone regeneration.

5.3 Swelling and Degradation Behavior

Swelling and degradation behavior basically dictate the performance and longevity of HA hydrogels in biological environments. These properties mandate how the hydrogel interacts with surrounding tissues, releases therapeutic agents, and eventually degrades in the body. Indeed, swelling studies provide information concerning the rate at which the hydrogel bulges and the possible structural stability in the physiological environment. This property is very essential for the application of hydrogels in drug delivery systems where the hydrogel should act as a reservoir of therapeutic molecules (Peppas et al., 2000).

Swelling behavior of HA hydrogels is influenced by their molecular weight, cross-linking density within the network, and the presence of either hydrophilic or hydrophobic groups inside the matrix. One of the typical ways of estimating swelling is derived from immersing a hydrogel in physiological solution, usually phosphate-buffered saline, and then measuring its change in weight or volume versus time. The swelling ratio defines the ratio of swollen weight or volume to initial weight or volume and gives a quantitative way to denote the swelling capacity of the hydrogel. For example, Heo et al. conducted research on the swelling property of dexamethasone-loaded HA hydrogels aimed at sustained delivery. Their work clearly demonstrated that one needs to very carefully control the swelling behavior of the hydrogel to achieve the desired controlled release of drugs; over-swelling could drive premature drug release, while too low swelling could hinder the therapeutic diffusion of the therapeutic molecules released from the hydrogel matrix (Caccavo et al., 2016; Heo et al., 2019).

Degradation of HA hydrogels is another crucial aspect to consider, as degradation products and rates directly impact their biocompatibility, cytotoxicity, and drug release kinetics. The understanding of degradation behavior for HA hydrogels is important to predict their shelf life and to consider proper, safe elimination from the body when the purpose would have been served by the HA hydrogels. Degradation studies are hence undertaken to predict the rate and extent of the breakdown of the hydrogel under physiological conditions, hence offering invaluable information for the design of hydrogels with predictable degradation profiles (Sahoo et al., 2008).

Typical *in vitro* degradation studies involve incubating the hydrogel in a solution that simulates physiological conditions, such as phosphate-buffered saline or a solution containing enzymes. During the degradation process, researchers measure the mass loss over time as well as changes in mechanical properties, including compressive modulus and tensile strength (Hahn et al., 2007). Thiolated HA hydrogels were also tested for degradation behavior by Ding et al. Those gels showed accelerated degradation in the presence of reducing factors like glutathione, which is highly abundant in the biological environment, especially in tumor tissues. This redox-responsive degradation also portends the basis for thiolated HA hydrogels in targeted drug delivery at tumor sites (Ding et al., 2012).

While *in vitro* degradation studies provide important details, the actual biological behavior of the hydrogel cannot be completely replicated in this controlled environment. Therefore, *in vivo* testing through animal models is crucial to evaluate the hydrogel's biocompatibility, interactions with surrounding tissues, and degradation kinetics. For instance, Halloway et al. investigated the *in vivo* degradation of BMP-2-loaded HA hydrogels implanted in a rat calvarial defect. Their findings demonstrated the biocompatibility of the hydrogels, with their gradual breakdown coinciding with bone regeneration. Notably, the degradation products of HA, such as oligosaccharides, are not known to be toxic and may even possess wound-healing properties (Holloway et al., 2014).

The characteristics and behaviors of swelling and degradation exhibited by HA hydrogels can be easily studied, which can guide the design and development of new biomaterials for tissue engineering and regenerative medicine applications.

6. Hyaluronic Acid Hydrogels: Biological Evaluation

Critical consideration during the appraisal of HA hydrogels for tissue engineering and regenerative medicine applications is rigorous biological assessment. This assessment is biconduitive in nature, combining both *in vitro* and *in vivo* processes. These studies will be about trying to fully understand cell behavior into the hydrogel, examining for its biocompatibility, and finally establishing its regenerative potential.

6.1 *In Vitro* Studies

In vitro studies provide a very controlled environment for hosting basic studies on the interaction of HA hydrogels with cells, particularly in studying cell viability, proliferation, and differentiation or biocompatibility.

6.1.1 Cell Viability and Proliferation

Cytocompatibility is based on the assessment of cells encapsulated within HA hydrogels. The viability of cultured cells can be determined using MTT, XTT, Live/Dead staining, or other similar staining methods; the data are generally quantitative regarding cell metabolic activities and membrane integrity, which are strong indicators of cell health and survival (Park et al., 2014). For instance, the HA hydrogels of Prestwich et al. (Prestwich, 2011) were shown fabulous cytocompatibility by finding close to 100% viability for the fibroblasts, using the MTT assay. Their results revealed that the HA-based hydrogels offered a supportive environment for cell survival.

Beyond the viability, cell proliferation of the cells on HA hydrogels must be determined if the hydrogel is going to support the proliferation of the new tissue. DNA synthesis assays, for example through the detection of BrdU incorporation, and also cell counting can be utilized to determine the rate at which cells are proliferating. Such studies can point out how much the scaffold will be supportive in terms of the cell proliferation. It was observed that in these studies the mesenchymal stem cells did may proliferate in RGD-derivatized HA hydrogels. This finding suggests the incorporation of bioactive cues into the hydrogel matrix, such as RGD peptides, in order to better the adhesion and propagation of cells (Bellis, 2011; Park et al., 2003).

6.1.2 Osteogenic Differentiation

This implies that their promotional potential in osteogenic differentiation by HA hydrogels is beyond their viability and proliferation, considering bone tissue engineering applications. Along with this, several methodologies are engaged to evaluate osteogenic differentiation through viewing markers of both early and late osteogenic differentiation. It is therefore common to measure the alkaline phosphatase activity when assessing the potential of osteoinduction in HA hydrogels, since enzyme activity serves as an early marker for the osteogenic differentiation process. These data further explains that increased ALP is indicative of cells committed to the osteogenic lineage (Hwang et al., 2023). This was demonstrated by Bai et al., wherein culturing with HAp-HA composite hydrogels increased the elicitation of a high ALP activity by MSCs, thereby presenting improved osteogenic differentiation (Bai et al., 2024).

Another important indicator of successful osteogenesis is calcium deposition—characteristic of mature osteoblasts and bone formation. Alizarin Red S and Von Kossa staining, together with quantitative calcium assays, are usually employed to assess and quantify the degree of mineralization in HA hydrogels (Xing et al., 2020). It was also found in one study that bone morphogenetic protein-2 loaded within HA hydrogels led to significantly high calcium deposition in such hydrogels, suggesting that these are potent hydrogels for bone tissue generation (Kang et al., 2011).

Molecular studies can also be done to understand the process of differentiation. For instance, quantitative analysis of expressions of osteogenic marker genes, such as RUNX2, osteocalcin, and collagen type I, can be conducted using qPCR and reverse transcription PCR. Lee et al. demonstrated that these osteogenic genes were upregulated in MSCs cultured in tyrosinase-crosslinked HA hydrogels, proving their osteoinductive potential (Lee et al., 2019).

6.1.3 Biocompatibility Assays

One of the major concerns when using HA hydrogels *in vivo* is their biocompatibility. Different assays are performed for the cytotoxic, hemocompatible, and immunogenic potentials of these hydrogels.

Cytotoxicity assays, such as lactate dehydrogenase release and annexin V/ propidium iodide staining, are gold standard inquisitional methods of cell membrane damage and apoptosis, respectively. These methods give accurate insights into the potential ability for HA hydrogels to induce cell death. For instance, Ding et al. demonstrated that the preparation exhibited excellent cytocompatibility by showing minimal LDH release and hence minimal cell membrane damage; therefore, it is safe for cell encapsulation (Wang et al., 2018).

For applications involving vascularization, the study of the interaction between HA hydrogels and blood components is always relevant. Test for the potential of the hydrogel to induce red blood cell rupture by hemolysis assays and platelet adhesion studies for thrombogenicity are required in vascular applications. Khadem et al., for instance, investigated the hemocompatibility of HA hydrogels containing silver nanoparticles and proved them applicable in wound healing since they exhibited very minimal hemolytic activities and platelet adhesion (Khadem et al., 2023).

Finally, understanding the inflammatory response of HA hydrogels is key to predicting their in vivo performance. This is easily achieved through assays of cytokine release by immune cells—for example, macrophages—through various available techniques, including enzyme-linked immunosorbent assays and flow cytometry. Pro-inflammatory and anti-inflammatory cytokines are quantified using these techniques, which give information on the immunomodulatory properties of these hydrogels. For example, the anti-inflammatory potential was shown by Reum et al., who reported that dexamethasone-loaded HA hydrogels decreased the levels of pro-inflammatory cytokines (Reum Son et al., 2015).

6.2 *In Vivo* Studies

In vivo studies on HA hydrogels are firmly implicated in the biological evaluation and enable the general understanding of their performance and safety in a living organism. Animal models involved in such studies fundamentally deal with bone regeneration, biocompatibility, and integration with host tissues.

6.2.1 Animal Models

Selection of the appropriate animal models for *in vivo* studies is therefore key to the translational findings. The choice of model depends on various factors, such as specific applications, size of bone defects, and the question asked.

One of the most commonly used is rodents, such as rats or mice, due to their lower cost, ease of handling, and genetic tools available. They may be used for preliminary testing of bone regeneration in small defects. Compared to rodents, rabbits possess larger bone structures that facilitate the creation of critical-sized defects and enable more comprehensive imaging and mechanical assessments. For instance, a study utilized a rabbit model to investigate the bone regeneration potential of 3D-printed hyaluronic acid hydrogels (Noh et al., 2019).

Thus, bigger animal models such as dogs or sheep are necessary for a closer evaluation with human-like scenarios of bony regeneration or when bigger defect sizes are required. However, their use has been confined by ethical consideration and also higher cost.

The method of implantation of the hydrogel has a direct impact for *in vivo* studies. Common techniques include:

- Subcutaneous implantation is when the hydrogel is placed under the skin. It has extensively been applied to basic studies on biocompatibility and early tissue responses.
- Intramuscular injection is a kind of direct injection of the hydrogel into the muscle for cell delivery tests or localized tissue regeneration.
- Direct implantation is likely the most translatable approach for bone tissue engineering. It enables direct assessment of bone regeneration within defect sites. For example, using this approach in a rat cranial defect model, Murphy and Atala showed that HA hydrogels are effective (Atala, 2002; Lyons et al., 2020; Stahl et al., 2021).

6.2.2 Bone Regeneration Assessment

Evaluation of bone repair in vivo greatly depends on noninvasive imaging techniques and mechanical testing for monitoring new bone formation and its functional integration.

X-ray radiography is a two-dimensional imaging modality that is used to examine bone structure and density for fracture healing and bone formation processes over time. Thus, three-dimensional images produced by use of micro-computed tomography supply great resolution in the study of bone architecture; therefore, it allows bone volume, mineral density, and trabecular structure to be evaluated in detail (Guldberg et al., 2004; Lima et al., 2009).

Magnetic resonance imaging provides information on soft tissue and bone marrow, which complements x-ray and micro-CT by visualizing the surrounding tissues and assessing vascularization (Majumdar, 2002).

The compressive strength measures the resistance of the regenerated bone to compression, hence survival of the loads. It would define the stiffness of the bone against resisting deformation under a state of stress. The fracture resistance is a measure of the energy amount needed to cause the fracture of a bone and could give some information about toughness of the bone (Bankoff, 2012).

6.2.3 Histological Analysis

Histological analysis is crucial for identifying the cellular and structural components of regenerated tissue and evaluating the integration between HA hydrogels and host tissues. Hematoxylin and eosin staining is a standard technique used to visualize general tissue morphology, differentiating the cell nucleus from the cytoplasm and staining the extracellular matrix pink. Masson's trichrome staining specifically stains collagen blue, enabling the assessment of collagen deposition and organization, which are critical for bone matrix formation (de Haas et al., 2014). The use of antibodies against specific proteins of interest provides insights into cellular activity and differentiation, such as the detection of osteocalcin or collagen type I. Confocal microscopy combined with fluorescent labeling allows for the visualization of specific cell types, proteins, or structures within the hydrogel and surrounding tissues. Live/dead staining can distinguish between living and dead cells, thereby providing information on the viability of cells within the hydrogel matrix. For instance, Kesti et al. employed fluorescence imaging to confirm the biocompatibility of HA and visualize the distribution and viability of cells inside the hydrogel matrix (Ho et al., 2022; Kesti et al., 2015).

In summary, *in vivo* studies are essential for assessing the clinical potential of HA hydrogels in bone tissue engineering. Such research provides critical parameters regarding bone regeneration, biocompatibility, and integration with host tissues, which can help optimize hydrogel properties and implantation strategies to achieve better outcomes.

7. Applications in Bone Tissue Engineering

Hyaluronic acid-based hydrogels thus present enormous potential within the bone tissue engineering arena. The perfect niche is formed with regard to biocompatibility, degradability, and providing a mechanistic base for cell growth and differentiation. This section highlights clinical applicability insights and potential that come from HA hydrogels, together with case studies and examples of their applications, and finally, all challenges and limitations existing with their employment.

7.1 Clinical Relevance and Potential

HA hydrogels present excellent scaffolds for bone regeneration because they could strongly simulate the natural extracellular matrix of bone tissue. They offer an excellent environment wherein osteogenic cells like mesenchymal stem cells can grow in number and differentiate into osteoblasts, which would then enable appropriate bone repair and regeneration.

One of the major advantages of HA hydrogels is their capability to embed and then release bioactive molecules in a controlled and temporal manner. This can be achieved through the incorporation of growth factors like bone morphogenetic proteins, which would enhance osteogenesis. For instance, BMP-2-loaded HA hydrogel systems have demonstrated enhanced bone healing in clinical settings by offering a controlled and sustained release for the growth factor to the target site.

HA is a naturally occurring polymer in the human body; therefore, it is highly biocompatible, inducing fewer adverse immune responses. Moreover, HA hydrogels are biodegradable and enable themselves to be gradually replaced without leaving behind harmful residues by newly formed bone tissue. What is particularly important about this property in tissue engineering applications in bone tissue could be related to its integration with host tissue over a very long time.

This trait makes HA hydrogels suitable for less invasive procedures. They can deliver by the use of minimally invasive techniques, reducing surgical trauma and improving patient outcomes, directly into the defect site. This feature is particularly useful for elderly patients or patients with compromised heath who have to go through an extensive surgical procedure..

7.2 Challenges and Limitations

Despite its very promising potential, there are some challenges and limitations associated with the use of HA hydrogels in bone tissue engineering.

One of the major drawbacks concerning HA hydrogels is their relatively low mechanical strength in comparison with native bone tissue. Of course, after modifications, stronger materials could be obtained, but still not enough for load-bearing applications. Investigations into the combination of HA with other materials, such as ceramics or synthetic polymers, are undertaken to compensate for this weakness.

Control over the HA hydrogels' degradation rate is indispensable and needs to be commensurate with new bone formation. If it is too fast, then tissue regeneration support can be undermined; if it happens at a pace that is too slow, then it might impede the integration of new tissue. In this direction, developing hydrogels having tunable degradation rates is quite important for their successful application.

Although HA is generally biocompatible, modifications and cross-linking processes may introduce new chemical groups that can cause an immune response. Ensuring the biocompatibility of modified HA hydrogels is thus very important to their safe application in a clinic. Rigorous testing for their biocompatibility and using biocompatible crosslinkers are necessary to reduce this risk.

Subsequently, the scalability and reproducibility of HA hydrogel fabrication processes will become crucial for clinical translation. Maintaining homogeneous quality and properties while producing large quantities is a challenge to wide diffusion in clinical use and achieving regulatory approval. Investigation of new fabrication technologies, including 3D printing and automated manufacturing methods, to meet these challenges is increasingly underway.

In particular, the manufacturing of HA hydrogels with advanced modifications can be really expensive. If the cost of their production is to be reduced while keeping up quality and effectiveness, this will greatly aid these materials in being more broadly applied in the clinic. Studies involving cheap methods and materials for preparation are underway. It is thus conclusive that HA-based hydrogels present enormous potential for application in bone tissue engineering owing to their biocompatibility, biodegradability, and capacity to support cell growth and differentiation. Challenges associated with mechanical strength, the rate of degradation, immune host response, scalability, and cost remain though, these materials are under continuous research and development to move the field forward. Case studies and clinical examples of HA hydrogel application in the promotion of bone regeneration further outline the great promise this material has for use in tissue engineering.

8. Future Perspectives and Emerging Trends

The field of hyaluronic acid-based hydrogels is becoming very dynamic; development and emerging trends open broad views on the use of these materials in the future engineering of bone tissue. This section reviewed recent improvements in hydrogel technology, their combination with other regenerative strategies, and potential applications for personalized medicine and tailored therapies.

8.1 Advances in Hydrogel Technology

Smart hydrogels are designed for the development of properties that, as a response to applied stimuli, undergo dynamic changes, whether it involves pH, temperature, or mechanical stresses. This makes them more functional for tissue application purposes. For example, the HA hydrogel, temperature-sensitive and also gelling at body temperature, provides less invasive delivery and more stable *in situ* formation for better integration with host tissues.

The incorporation of nanoparticles into HA hydrogels may increase the mechanical strength, bioactivity, and functional features to a great extent. Nanocomposites, for example, based on hydroxyapatite nanoparticles or carbon nanotubes give better mechanical support to make it more biomimetic with respect to natural bone. This could also imbibe new functionalities to the construct such as antimicrobial properties or targeted drug delivery.

Out of all the bioprinting technologies available today, HA hydrogels can make a reality the fabrication of 3D complex structures with high precision. Given the ability to create patient-specific scaffolds possessing complex geometries and gradients in mechanical/bioactive properties, this technology permits one to design constructs that very closely replicate natural bone tissue architecture and significantly enhances the effectiveness of regeneration.

8.2 Integration with Other Regenerative Approaches

By the combination of HA hydrogels with gene therapy, there has been very great potential for bone tissue engineering. Vectors of gene delivery can be combined with HA hydrogels for appropriate means of osteogenesis promotion through sustained release of growth factors or therapeutic genes and tissue repair.

One of the latest trends in tissue engineering is using scaffold-free approaches where the hydrogels of HA would induce the formation of cell aggregates or spheroids and allow these aggregates to self-assemble into tissue structures. In this way, it can obviate potential scaffold-related complications and let natural tissue architecture form.

Deepening our understanding of the immune response to the HA hydrogels and modulating this response will permit improving the biocompatibility and therapeutic outcomes. Another emerging area in regenerative medicine involves investigation into immunomodulating hydrogels, which would help tailor a positive healing environment and reduce inflammation.

8.3 Personalized Medicine and Tailored Therapies

The tendency in tissue engineering is toward the application of patient-specific data, like imaging and genetic information, for the design and construction of HA hydrogel scaffolds in a patient-specific manner. Personalized scaffolds improve integration and functionality of the regenerated tissue, thus yielding better clinical results.

For instance, targeted delivery at an injured or diseased site can be engineered with HA hydrogels for delivery of drugs or bioactive molecules. Such a targeted drug delivery system could increase therapeutic efficacy while reducing the side effects of these therapies against bone-related conditions.

Perhaps the most important research is the development of HA hydrogels that will interact with the biology of the patient at a molecular level to enhance tissue regeneration. Being developed now, such biomaterials will provide all the necessary signals for cellular growth, differentiation, and functional tissue formation and form an important part of the personalized regenerative therapies.

9. Conclusion

This chapter has provided an overview of hyaluronic acid-based hydrogels for bone tissue engineering applications. Hyaluronic acid-based hydrogels represent one of the most promising biomaterials in bone tissue engineering today because they possess a unique combination of properties:

- HA is present in the human body, so there is little acceptance upon implantation, hence decreasing the possibility of immune rejection
- HA biodegrades into non-toxic by-products which are naturally processed by the body.
- The synthesis and fabrication methods can be changed to vary the mechanical strength and degradation rate of HA hydrogels.
- HA hydrogels support attachment, proliferation, and differentiation of cells—all events associated with the technique of bone regeneration.

In vitro and *in vivo* studies have proven HA hydrogels to be effective for bone regeneration. These studies have exposed the potentials of HA hydrogels in many clinical applications, including:

- Bone defect repair: HA hydrogels help as scaffolds guiding the self-healing in bone tissue at the injury location.
- Tissue integration enhancement: A number of studies have investigated the potential of HA hydrogels in enhancing the integration of implanted materials with the surrounding host tissue.

References

Aguilar, G. V. 2018. Introductory chapter: a brief semblance of the sol-gel method in research. Sol-Gel Method-Design and Synthesis of New Materials with Interesting Physical, Chemical and Biological Properties.

Alameddine, R. S., L. Hamieh and A. Shamseddine. 2014. From sprouting angiogenesis to erythrocytes generation by cancer stem cells: evolving concepts in tumor microcirculation. BioMed Research International 2014(1): 986768.

Al-Fotawi, R. and W. Fallatah. 2023. Revascularization and angiogenesis for bone bioengineering in the craniofacial region: a review. Journal of Materials Science: Materials in Medicine 34(6): 30.

Allison, D. D. and K. J. Grande-Allen. 2006. Hyaluronan: a powerful tissue engineering tool. Tissue Engineering 12(8): 2131–2140.

Alqahtani, A. M. 2023. Guided tissue and bone regeneration membranes: a review of biomaterials and techniques for periodontal treatments. In Multidisciplinary Digital Publishing Institute (Vol. 15, Issue 16, p. 3355). doi: 10.3390/polym15163355.

Atala, A. 2002. Tissue engineering for the correction of deficits in body structure and function. E-Biomed: The Journal of Regenerative Medicine 3(1): 1–6.

Avila, L. Z., D. A. Gianolio, P. A. Konowicz, M. Philbrook, M. R. Santos and R. J. Miller. 2008. Drug delivery and medical applications of chemically modified hyaluronan. pp. 333–357. In Carbohydrate Chemistry, Biology and Medical Applications. Elsevier.

Azevedo, A. M., V. C. Martins, D. M. F. Prazeres, V. Vojinovic, J. M. S. Cabral and L. P. Fonseca. 2003. Horseradish peroxidase: a valuable tool in biotechnology. Biotechnology Annual Review 9(3): 1387–2656.

Bae, J. W., J. H. Choi, Y. Lee and K. D. Park. 2015. Horseradish peroxidase-catalysed in situ-forming hydrogels for tissue-engineering applications. Journal of Tissue Engineering and Regenerative Medicine 9(11): 1225–1232.

Bae, M. S., J.-Y. Ohe, J. B. Lee, D. N. Heo, W. Byun, H. Bae et al. 2014. Photo-cured hyaluronic acid-based hydrogels containing growth and differentiation factor 5 (GDF-5) for bone tissue regeneration. Bone 59: 189–198.

Bai, Y., Z. Wang, X. He, Y. Zhu, X. Xu, H. Yang et al. 2024. Application of bioactive materials for osteogenic function in bone tissue engineering. Small Methods 2301283.

Ballini, A., S. Cantore, S. Capodiferro and F. R. Grassi. 2009. Esterified hyaluronic acid and autologous bone in the surgical correction of the infra-bone defects. International Journal of Medical Sciences 6(2): 65.

Bankoff, A. D. P. 2012. Biomechanical characteristics of the bone. Human Musculoskeletal Biomechanics 61: 86.

Baron, R. 2008. Anatomy and ultrastructure of bone-histogenesis, growth and remodeling. Diseases of Bone and Mineral Metabolism. Endotext. Com2008.

Basara, G., X. Yue and P. Zorlutuna. 2019. Dual crosslinked gelatin methacryloyl hydrogels for photolithography and 3D printing. Gels 5(3): 34.

Beauvais, S., O. Drevelle, J. Jann, M.-A. Lauzon, M. Foruzanmehr, G. Grenier et al. 2016. Interactions between bone cells and biomaterials: An update. Front Biosci (Schol Ed) 8: 227–263.

Bellis, S. L. 2011. Advantages of RGD peptides for directing cell association with biomaterials. Biomaterials 32(18): 4205–4210.

Bhattarai, R. S., R. D. Bachu, S. H. S. Boddu and S. Bhaduri. 2018. Biomedical applications of electrospun nanofibers: Drug and nanoparticle delivery. Pharmaceutics 11(1): 5.

Bishop, E. S., S. Mostafa, M. Pakvasa, H. H. Luu, M. J. Lee, J. M. Wolf et al. 2017. 3-D bioprinting technologies in tissue engineering and regenerative medicine: Current and future trends. Genes & Diseases 4(4): 185–195.

Bokov, D., A. Turki Jalil, S. Chupradit, W. Suksatan, M. Javed Ansari, I. H. Shewael et al. 2021. Nanomaterial by sol-gel method: synthesis and application. Advances in Materials Science and Engineering 2021(1): 5102014.

Borzacchiello, A., L. Russo, B. M. Malle, K. Schwach-Abdellaoui and L. Ambrosio. 2015. Hyaluronic acid based hydrogels for regenerative medicine applications. BioMed Research International 2015(1): 871218.

Buckley, C., T. R. Montgomery, T. Szank, B. A. Murray, C. Quigley and I. Major. 2023. Modification of hyaluronic acid to enable click chemistry photo-crosslinking of hydrogels with tailorable degradation profiles. International Journal of Biological Macromolecules 240: 124459.

Burdick, J. A. and G. D. Prestwich. 2011. Hyaluronic acid hydrogels for biomedical applications. Advanced Materials 23(12): H41–H56.

Caccavo, D., S. Cascone, G. Lamberti and A. A. Barba. 2018. Hydrogels: experimental characterization and mathematical modelling of their mechanical and diffusive behaviour. Chemical Society Reviews 47(7): 2357–2373.

Caccavo, D., S. Cascone, G. Lamberti, A. A. Barba and A. Larsson. 2016. Swellable hydrogel-based systems for controlled drug delivery. Smart Drug Delivery System 237–303.

Caon, I., A. Parnigoni, M. Viola, E. Karousou, A. Passi and D. Vigetti. 2021. Cell energy metabolism and hyaluronan synthesis. Journal of Histochemistry & Cytochemistry 69(1): 35–47.

Carvalho, A. M., R. L. Reis and I. Pashkuleva. 2023. Hyaluronan receptors as mediators and modulators of the tumor microenvironment. Advanced Healthcare Materials 12(5): 2202118.

Casale, J. and J. S. Crane. 2019. Biochemistry, glycosaminoglycans.

Censi, R., P. J. Fieten, P. di Martino, W. E. Hennink and T. Vermonden. 2010. *In situ* forming hydrogels by tandem thermal gelling and Michael addition reaction between thermosensitive triblock copolymers and thiolated hyaluronan. Macromolecules 43(13): 5771–5778.

Chang, K.-H., H.-T. Liao, & J.-P. Chen. 2013. Preparation and characterization of gelatin/hyaluronic acid cryogels for adipose tissue engineering: *In vitro* and *in vivo* studies. Acta Biomaterialia 9(11): 9012–9026.

Choi, S., H. Ahn and S. Kim. 2022. Tyrosinase-mediated hydrogel crosslinking for tissue engineering. Journal of Applied Polymer Science 139(14): 51887.

Conduah, A. H., C. L. Baker III and C. L. Baker Jr. 2009. Managing joint pain in osteoarthritis: safety and efficacy of hylan GF 20. Journal of Pain Research 87–98.

Cook, M. T., P. Haddow, S. B. Kirton and W. J. McAuley. 2021. Polymers exhibiting lower critical solution temperatures as a route to thermoreversible gelators for healthcare. Advanced Functional Materials 31(8): 2008123.

D'Amora, U., A. Ronca, M. G. Raucci, S. M. Dozio, H. Lin, Y. Fan et al. 2019. *In situ* sol-gel synthesis of hyaluronan derivatives bio-nanocomposite hydrogels. Regenerative Biomaterials 6(5): 249–258.

de Haas, H. J., E. Arbustini, V. Fuster, C. M. Kramer and J. Narula. 2014. Molecular imaging of the cardiac extracellular matrix. Circulation Research 114(5): 903–915.

Deshmukh, K., T. Kovářík, T. Křenek, D. Docheva, T. Stich and J. Pola. 2020. Recent advances and future perspectives of sol–gel derived porous bioactive glasses: a review. RSC Advances 10(56): 33782–33835.

Dicker, K. T., L. A. Gurski, S. Pradhan-Bhatt, R. L. Witt, M. C. Farach-Carson and X. Jia. 2014. Hyaluronan: A simple polysaccharide with diverse biological functions. Acta Biomaterialia 10(4): 1558–1570. doi: https://doi.org/10.1016/j.actbio.2013.12.019.

Ding, J., R. He, G. Zhou, C. Tang and C. Yin. 2012. Multilayered mucoadhesive hydrogel films based on thiolated hyaluronic acid and polyvinylalcohol for insulin delivery. Acta Biomaterialia 8(10): 3643–3651.

Dogan, M., O. Karatay, D. Cokeliler and I. C. Kocum. 2009. Design and implementation of electrospinning system. 2009 14th National Biomedical Engineering Meeting 1–4.

Duan, B., L. A. Hockaday, E. Kapetanovic, K. H. Kang and J. T. Butcher. 2013. Stiffness and adhesivity control aortic valve interstitial cell behavior within hyaluronic acid based hydrogels. Acta Biomaterialia 9(8): 7640–7650.

Duarte, L., C. R. Matte, C. V. Bizarro and M. A. Z. Ayub. 2020. Transglutaminases: part I—origins, sources, and biotechnological characteristics. World Journal of Microbiology and Biotechnology 36: 1–18.

Einhorn, T. A. 1996. The bone organ system: form and function. Osteoporosis.

Fakhari, A. and C. Berkland. 2013. Applications and emerging trends of hyaluronic acid in tissue engineering, as a dermal filler and in osteoarthritis treatment. Acta Biomaterialia 9(7): 7081–7092.

Falcone, S. J., D. M. Palmeri and R. A. Berg. 2006. Rheological and cohesive properties of hyaluronic acid. Journal of Biomedical Materials Research Part A: An Official Journal of The Society for

Biomaterials, The Japanese Society for Biomaterials, and The Australian Society for Biomaterials and the Korean Society for Biomaterials 76(4): 721–728.

Fallacara, A., E. Baldini, S. Manfredini and S. Vertuani. 2018. Hyaluronic acid in the third millennium. Polymers 10(7): 701.

Feng, Q., S. Lin, K. Zhang, C. Dong, T. Wu, H. Huang et al. 2017. Sulfated hyaluronic acid hydrogels with retarded degradation and enhanced growth factor retention promote hMSC chondrogenesis and articular cartilage integrity with reduced hypertrophy. Acta Biomaterialia 53: 329–342.

Florea, D. A., C. Chircov and A. M. Grumezescu. 2020. Hydroxyapatite particles—directing the cellular activity in bone regeneration processes: an up-to-date review. Applied Sciences 10(10): 3483.

França, C. G., D. G. Villalva and M. H. A. Santana. 2021. Oxi-HA/ADH hydrogels: a novel approach in tissue engineering and regenerative medicine. Polysaccharides 2(2): 477–496.

Gan, Z., X. Qin, H. Liu, J. Liu and J. Qin. 2023. Recent advances in defined hydrogels in organoid research. Bioactive Materials 28: 386–401.

Geever, L. M., D. M. Devine, M. J. D. Nugent, J. E. Kennedy, J. G. Lyons, A. Hanley et al. 2006. Lower critical solution temperature control and swelling behaviour of physically crosslinked thermosensitive copolymers based on N-isopropylacrylamide. European Polymer Journal 42(10): 2540–2548.

Gottschalk, J., M. Aßmann, J. Kuballa and L. Elling. 2022. Repetitive synthesis of high-molecular-weight hyaluronic acid with immobilized enzyme cascades. ChemSusChem 15(9): e202101071.

Grieco, M., O. Ursini, I. E. Palamà, G. Gigli, L. Moroni and B. Cortese. 2022. HYDRHA: Hydrogels of hyaluronic acid. New biomedical approaches in cancer, neurodegenerative diseases, and tissue engineering. Materials Today Bio 17: 100453.

Grosso, R. and M.-V. de-Paz. 2021. Thiolated-polymer-based nanoparticles as an avant-garde approach for anticancer therapies—reviewing thiomers from chitosan and hyaluronic acid. Pharmaceutics 13(6): 854.

Guarino, V., M. G. Raucci, A. Ronca, V. Cirillo and L. Ambrosio. 2014. Multifunctional scaffolds for bone regeneration. pp. 95–117. In Bone Substitute Biomaterials. Elsevier.

Guldberg, R. E., A. S. P. Lin, R. Coleman, G. Robertson and C. Duvall. 2004. Microcomputed tomography imaging of skeletal development and growth. Birth Defects Research Part C: Embryo Today: Reviews 72(3): 250–259.

Hahn, S. K., J. K. Park, T. Tomimatsu and T. Shimoboji. 2007. Synthesis and degradation test of hyaluronic acid hydrogels. International Journal of Biological Macromolecules 40(4): 374–380.

Haider, A., S. Haider, S. S. Han and I.-K. Kang. 2017. Recent advances in the synthesis, functionalization and biomedical applications of hydroxyapatite: a review. Rsc Advances 7(13): 7442–7458.

Hamlet, S. M., C. Vaquette, A. Shah, D. W. Hutmacher and S. Ivanovski. 2017. 3-Dimensional functionalized polycaprolactone-hyaluronic acid hydrogel constructs for bone tissue engineering. Journal of Clinical Periodontology 44(4): 428–437.

Han, C., H. Zhang, Y. Wu, X. He and X. Chen. 2020. Dual-crosslinked hyaluronan hydrogels with rapid gelation and high injectability for stem cell protection. Scientific Reports 10(1): 14997.

Henderson, T. M. A., K. Ladewig, D. N. Haylock, K. M. McLean and A. J. O'Connor. 2015. Formation and characterisation of a modifiable soft macro-porous hyaluronic acid cryogel platform. Journal of Biomaterials Science, Polymer Edition 26(13): 881–897.

Hendi, A., M. Umair Hassan, M. Elsherif, B. Alqattan, S. Park, A. K. Yetisen et al. 2020. Healthcare applications of pH-sensitive hydrogel-based devices: a review. International Journal of Nanomedicine 3887–3901.

Henkel, J., M. A. Woodruff, D. R. Epari, R. Steck, V. Glatt, I. C. Dickinson et al. 2013. Bone regeneration based on tissue engineering conceptions—a 21st century perspective. Bone Research 1(1): 216–248.

Heo, J. Y., J. H. Noh, S. H. Park, Y. B. Ji, H. J. Ju, D. Y. Kim et al. 2019. An injectable click-crosslinked hydrogel that prolongs dexamethasone release from dexamethasone-loaded microspheres. Pharmaceutics 11(9): 438.

Hickey, T., D. Kreutzer, D. J. Burgess and F. Moussy. 2002. Dexamethasone/PLGA microspheres for continuous delivery of an anti-inflammatory drug for implantable medical devices. Biomaterials 23(7): 1649–1656.

Ho, T.-C., C.-C. Chang, H.-P. Chan, T.-W. Chung, C.-W. Shu, K.-P. Chuang et al. 2022. Hydrogels: properties and applications in biomedicine. Molecules 27(9): 2902.

Holloway, J. L., H. Ma, R. Rai and J. A. Burdick. 2014. Modulating hydrogel crosslink density and degradation to control bone morphogenetic protein delivery and *in vivo* bone formation. Journal of Controlled Release 191: 63–70.

Hosseinkhani, M., D. Mehrabani, M. H. Karimfar, S. Bakhtiyari, A. Manafi and R. Shirazi. 2014. Tissue engineered scaffolds in regenerative medicine. World Journal of Plastic Surgery 3(1): 3.

Hwang, C., Y. Min, Y.-J. Seong, D.-E. Kim, H.-E. Kim and S.-H. Jeong. 2019. Enhanced biolubrication on biomedical devices using hyaluronic acid-silica nanohybrid hydrogels. Colloids and Surfaces B: Biointerfaces 184: 110503.

Hwang, H. S. and C.-S. Lee. 2023. Recent progress in hyaluronic-acid-based hydrogels for bone tissue engineering. Gels 9(7): 588.

Itano, N. and K. Kimata. 2009. Altered hyaluronan biosynthesis in cancer progression. Hyaluronan in Cancer Biology 171–185.

Jiang, G., C. Liu, X. Liu, G. Zhang, M. Yang and F. Liu. 2009. Construction and properties of hydrophobic association hydrogels with high mechanical strength and reforming capability. Macromolecular Materials and Engineering 294(12): 815–820.

Jin, R., L. S. M. Teixeira, A. Krouwels, P. J. Dijkstra, C. A. Van Blitterswijk, M. Karperien et al. 2010. Synthesis and characterization of hyaluronic acid–poly (ethylene glycol) hydrogels via Michael addition: An injectable biomaterial for cartilage repair. Acta Biomaterialia 6(6): 1968–1977.

Jun, I., H.-S. Han, J. R. Edwards and H. Jeon. 2018. Electrospun fibrous scaffolds for tissue engineering: Viewpoints on architecture and fabrication. International Journal of Molecular Sciences 19(3): 745.

Kamata, H., X. Li, U. Chung and T. Sakai. 2015. Design of hydrogels for biomedical applications. Advanced Healthcare Materials 4(16): 2360–2374.

Kang, S.-W., J.-S. Kim, K.-S. Park, B.-H. Cha, J.-H. Shim, J. Y. Kim et al. 2011. Surface modification with fibrin/hyaluronic acid hydrogel on solid-free form-based scaffolds followed by BMP-2 loading to enhance bone regeneration. Bone 48(2): 298–306.

Kesti, M., M. Müller, J. Becher, M. Schnabelrauch, M. D'Este, D. Eglin et al. 2015. A versatile bioink for three-dimensional printing of cellular scaffolds based on thermally and photo-triggered tandem gelation. Acta Biomaterialia 11: 162–172.

Khadem, E., M. Kharaziha and S. Salehi. 2023. Colorimetric pH-responsive and hemostatic hydrogel-based bioadhesives containing functionalized silver nanoparticles. Materials Today Bio 20: 100650.

Khunmanee, S., Y. Jeong and H. Park. 2017. Crosslinking method of hyaluronic-based hydrogel for biomedical applications. Journal of Tissue Engineering 8: 2041731417726464.

Kim, H. H., J. B. Park, M. J. Kang and Y. H. Park. 2014. Surface-modified silk hydrogel containing hydroxyapatite nanoparticle with hyaluronic acid–dopamine conjugate. International Journal of Biological Macromolecules 70: 516–522.

Kim, I. L., R. L. Mauck and J. A. Burdick. 2011. Hydrogel design for cartilage tissue engineering: a case study with hyaluronic acid. Biomaterials 32(34): 8771–8782.

Kim, I. Y., M. K. Yoo, B. C. Kim, I. Y. Park, H. C. Lee and C. S. Cho. 2008. Thermogelling behaviors of poly (caprolactone-b-ethylene glycol-b-caprolactone) triblock copolymer in the presence of hyaluronic acid. Journal of Polymer Science Part A: Polymer Chemistry 46(11): 3629–3637.

Kim, J.-W., Y.-S. Han, H.-M. Lee, J.-K. Kim and Y.-J. Kim. 2021. Effect of morphological characteristics and biomineralization of 3D-printed gelatin/hyaluronic acid/hydroxyapatite composite scaffolds on bone tissue regeneration. International Journal of Molecular Sciences 22(13): 6794.

Kim, S.-H., S.-H. Lee, J.-E. Lee, S. J. Park, K. Kim, I. S. Kim et al. 2018. Tissue adhesive, rapid forming, and sprayable ECM hydrogel via recombinant tyrosinase crosslinking. Biomaterials 178: 401–412.

King, V. R., A. Alovskaya, D. Y. T. Wei, R. A. Brown and J. V. Priestley. 2010. The use of injectable forms of fibrin and fibronectin to support axonal ingrowth after spinal cord injury. Biomaterials 31(15): 4447–4456.

Kurniawansyah, I. S., T. Rusdiana, I. Sopyan, I. F. Desy Arya, H. A. Wahab and D. Nurzanah. 2023. Comparative study of *in situ* gel formulation based on the physico-chemical aspect: Systematic review. Gels 9(8): 645.

Lamberti, G., A. A. Barba, S. Cascone, A. Dalmoro and D. Caccavo. 2016. An engineering point of view on the use of the hydrogels for pharmaceutical and biomedical applications. pp. 201–230. In Emerging Concepts in Analysis and Applications of Hydrogels. Sutapa Biswas Majee.

Lee, F., J. E. Chung and M. Kurisawa. 2009. An injectable hyaluronic acid–tyramine hydrogel system for protein delivery. Journal of Controlled Release 134(3): 186–193.

Lee, H., J. Kim, C. Hwang, H. Kim and S. Jeong. 2018. Strategy for preparing mechanically strong hyaluronic acid–silica nanohybrid hydrogels via *in situ* sol–gel process. Macromolecular Materials and Engineering 303(9): 1800213.

Lee, H.-Y., H.-E. Kim and S.-H. Jeong. 2019. One-pot synthesis of silane-modified hyaluronic acid hydrogels for effective antibacterial drug delivery via sol–gel stabilization. Colloids and Surfaces B: Biointerfaces 174: 308–315.

Lee, J. E., Y. Yin, S. Y. Lim, E. S. Kim, J. Jung, D. Kim et al. 2019. Enhanced transfection of human mesenchymal stem cells using a hyaluronic acid/calcium phosphate hybrid gene delivery system. Polymers 11(5): 798.

Lee, K.-S., H.-A. Kim, Y.-S. Kim, H.-K. You and H.-S. Shin. 2002. The effect of high molecular hyaluronic acid on bone formation in human fetal osteoblasts. The Journal of the Korean Academy of Periodontology 32(3): 589–602.

Levett, P. A., F. P. W. Melchels, K. Schrobback, D. W. Hutmacher, J. Malda and T. J. Klein. 2014. A biomimetic extracellular matrix for cartilage tissue engineering centered on photocurable gelatin, hyaluronic acid and chondroitin sulfate. Acta Biomaterialia 10(1): 214–223.

Li, D., Y. Wang and Y. Xia. 2004. Electrospinning nanofibers as uniaxially aligned arrays and layer-by-layer stacked films. Advanced Materials 16(4): 361–366.

Li, L., Y. He, X. Zheng, L. Yi and W. Nian. 2021. Progress on preparation of pH/temperature-sensitive intelligent hydrogels and applications in target transport and controlled release of drugs. International Journal of Polymer Science 2021(1): 1340538.

Li, R., Z. Cai, Z. Li, Q. Zhang, S. Zhang, L. Deng et al. 2017. Synthesis of *in-situ* formable hydrogels with collagen and hyaluronan through facile Michael addition. Materials Science and Engineering: C 77: 1035–1043.

Lima, I., J. T. de Assis and R. T. Lopes. 2009. Three-dimensional conic beam X-ray microtomography in bone quality. Spectrochimica Acta Part B: Atomic Spectroscopy 64(11–12): 1173–1179.

Lin, H., C. Yin, A. Mo and G. Hong. 2021. Applications of hydrogel with special physical properties in bone and cartilage regeneration. Materials 14(1): 235.

Lin, P., S. Ma, X. Wang and F. Zhou. 2015. Molecularly engineered dual-crosslinked hydrogel with ultrahigh mechanical strength, toughness, and good self-recovery. Advanced Materials 27(12): 2054–2059.

Liu, T., X. Peng, Y. Chen, Q. Bai, C. Shang, L. Zhang et al. 2018. Hydrogen-bonded polymer–small molecule complexes with tunable mechanical properties. Macromolecular Rapid Communications 39(9): 1800050.

Liu, X., G. Zhang, C. Hou, H. Wang, Y. Yang, G. Guan et al. 2013. Vascularized bone tissue formation induced by fiber-reinforced scaffolds cultured with osteoblasts and endothelial cells. BioMed Research International 2013(1): 854917.

Liu, Y., W. He, Z. Zhang and B. P. Lee. 2018. Recent developments in tough hydrogels for biomedical applications. Gels 4(2): 46.

Liu, Z., S. Ramakrishna and X. Liu. 2020. Electrospinning and emerging healthcare and medicine possibilities. APL Bioengineering 4(3).

Luo, Y., J. Tan, Y. Zhou, Y. Guo, X. Liao, L. He et al. 2023. From crosslinking strategies to biomedical applications of hyaluronic acid-based hydrogels: A review. International Journal of Biological Macromolecules 231: 123308.

Lyons, J. G., M. A. Plantz, W. K. Hsu, E. L. Hsu and S. Minardi. 2020. Nanostructured biomaterials for bone regeneration. Frontiers in Bioengineering and Biotechnology 8: 922.

Majumdar, S. 2002. Magnetic resonance imaging of trabecular bone structure. Topics in Magnetic Resonance Imaging 13(5): 323–334.

Mero, A. and M. Campisi. 2014. Hyaluronic acid bioconjugates for the delivery of bioactive molecules. Polymers 6(2): 346–369.

Mohit, E., M. Tabarzad and M. A. Faramarzi. 2020. Biomedical and pharmaceutical-related applications of laccases. Current Protein and Peptide Science 21(1): 78–98.

Mori, M., M. Yamaguchi, S. Sumitomo and Y. Takai. 2004. Hyaluronan-based biomaterials in tissue engineering. Acta Histochemica et Cytochemica 37(1): 1–5.

Noh, I., N. Kim, H. N. Tran, J. Lee and C. Lee. 2019. 3D printable hyaluronic acid-based hydrogel for its potential application as a bioink in tissue engineering. Biomaterials Research 23(1): 3.
Orban, J. M., K. G. Marra and J. O. Hollinger. 2002. Composition options for tissue-engineered bone. Tissue Engineering 8(4): 529–539.
Pang, X., H. Li, F. Guan and X. Li. 2018. Multiple roles of glycans in hematological malignancies. Frontiers in Oncology 8: 364.
Pape, H. C., A. Evans and P. Kobbe. 2010. Autologous bone graft: properties and techniques. Journal of Orthopaedic Trauma 24: S36–S40.
Pardue, E. L., S. Ibrahim and A. Ramamurthi. 2008. Role of hyaluronan in angiogenesis and its utility to angiogenic tissue engineering. Organogenesis 4(4): 203–214.
Park, H. S., S. Y. Lee, H. Yoon and I. Noh. 2014. Biological evaluation of micro-patterned hyaluronic acid hydrogel for bone tissue engineering. Pure and Applied Chemistry 86(12): 1911–1922.
Park, Y. D., N. Tirelli and J. A. Hubbell. 2003. Photopolymerized hyaluronic acid-based hydrogels and interpenetrating networks. Biomaterials 24(6): 893–900.
Patel, J. M., K. S. Saleh, J. A. Burdick and R. L. Mauck. 2019. Bioactive factors for cartilage repair and regeneration: improving delivery, retention, and activity. Acta Biomaterialia 93: 222–238.
Peppas, N. A., P. Bures, W. S. Leobandung and H. Ichikawa. 2000. Hydrogels in pharmaceutical formulations. European Journal of Pharmaceutics and Biopharmaceutics 50(1): 27–46.
Pérez-Luna, V. H. and O. González-Reynoso. 2018. Encapsulation of biological agents in hydrogels for therapeutic applications. Gels 4(3): 61.
Petrey, A. C. and C. A. de la Motte. 2014. Hyaluronan, a crucial regulator of inflammation. Frontiers in Immunology 5: 101.
Petta, D., U. D'amora, L. Ambrosio, D. W. Grijpma, D. Eglin and M. D'este. 2020. Hyaluronic acid as a bioink for extrusion-based 3D printing. Biofabrication 12(3): 032001.
Prestwich, G. D. 2011. Hyaluronic acid-based clinical biomaterials derived for cell and molecule delivery in regenerative medicine. Journal of Controlled Release 155(2): 193–199.
Qiao, Y., S. Xu, T. Zhu, N. Tang, X. Bai and C. Zheng. 2020. Preparation of printable double-network hydrogels with rapid self-healing and high elasticity based on hyaluronic acid for controlled drug release. Polymer 186: 121994.
Rahar, S., G. Swami, N. Nagpal, M. A. Nagpal and G. S. Singh. 2011. Preparation, characterization, and biological properties of β-glucans. Journal of Advanced Pharmaceutical Technology & Research 2(2): 94–103.
Raia, N. R., B. P. Partlow, M. McGill, E. P. Kimmerling, C. E. Ghezzi and D. L. Kaplan. 2017. Enzymatically crosslinked silk-hyaluronic acid hydrogels. Biomaterials 131: 58–67.
Reddy, M. S. B., D. Ponnamma, R. Choudhary and K. K. Sadasivuni. 2021. A comparative review of natural and synthetic biopolymer composite scaffolds. Polymers 13(7): 1105.
Reneker, D. H. and A. L. Yarin. 2008. Electrospinning jets and polymer nanofibers. Polymer 49(10): 2387–2425.
Reum Son, A., D. Y. Kim, S. Hun Park, J. Yong Jang, K. Kim, B. Ju Kim et al. 2015. Direct chemotherapeutic dual drug delivery through intra-articular injection for synergistic enhancement of rheumatoid arthritis treatment. Scientific Reports 5(1): 14713.
Ruel-Gariepy, E. and J.-C. Leroux. 2004. *In situ*-forming hydrogels—review of temperature-sensitive systems. European Journal of Pharmaceutics and Biopharmaceutics 58(2): 409–426.
Ruppert, S. M., T. R. Hawn, A. Arrigoni, T. N. Wight and P. L. Bollyky. 2014. Tissue integrity signals communicated by high-molecular weight hyaluronan and the resolution of inflammation. Immunologic Research 58: 186–192.
Rutledge, G. C. 2008. Processes and applications of electrostatic fiber formation. Journal of Physics: Conference Series 142(1): 012026.
Rutledge, G. C. and S. V. Fridrikh. 2007. Formation of fibers by electrospinning. Advanced Drug Delivery Reviews 59(14): 1384–1391.
Sahoo, S., C. Chung, S. Khetan and J. A. Burdick. 2008. Hydrolytically degradable hyaluronic acid hydrogels with controlled temporal structures. Biomacromolecules 9(4): 1088–1092.
Santos, J. R., N. M. Alves and J. F. Mano. 2010. New thermo-responsive hydrogels based on poly (N-isopropylacrylamide)/hyaluronic acid semi-interpenetrated polymer networks: swelling properties and drug release studies. Journal of Bioactive and Compatible Polymers 25(2): 169–184.

Savani, R. C., G. Cao, P. M. Pooler, A. Zaman, Z. Zhou and H. M. DeLisser. 2001. Differential involvement of the hyaluronan (HA) receptors CD44 and receptor for HA-mediated motility in endothelial cell function and angiogenesis. Journal of Biological Chemistry 276(39): 36770–36778.

Sayed, M. M. El. 2023. Production of Polymer Hydrogel Composites and Their Applications. In Springer Science+Business Media (Vol. 31, Issue 7, pp. 2855–2879). doi: 10.1007/s10924-023-02796-z.

Seliktar, D. 2012. Designing cell-compatible hydrogels for biomedical applications. Science 336(6085): 1124–1128.

Seol, D., M. J. Magnetta, P. S. Ramakrishnan, G. L. Kurriger, H. Choe, K. Jang et al. 2013. Biocompatibility and preclinical feasibility tests of a temperature-sensitive hydrogel for the purpose of surgical wound pain control and cartilage repair. Journal of Biomedical Materials Research Part B: Applied Biomaterials 101(8): 1508–1515.

Serban, M. A., G. Yang and G. D. Prestwich. 2008. Synthesis, characterization and chondroprotective properties of a hyaluronan thioethyl ether derivative. Biomaterials 29(10): 1388–1399.

Shin, J., J. S. Lee, C. Lee, H. Park, K. Yang, Y. Jin et al. 2015. Tissue adhesive catechol-modified hyaluronic acid hydrogel for effective, minimally invasive cell therapy. Advanced Functional Materials 25(25): 3814–3824.

Shraddha, Shekher, R., S. Sehgal, M. Kamthania and A. Kumar. 2011. Laccase: microbial sources, production, purification, and potential biotechnological applications. Enzyme Research 2011(1): 217861.

Sivaraj, K. K. and R. H. Adams. 2016. Blood vessel formation and function in bone. Development 143(15): 2706–2715.

Slevin, M., J. Krupinski, J. Gaffney, S. Matou, D. West, H. Delisser et al. 2007. Hyaluronan-mediated angiogenesis in vascular disease: uncovering RHAMM and CD44 receptor signaling pathways. Matrix Biology 26(1): 58–68.

Snetkov, P., S. Morozkina, M. Uspenskaya and R. Olekhnovich. 2019. Hyaluronan-based nanofibers: Fabrication, characterization and application. Polymers 11(12): 2036.

Soranzo, C., D. Renier and A. Pavesio. 2004. Synthesis and characterization of hyaluronan-based polymers for tissue engineering. Biopolymer Methods in Tissue Engineering 25–39.

Stahl, A. and Y. P. Yang. 2021. Regenerative approaches for the treatment of large bone defects. Tissue Engineering Part B: Reviews 27(6): 539–547.

Sturabotti, E., S. Consalvi, L. Tucciarone, E. Macrì, V. Di Lisio, I. Francolini et al. 2022. Synthesis of novel hyaluronic acid sulfonated hydrogels using safe reactants: a chemical and biological characterization. Gels 8(8): 480.

Suwan, J., A. Torelli, A. Onishi, J. S. Dordick and R. J. Linhardt. 2012. Addressing endotoxin issues in bioengineered heparin. Biotechnology and Applied Biochemistry 59(6): 420–428.

Sze, J. H., J. C. Brownlie and C. A. Love. 2016. Biotechnological production of hyaluronic acid: a mini review. 3 Biotech, 6: 1–9.

Tan, H., H. Li, J. P. Rubin and K. G. Marra. 2011. Controlled gelation and degradation rates of injectable hyaluronic acid-based hydrogels through a double crosslinking strategy. Journal of Tissue Engineering and Regenerative Medicine 5(10): 790–797.

Tan, H. and K. G. Marra. 2010. Injectable, biodegradable hydrogels for tissue engineering applications. Materials 3(3): 1746–1767.

Toole, B. P. 2004. Hyaluronan: from extracellular glue to pericellular cue. pp. 528–539. In Nature Portfolio (Vol. 4, Issue 7). doi: 10.1038/nrc1391.

Trombino, S., C. Servidio, F. Curcio and R. Cassano. 2019. Strategies for hyaluronic acid-based hydrogel design in drug delivery. Pharmaceutics 11(8): 407.

Volpi, N., J. Schiller, R. Stern and L. Soltes. 2009. Role, metabolism, chemical modifications and applications of hyaluronan. Current Medicinal Chemistry 16(14): 1718–1745.

Walimbe, T., A. Panitch and P. M. Sivasankar. 2017. A review of hyaluronic acid and hyaluronic acid-based hydrogels for vocal fold tissue engineering. Journal of Voice 31(4): 416–423.

Wang, G., X. Cao, H. Dong, L. Zeng, C. Yu and X. Chen. 2018. A hyaluronic acid based injectable hydrogel formed via photo-crosslinking reaction and thermal-induced diels-alder reaction for cartilage tissue engineering. Polymers 10(9): 949.

Weigel, P. H., V. C. Hascall and M. Tammi. 1997. Hyaluronan synthases*. Journal of Biological Chemistry 272(22): 13997–14000. doi: https://doi.org/10.1074/jbc.272.22.13997.

Weigel, P. H. 2015. Hyaluronan synthase: the mechanism of initiation at the reducing end and a pendulum model for polysaccharide translocation to the cell exterior. International Journal of Cell Biology 2015(1): 367579. doi: https://doi.org/10.1155/2015/367579.

Xing, F., C. Zhou, D. Hui, C. Du, L. Wu, L. Wang et al. 2020. Hyaluronic acid as a bioactive component for bone tissue regeneration: Fabrication, modification, properties, and biological functions. Nanotechnology Reviews 9(1): 1059–1079.

Xu, X., A. K. Jha, D. A. Harrington, M. C. Farach-Carson and X. Jia. 2012. Hyaluronic acid-based hydrogels: from a natural polysaccharide to complex networks. Soft Matter 8(12): 3280–3294.

Yang, Y., C. He, E. Dianyu, W. Yang, F. Qi, D. Xie et al. 2020. Mg bone implant: Features, developments and perspectives. Materials & Design 185: 108259.

Yao, W., M. Chen, X. Dou, H. Jin, X. Zhang, Y. Zhu et al. 2019. Unravel a neuroactive sHA sulfation pattern with neurogenesis activity by a library of defined oligosaccharides. European Journal of Medicinal Chemistry 163: 583–596.

Yeom, J., B. W. Hwang, D. J. Yang, H.-I. Shin and S. K. Hahn. 2014. Effect of osteoconductive hyaluronate hydrogels on calvarial bone regeneration. Biomaterials Research 18(1): 8.

Yuan, J., P. Maturavongsadit, K. Metavarayuth, J. A. Luckanagul and Q. Wang. 2019. Enhanced bone defect repair by polymeric substitute fillers of multiarm polyethylene glycol-crosslinked hyaluronic acid hydrogels. Macromolecular Bioscience 19(6): 1900021.

Zhao, H., Y. Zhang, Y. Liu, P. Zheng, T. Gao, Y. Cao et al. 2021. *In situ* forming cellulose nanofibril-reinforced hyaluronic acid hydrogel for cartilage regeneration. Biomacromolecules 22(12): 5097–5107.

Zhao, S., M. Cao, H. Li, L. Li and W. Xu. 2010. Synthesis and characterization of thermo-sensitive semi-IPN hydrogels based on poly (ethylene glycol)-co-poly (ε-caprolactone) macromer, N-isopropylacrylamide, and sodium alginate. Carbohydrate Research 345(3): 425–431.

Zheng, L., S. Zhang, Z. Ying, J. Liu, Y. Zhou and F. Chen. 2020. Engineering of aerogel-based biomaterials for biomedical applications. International Journal of Nanomedicine 2363–2378.

Zhu, L., D. Luo and Y. Liu. 2020. Effect of the nano/microscale structure of biomaterial scaffolds on bone regeneration. International Journal of Oral Science 12(1): 6.

CHAPTER 7

Advances in Starch-Based Bio Nanocomposites

Pioneering Sustainable Solutions for Biomedical Applications and Tissue Engineering

Ayushi Bhatnagar and *Gangotri Pemawat**

1. Introduction

Escalating environmental concerns have spurred academic and modern efforts towards crafting "green materials" sourced from natural reservoirs. There's a growing push for developing biodegradable materials from sustainable origins, easily tailored to specific applications. Natural polymers like starch, lignin, cellulose, and proteins are gaining traction in the plastics industry, aiming to repurpose agricultural by-products. Biodegradable polymers are increasingly seen as viable substitutes for petroleum-based plastics, driven by mounting awareness of environmental risks. Within the vibrant sphere of biomedical research, starch emerges as a remarkable polysaccharide with unrivaled versatility and promise. Revered for its intricate molecular structure, governed by amylose and amylopectin, starch holds vast potential across myriad applications, notably in tissue engineering and bio composite fabrication. However, despite its inherent merits, starch encounters hurdles in direct application, grappling with mechanical limitations and environmental susceptibilities. In response, researchers have embarked on an ambitious quest into the realm of bio

Department of Chemistry, University College of Science, Mohanlal Sukhadia University, Udaipur, Rajasthan, India 313001.

* Corresponding author: drgpemawat@mlsu.ac.in

DOI: 10.1201/9781003518792-7

nanocomposites, leveraging the prowess of nanotechnology to enhance starch's properties. Concurrently, the field of tissue engineering witnesses' remarkable strides, with starch-based hydrogels emerging as deciding platforms for orchestrating cellular responses and fostering tissue regeneration. This convergence of disciplines signifies a transformative era in biomedical science, promising groundbreaking opportunities for innovation in regenerative medicine and tissue repair. Fuelling this progress are innovative initiatives exploring diverse formulations and applications of starch-based materials, from novel nanocomposite blends to customized hydrogel scaffolds. These advancements hold immense potential for overcoming the limitations of conventional approaches, ushering in more effective strategies for tissue regeneration. As interdisciplinary collaborations flourish, the integration of starch-based materials into biomedical solutions continues to evolve, pushing the boundaries of regenerative medicine and tissue repair. With ongoing advancements in nanotechnology and tissue engineering, the horizon is brimming with exciting possibilities for harnessing starch's full potential in addressing critical healthcare needs and elevating patient outcomes.

2. Starch

Starch emerges as an extraordinary polysaccharide, standing at the forefront of material innovation with its boundless potential (Whistler et al., 2012). Its principal constituents, amylose and amylopectin, serve as the cornerstones of its structural elegance and functional versatility. Amylose, characterized by its linear assembly of D-glucose units interconnected through α-d-(1,4) bonds, juxtaposes the intricate branching pattern of amylopectin, which intertwines α-d-(1,4) and α-d-(1,6) linkages in a complex web of molecular architecture (Figure 1) (Subroto et al., 2020). The chemical composition of starches sourced from different origins exhibits notable variations, encompassing differences in α-glucans, moisture content, lipid profiles, protein composition, and phosphorylated residues (Ahmad et al., 2020).

Moreover, genetic and environmental factors contribute to the diversity observed in the size and shape of starch granules, which in turn are influenced by the content, structure, and spatial arrangement of amylose and amylopectin (Vamadevan and Bertoft, 2020). The ratio of amylose to amylopectin plays a key role in dictating the thermal, mechanical, and barrier properties of starch.

Despite its potential, the practical applications of starch have been hindered by several limitations, including inherent brittleness, heightened susceptibility to water, inadequate gas and moisture barrier properties, proneness to retrogradation, elevated viscosity, and limited solubility. To address these challenges, researchers have explored various strategies such as the incorporation of plasticizers, chemical modifications, and the integration of nanofillers like starch nanoparticles, nanoparticles, nanoclays, and nanofibers (Gamage et al., 2022). These approaches aim to enhance the performance and expand the utility of starch-based materials by mitigating their shortcomings and unlocking their full potential across diverse applications. The remarkable attributes of starch, spanning from its innate non-toxicity to its impeccable biocompatibility and biodegradability, coupled with its desirable mechanical prowess, underscore its aptitude for crafting starch-based

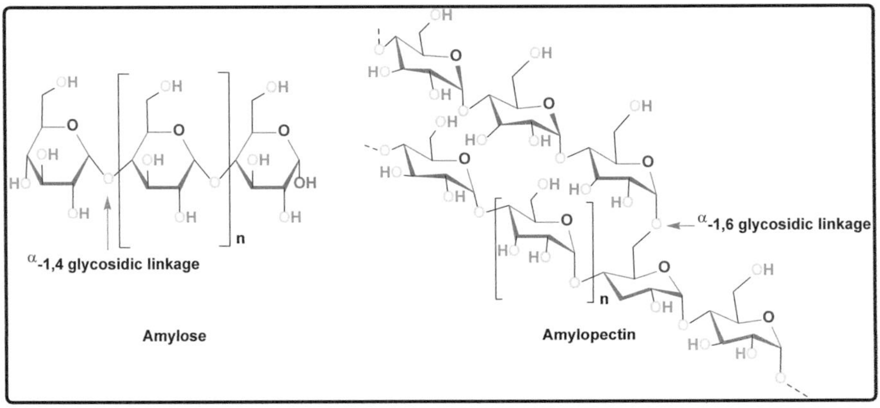

Figure 1. Chemical structure of Amylose and Amylopectin.

bionanocomposites (Onyeaka et al., 2022). These materials herald a new era in biomedical engineering, poised to revolutionize drug delivery systems, usher in breakthroughs in bone regeneration methodologies, redefine the frontiers of tissue engineering, and pioneer the development of advanced hydrogel technologies. While carbon nanotubes (CNTs) have historically captured the imagination of researchers in the medical domain, serving as potent reinforcements capable of catalyzing bone formation, the spotlight is now shifting towards unlocking the untapped potential of starch in similar realms of application (Calori et al., 2020). Noteworthy is the revelation that amylopectin presents a stretched conformation, a stark contrast to the linear arrangement observed in amylose. The labyrinthine intricacies of starch's molecular architecture continue to captivate researchers, driving relentless exploration into its manifold applications and synergistic possibilities with other biomaterials in the ever-evolving landscape of biomedical science and engineering (Vega et al., 2022).

2.1 Starch Bionanocomposites

Bionanocomposites emerge as a cutting-edge category of nanostructured materials, featuring at least one component of biological origin and particles with nanoscale dimensions. Analogous to traditional nanocomposites, these hybrids offer the promise of enhanced structural and functional characteristics, opening avenues in specialized domains such as regenerative medicine and everyday applications like eco-friendly materials (Joseph et al., 2020). The inclusion of biopolymers is particularly intriguing due to their minimal environmental impact, versatility in chemical and structural configurations, and responsiveness to molecular interactions and biological cues (Mary et al., 2022). Conversely, the presence of an inorganic phase offers potential benefits such as bolstered thermal and physical stability, along with imparting functional traits like conductivity, optical clarity, and magnetic properties. The development of bio nanocomposites necessitates interdisciplinary collaboration across fields exploring biomineralization processes, bioinspired materials, and biomimetic systems (Choi et al., 2021). This convergence of expertise holds promise for advancing the frontier of materials science, propelling the innovation of sustainable, multifunctional materials with diverse applications in both specialized and mainstream contexts. Starch stands out as a plentiful, biodegradable, cost-effective, and renewable biopolymer (Singh et al., 2024). Despite these merits, research indicates that its mechanical strength and water resistance fall short compared to petroleum-based plastics. To address this limitation, nanofillers like nanocellulose are introduced to enhance its properties (Zarski et al., 2021). Nanocellulose garners significant attention owing to its commendable physical and mechanical attributes The direct application of starch is often hindered by its subpar mechanical characteristics, limited long-term stability, and heightened sensitivity to water. To overcome these constraints, significant attention has been directed towards the incorporation of diverse nanofillers such as particles, fibers, and layers into the starch matrix to fabricate nanocomposites. Starch is frequently blended with natural or synthetic polymers or infused with various nanomaterials to bolster its mechanical, physical, and barrier attributes (Thakur et al., 2022). Nanoparticles, in particular, display unique physical, mechanical, optical, magnetic, and electrical characteristics due to their increased surface area relative to volume. The size of the combined nanofillers plays a crucial role in shaping the ultimate properties of the nanocomposites, distinguishing them from conventional micro composites (Fu et al., 2019). This unique characteristic allows nanometer-sized fillers to establish molecular bridges within the starch matrix, thereby augmenting the properties of the resulting nanocomposites (Salmon and Catala, 2018). These enhancements encompass improved mechanical performance, enhanced thermal stability, heightened resistance to moisture and oxygen permeation, and accelerated biodegradation rates. Moreover, nanocomposites offer myriad advantages over their conventional counterparts. By incorporating relatively small amounts of fillers (typically less than 15%), nanocomposites achieve desired properties without adding excessive weight, thereby preserving the composite's structural integrity (Jamróz et al., 2019). Furthermore, the diminutive size of nanofillers relative to the wavelength of light ensures minimal light scattering, thereby enhancing the optical transparency of the composite. In light of these advancements, bio nanocomposites emerge as a promising avenue for fortifying the mechanical and barrier properties of starch-based materials, heralding a new era of multifunctional and sustainable composite materials (Guchait et al., 2022). Amylose and amylopectin levels in starch derived from different sources are tabulated in Table 1.

2.2 Amylose and Amylopectin Levels in Starch Derived from Different Sources

2.2.1 Tissue Engineering

Tissue engineering has emerged from the realm of biomaterials development, encompassing the integration of scaffolds, cells, and biologically active compounds to fabricate functional tissues. The overarching aim is to fabricate constructs capable of restoring, preserving, or enhancing damaged tissues or entire organs (Figure 2) (Eldeeb et al., 2022). Notably, artificial skin and cartilage stand as FDA-approved

Table 1. Amylose and amylopectin levels in starch derived from different sources.

S. No.	Starch Source	Amylose %	References
1.	Arrowroot	35.52	(Nogueira et al., 2018)
2.	Sweet potato	30.4	(Zhou et al., 2015)
3.	Maize (high amylose content)	35.5-65.8	(Zhong et al., 2020)
4.	Maize (normal)	22.7-28.9	(Schirmer et al., 2013)
5.	Banana peel	25.7	(Lemos et al., 2018; Li et al., 2018)
6.	Banana pulp	16.36	(Thanyapanich et al., 2021)
7.	Potato	18.6	(Lemos et al., 2018)
8.	Rice	0.1-28.7	(Govindaraju et al., 2022)
9.	Wheat	6.2-22.8	(Chisenga et al., 2019; Schirmer et al., 2013)
10.	Cassava	2.5-32.12	(Chisenga et al., 2019)
11.	Corn	0.9-79.5	(Luchese et al., 2017)

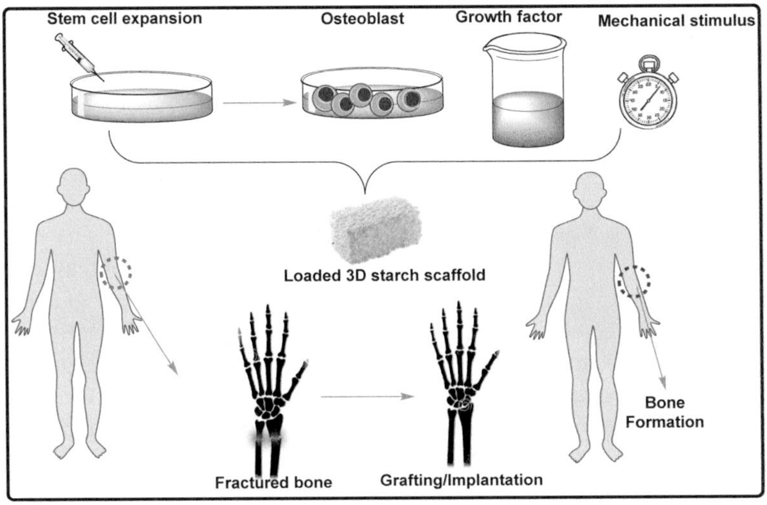

Figure 2. Bone tissue scaffold implantation.

examples of engineered tissues. The process of tissue regeneration is intricate, involving the intricate interplay of various cellular and molecular elements. A fundamental strategy in tissue engineering involves the incorporation of growth factors, cytokines, and other bioactive agents into scaffolds to augment tissue regeneration (Esdaille et al., 2021).

In the field of tissue engineering, starch-based hydrogels stand out as versatile platforms, functioning as scaffolds or matrices to support cell proliferation, tissue regeneration, and the controlled release of bioactive substances (Han et al., 2020). The controlled release of these compounds from hydrogels initiates a sequential cascade of cellular events, ultimately leading to the formation of fully functional tissues. Additionally, the mechanical strength and degradation rates of starch-based hydrogels can be finely tuned to meet the specific requirements of various tissues and organs (Lee and Hwang, 2023). This versatility underscores the potential of starch-based hydrogels as customizable platforms for advancing the field of tissue engineering, paving the way for innovative therapeutic interventions and regenerative medicine approaches. In a synchronous undertaking, several studies showed that hydroxyapatite (HA) stands as a highly esteemed biomaterial in orthopedics and dentistry, cherished for its outstanding biocompatibility, bioactivity, and osteoconductivity. The integration of nano-HA further enhances scaffold functionality by amplifying surface area. Starch, recognized for its biodegradability and biocompatibility, emerges as a valuable component in tissue engineering scaffold composites. Cytotoxicity evaluations conducted on starch/HA scaffolds affirm their benign nature toward mammalian cells. Additionally, starch infusion addresses the brittleness often associated with HA scaffolds. This improvement stems from the helical structure of amylose within starch, which, when stretched, forms an expansive network characterized by a hydrophilic exterior surface and hydrophobic interior cavities (Roslan et al., 2016). This network intertwines with HA nanoparticles, fostering adhesive interactions that fortify the HA scaffolds through interlocking mechanisms. In a study conducted by Roslan and colleagues, the potential suitability of Balik Wangi rice starch as a biomaterial for fabricating bone tissue scaffolds was investigated. The researchers employed a combination of solvent casting and salt leaching techniques to fabricate the scaffolds, and their analysis revealed distinct trends in scaffold density, porosity, and pore size (Roslana et al., 2016). The porosities of the scaffolds, as determined in this study, fell within the range typically observed in spongy bone, spanning approximately 30% to 90%. Moreover, the density values obtained closely resembled those of trabecular bone. Interestingly, an increase in the percentage of starch in the scaffold formulation appeared to correlate with higher porosity values, while concurrently reducing scaffold density. However, this trend was observed only up to a starch percentage of 70 wt%, beyond which further increases compromised the structural integrity of the scaffold. Additionally, the researchers noted that higher salt ratios led to an augmentation in scaffold porosity. In a similar study reported by Beh et al. suggested the utilization of the solvent casting and particulate leaching technique, that helped in fabrication of scaffolds comprised of cornstarch and nano-hydroxyapatite (n-HAp). These porous composites, featuring varying cornstarch compositions, were synthesized, aiming to explore their suitability for bone tissue engineering applications. The findings indicated that by adjusting the cornstarch content, it was possible to produce highly porous cornstarch/n-HAp scaffolds possessing satisfactory mechanical properties. These results suggest the potential of such scaffolds for use in bone tissue engineering, offering promising avenues for the development of innovative biomaterials in this field (Beh et al., 2020). Torres and colleagues utilized an innovative microwave processing method to create biodegradable scaffolds for tissue engineering. They used a range of starch-based polymers, including potato, sweet potato, corn starch, and non-isolated amaranth and quinoa starch. These scaffolds had porous structures, with water and glycerol acting as plasticizers for the different types of starch (Torres et al., 2007). The compressive mechanical properties of the scaffolds were assessed using indentation tests, which showed a notable correlation between indentation stress and average pore size. Additionally, studies performed in simulated body fluid allowed for the evaluation of *in vitro* bioactivity, degradability, and changes in surface morphology of the scaffolds (Torres and Arce, 2015).

2.2.2 Starch-based Hydrogels

Starch-based hydrogels demonstrate considerable versatility as scaffolds or matrices, promoting cell growth, tissue regeneration, and the controlled release of bioactive substances. Serving as a supportive framework, they enhance cell proliferation and tissue renewal (Figure 3). Their porous structure allows for the essential exchange of nutrients and oxygen, crucial for cellular functions. Additionally, hydrogels can be tailored to mimic the mechanical properties of specific tissues, thereby optimizing the conditions for cell adhesion, growth, and differentiation. In a pioneering study, Mao et al. explored wound care innovation with the introduction of an impressive oxidized starch/gelatin-based shape memory hydrogel (OSG). This innovative formulation acts as a self-contracting wound dressing designed to expedite the noninvasive closure of wounds. The key element of the OSG hydrogel is the thermo-reversible gelatin component, which serves as a shape memory switch, enhancing both structural stability and mechanical properties. Impressively, the OSG exhibited an effective thermo-responsive nature, manifesting its shape memory prowess at a balmy 38°C. Its mechanical robustness, coupled with tissue compatibility, positioned it as a promising candidate for facilitating wound tissue remodeling. Histological scrutiny further unveiled the therapeutic potential of OSG, showcasing thicker epidermal and dermal layers in treated wounds, indicative of enhanced tissue reconstruction (Mao et al., 2020).

Similarly In a complementary endeavor, Pal et al. developed a starch-based hydrogel by carefully crosslinking polyvinyl alcohol with a starch suspension, using glutaraldehyde as the crosslinking agent. This innovative hydrogel exhibited significant strength while minimizing toxicity concerns. Remarkably, the tensile strength of the hydrogel membrane was comparable to that of natural skin, making it a promising option for use as artificial skin in wound management. Beyond simply covering wounds, this membrane was designed to deliver essential nutrients and healing factors directly to the wound site, thus accelerating the healing process. This comprehensive approach represents a significant advancement in wound care, offering the potential for faster recovery and better patient outcomes (Pal et al., 2006).

3. Assorted Starch Bio Composite Blends

Mahdieh and her team creatively developed a nanocomposite biomaterial by integrating a TPS/EVOH matrix with nano forsterite reinforcement through extrusion and injection molding methods. Vitamin E was crucial as a thermal stabilizer during the melt compounding process. The inclusion of Mg^{2+} in forsterite provided a pH buffering effect, which notably enhanced the cell viability of the biomaterial. Furthermore, fibroblast cells demonstrated robust adhesion to the sample surfaces, confirming the non-cytotoxic nature of these biomaterials and their suitability for

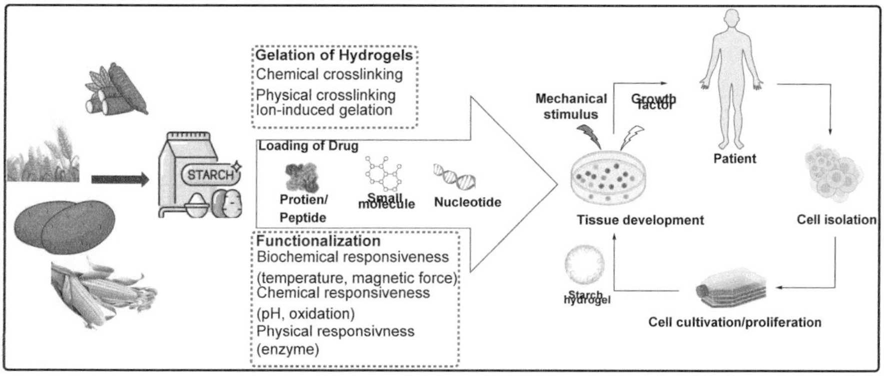

Figure 3. Illustration of tissue engineering using starch-based hydrogel.

supporting the growth and proliferation of fibroblast cells (Pal et al., 2006). In a parallel pursuit, Mirab et al. explored the promise of starch-based scaffolds as viable biomaterials for bone tissue engineering. Their research presented a highly porous nanocomposite scaffold created by combining starch with polyvinyl alcohol (PVA) and integrating various bio-additives such as citric acid, cellulose nanofibers, and hydroxyapatite (HA) nanoparticles. The fabrication process utilized a blend of unidirectional and cryogenic freeze-casting techniques, followed by freeze-drying. The study revealed that cellulose and HA nanofillers had differing impacts on scaffold biodegradation rates, with cellulose decelerating degradation and HA accelerating it. Results from MTT assays indicated the scaffold's excellent cytocompatibility with osteoblast cells. Notably, the scaffold provided an excellent platform for cell adhesion, underscoring its potential for bone tissue regeneration applications (Mirab et al., 2018). Similarly, Nasrabadi and colleagues have reported the creation of starch/cellulose nanofiber composites tailored for cartilage tissue engineering. These composites are characterized by optimal porosity, pore size, mechanical strength, and biodegradability. Through a meticulous approach combining film casting, salt leaching, and freeze-drying techniques, porous thermo-plastic starch-based composites were crafted. The outcomes underscored the efficacy of the fabricated scaffold, as evidenced by the successful cultivation of isolated rabbit chondrocytes. Notably, the incorporation of nanofibers into the starch matrix was found to significantly enhance cell attachment and proliferation, emphasizing the scaffold's potential for fostering cartilage regeneration (Nasri-Nasrabadi et al., 2014).

4. Conclusion

In essence, starch-based bio nanocomposites are carving a significant niche in the realm of biomaterials, especially within biomedical applications like tissue engineering. By capitalizing on the inherent properties and widespread availability of starch, researchers are pioneering innovative material solutions. Despite challenges such as brittleness and water susceptibility, advancements like the incorporation of nanofillers are significantly enhancing the mechanical strength, thermal stability, and biocompatibility of these materials. The emergence of starch-based hydrogels as adaptable scaffolds for tissue regeneration, along with promising developments in scaffold fabrication and composite materials, underscores the versatility and potential of starch-based bio nanocomposites. Interdisciplinary collaboration remains crucial in driving forward the development of these materials, fostering breakthroughs in regenerative medicine and tissue repair. As the commercialization of bio nanocomposites expands, their role in addressing environmental concerns and promoting sustainability across various industries is becoming increasingly significant as it highlights the transformative impact of starch-based bio nanocomposites on biomedical research and sustainable material development, emphasizing their pivotal contribution to the future of tissue engineering and beyond.

References

Ahmad, S., K. Manzoor, M. Ahmad, R. Purwar and S. Ikram. 2020. Starch-based bionanocomposites. pp. 157–171. *In*: Mahmood Zia, K., F. Jabeen, M. Anjum and S. Ikram (eds.). Bionanocomposites: Green Synthesis and Applications. Elsevier.

Beh, C., E. Cheng, N. Mohd Nasir, M. Abdul Majid, M. Mohd Roslan, K. You and M. Ridzuan. 2020. Fabrication and characterization of three-dimensional porous cornstarch/n-HAp biocomposite scaffold. Bull. Mater. Sci. 43: 1–9.

Calori, I. R., G. Braga, P. de Jesus, H. Bi and A. C. Tedesco. 2020. Polymer scaffolds as drug delivery systems. Eur. Poly. J. 129: 109621.

Choi, G., N. S. Rejinold, H. Piao and J.-H. Choy. 2021. Inorganic–inorganic nanohybrids for drug delivery, imaging and photo-therapy: recent developments and future scope. Chem. Sci. 12(14): 5044–5063.

Eldeeb, A. E., S. Salah and N. A. Elkasabgy. 2022. Biomaterials for tissue engineering applications and current updates in the field: A comprehensive review. Aaps. Pharmsci. Tech. 23(7): 267.

Esdaille, C. J., K. S. Washington and C. T. Laurencin. 2021. Regenerative engineering: A review of recent advances and future directions. Regenerative Medicine 16(5): 495–512.

Fu, S., Z. Sun, P. Huang, Y. Li and N. Hu. 2019. Some basic aspects of polymer nanocomposites: A critical review. Nano Mater. Sci. 1(1): 2–30.

Gamage, A., P. Thiviya, S. Mani, P. G. Ponnusamy, A. Manamperi, P. Evon et al. 2022. Environmental properties and applications of biodegradable starch-based nanocomposites. Poly. 14(21): 4578.

Guchait, A., A. Saxena, S. Chattopadhyay and T. Mondal. 2022. Influence of nanofillers on adhesion properties of polymeric composites. ACS Omega 7(5): 3844–3859.

Han, F., J. Wang, L. Ding, Y. Hu, W. Li, Z. Yuan et al. 2020. Tissue engineering and regenerative medicine: achievements, future, and sustainability in Asia. Front. Bioengin. Biotech. 8: 83.

Jamróz, E., P. Kulawik and P. Kopel, 2019. The effect of nanofillers on the functional properties of biopolymer-based films: A review. Poly. 11(4): 675.

Joseph, B., S. Krishnan, V. Sagarika, A. Tharayil, N. Kalarikkal and S. Thomas. 2020. Bionanocomposites as industrial materials, current and future perspectives: a review. Emerg. Mater. 3: 711–725.

Lee, C.-S. and H. S. Hwang. 2023. Starch-based hydrogels as a drug delivery system in biomedical applications. Gels 9(12): 951.

Luchese, C. L., N. Sperotto, J. C. Spada and I. C. Tessaro. 2017. Effect of blueberry agro-industrial waste addition to corn starch-based films for the production of a pH-indicator film. Int. J. Biol. Macromol. 2017 Nov; 104(Pt A): 11–18. doi: 10.1016/j.ijbiomac.2017.05.149.

Mao, Q., O. Hoffmann, K. Yu, F. Lu, G. Lan, F. Dai et al. 2020. Self-contracting oxidized starch/gelatin hydrogel for noninvasive wound closure and wound healing. Mater. Des. 194: 108916.

Mary, S. K., R. R. Koshy, R. Arunima, S. Thomas and L. A. Pothen. 2022. A review of recent advances in starch-based materials: Bionanocomposites, pH sensitive films, aerogels and carbon dots. Carbohydrate Poly. Techn. Appl. 3: 100190.

Mirab, F., M. Eslamian and R. Bagheri. 2018. Fabrication and characterization of a starch-based nanocomposite scaffold with highly porous and gradient structure for bone tissue engineering. Biomed. Phy. Engine. Exp. 4(5): 055021.

Nasri-Nasrabadi, B., M. Mehrasa, M. Rafienia, S. Bonakdar, T. Behzad and S. Gavanji. 2014. Porous starch/cellulose nanofibers composite prepared by salt leaching technique for tissue engineering. Carb. Poly. 108: 232–238.

Onyeaka, H., K. Obileke, G. Makaka and N. Nwokolo. 2022. Current research and applications of starch-based biodegradable films for food packaging. Polymers 14(6): 1126.

Pal, K., A. Banthia and D. Majumdar. 2006. Characterization of prepared corn starch-based hydrogel membranes. J. Appl. Biomater. Biomech. 4(1): 38–44.

Robyt, J. F. 2012. Enzymes in hydrolysis and synthesis of starch. pp. 87–118. *In*: Whistler, R. L., J. N. BeMiller and E. F. Paschall (eds.). Starch: Chemistry and Technology. Academic Press.

Roslan, M. R., N. M. Nasir, E. Cheng and N. Amin. 2016. Tissue engineering scaffold based on starch: A review. Proc. Inter. Confer. Electric. Electron. Optim. Techn. 1857–1860.

Roslana, M. R., N. Nasirb, E. Cheng and N. Mamat. 2016. The characterization of nano HA-Balik Wangi rice starch tissue engineering scaffold. Int. J. Mech. Mechatron. Eng. 16: 36–41.

Salmon, L. and L. Catala. 2018. Spin-crossover nanoparticles and nanocomposite materials. Comptes Rendus Chimie 21(12): 1230–1269.

Singh, P., V. K. Pandey, R. Singh, K. Singh, K. K. Dash and S. Malik. 2024. Unveiling the potential of starch-blended biodegradable polymers for substantializing the eco-friendly innovations. J. Agri. Food Res. 101065.

Subroto, E., G. Jeanette, Y. Meiyanasari, I. Luwinsky and S. Baraddiaz. 2020. Review on the analysis methods of starch, amylose, amylopectinin food and agricultural products. Inter. J. Emerg. Trends Engine. Res. 8(7): 3519–3524.

Thakur, M., M. Chandel, A. Rani and A. Sharma. 2022. Introduction to biorenewable nanocomposite materials: methods of preparation, current developments, and future perspectives. pp. 1–24. *In*: Pathania, D. and L. Singh (eds.). Biorenewable Nanocomposite Materials, Vol. 2: Desalination and Wastewater Remediation. ACS Publications.

Torres, F. G. and D. Arce. 2015. Starch-based nanocomposites for biomedical applications. Biodegrad. Polym. Nanocomp. Adv. Biomed. Appl. 73–94.

Torres, F. G., A. R. Boccaccini and O. P. Troncoso. 2007. Microwave processing of starch-based porous structures for tissue engineering scaffolds. J. Appl. Poly. Sci. 103(2): 1332–1339.

Vamadevan, V. and E. Bertoft. 2020. Observations on the impact of amylopectin and amylose structure on the swelling of starch granules. Food Hydrocoll. 103: 105663.

Vega, Y. A., A. A. Arrieta, M. F. Acosta-Humánez, A. Consuegra, M. Palencia, E. Combatt et al. 2022. Starch and its potential applications as a biomaterial for hydrogels with agricultural purposes: a mini-review. J. Sci. Tech. Appl. 18: 1–10.

Zarski, A., K. Bajer and J. Kapuśniak. 2021. Review of the most important methods of improving the processing properties of starch toward non-food applications. Poly. 13(5): 832.

CHAPTER 8

The Role of Alginate in Scaffold Design for Tissue Engineering Applications

Jila Majidi,[1,2,†] *Mohammadamin Momeny,*[3,†] *Mohadese Saraei,*[4] *Mohammad Zahedi,*[4] *Saghar Hoseinzadeh,*[4] *Sanaz Alizadeh,*[5,*] *Narendra Pal Singh Chauhan*[6,*] and *Mazaher Gholipourmalekabadi*[1,4,7,*]

1. Introduction

Natural polymers have a consistent interest in biomedical applications. Polysaccharides are a significant class of naturally occurring macromolecular materials in various plants, animals, and microorganisms. They are crucial for biological processes such as molecular recognition of the immune system, cell adhesion, and signal transduction (Dong and Guo, 2021). Polysaccharides are composed of O-glycosidic bonds that hold together monosaccharides (sugars), and their physical properties, like

[1] Cellular and Molecular Research Center, Iran University of Medical Sciences, Tehran, Iran.
[2] Department of Tissue Engineering & Regenerative Medicine, Faculty of Advanced Technologies in Medicine, Tehran, Iran.
[3] Department of Chemical and Petroleum Engineering, Sharif University of Technology, Tehran, Iran.
[4] Department of Medical Biotechnology, Faculty of Allied Medicine, Iran University of Medical Sciences, Tehran, Iran.
[5] Burn Research Center, Iran University of Medical Science, Tehran, Iran.
[6] Department of Chemistry, Faculty of Science, Bhupal Nobles' University, Udaipur-313002, Rajasthan, India.
[7] NanoBiotechnology and Regenerative Medicine Innovation Group, Noavarn Salamat ZHINO (PHC), Tehran, Iran.
* Corresponding authors: s.alizadeh@rsct.ir; alizadehsannaz@gmail.com, narendrapalsingh14@gmail.com, mazaher.gholipour@iums.ac.ir; mazaher.gholipour@gmail.com
† Equally contributed to this study.

DOI: 10.1201/9781003518792-8

solubility and viscosity, are affected mainly by monosaccharide composition, chain forms, and molecular weight. Polysaccharides are derived from various resources, including higher plants (cellulose, pectin, starch), animals (hyaluronan, chondroitin, heparin, chitin, and chitosan), and algae (alginate, agarose, carrageenan) (Rezvani Ghomi et al., 2019). Sodium alginate (SA) is widely used in biological repair and soft tissue regeneration due to its low cost, high biocompatibility, and rapid and moderate cross-linking (Borda et al., 2016). SA (NaC6H7O6) is a linear polysaccharide, a derivative of alginic acid, consisting of 1,4-β-d-mannuronic acids (M) and α-l-guluronic acid (G) (Nazarnezhada et al., 2020; Majidi Ghatar et al., 2023). SA is a component of the cell wall of algae and contains approximately 30–60% alginic acid. Converting alginic acid to SA makes it water-soluble and more accessible to extract (Nazarnezhada et al., 2020; Alizadeh et al., 2024). Linear polysaccharide (Alizadeh et al., 2024), hydrophilic (Rastogi and Kandasubramanian, 2019), pH-sensitive (Zhang and Zhao, 2020), aquaphilic (Reig-Vano et al., 2021), bio-resorbable (Brumberg et al., 2021), non-immunogenic (Koehler et al., 2018; Op't Veld et al., 2020), biocompatible (Koehler et al., 2018), and inexpensive polyanionic copolymers (Wang and Lu, 2023; Jabbar et al., 2024), are all characteristics of SA. This polymer exhibits film-forming properties, chelating properties, and a high-water viscosity (Ahmad et al., 2021). The present polysaccharide has potential applications in tissue engineering (Sudhakar et al., 2017), gene transfer (Senturk Parreidt et al., 2018), and wound dressing (Senturk Parreidt et al., 2018; Varaprasad et al., 2020).

Besides using sodium alginate in gene delivery, it is crucial in anti-tumor chemotherapy, anti-viral drug delivery, and recognition of cancers like protease cancer (Hurtado et al., 2022; Jahani et al., 2024). Using sodium alginate in tissue engineering as a multipurpose scaffold, *in vivo* wound healing, chronic wounds (Teo and Ramakrishna, 2006) and other severe injuries, healing applications can be more achievable in soft tissue engineering and bone tissue engineering (Mallakpour et al., 2021; Deptuła et al., 2023). Low drug loading, hydrophilicity, microbial degradation, weak mechanical strength, and sodium alginate cell adhesion are considered sodium alginate defects (Negahdari et al., 2024; Teo and Ramakrishna, 2006). Due to these drawbacks, sodium alginate has been mixed with different natural and synthetic components, strengthening its properties (Ahmad et al., 2021). Sodium Alginate-based biomaterials have been utilized to form hydrogels, sponges, fibers, films, foams, 3D bioprinting, beads, etc. (Wang and Lu, 2023; Zhang and Zhao, 2020). In this chapter, the application of several features of sodium alginate-based composites, like hydrogel, beads, films and other forms in tissue engineering have been introduced.

2. Properties of Alginate

Alginate is characterized by its biocompatibility, biodegradability, and non-toxicity, making it suitable for various applications. It has a high viscosity and exhibits shear-thinning behavior, which allows it to function effectively as a thickening, gelling, and stabilizing agent. These attributes enable alginate to be used in drug delivery systems and wound dressings (Gheorghita Puscaselu et al., 2020; Kothale et al., 2020). Alginate biosynthesis entails the oxidation of a carbon substrate to acetyl-CoA, which then enters the TCA cycle and is transformed into fructose-6-phosphate by gluconeogenesis. Fructose-6-phosphate undergoes a series of biochemical transformations to be turned into GDP-mannuronic acid, which serves as a precursor for alginate production. The biosynthetic process may be delineated into four stages: (1) production of the GDP-mannuronic acid precursor; (2) transport to the cytoplasmic membrane and polymerization into polymannuronic acid; (3) transfer to the periplasm and subsequent modification; and (4) export through the outer membrane. Post-polymerization modification of alginates transpires at step (3), whereby polymannuronic acid undergoes acetylation at the O-2 and O-3 sites by several transacetylases. A group of epimerase enzymes then executes epimerization to transform some non-acetylated M residues into G residues. Ultimately, alginate is expelled from the cell via transmembrane porins (Pawar and Edgar, 2012). Alginate has two essential components: D-mannuronate and L-guluronate, which are linked by a $1 \rightarrow 4$ connection. Dmannuronate is the principal constituent of alginate, and upon fractional precipitation with calcium and magnesium ions, alginate is considered a block copolymer of natural origin. The ratio of mannuronate to guluronate fluctuates based on the source of occurrence. Recently, over 200 distinct kinds of alginates have been found and isolated from nature. The alginate family comprises linear copolymers including segments of (1, 4)-linked α-L-guluronate (G) and β-D-mannuronate (M). The distinct blocks of alginate are organized sequentially as G residues (GGGGGG), followed by M residues (MMMMMM), and then alternating between G and M residues (GMGMGM). The G-block concentration in Laminaria hyperborean stems is around 60%, while in other commercially available alginates, it ranges from 14.0% to 31.0%. It was anticipated that just the G-blocks inside the alginate would engage in intermolecular covalent cross-linking with certain divalent cations, such as Ca^{2+}, hence promoting hydrogel formation. Consequently, the dimensions of the G-block, molecular weight, and M/G ratio are critical determinants influencing the physical characteristics and the potential for hydrogel formation in alginate. The alginate derived from various sources has distinct chemical structures. The physical features of alginate substantially influenced the pace of gel formation, the drug release rate from the gel, and the functionality of cells contained inside the alginate gels (Sahoo and Biswal, 2021; Hecht and Srebnik, 2016; Najdahmadi et al., 2018).

3. Sodium Alginate-Based Composites for Tissue Engineering

SA is a natural polysaccharide derived from brown algae such as kelp or Sargassum. This biomaterial is both economical and biocompatible. It can crosslink quickly and gently, which makes it a popular component in soft tissue regeneration and repair processes. SA is a versatile and promising biomaterial because of its many special qualities. It is mucoadhesive, non-toxic, water soluble, biodegradable, immunogenic, and pH-sensitive. It also shows transparency, easy gelation, and chelating ability. It can be used in a variety of applications, such as tissue engineering and medication administration, thanks to these qualities. SA is a very adaptable hydrogel that has gained prominence and remarkable status in the field of soft tissue scaffold development. Because of its pH-responsive nature, SA-containing hydrogels release drugs slowly in stomach settings. Its applications for dressings and wound healing are especially remarkable. SA has shown promise in supporting neuro-regenerative therapies, creating multifunctional scaffolds for tissue engineering (soft tissue, bone tissue, and 3D bioprinting), and allowing the transplantation of OE-MSCs.

Moreover, it has demonstrated effectiveness in the treatment of alkali burns and other serious eye injuries, *in vivo* wound healing, chronic and irregular wounds, and mechanical filling for cartilage regeneration. Despite being used in many pharmaceutical goods, sodium alginate's industrial utilization is restricted because of several issues. Bloating, diarrhea, and nausea are gastrointestinal adverse effects that it may produce. Various synthetic and natural polymers have been combined with SA to improve and alleviate these problems. By enhancing SA's compatibility and usefulness, this method seeks to increase its potential for a wider range of industrial uses.

A wound dressing is essential to wound care because it is a physical barrier to keep the wound from worsening or becoming infected. Wound dressings not only keep bacteria and other pathogens out; they also promote collagen synthesis, reepithelialization, hypoxia, and a decrease in wound bed pH, which hasten wound healing and lessen the likelihood of infection (Dong and Guo, 2021). Previously, natural materials, including leaves, spider webs, honey, cotton bandages, and gauze, were used as dressings (Rezvani Ghomi et al., 2019). Foams, hydrocolloids, ahydrogels are examples of modern wound dressings that have emerged alongside advances in medical care. While gauze and other passive dressings can cover and absorb wound exudates, they can cling to freshly formed granulation and be painful to remove. Furthermore, conventional dressings do not possess any antimicrobial properties (Borda et al., 2016). In addition to protecting the wound from further harm and keeping it wet, multipurpose wound dressings can speed up the healing process by affecting different steps in the repair process. To suppress infection and cell growth, it is common practice to utilize several medicines, iron nanoparticles, or antimicrobial peptides (such as host-defense peptides) all at once (Nazarnezhada et al., 2020).

3.1 Hydrogel

Cross-linked polymers with a water content of above 90% are called hydrogels, and they can be produced synthetically or organically. Hydrogels' three-dimensional structure is mostly dependent on cross-linking, whether it be covalent or physical. This structure enables hydrogels to absorb large volumes of water due to its hydrophilic porous nature. Because of these special qualities, hydrogels are a good choice for dressings since they are pliable, non-adhesive, and resemble live tissue. They can add moisture to the area where the wound is located and produce a damp atmosphere that encourages cell migration, which can hasten the healing process. One of the assertions made by the U.S. FDA during the approval of hydrogel products is their ability to hydrate and maintain a moist environment effectively. Hydrogels have the additional benefit of providing wound cooling and pain reduction, making them particularly useful for treating burns or painful wounds (Brumberg et al., 2021). Furthermore, this topical administration method enhances effectiveness, particularly in treating wounds, and reduces the potential for drug toxicity or adverse reactions. Hydrogels have superior attributes that render them the optimal selection for wound dressings compared to alternative materials (Koehler et al., 2018; Op't Veld et al., 2020). Alginate hydrogel possesses both the hydrophilic properties of a hydrogel and additional characteristics such as biodegradability, moisture permeability, and exudate absorption. Due to their nontoxic nature and high water absorption capacity, alginate hydrogels are superior to typical medical dressings (Rastogi and Kandasubramanian, 2019). They can create a hydrogel network on the wound surface, which helps maintain a moist environment (Zhang and Zhao, 2020). It is crucial to prevent alginate hydrogels from drying or becoming saturated with exudate. These hydrogels exhibit non-adhesive properties to wound tissues, ensuring that their removal does not harm the wound surface. However, the dehydrated hydrogel has the potential to stick to the wound surface, causing harm upon removal. In addition, alginate tends to gather the exudate towards the intact skin adjacent to the incision, a phenomenon referred to as "lateral wicking." This underscores the importance of careful dressing management (Reig-Vano et al., 2021).

Functionalizing hydrogels with bioactive molecules, nanoparticles, or growth factors significantly enhances their performance by improving structural stability and promoting biological activity. This process improves the mechanical properties of hydrogels, making them more stable and durable for various applications. Crosslinked hydrogels can withstand physiological conditions better than non-crosslinked counterparts, which is crucial for maintaining their integrity in tissue engineering applications (Mitrovic et al., 2023; López-Gutierrez et al., 2023). Functionalized hydrogels create a favorable microenvironment that mimics the extracellular matrix (ECM), facilitating cell migration, adhesion, and growth. For instance, peptide-functionalized hydrogels have been shown to enhance angiogenesis and ECM remodeling, which are vital for effective wound healing (Li et al., 2023; López-Gutierrez et al., 2023).

A study explored the use of an injectable SA and collagen (Col) hydrogel to enhance stem cell therapy for wound healing. Figure 1 displays SEM pictures of the hydrogel. Designed to improve the retention and effectiveness of human umbilical cord-derived mesenchymal stem cells (hUC-MSCs), the SA/Col hydrogel was shown to be biodegradable, injectable, and minimally immunogenic. When applied to a full-thickness skin wound model, the SA/Col hydrogel with hUC-MSCs significantly accelerated wound closure by promoting granulation tissue formation, collagen deposition, and angiogenesis. This combination increased the production of VEGF and TGF-β1, while reducing inflammatory markers TNF-α and IL-1β and boosting anti-inflammatory cytokines IL-4 and IL-10. Additionally, SA/Col with hUC-MSCs helped suppress the NLRP3 inflammasome pathway, which is associated with inflammation (Zhang et al., 2021). Another study introduced a thermosensitive hydrogel membrane for wound healing, primarily composed of SA, poloxamer 407, pluronic F-127, and polyvinyl alcohol (PVA). SA is critical in creating a biocompatible scaffold, enhancing moisture retention, and providing structural integrity through cross-linking. The hydrogel's ability to swell and retain water supported gradual drug release, with amikacin embedded to offer antibacterial properties. The hydrogel effectively inhibits *S. aureus* and *P. aeruginosa*, critical pathogens in wound infections. *In vivo* testing on excisional wound models in

Figure 1. SEM images of hydrogel membrane at 5 µm and 2 µm magnifications (Abbasi et al., 2020).

animals showed faster wound closure, enhanced re-epithelization, and collagen deposition within 21 days (Abbasi et al., 2020). Satish et al. (Satish et al., 2019) focused on a lyophilized hydrogel wound dressing composed of SA, gelatin, and PVA, designed for chronic wound management, especially in exudate-heavy wounds. The primary role of SA in this scaffold is to enhance moisture retention, exudate absorption, and maintain an optimal healing environment, leveraging its natural gel-forming properties and biocompatibility. Triiodothyronine (T3), a thyroid hormone known for its regenerative effects, is incorporated into the scaffold to support wound healing. The hydrogel's properties allow it to release T3, which promotes cellular proliferation, migration, and angiogenesis. *In vivo* studies on rats with full-thickness wounds demonstrated that the SA-based scaffold promotes faster wound closure, increased collagen deposition, and reduced scar formation.

3.2 Aerogels

The discovery of aerogels dates back to 1931 when Kistler published the first article on the subject in Nature, titled "Coherent expanded aerogels and jellies". Kistler's groundbreaking work involved the successful synthesis of aerogels from silica, achieved through the condensation of sodium metasilicate. He later expanded his research to include aerogels made from alumina, tungsten, nickel tartrate, cellulose, and gelatin (Lázár et al., 1931). Aerogels are a fascinating type of synthesized solid materials. Their appearance attracts attention due to the way light is dispersed through the material, giving them a translucent and smoked look similar to that of glass in the case of silica aerogels. Some such as aerogels prepared from (bio)polymers look like soft, thin foams while others such as carbon-based aerogels look like crunchy black bodies. They feel fragile and extremely light to the touch because they are formed by up to 90%–99% of gas or air (Montes and Maleki, 2020). Aerogel is a synthetic porous ultralight material derived from a gel, in which the liquid component of the gel has been replaced with a gas; for example, graphene aerogels are so light that they can rest on top of a grass leaf. The combination of high porosity and extremely small pores provides aerogels with their extreme properties: solid with extremely low density and low thermal conductivity (Thapliyal and Singh, 2014).

Due to its biocompatibility, biodegradability, abundance, and low cost, SA-based aerogels have attracted significant interest for various applications. SA-based aerogels have several advantages: high porosity, low density, biocompatibility, and biodegradability (Wang and Lu, 2023). The extensive pore network facilitates efficient adsorption and transport of substances. Their minimal weight renders them suitable for applications prioritizing weight reduction. Derived from natural sources, SA-based aerogels often exhibit biocompatibility, suggesting potential in biomedical fields. Furthermore, their susceptibility to degradation by natural processes aligns with environmental sustainability goals (Jabbar et al., 2024). SA-based aerogels exhibit unique properties that make them well-suited for various applications. Their large surface area and porous nature enable them to efficiently adsorb pollutants, heavy metals, and dyes from water and air. Additionally, their low thermal conductivity and density make them promising candidates for thermal insulation. Their biocompatibility and ability to control drug release are advantageous in the pharmaceutical field. The porous structure and biocompatibility also contribute to their potential in tissue engineering by supporting cell growth and tissue regeneration. Furthermore, their use as electrode materials in energy storage devices highlights their versatility and potential for technological advancement (Ahmad et al., 2021). SA-based aerogels exhibit unique properties that make them well-suited for various applications. Their large surface area and porous nature enable them to efficiently adsorb pollutants, heavy metals, and dyes from water and air. Additionally, their low thermal conductivity and density make them promising candidates for thermal insulation. Their biocompatibility and ability to control drug release are advantageous in the pharmaceutical field. The porous structure and biocompatibility also contribute to their potential in tissue engineering by supporting cell growth and tissue regeneration.

A study focused on the development of highly porous two-component alginate-chitosan aerogels designed for controlled drug delivery systems. Utilizing sol-gel and supercritical fluid technologies, the researchers created aerogels based on the sodium alginate–chitosan interpolymer complex. The investigation examined how varying the ratios of sodium alginate and chitosan influenced the aerogels' textural properties and water uptake capacity. The study specifically evaluated the release kinetics of levomycetin (LC), an antibiotic effective against various Gram-positive and Gram-negative bacteria, from the aerogel matrix. Results showed up to 95% of the antibiotic could be released within four hours, demonstrating a prolonged release profile (Gorshkova et al., 2021).

Guo et al. (Guo et al., 2022) presented the development of a novel sodium alginate-based aerogel, termed SA@5,10,15,20-Tetra (4-aminophenyl) porphyrin (TPAPP)@ phenylboronic acid (PBA), designed for use as an antibacterial photodynamic wound dressing. The aerogel was synthesized by covalently modifying sodium alginate with photosensitizers and phenylboronic acid (PBA), followed by cross-linking with calcium ions. This innovative approach aimed to enhance the solubility of the photosensitizers and improve their efficacy in photodynamic therapy (aPDT) against bacterial infections, particularly *Staphylococcus aureus*. The resulting aerogel exhibited a high porosity, which is beneficial for rapid hemostasis and absorption of blood exudates. The PBA modification allowed the aerogel to effectively capture *S. aureus* through reversible covalent bonding, significantly enhancing the aPDT effectiveness by shortening the distance between bacteria and photosensitizer molecules. *In vitro* assays demonstrated that the SA@TPAPP@PBA aerogel achieved an impressive antibacterial rate of approximately 90.7% against *S. aureus* under 660 nm laser irradiation. The study also highlighted the aerogel's excellent hemostatic properties and biosafety, making it suitable for pre-hospital care applications. Another study investigated the creation of highly porous collagen-alginate aerogels, with or without graphene oxide (GO), intended for use as scaffolds in tissue engineering. The aerogels were synthesized using a combination of collagen and sodium alginate, followed by a supercritical drying process to enhance their structural integrity. Characterization of the aerogels revealed a highly porous interconnected network, which is essential for promoting cell migration, attachment, and proliferation. The results indicated that the collagen-alginate scaffolds effectively supported osteoblast attachment and proliferation, demonstrating their potential as a suitable platform for tissue engineering applications. In contrast, the incorporation of GO into the scaffolds appeared to hinder cell attachment and proliferation, suggesting that further research is needed to understand the interactions between cells and GO within the aerogel structure. The study emphasizes the advantages of using hybrid biomaterials like collagen and alginate to mimic the extracellular matrix (ECM) of native tissues, providing a three-dimensional environment conducive to cellular activities (Muñoz-Ruíz et al., 2019).

3.3 Beads

Alginate beads, known for their biodegradability, undergo natural decomposition either within the body or the environment, making them highly suitable for various applications that necessitate the controlled release of substances. The unique characteristic of these beads lies in their adjustable porosity, allowing for precise modifications to the release rate of the encapsulated substance. The versatility of sodium alginate beads has led to their widespread utilization across many fields. One such application involves the controlled dispensation of therapeutic agents such as drugs, proteins, and nucleic acids, facilitating targeted and sustained release formulations for enhanced efficacy (Lee et al., 2013). Furthermore, these alginate beads find significant use in the encapsulation of cells for purposes like transplantation, tissue engineering, and cell-based therapies, showcasing their adaptability and utility in the biomedical domain. Beyond the medical realm, these beads are also employed in the food industry to encapsulate flavors, probiotics, and nutrients, thereby improving the functionality and appeal of various food products. The diverse applications of alginate beads underscore their importance and relevance in modern scientific and industrial settings, highlighting their pivotal role in enabling innovative solutions for controlled substance release. Particularly noteworthy is their potential in cancer treatment.

Alginate beads can encapsulate various cell types, providing a protective environment that enhances cell viability and function. This is particularly beneficial for developing bioartificial organs and regenerative therapies. The porous structure of alginate beads facilitates nutrient and waste exchange, essential for maintaining cell health within the scaffold. This property is vital when designing scaffolds for hard and soft tissue regeneration. Alginate can be combined with other materials, such as collagen or graphene oxide, to improve mechanical strength and promote better cell adhesion. These hybrid scaffolds exhibit enhanced properties suitable for specific tissue engineering applications. Alginate's shear-thinning properties make it an excellent bioink for 3D printing technologies, allowing for the creation of complex tissue structures that closely mimic native tissues. Alginate-based scaffolds have been successfully applied in regenerating various tissues, including cartilage, bone, and skin. Their ability to support cell growth while allowing for controlled drug release enhances their therapeutic potential (Al-barudi et al., 2024; Ouwerx et al., 1998). Tarun et al. (Tarun et al., 2016) explored the development of a novel bone tissue engineering (BTE) implant utilizing calcium alginate beads arranged in a hexagonal close-packed (HCP) structure. This unique architecture enhances the osteogenic micro-architecture necessary for effective bone regeneration. The alginate beads were designed to facilitate the spatio-temporal distribution of therapeutic molecules, such as growth factors and drugs, within the scaffold. The implant demonstrated a significant increase in compressive modulus upon incubation in simulated body fluid (SBF), indicating improved mechanical properties suitable for *in vivo* applications. The study also highlighted the tunability of drug release kinetics by altering the spatial arrangement of drug-loaded beads within the implant. *In vitro* experiments using osteoblast cells (MG-63) confirmed the osteoconductive nature of the scaffold, as evidenced by increased cell viability, cell cycle progression, and expression of differentiation markers such as alkaline phosphatase (ALP), runx2, and collagen type I. Furthermore, when tested *in vivo*, implants loaded with vascular endothelial growth factor (VEGF) successfully induced angiogenesis in a mouse model. A study highlighted the use of calcium/cobalt (Ca/Co) alginate beads as a promising scaffold for cartilage tissue engineering. Articular cartilage injuries, challenging to repair due to the tissue's limited self-healing ability, often result in chronic health issues. Traditional regeneration methods for cartilage rely on costly growth factors, but this study proposes an alternative approach that harnesses the synergy between alginate and Co. The Ca/Co alginate beads mimic the natural cartilage environment, supporting chondrogenic differentiation through their structure, which allows for effective cell encapsulation and nutrient diffusion. The beads are produced using a simple dripping technique, which is not only cost-effective but also reproducible and gentle on cells, thus ensuring high viability of the encapsulated human adipose-derived mesenchymal stem cells. This technique creates a tissue-like matrix, fostering the development of chondroprogenitor cells in a growth-factor-free setting (Focaroli et al., 2016). Another study explored the effects of crocin, a saffron-derived compound with known anti-tumor properties, on breast cancer cells (MCF-7) using both two-dimensional (2D) and three-dimensional (3D) cell culture models. Recognizing the limitations of 2D cultures in replicating the *in vivo* tumor environment, researchers developed a 3D cell model using calcium alginate/gelatin (CAG) beads. The findings showed that crocin effectively inhibits MCF-7 cell proliferation in both models, with effects that are time- and concentration-dependent. CAG beads, which were selected based on their enhanced permeability, biocompatibility, and reduced shrinkage, facilitated cell adhesion and colony formation more effectively than calcium alginate (CA) beads alone. Gelatin's integration into the alginate beads improved cell proliferation and adhesion by altering the beads' cross-linking, which supports cell migration and colony growth through internal bead pores. These findings underline the potential of CAG beads as 3D scaffolds for studying anti-tumor agents like crocin (Zhu et al., 2021). Zhou et al. (Zhou and Xu, 2011) developed fast-degrading oxidized alginate-fibrin microbeads to deliver stem cells for tissue regeneration, especially in bone engineering. The microbeads degrade quickly, releasing cells by day 4 and promoting rapid cell proliferation and osteogenic differentiation, which is essential for bone formation. The addition of 0.1% fibrin optimized the degradation rate without compromising the microbead structure. By day 21, cells released from these microbeads showed significantly higher density and mineralization levels than cells from slower-degrading beads. When used in injectable scaffolds, the microbeads release cells throughout the scaffold while creating macropores that enhance cell migration and nutrient flow. This system showed strong potential for applications in bone tissue engineering and injectable biomaterials for tissue repair. Researchers created an innovative construct, hUCMSC-microbead-AM, that wraps hUCMSCs within alginate-fibrin microbeads embedded in an Arg-Gly-Asp (RGD)-modified alginate matrix (AM). The hUCMSC-microbead-AM construct demonstrated enhanced cell viability and myogenic differentiation relative to the conventional technique of seeding cells within a hydrogel matrix. The novel methodology markedly enhanced cell survival and myogenic differentiation, in contrast to the conventional way of encapsulating cells within the hydrogel matrix devoid of microbeads. The novel hUCMSC-microbead-AM construct exhibited superior cell viability, improved spreading morphology, and significantly enhanced myogenic expressions. This design holds potential for applications in dental, craniofacial, and skeletal muscle engineering. The project seeks to investigate methods for generating muscle tissue to address the lip muscle deficiency in infants born with cleft lip and palate. The technique outlined above may prove advantageous for several people with congenital and acquired muscular deficits and diseases (Liu et al., 2012).

3.4 Film

In recent years, there has been a significant increase in research on renewable and biodegradable materials, intending to replace conventional polymers generated from fossil fuels with biopolymers. Natural substances, including proteins, lipids, and polysaccharides, are the source of biopolymers (Senturk Parreidt et al., 2018; Azeredo and Waldron, 2016). Biopolymers are fully biodegradable and have the potential to decompose rapidly in the environment. Alginate is one of the polysaccharides that have generated commercial attention due to its availability, affordability, high nutritional content, biodegradability, biocompatibility, low oxygen permeability, and non-toxicity (Varaprasad et al., 2020). However, alginate-based films have poor mechanical properties and high water vapor permeability; raising these attributes to the level of Synthetic equivalents will be challenging. Alginate–alginate chains have intermolecular solid interactions, so it is imperative to add an appropriate plasticizer to the formulation to reduce them. This is because alginate-based films are very stiff and are not very flexible (Varaprasad et al., 2020). These films and coatings improve fruit and vegetable quality by lowering shrinkage, moisture loss, oxidative rancidity, oil absorption, flavor loss, and color loss (Hurtado et al., 2022). Rosellini et al. (Rosellini et al., 2009) sought to combine the benefits of gelatin and alginate films, two natural polymers, in varying weight ratios for cardiac tissue engineering. Good miscibility and interactions between the functional groups of pure biopolymers were found by physicochemical characterizations using techniques such as differential scanning calorimetry, Fourier transforms infrared spectroscopy and thermogravimetric analysis. Tests of swelling and degradation revealed that blends with a higher gelatin content and PBS had greater weight losses and swelling degrees. These findings pointed to a hydrolytic breakdown mechanism and improved stability in cell culture media. All blends containing more than 60% gelatin demonstrated strong cell proliferation in cell culture assays using C_2C_{12} myoblasts; the mix with the greatest response was the alginate/gelatin 20:80. The only mix in which cell differentiation was seen was this one. According to the findings, the 20:80 combination of alginate and gelatin is a good material to use for creating scaffolds for cardiac tissue engineering. In a study, chitosan (C), alginate (A), and hyaluronate (H) complexes are evaluated as scaffolds for tissue engineering. The complexes' surfaces may be altered by a protein called Arg-Gly-Asp (RGD), which has a cellulose-binding domain (RGD). These complexes can be cast into films or created as scaffolds. Immortalized rat chondrocytes and primary articular chondrocytes obtained from rabbits are used in cytocompatibility studies. The development of neocartilage in cell-seeded scaffolds is investigated in rabbits and *in vitro*. R improves cell adhesion and proliferation in C-A and C-A-H complex films, according to the results. With H and R, complex C1A1HR (C:A = 1:1) performs better than other formulations. When C1A1HR constructs are implanted into rabbit knee cartilage lesions, the defects are fully fixed in six months, but partial healing is shown after one month when chondrocyte scaffolds are implanted. The study suggests that although early inflammation might not have an impact on the subsequent healing process, C1A1HR is a promising tissue-engineering scaffold for cartilage regeneration (Hsu et al., 2004).

The fabrication and characterization of thin hydrogel films composed of varying ratios of alginate and aloe vera gel were the main objectives of Pereira et al. (Pereira et al., 2013). The films' mechanical, chemical, thermal, and contact angle characteristics as well as light transmission were assessed. The films' water solubility and swelling behavior were examined after they were created by crosslinking with a 5% calcium chloride solution. Aloe vera enhanced the films' transparency and thermal stability, according to the results, which made them appropriate for use on the skin. After being submerged in distilled water for 24 h, the crosslinked films became insoluble, indicating that their mechanical qualities were sufficient for skin application. The amount of aloe vera and the pH of the solution affect how the films absorb water and swell. According to the research, alginate/aloe vera films offer qualities that make them acceptable for drug delivery and wound healing applications.

In a different research, titania's surface was altered by applying a self-assembled, chemically crosslinked biopolymer film composed of collagen and alginate. In terms of morphology and mechanical characteristics, this biofilm closely resembles the extracellular matrix found in bone. Carbodiimide chemistry and the layer-by-layer approach are used to create the biofilms, which have a compact and stable structure. The alginate-terminated coating exhibited greater biological activity and improved adherence to the coated surface in osteoblast MC3T3-E1 cells. According to the findings, further biological research should be done to determine how alginate absorption and enhanced osteoblast activity are related. Compared to bare titanium, the coated surfaces exhibit more notable bioactive qualities and have a great chance of surviving bodily implant rejections (Gregurec et al., 2016). Kumar et al. (Kumar et al., 2019) demonstrated a keratin-based (KE) tissue regenerating alternative derived from leftover goat foot that was combined with sodium alginate and gelatin (gel) to enhance its mechanical and biological qualities. The biofilm has good swelling and antibacterial qualities, as shown by FTIR, XRD, TG-DTA curve, and SEM examination. Fibroblast cell lines also exhibit outstanding biocompatibility. The KE-gel-SA combination of biofilms maintains oxygen permeability, inhibits infection at wound sites, and facilitates simple tissue regeneration processes. The biofilms show enhanced cell proliferation, tensile strength, thermal nature, surface shape, and physiochemical properties. For wound therapy, Ca^{2+}-Alg films were quickly and precisely created using an electrodeposition technique by Liu et al. (Liu, Liu, et al., 2017). The growth and wound-healing characteristics of the film were found to be correlated with the electrode's charge transfer. Wounds treated with this electro-fabricated sheet closed more quickly than wounds left untreated, according to *in vivo* experiments. The newly created dermal tissue was made up of stratified and rearranged epithelial layers, fully grown hair follicles, sebaceous glands, aligned collagen, and connective tissue. The electro-fabricated film was a viable replacement for wound care dressings since it demonstrated a greater rehabilitative impact than untreated groups.

3.5 Bioprinting

Bioprinting is an advanced technology that combines principles of 3D printing with biological materials to create living tissues and potentially entire organs. This innovative process utilizes bio-inks, which are mixtures of living cells and biomaterials, to construct three-dimensional structures that can mimic the natural architecture and functions of human tissues. Bioprinting is an advanced technology that combines principles of 3D printing with biological materials to create living tissues and potentially entire organs. This innovative process utilizes bio-inks, which are mixtures of living cells and biomaterials, to construct three-dimensional structures that can mimic the natural architecture and functions of human tissues. 3D bioprinting involves several key components: biomaterials, cells, and growth factors. The live cells are integrated with biocompatible materials, allowing for the design of tissue or organ models *in situ* for various biomedical applications (Tripathi et al., 2023). There are different kinds of 3D bioprinting methods. Extrusion-based bioprinting, often referred to as material extrusion or MEX-bioprinting, is a significant technique in tissue engineering and regenerative medicine. This method involves the layer-by-layer deposition of bio-inks, which are combinations of biomaterials and living cells, to create three-dimensional (3D) constructs that replicate the architecture and properties of native tissues. The process of MEX-bioprinting typically utilizes specialized printers that extrude bio-inks through a nozzle, allowing for precise control over the placement of materials (Rossi et al., 2024). Another 3D bioprinting method is Inkjet-based bioprinting. It is an innovative technology that has gained traction in tissue engineering due to its ability to produce precise, high-speed droplets of bio-inks—combinations of biomaterials and living cells. This method allows for the non-contact deposition of these bioinks, facilitating the creation of complex three-dimensional (3D) structures that can closely mimic the architecture and properties of native tissues (Park et al., 2023; Li et al., 2020). There are also other methods of 3D bioprinting, such as stereolithography, selective laser sintering, and fused deposition modeling (Beheshtizadeh et al., 2022).

The 3D bio-printed alginate scaffolds were created using the 3D bioprinting technique to create 3D scaffolds for CTE. Using 3D Bioprinting, complex circumstances that resemble those seen in the human body may be simulated by suspending cells in a regulated environment. Environmental cues, both chemical and physical, especially those provided by the extracellular matrix (ECM), significantly impact cell biology. Prior studies have exhibited the efficacy of bio-inks Alginate in the extrusion of bioprinting techniques for creating three-dimensional scaffolds (Jahani et al., 2024). Alginate's exceptional printability, biocompatibility, almost inexpensive cost, and low toxicity make it a popular choice for 3D bioprinting technology. It also gels quickly in one step when combined with a Ca^{2+} cross-linker (Mallakpour et al., 2021).

Alginate's mechanical properties and ability to support cell growth make it suitable for creating scaffolds that replicate cartilage. Studies have shown that alginate-based constructs can promote chondrogenesis and maintain cell viability during and after the printing process (Axpe and Oyen, 2016; Jahani et al., 2024). Nguyen et al. (Nguyen et al., 2017) investigated the promising role of alginate-based bioinks in 3D bioprinting for cartilage tissue engineering using induced pluripotent stem cells (iPSCs). Researchers compared nanofibrillated cellulose (NFC) composite bioinks: NFC with alginate (NFC/A) and NFC with hyaluronic acid (NFC/HA). The NFC/A (60/40 dry weight ratio) bioink demonstrated superior performance by maintaining the pluripotency of iPSCs after bioprinting and supporting their differentiation into hyaline-like cartilage over five weeks. Key findings showed that iPSCs in NFC/A constructs maintained high cell viability, increased cell proliferation, and expressed collagen type II, a hallmark of cartilage tissue. Importantly, the absence of OCT4 protein (linked to pluripotency and tumorigenicity) post-differentiation underscored the safety of this approach for clinical use. The co-culture with irradiated chondrocytes further enhanced iPSC differentiation through cell-to-cell interactions and growth factors like BMP2 and GDF5. The NFC/A bioink provided a mechanically stable and non-cytotoxic environment for bioprinting. It facilitated sustained growth and differentiation of iPSCs into functional cartilage, paving the way for regenerative therapies in osteoarthritis and joint repair. Using a molecular modeling and experimental approach, a study sought to identify the ideal SA/PVA mix for bioprinting cartilage scaffolds. The findings indicated that SA and PVA were compatible. The system became smaller and denser with the addition of PVA, and the hydrogel's pore shape was ideal for chondrocyte growth. As the PVA percentage increased, the forming quality deteriorated; nevertheless, the forming quality was comparatively high when the PVA level was within 30%. Because they had the highest tensile strength and the ideal pore shape for chondrocyte proliferation, the mix hydrogel precursors of 8SA/2PVA were thought to be the most appropriate for 3D printing cartilage scaffolds. This study offers useful advice for creating hydrogels using a SA/PVA combination for 3D cartilage scaffolds (Wei et al., 2022). Another study used additive manufacturing (AM) with a multi-head deposition system (MHDS) to offer a novel 3D cell-printed scaffold for cartilage regeneration. Polycaprolactone (PCL) and alginate hydrogel encapsulated in chondrocyte cells were used to generate the scaffolds. Alginate concentrations and the cell dispensing procedure were established, and *in vitro* biochemical tests revealed increased extracellular matrix (ECM) synthesis. After being implanted in female naked mice, the scaffolds demonstrated increased cartilage tissue and the creation of type II collagen fibrils. The use of 3D-printed scaffolds as an experimental tissue model for cell biology, drug screening, and drug discovery research will have a big influence on regenerative medicine. The potential of the created 3D-engineered tissue structures will be investigated in animal models devoid of blood vessels (Kundu et al., 2015).

Researchers have explored alginate's potential in fabricating bone grafts. Its ability to be combined with other materials enhances its mechanical strength and bioactivity, making it a promising option for bone tissue engineering (Datta, 2023; Ahmadi Soufivand et al., 2023). A novel formulation of alginate, polyvinyl alcohol (PVA), and hydroxyapatite (HA) hydrogel has been created for the 3D bioprinting of mouse calvaria 3T3-E1 (MC3T3) cells into scaffolds with good shape fidelity. By adjusting the concentrations of alginate, phosphate, calcium, and the PVA-HA solution, the hydrogel's rheological characteristics were improved. The primary mineral found in natural bone, HA, was added to provide an environment that was conducive to the formation of new bone. The MC3T3 cells showed good vitality, and the 3D-printed scaffolds held up well in α-MEM cell culture conditions for 14 days. A potential option for correcting customized bone abnormalities is an alginate-PVA-HA mixture that is osteoconductive and biodegradable. The capacity of the encapsulated MC3T3 cells to differentiate and produce new bone will be examined in further *in vitro* research. This alginate-PVA-HA hydrogel composition might be printed to correct faults individually and encourage consistent healing (Bendtsen et al., 2017). Another research used the Cellink® BioX printer to create a 3D-bioprintable scaffold with β-tricalcium phosphate (β-TCP) and alginate. Printability, swelling behavior, degradability, and element analysis were all assessed for the scaffolds. The best method for creating 3D acellular bio-printed scaffolds, according to the study, was 10% alginate/β-TCP bioink, which showed improved alkaline phosphatase activity, cell survival, and physical characteristics. The particular application, planned usage, and desired features of the biomaterial composite determine whether combining alginate with β-TCP is successful. This acellular bioink's affordability supports its use in intrabony defect regeneration methods and provides a workable approach for customized bone regeneration. However, while thinking about these biomaterials' application in the body, one must take into account their immunological effects (Wu et al., 2024). The effects of bioactive glass (BaG) and wood-based cellulose nanofibrils (CNF) on the rheological characteristics of gelatin-alginate bio-inks and the early reactions of bone cells embedded in these inks were assessed by Ojansivu et al. (Ojansivu et al., 2019). The findings demonstrated that whereas BaG increased viscosity and reduced cell survival and proliferation, CNF enhanced the flow behavior of hydrogels. Applications in bone tissue engineering might greatly benefit from the inclusion of CNF and BaG in gelatin-alginate bioinks. Additionally, the study discovered that the bioink's viscosity affected cell survival and proliferation, indicating that the bioink should be customized for each kind of cell. There is a lot of potential for 3D bioprinting mesenchymal stem cells for bone tissue engineering applications since the addition of BaG and CNF to the gelatin-alginate mixture improved their printability and osteogenic activity.

A research created scaffolds for bone tissue engineering utilizing chitosan/alginate/hardystonite (Cs/Alg/HD). The bioactive HD powder was incorporated into CS/Alg at varying proportions, yielding well-structured porous scaffolds. The CsAlg-HD70 scaffold exhibited the greatest yield strength, elastic modulus, degradation rate, and bioactivity. The incorporation of HD particles enhanced cell survival and attachment, achieving 150% by the seventh day. The scaffolds, particularly those containing a greater proportion of HD powders, exhibited favorable *in vitro* performance (Figure 2) (Mohandesnezhad et al., 2023).

Alginate has been used to develop vascular networks within printed tissues, which is essential for nutrient delivery and waste removal in larger constructs (Axpe and Oyen, 2016). By producing scaffolds with or without cells, 3D printing is a promising tissue engineering approach that may be used to treat cardiovascular disorders. A new rotational forming tool for creating vessel-like structures using an alginate-gelatin solution is presented in another research. To examine motion synchronization parameters and test the solution, the researchers put out a theoretical model. The findings demonstrate that while the thickness of the vessel-like structures may be altered by adjusting the parameters, the theoretical model is unable to predict it with any degree of accuracy. To estimate the actual thickness in 3D printing, the optimum thickness parameters have been modified. With greater forming accuracy and faster molding speed than deposition modeling, the study shows that rotary forming is feasible for 3D printing. The researchers carefully adjusted the parameters under various pressures and nozzle movement speeds to guarantee the model's correctness. Angiogenesis in tissue engineering and organ regeneration may be possible with this method, which might help cure cardiovascular conditions (Liu, Zhou et al., 2017).

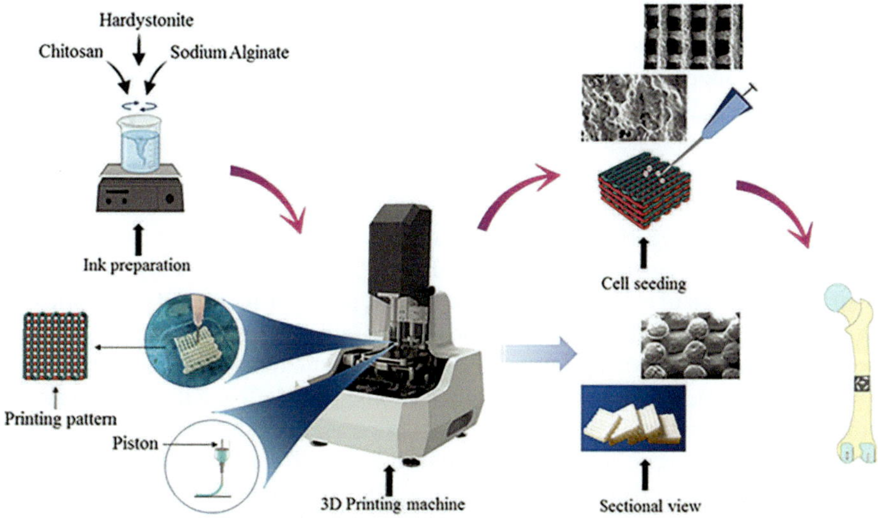

Figure 2. Graphical abstract depicting the procedures for synthesizing a 3D-printed scaffold, encompassing cell implantation and preparation for *in vivo* testing (Mohandesnezhad et al., 2023).

3.6 Electrospinning

Because electrospinning can produce micro- or nanofibers from a wide variety of materials, it has attracted much attention in the last 20 years. An electrical potential is used during this procedure to create an electric field between the syringe and the collector. A jet is produced by charging the polymer droplet at the syringe's tip, which surface tension holds in place. The fibers with diameters ranging from a few nanometers to a few microns are then produced by stretching and extending this jet (Yeo and Kim, 2015; Jadbabaei et al., 2021). The main factors limiting the electrospinnability of alginate solutions have been the confirmation of this structure and its polyelectrolytic nature (Tahermanesh et al., 2024; Nolasco et al., 2022). The presence of polyelectrolytes complicates the electrospinning of alginate because they increase electrical conductivity. Additionally, the limited spin potential is further impacted by the lack of chain entanglements and gelation at low concentrations (i.e., both high surface tension and below the entanglement point). Numerous investigations have concentrated on enhancing alginate's spinning properties by mixing it with water-soluble polymers, establishing a co-solvent system, and modifying the alginate itself (Osfouri et al., 2020).

There is a worldwide public health emergency with the scarcity of premium donor cornea; alternate donor tissue is thus required. Although they lack the fibrous structure of natural tissue, hydrogels which have a lot of water are employed in ocular tissue regeneration. Mechanical characteristics of compliant hydrogels can be improved using electrospun nanofiber reinforcement (Tonsomboon and Oyen, 2013). A study introduced innovative hybrid scaffolds composed of cell-laden electrospun fibers and poly(ε-caprolactone) microstructures for tissue engineering applications. The electrospun fibrous architecture must provide sufficient cell survival and consistent fiber development. The ideal parameters for the cell-electrospinning process were established by modifying alginate concentration and electric field intensity. The amalgamation of cell-laden bioink (2×10^5 MG63 cells/mL, 2 wt% alginate, 2 wt% poly(ethylene oxide), and 0.7 wt% lecithin) with a 0.16 kV/mm electric field exhibited the greatest cell vitality and fiber production. The cells incorporated within the fibrous matrix were viable and underwent proliferation (Figure 3). They exhibit enhanced mechanical characteristics relative to pure cell-laden fibrous mats. *In vitro* cellular activities were evaluated to determine their viability as biomedical scaffolds for tissue regeneration (Yeo and Kim, 2015). In another study a composite scaffold composed of alginate, chitosan, collagen, and hydroxyapatite was fabricated by electrospinning methods. The composite scaffold markedly decreased collagen degradation by 35% over 10 days in collagenase solution, indicating its potential as a scaffold for bone tissue engineering applications. The scaffold's three-dimensional porous architecture facilitates cellular penetration and proliferation. After 10 days in SBF and collagenase solution, the collagen release from the scaffolds was 17% and 2% compared to that from the collagen film. This indicates that employing a chitosan alginate nanofibrous mat as the scaffold framework might reduce scaffold degradation relative to traditional collagen scaffolds, rendering it appropriate for bone tissue regeneration (Yu et al., 2013).

SA is a biodegradable and biocompatible polymer with significant potential in tissue engineering and regenerative medicine. Nevertheless, its limited spinnability poses a barrier for nanofiber-based scaffolds. Jadbabaei et al. (Jadbabaei et al., 2021) sought to enhance the spinnability of SA in conjunction with polyvinyl alcohol (PVA) via response surface methods based on a central composite design. Scanning electron microscopy and Fourier transform infrared spectroscopy analyzed the completed nanofiber scaffolds to assess degradability, swelling, tensile strength, porosity, nanofiber diameter, contact angle, and cytotoxicity. The optimal ratio

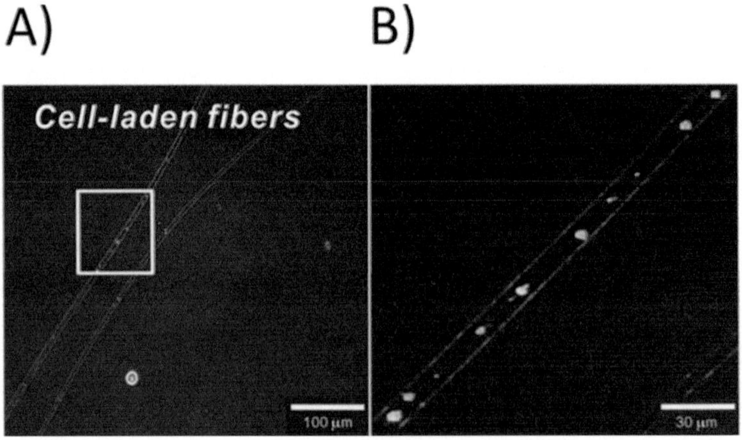

Figure 3. (A) and (B) Fluorescence pictures depicting living cells within the electrospun alginate-based fibers (Yeo and Kim, 2015).

for SA:PVA was determined to be 1:1.6.5, which was manipulable under diverse process settings. The engineered scaffolds exhibited favorable physical, chemical, mechanical, and biological properties. The engineered scaffolds with the optimal ratio under the specified circumstances demonstrate strong biological compatibility for skin tissue engineering. A research examined the biocompatibility and swelling behavior of nanofibers upon the incorporation of alginates into a polymer solution. Glutaraldehyde was utilized to enhance nanofibers, and improved cross-linking was attained by amalgamating it with the polymer solution prior to electrospinning. Dexpanthenol (Sex) was included into the polymeric composition of PVA and SA by a mixing technique. The resultant material served as the foundation for the nanofibers, with drug release regulated by several shells. The incorporation of 1% (w/v) chitosan on the shell side of nanofibers enhanced the regulation of drug release. The release of the medication from dexpanthenol-loaded wound dressings adhered to the Fickian diffusion process as described by the Korsmeyer-Peppas model. MTT measurement and cell culture demonstrated that dexpanthenol-loaded PVA/SA/Triton-Chitosan nanofibers exhibited nontoxicity towards fibroblast cells and positively influenced cellular adhesion and shape. The research also synthesized nanofibers with a core including Triton X-100, PVA, SA, and Dex, encased in different shells by the electrospinning method. The incorporation of SA into the polymer solution augmented the vacancy percentage and swelling of the resultant nanofibers, hence enhancing their water-absorption capacity. The nanofibers with 1% w/v CS shell improved drug release control, whereas the PVA/SA/Triton-CS Dex-loaded nanofibers promoted fibroblast cell attachment without inducing toxicity. Consequently, PVA/SA/Triton-CS nanofibers exhibit suitable biocompatibility and cell adhesion, rendering them advantageous for wound healing and the enhancement of particular tissue regeneration (Najafiasl et al., 2020).

3.6.1 Pure Alginate

The electrospinnability of the alginate solution without carrier polymers has only been studied in a few researches. Fang et al. (2023) looked at the ability to electrospun pure SA in their investigation. However, because there were no entanglements between the molecular chains, the collector only collected irregular droplets. They injected calcium chloride to promote the interlocking of chains, which led to the development of long, discontinuous fibers. The research claims that the addition of calcium ions (Ca^{2+}) enhanced molecular chain interaction by causing SA to form a gel. Alginate was then easier to electrospun as a result.

Similarly, Saquing et al. (Saquing et al., 2013) observed the formation of droplets exclusively, independent of the rise in alginate concentration ranging from 1 to 6 weight percent. There is more to alginate solution's electrospin challenges than just a deficiency in chain entanglements. Surface tension and viscosity are critical factors hindering the production of defect-free alginate fibers. Even though the surface tension was cut in half, the researchers found that adding Triton X-100, a non-ionic surfactant, caused the size of the collected droplets to decrease. Alginate's molecular structure, surface tension, conductivity, and solution viscosity are some variables that affect its electrospinnability. Another factor contributing to this phenomenon is gelation at a relatively low concentration before the stable jet forms. Numerous techniques have been used up to this point to improve alginate's spinnability.

3.6.2 Co-Solvent Systems

Co-solvent systems consist of one or two extra solvents, often comprising less than 20% of the total weight compared to the solvent that dissolves the polymer. Previous studies described the production of alginate nanofibers using a solvent system solution by electrospinning process (Nie et al., 2008). The co-solvent solution consisted of the highly polar substances glycerol and water. Nanofibers with an average diameter ranging from 120 to 300 nm were obtained by increasing the concentration from 1.6 to 2.4 weight percent. To crosslink the fibers, the solution was electrospun immediately into a coagulation bath made of ethanol and a 10-weight percent aqueous $CaCl_2$ solution. Glycerol was found to be the agent that increased flexibility and encouraged chain entanglements to improve alginate's electrospinnability. In addition, the solution's viscosity increased while its conductivity and surface tension decreased. Furthermore, it was demonstrated that by lowering surface tension and electrical conductivity, the co-solvent solution improved alginate's ability to spin (Fang et al., 2011). A complex solvent system comprising water, ethanol, and dimethylformamide (in the ratio of 75/15/10) with 1.5 weight percent calcium chloride was used to create a continuous fluid jet.

3.7 Microspheres

Because of their exceptional mechanical properties, biodegradability, biocompatibility, and ease of processing, alginates have been approved by the FDA for use in drug delivery. These attributes have also contributed to their lengthy history and present status as popular biomaterials. Fewer research has been done on other biomacromolecules than on alginate microspheres as a basis for drug carriers. Numerous recent developments in the application of alginate-based microsphere delivery technologies and instruments have been documented in several biomedical domains, such as immunology, vaccines, diabetes, cancer, TB, orthopedics, gene therapy, stem cell research, vascular tissue engineering, protein engineering, and other medical specialties. Everyone working on the research, therapy, or diagnosis (Uyen et al., 2020). In a study human bone marrow mesenchymal stem cells (hBMMSCs) was encapsulated in alginate microspheres with murine anti-BMP2 monoclonal antibodies (mAb) found that anti-BMP2 mAbs bind BMP2 ligands, boosting osteogenesis. This could be a potential technique for healing and regeneration of lesions in the cranium, axial bone, and appendix. The study suggests that the inclusion of anti-BMP2 mAb could improve osteo-differentiation *in vivo* (Moshaverinia et al., 2013).

The creation of hydroxyapatite/sodium alginate/chitosan (HA/SA/CS) composite microspheres, which have superior biocompatibility for certain biomedical applications, is the main goal of the study that published by Bi et al. (Bi et al., 2019). Using an emulsion crosslink process and calcium ions as a cross-linking agent, the HA/SA/CS composite microspheres were effectively created. According to the findings, HA/SA/CS composite microspheres outperformed HA nanoparticles in terms of drug loading and encapsulation efficiency. Doxorubicin-loaded HA/SA/CS composite microspheres demonstrated good blood and cell compatibility as well as strong pH-sensitive drug-release capabilities. In comparison to HA nanoparticles and HA/SA composite microspheres, they also demonstrated superior cell attachment and proliferation capabilities. The cell proliferation experiment showed that the proliferation capacity of MG-63 cells on HA/SA/CS microspheres increased with prolonged incubation. The three-dimensional network structure of HA/SA microspheres was more advantageous for cell attachment than individual HA nanoparticles. The adhesion ability of HA/SA/CS microspheres to osteoblasts was significantly enhanced compared to HA nanoparticles and HA/SA microspheres. The rough surface and three-dimensional network structure of HA/SA/CS microspheres were the primary factors enhancing the adhesion, migration, and proliferation of osteoblasts on the microspheres. The results indicate that HA/SA/CS composite microspheres facilitate the adhesion of MG-63 cells to their surface and exhibit a strong affinity for MG-63 cells (Figure 4).

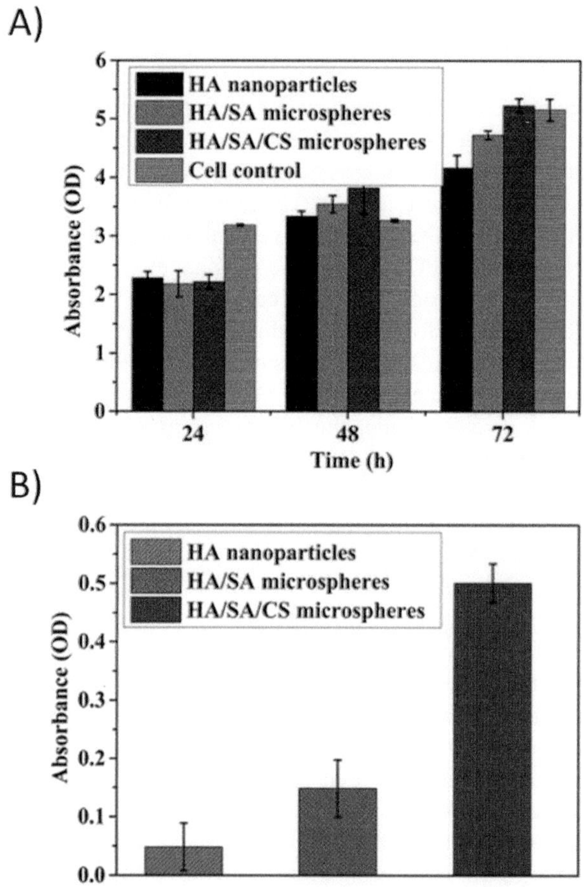

Figure 4. Results of the cell proliferation and cell attachment experiment (Bi et al., 2019).

For use in brain tissue creation, a novel biomimetic scaffold based on gelatin was created. The scaffold was made by combining thermally induced phase separation with porogen leaching, and it was merged with alginate microspheres filled with nerve growth factor (NGF). When the concentration of gelatine reached at least 7.5%, the nanofibrous matrix was produced. NGF was encapsulated into the microspheres with 85% and 100% efficiency, respectively, and the scaffold's modulus value was comparable to that of brain tissue. The poly(L-lysine) coating and alginate concentration regulated the rate of NGF release. Neurite extension of PC12 cells was triggered by the nanofibrous structure and the regulated release of NGF from the microspheres, indicating that the released NGF was in an active state. Because of their mechanical and topologic characteristics as well as their pore structure that promotes cell growth and differentiation, the scaffolds may find use in brain tissue engineering (Büyüköz et al., 2018).

Protein-loaded microspheres for long-term protein delivery in scaffolds for tissue engineering have been studied. Fluorescently tagged immunoglobulin G (IgG) was loaded into alginate microspheres, which were made utilizing a small-scale water-in-oil emulsion process. The force produced by mechanical stirring during emulsion has an impact on the size of the microsphere. IgG release from microspheres may be prolonged, according to protein release tests, and the rate of protein release could be further slowed by an additional chitosan coating. In order to obtain sustained and extended protein delivery in tissue engineering applications, the work shows that protein-loaded microspheres may be made with a regulated release profile and conserved biological activity. These microspheres can then be integrated into scaffolds. Additionally, the study discovered that altering the stirring speed had no discernible effect on the IgG release patterns (Zhai et al., 2013).

The use of human mesenchymal stem cells (hMSCs) in alginate microspheres modified with RGD peptide for cardiac healing was examined in one research. The RGD-modified alginate enhanced cell adhesion, proliferation, and the expression of angiogenic growth factors, according to the *in vitro* investigation. When compared to the PBS control or cells alone, the hMSC group demonstrated much higher cell survival. The efficiency of an RGD-alginate microsphere system for cardiac healing and stem cell delivery to the myocardium was also shown in the study. This demonstrates the promise of stem cell treatment and synthetic biomaterials in cardiac regeneration and repair (Yu et al., 2010).

The objective of the Soran et al. (Soran et al., 2012) study was to create a scaffold-microsphere system for periodontal tissue creation. Alginate microspheres containing bone morphogenetic protein-6 (BMP-6) were produced under ideal electrospraying conditions. These microspheres were added to chitosan gels to create a new scaffold with preserved pore diameters and interconnectivity. The controlled release of BMP-6 from chitosan scaffolds was seen over seven days. Research employing rat mesenchymal stem cells in cell culture has shown that osteogenic differentiation was significantly impacted by the controlled injection of BMP-6. The color shift of BMP-6, which contained chitosan scaffolds loaded with alginate microspheres, was more pronounced than the matrix calcification seen in the scaffolds by the von Kossa study. The integrated method enhanced the osteoblastic growth of rat mesenchymal stem cells derived from bone marrow by controlling the release of BMP-6. One may consider this technique to be an appropriate medium for periodontal tissue engineering applications.

In a lengthy bone defect, research looked at how endothelial cells could control the osteogenic capacity of osteoprogenitor cells both *in vitro* and *in vivo*. Alginate microspheres, a naturally occurring polymer utilized for cell encapsulation, were employed in another study to immobilize human osteoprogenitors (HOP). The findings revealed that co-immobilized HOPs and HUVECs had mineralization deposits, elevated cell metabolic activity, and upregulated alkaline phosphatase and osteocalcin gene expression. The study also discovered that mineralization was considerably enhanced when HOPs and HUVECs were co-immobilized. The results indicate that a promising approach to bone tissue creation is the co-immobilization of osteogenic and endothelial cells within RGD-grafted alginate microspheres (Grellier et al., 2009).

3.8 Microcarriers

In bioreactor systems, microcarriers serve as support matrices that promote cell development. To facilitate cell adhesion and subsequent cell proliferation, growth media is suspended in a fine mesh of microscopic solid particles onto which cells are cultured (Safonova et al., 2021).

Bioreactor microcarriers boost cell proliferation (Chen et al., 2020). Cells can adhere to and grow on the surface of tiny solid particles suspended in a growth medium. The first evidence of microcarriers was found in 1967 by van Wezel and associates. New microcarrier designs and surface changes have expanded their clinical and *ex vivo* applications (Derakhti et al., 2019). Basic microcarriers are 100–300-micron beads with sufficient density to remain suspended while stirring. Numerous microcarriers, such as glass, collagen, DEAE-dextran, acrylamide, polystyrene, and alginate, have been produced and sold (Ding et al., 2022). Alginic acid is a widely used biomaterial for drug delivery and cell encapsulation. Cartilage cells grown on alginate can create an ECM that resembles natural cartilage. The surfaces of these materials can be changed to influence the adhesion of proteins and cells. Fish, birds, insects, and mammal cells were successfully cultured in microcarriers (Kulchar et al., 2021). One can cultivate primary cells, diploid cell strains, altered or established cell lines, hybrid cell lines, and tumor cells on microcarriers. Cell types and applications are the basis for choosing microcarriers (Dias et al., 2017).

In a study, the novel composite microcarriers for the adhesion and proliferation of mesenchymal stem cells (MSCs) were investigated. To cause a conformational shift in silk, alginate microcarriers were made, introduced to a solution of silk fibroin, and then treated with ethanol. The uniformity, coating stability, and size distribution of the microcarriers were assessed. The suitability and *in vitro* cytocompatibility of MSC delivery devices were examined. The findings demonstrated reliable and repeatable production procedures, stable silk/alginate microcarriers with spherical shapes, and a uniform layer of fibroin covering the surface. After three days of culture, MSCs quickly clung to the microcarrier surface and covered it. The created 3D culture system maintained the metabolic activity and ability for multi-lineage differentiation of MSCs, which makes it appropriate for use in regenerative medicine, enhanced treatment, and cell delivery. The findings imply that the technical procedure is reliable for creating fibroin/alginate microcarriers on a laboratory scale, which promotes the adhesion and growth of MSCs (Perteghella et al., 2017).

Li et al. (Li et al., 2016) introduced a collection of alginate/PEG (AL/PEG) semi-interpenetrating network microcarriers for the proliferation and extraction of human umbilical cord blood mesenchymal stem cells (hUCB-MSCs) (Figure 5). The AL/PEG system comprises chemically cross-linked alginate networks created by the reaction between carboxylic acid and cysteamine. PEG is utilized to regulate the breakdown of microcarriers, while chitosan and biocompatible fibronectin are applied to the surface to enhance mechanical strength. The dimensions of AL/PEG microcarriers can be modified by altering the composition and molecular weight of PEG. The microcarriers may be destroyed by the breakage of disulfide cross-linkages, and the self-renewal of collected cells is examined. The shape and dimensions of AL/PEG microcarriers are contingent upon the PEG concentration, its molecular weight, and the surface modifications applied. AL/PEG microcarriers facilitate the adhesion and proliferation of hUCB-MSCs and may be collected by breakdown, minimizing cellular damage associated with conventional proteolytic enzyme methods. The synthesized AL/PEG-based microcarriers are anticipated to hold significant use in large-scale stem cell cultivation for tissue engineering and cellular transplantation.

Mesenchymal stem cells (MSCs) are promising therapeutic possibilities owing to their capacity to develop into many cell types. Nonetheless, the disparity between availability and demand necessitates other expansion strategies. A hydrogel microcarrier composed of genipin cross-linked alginate–chitosan beads was created, demonstrating a 26% increase in cell adhesion and a twofold proliferation rate relative to the commercial microcarrier Cytodex 1. The cells readily separated from the hydrogel beads without necessitating prolonged incubation or vigorous agitation during harvesting. The research found that genipin cross-linked alginate–chitosan microcarriers may serve as a viable substitute for commercial microcarriers in the proliferation of MSCs (Chui et al., 2019).

Lehnert et al. (Lehnert and Sikorski, 2022) presented a novel approach for creating 3D constructions including different zones of elevated cell densities without the need for complex equipment. The procedure entails embedding cells within a sacrificial alginate matrix, facilitating the integration of cell-specific biopolymers such as collagen to create a microscale, spatially limited, interpenetrating network.

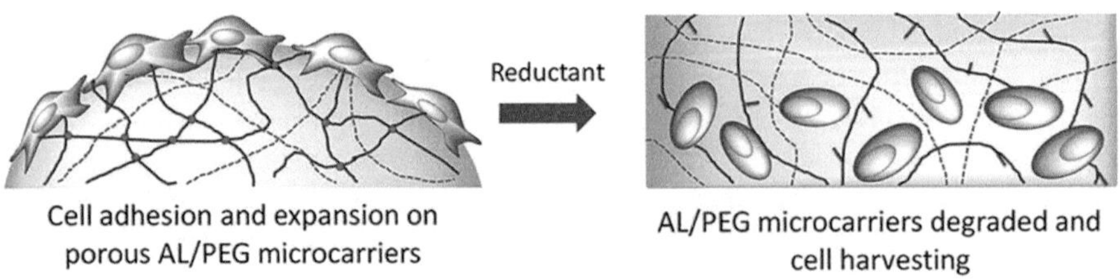

Figure 5. Schematic illustration of the detachment processes of hUCB-MSCs triggered by the degradation of AL/PEG microcarriers (Li et al., 2016).

A macroscale layered structure can be attained by integrating these cell-laden microspheres within thin collagen layers. Cells are locally freed by de-gelling the alginate matrix, and their adhesion to the collagen hydrogel layers is examined. The mouse pre-osteoblast cell line MC3T3-E1 demonstrates that cell migratory patterns vary according on the initial composition of the alginate microspheres. The manufactured structures do not diminish in size with prolonged culture duration, facilitating extended examination of artificially produced constructs. Future research may involve including more ECM-related biopolymers, introducing supplementary microspheres with a secondary cell population, or creating gradients by layering several inherently distinct materials. The microsphere microstructure may be altered to form cell-specific compartments holding biochemical molecules, so establishing a chemical gradient within the system.

4. Commercial Alginate-Based Biomaterials Product with Application in Tissue Engineering-Regenerative Medicine (TE-RM)

Given the successful outcomes of various alginate composite formulations in different TE-RM domains, several commercial products, particularly those used as wound dressings, have been developed. Table 1 lists a number of these products.

5. Clinical Trial Overview

Information specific to commerce is also covered in this chapter. It also details a number of clinical trials and provides an overview of the uses of alginate-based composites in biomedicine. ClinicalTrials.gov data is used in Table 2 and text to describe clinical trials that will take place before 2021. The clinical trial aimed to determine the safety and effectiveness of using alginate hydrogel to reconstruct the left ventricle as a treatment for heart failure. The alginate hydrogel was designed to be implanted into the myocardium as an artificial filler to prevent ventricular expansion and restore favorable heart shape and size. It was expected that the structure and function of the heart would improve, thereby reducing symptoms and enhancing patients' quality of life. Six months post-procedure, the patient's clinical condition improved significantly, and better LV ejection fraction reduced end-diastolic and end-systolic volume indices were observed (NCT04781660) (Wang et al., 2023). Tzu Chi University conducted a study investigating the effects of Calcium alginate silver dressing (CA ALG silver dressing) compared to conventional wound dressings in patients with pressure injuries (PI) residing in long-term care institutions in promoting wound healing for chronic wounds. Study results provided evidence-based guidance for ALG wound dressings in long-term care facilities and improved patient care (NCT05667831). Another study, A Single Center Trial of Donor Site Wound Dressings after split thickness skin grafting (STSGs) was initiated in 2023 by Dow University of Health Sciences in Pakistan. It aimed to evaluate various dressing types for donor site wounds from STSG. Following 12 weeks, the patient's conditions were reassessed, with the following dressing groups assessed: film, alginate, gauze, hydro fiber, hydrocolloid, and silicone. Different dressing

Table 1. Some commercially available ALG products and their properties with application in TE-RM.

Brand name	Manufacturer	Origin	Type	Properties	USFDA 510(k) Number	Ref
ALGS6 Ag	Foshan United Medical Technologies	China	Dressing	• Ca ALG fiber, Lyocell fiber, and 1.7% ionic silver • Controls minor bleeding • Antimicrobial efficiency • High absorption and fluid retention	K172570	(Barbu et al., 2021)
Calgitrol™ or SilverSite™	B. BRAUN VET CARE GMBH	Germany	Dressing	• Ag ALG—Ca ALG—polyurethane • Management of moderately to heavily exuding wounds • High absorption capacity • Atraumatic dressing removal	K041268	(Barbu et al., 2021)
SeaSorb®	Coloplast Corp.	United States	Dressing	• Na ALG/Ca ALG—polyethylene	K983519	(Szekalska et al., 2016; Barbu et al., 2021)
Medihoney®	Derma Sciences	United States	Gel, hydrogel and paste	• Ca ALG—Manuka honey • High osmotic potential and low pH helps to decrease protease activity • Antibacterial effect • Suitable for Lightly to moderately exuding wounds	K150985	(Barbu et al., 2021)
Flaminal Forte® gel	Flenhealth	Belgium	Gel	• ALG—PEG matrix combined with a unique antimicrobial enzyme system (GLG — glucose oxidase and lactoperoxidase, stabilized by guaiacol • Intended for moderately to highly exuding wounds • Debrides the wound and manages moisture balance. • Antimicrobial properties	-	(Szekalska et al., 2016; Barbu et al., 2021)
Hyalogran®	FIDIA FARMACEUTICI S.p.A.	Italy	Dressing	• Hyaluronic acid ester—Na ALG • Absorbs wound exudate and transforms into a soft gel • Management slougy and necrotic wounds	-	(Szekalska et al., 2016; Barbu et al., 2021)

Algivon®	Advancis Medical	United Kingdom	Dressing	• Ca ALG -Manuka honey • Alginate fibers gradually release the manuka honey • Provides antimicrobial properties and optimal moisture balance • Suitable for the therapy of moderately to severely leaking wounds, including as leg ulcers, pressure ulcers, diabetic foot ulcers, surgical wounds, and traumatic wounds. both a main and secondary dressing	-	(Wang et al., 2015; Szekalska et al., 2016)
Algidex Ag+	DeRoyal	United States	Foam, Gel Pre-Cut Trach Hydrogel, Gauze, I.V.Dressing	• Ionic silver with ALG and maltodextrin • Maintaining an optimal moisture level • Sustains an antimicrobial barrier for up to seven days	K173072	(Hartzell et al., 2014; Ousey et al., 2016)
Aquacel™ Ag EXTRA™	Convatec	United Kingdom	Hydrofiber dressing	• Na ALG CMC—Ag+ • Moderately to highly exuding wounds • Sustained antimicrobial activity for up to 14 days	K173675	(Hurlow, 2012; Joergensen, Blaise et al., 2022)
Biatain® Alginate Ag	Coloplast Corp.	United States	Dressing	• Ca ALG, CMC, and an ionic silver complex • Moderately to heavily exuding infected wounds • Highly absorbent • Antimicrobial effect	K231953	(Hurlow, 2012; Joergensen et al., 2022)
Guardix-SG®	Genewel	Korea	Thermosensitive dressing	• Na ALG- poloxamer- CaCl • Thermosensitive Adhesion Barrier • Non-pyrogenic and nontoxic • Be used as an adjunct to abdominal surgery, spine surgery, and thyroidectomy	-	(Kim et al., 2019; Barbu et al., 2021)
Purilon Gel®	Coloplast Corp.	United States	Hydrogel	• Consists of Na CMC—Ca ALG • Hydrates necrotic tissue and absorbs excess exudate • Used in conjunction with a secondary dressing. • Used for dry and sloughy necrotic wounds, non-infected diabetic foot ulcers, 1st and 2nd-degree burns, and pressure and venous ulcers	K971597	(Szekalska et al., 2016)

Table 1 contd. ...

...Table 1 contd.

Brand name	**Manufacturer**	**Origin**	**Type**	**Properties**	**USFDA 510(k) Number**	**Ref**
KALTOSTAT®	Conva tec	United Kingdom	Gel-fiber matrix	• It is Composed of Ca-Na ALG fiber • Absorbs exudate and helps control minor bleeding • Suitable for moderately to heavily exuding wounds	K921009	(Barbu et al., 2021)
Luofucon®	foryoumedical	China	Gelling fiber	• Consisting dressings composed of Ca ALG—Ag and Ca ALG - PHMB • Highly absorbent • Antibacterial properties	K172554	(Barbu et al., 2021)
Fibracol™	Solventum	United States	Gelling fiber	• Structural support of collagen with the exudate management of Ca ALG • Non-adherent upon removal • Maintains integrity and a moist wound environment • Conducive to granulation tissue formation and epithelialization	K925548	Szekalska et al., 2016; Barbu et al., 2021)
Silvercel™	Solventum	United States	Dressing	• Containing a high G (guluronic acid) ALG, CMC and Ag-coated nylon fibers • High fluid absorption from moderately to heavily exuding wounds • Antimicrobial action	K081363	(Barbu et al., 2021)
Saf-gel®	Coloplast	United States	Gel	• Composed of Na CMC—Ca ALG • Using it in conjunction with a secondary dressing • Absorbs excess exudate, effective autolytic debridement • Used for necrotic and sloughy wounds	K971597	Szekalska et al., 2016; Barbu et al., 2021)

Suprasorb® A—Ag	Lohmann & Rauscher	Germany	Rope and dressing	• Contains Ca ALG Dressing with Silver • Creates a moist wound environment. for heavily discharged or infected wounds • Antibacterial properties • Promotes healing by high exudate absorption	K183208	(Ousey et al., 2016)
Tromboguard®	TRICOMED S.A.	Poland	Dressing	• CS, NaALG/Ca ALG with the addition of silver salt • Rapid hemostasis • Promotes clot formation and the wound-healing process in the treatment of traumatic and postoperative wounds	-	(Szekalska et al., 2016; Barbu et al., 2021; Strużyna et al., 2024)
Comfeel Plus™	Coloplast Corp	United States	Hydrocolloid dressing	• Na-CMC/ Ca-ALG • Management of low to moderately exuding wounds • Can be used as a secondary dressing.	K951325	https://www.coloplast.co.za/global/south%20africa/coloplast%20wound%20care%20catalogue%202022.pdf.
Sorbalgon® Ag	Hartmann	United States	Dressing	• It is Composed of Ca ALG with silver • For treatment of any external wounds autolytic debridement Highly absorbent • Antimicrobial properties up to 7 days	-	(Lin et al., 2012)
Progenix putty®	Medtronic Sofamor Danek	United States	Putty form	• DBM in Type-1 bovine collagen Na ALG • A bone graft substitute and bone void filler, remodels into host bone, making it an effective choice for bone regeneration	K080462	(Szekalska et al., 2016)

Table 1 contd. ...

...Table 1 contd.

Brand name	Manufacturer	Origin	Type	Properties	USFDA 510(k) Number	Ref
AlgiMatrix®	Thermo Fisher Scientific	United States	Sponge	• Lyophilized ALG gel • 3D cell culture scaffold • Supports the formation of 3D cellular aggregates • For a broad range of procedures • Chemically defined, animal-origin-free material • Easy cell visualization • Non-enzymatic cell	-	(Brancato et al., 2020)
NovaMatrix®	Novamatrix	Norway	3D cell culture scaffold	• Ultrapure Na ALG and Na hyaluronate • Gentle cell immobilization • For use in drug delivery, tissue engineering, cell encapsulation, and medical devices	-	(Lee et al., 2012)
NOVATACH™	Novamatrix	Norway	3D cell culture scaffold	• Peptide-coupled ALG • Helps biological interactions between ALG hydrogels and cells. • For use in various biomedical contexts; cell therapy and advanced drug delivery, 3D bioprinting, Tissue engineering	-	(Pahlevanzadeh et al., 2020)
Hygedent	HYGEDENT INC	China	Powder	• Creates models for the production of restorative prosthetic devices • High tear resistance and good elastic properties	K140074	(Omo et al., 2025)

Abbreviations: United States Food and Drug Administration (USFDA); Chitosan (CS); Calcium alginate (Ca ALG); carboxymethylcellulose (CMC); Sodium alginate (Na ALG); Alginate silver (ALG Ag); poly hexamethylene biguanide (PHMB); polymers polyethylene glycol (PEG); Demineralized bone matrix (DBM).

Table 2. A brief overview of Clinical Trials findings on ALG-based combinations in TE-RM.

Study Title	ClinicalTrials.gov Identifier	Purpose	Study Type Model	Randomization	Recruitment status	Ref
The Safety Evaluation of Silver Alginate (Algidex™) Dressing in Very Low Birth Weight (VLBW) Infants Baylor University Medical Center, United States	NCT00593684	Prevention	Interventional Parallel Assignment	Randomized	Completed 2008	(Khattak et al., 2010)
Medical and Economic Evaluation of FORESEAL Versus the Current Therapeutic Approach (Stapling Alone or Associated with Tissue Sealant) in Terms of Air Leakage Duration After Lung Resection for Cancer. Hôpitaux de Paris	NCT00925444	Treatment	Interventional Parallel Assignment	Randomized	Completed 2009	(Wong et al., 2024)
Safety and Efficacy of Inhaled OligoG CF-5/20 for the Treatment Cystic Fibrosis. Merthyr Tydfil, United Kingdom	NCT00970346	Treatment	Interventional Single Group Assignment	Randomized	Completed 2009	(Tøndervik et al., 2014)
Restore Calcium Alginate Dressing, Silver vs Aquacel Ag in the Treatment of Critically Colonized Venous Leg Ulcers. Hollister Wound Care LLC, United States	NCT01396304	Treatment	Interventional Parallel Assignment	Randomized	Terminated (slow accrual) 2011	(Moore, 2013)
A Randomized, Controlled Study to Evaluate Algisyl-LVR™ as a Method of Left Ventricular Augmentation for Heart Failure (AUGMENT-HF). LoneStar Heart, Inc., United States	NCT01311791	Treatment	Interventional Parallel Assignment	Randomized	Completed 2011–2017	(Lee et al., 2015)
Open-label Investigation of the Safety and Efficacy of DIABECELL in Patients with Type 1 Diabetes Mellitus, Living Cell Technologies, New Zealand	NCT01736228	Treatment	Interventional Single Group Assignment	Non-Randomized	Completed 2012	(Arutyunyan et al., 2020)
Open-label Investigation of the Safety and Clinical Effects of NTCELL in Patients with Parkinson's Disease. Auckland City Hospital, New Zealand	NCT01734733	Treatment	Interventional Single Group Assignment	N/A	Completed 2012	(Snow et al., 2019)
To Evaluate the Effect of Contact Lens with Alginic Acid (7 Days) in Patients with Dry Eye. Singapore National Eye Centre Singapore Eye Research Institute, Singapore	NCT01918410	Treatment	Interventional Parallel Assignment	Randomized	Completed 2013	(Chong et al., 2016)

Table 2 contd. ...

...Table 2 contd.

Study Title	ClinicalTrials.gov Identifier	Purpose	Study Type Model	Randomization	Recruitment status	Ref
A Phase IIb Study of OligoG in Subjects with Cystic Fibrosis (SMR-2984), Germany	NCT02157922	Treatment	Interventional Crossover Assignment	Randomized	Completed 2014	(Chong et al., 2016)
A Study of OligoG in Cystic Fibrosis Subjects with Burkholderia spp. Infection, Germany	NCT02453789	Treatment	Interventional Crossover Assignment	Randomized	Completed 2015	(Weiser et al., 2021)
Investigation of the Safety and Efficacy of NTCELL® [Immunoprotected (Alginate-Encapsulated) Porcine Choroid Plexus Cells for Xenotransplantation] in Patients with Parkinson's Disease, Auckland City Hospital, New Zealand	NCT02683629	Treatment	Interventional Parallel Assignment	Randomized	Completed 2016	(Snow et al., 2019)
Hydrogel Injection to Assist Endoscopic Submucosal Dissection, National Cheng-Kung University Hospital, Taiwan	NCT03321396	Treatment	Interventional Single Group Assignment	N/A	Completed 2017	(Wang et al., 2023)
Urothelium Tissue Engineering Using Bladder Mucosa from Transurethral Resection of Prostate. Central Hospital, France	NCT03698721	Treatment	Observational	-	Not yet recruiting 2018	(Keshel et al., 2020)
Alginate Dressings Versus Gauge Dressings After Pilonidal Cyst Resection: Examination of the Quality of Life, Larissa University Hospital, Greece	NCT03757572	Treatment	Interventional Parallel Assignment	Randomized	Terminated 2018	(Mamaloudis et al., 2022)
A Phase 2b Randomized, Double-blind, Parallel-group Study of Alginate Oligosaccharide (OligoG) Dry Powder Inhalation in Addition to Standard of Care Compared to Placebo in Addition to Standard of Care in Patients With Cystic Fibrosis (CF), AlgiPharma AS, Australia	NCT03822455	Treatment	Interventional Parallel Assignment	Randomized	Unknown 2019	(Jaques et al., 2020)

groups took an average time to complete wound healing and re-epithelialization. The fastest recovery times were seen with hydro fiber and silicone dressings.

Perturbation, pain, scarring, or patient satisfaction did not differ significantly (NCT06044519) (Ismail et al., 2023). Lohmann & Rauscher's company aimed to evaluate the performance and safety of Suprasorb® A + Ag (CA ALG silver dressing) in treating wounds at risk of infection or infected wounds. Ninety-nine patients with various types of wounds participated in this study, and this conventional multicenter, open, single-arm cohort study was conducted for three weeks (NCT05646121) (Clegg et al., 2016). A study on the use of ALG microbeads with this subject Hepatocyte microbeads for Acute Liver Failure (HELP), a combination of hepatocytes and mesenchymal stromal cells co-encapsulated in peptide-alginate microbeads is novel cell therapy for the treatment of patients with acute liver failure leading to liver transplantation or regeneration (NCT05491135) (Fitzpatrick et al., 2023). The α-Mangostin Hydrogel Film with chitosan-ALG base study looks at the efficacy and security of a new therapy for people with recurrent acute salivary syndrome (RAS). α-Mangostin is administered via a drug delivery system consisting of chitosan and SA. The purpose of the trial is to assess the hydrogel film's safety and effectiveness in lowering pain and accelerating recovery.

The PMS Study of Silver I Alginate Non-Woven Dressing (Hydro-ALG), an open-label, multicenter, single-arm clinical research, attempts to validate the continuous safety and efficacy of this wound dressing in both acute and chronic wounds. On STSG donor site wounds, several dressing types, such as ALG, film, gauze, hydrocolloid, and hydro-fiber silicon, were applied at random (NCT05690685) (Vitamia et al., 2024).

6. Future Perspectives

Due to unique biological attributes, naturally derived polymers play a crucial role in biomedical and pharmaceutical industries. The significant progress associated with tissue engineering and regenerative medicine (TE-RM) has led to using renewable, cost-effective biomaterials. Alginate-based biomaterials exhibit remarkable and outstanding characteristics, such as non-toxicity, non-immunogenicity, high biological compatibility, biodegradability, slow degradation, and abundant availability (Zdiri et al., 2022; Sahoo and Biswal, 2021). High viscosity, low solubility, low cell adhesion, poor mechanical stability, hydrophilicity, and instability during heat treatment are just a few of the disadvantages and restrictions associated with alginate alone (Farshidfar et al., 2023; Ahmad et al., 2021).

To overcome these restrictions and enhance their qualities as promising candidates in a variety of derivatives, such as hydrogels, composites, films, aerogels, microspheres, and beads, it has been mixed with a variety of natural or synthetic polymers. Antimicrobial agents, wound dressings, tissue regeneration, drug and gene delivery, cancer therapy, and biosensors are just a few of the uses for these. The complex process of wound healing may require individualized care. Alginate-based compound modifications are moving quickly toward becoming effective and targeted drug delivery vehicles for the healing of chronic wounds and wound dressings, as well as innovative options for wound healing due to their hemostatic and antibacterial qualities. With the goal of creating the perfect wound dressing, the number of FDA-approved commercial wound dressings is growing quickly.

On the other hand, 3D bioprinting is an emerging technology with promising results in the TE-RM field. Alginate-based combinations are used extensively as bio-inks due to printability, low cost, ease of gelation, and biocompatibility. Blending it with other materials to overcome its limitations and reach an ideal bio-ink holds significant promise in advancing TE-RM applications. It is expected to reach the clinical use stage in the coming years. This versatile biopolymer improves sustained delivery, rate of biodistribution/dissolution, and targeting properties. They can promote extracellular matrix production and offer 3D support for cellular growth, control the release of tissue induction factors like growth factors and morphogenetic proteins for *in vivo* cell delivery, and adapt to chemical changes with adhesion ligands facilitating and accelerating the fabrication and regeneration of a wide range of organs and tissues. Due to these remarked features, they are considered a new therapeutic strategy in bone and cartilage, corneal, nerve, pancreas, skeletal muscle, liver, coronary, and peripheral arterial regeneration in a minimally invasive way and have been regarded as cutting-edge technology in TE-RM. Overall, research and production in this field are expected to increase significantly. More studies and reviews in Clinical translational and systematic *in vivo* can assist in achieving effective alginate-based composites for applications in TE-RM.

References

Abbasi, Asma Riaz, Muhammad Sohail, Muhammad Usman Minhas, Touba Khaliq, Mubeen Kousar, Shahzeb Khan et al. 2020. Bioinspired sodium alginate based thermosensitive hydrogel membranes for accelerated wound healing. International Journal of Biological Macromolecules 155: 751–65.

Ahmad, Awais, N. M. Mubarak, Fakiha Tul Jannat, Tayyaba Ashfaq, Carlo Santulli, Muhammad Rizwan et al.. 2021. A critical review on the synthesis of natural sodium alginate based composite materials: An innovative biological polymer for biomedical delivery applications. Processes 9: 137.

Ahmadi Soufivand, Anahita, Jessica Faber, Jan Hinrichsen and Silvia Budday. 2023. Multilayer 3D bioprinting and complex mechanical properties of alginate-gelatin mesostructures. Scientific Reports 13: 11253.

Al-barudi, Amenah, Genada Sinani and Zeynep Ulker. 2024. Biodegradable polysaccharide aerogels based on tragacanth and alginate as novel drug delivery systems. Journal of Sol-Gel Science and Technology 109: 748–56.

Alizadeh, Sanaz, Jila Majidi, Mozhgan Jahani, Zahra Esmaeili, Zeinab Nokhbedehghan, Zahra Aliakbar Ahovan et al. 2024. Engineering of a decellularized bovine skin coated with antibiotics-loaded electrospun fibers with synergistic antibacterial activity for the treatment of infectious wounds. Biotechnology and Bioengineering 121: 1452–63.

Axpe, Eneko and Michelle L. Oyen. 2016. Applications of alginate-based bioinks in 3D bioprinting. International Journal of Molecular Sciences 17: 1976.

Azeredo, Henriette M. C. and Keith W. Waldron. 2016. Crosslinking in polysaccharide and protein films and coatings for food contact—A review. Trends in Food Science & Technology 52: 109–22.

Beheshtizadeh, Nima, Mahmoud Azami, Hossein Abbasi and Ali Farzin. 2022. Applying extrusion-based 3D printing technique accelerates fabricating complex biphasic calcium phosphate-based scaffolds for bone tissue regeneration. Journal of Advanced Research 40: 69–94.

Bendtsen, Stephanie T., Sean P. Quinnell and Mei Wei. 2017. Development of a novel alginate-polyvinyl alcohol-hydroxyapatite hydrogel for 3D bioprinting bone tissue engineered scaffolds. Journal of Biomedical Materials Research Part A 105: 1457–68.

Bi, Yong-guang, Zi-ting Lin and Shi-ting Deng. 2019. Fabrication and characterization of hydroxyapatite/sodium alginate/chitosan composite microspheres for drug delivery and bone tissue engineering. Materials Science and Engineering: C 100: 576–83.

Borda, Luis J., Flor E. Macquhae and Robert S. Kirsner. 2016. Wound dressings: a comprehensive review. Current Dermatology Reports 5: 287–97.

Brancato, V., J. M. Oliveira, V. M. Correlo, R. L. Reis and S. C. Kundu. 2020. Could 3D models of cancer enhance drug screening? Biomaterials 2020 Feb; 232: 119744. doi: 10.1016/j.biomaterials.2019.119744.

Brumberg, Valentin, Tatiana Astrelina, Tatiana Malivanova and Alexander Samoilov. 2021. Modern wound dressings: Hydrogel dressings. Biomedicines 9: 1235.

Büyüköz, Melda, Esra Erdal and Sacide Alsoy Altinkaya. 2018. Nanofibrous gelatine scaffolds integrated with nerve growth factor-loaded alginate microspheres for brain tissue engineering. Journal of Tissue Engineering and Regenerative Medicine 12: e707–e19.

Chen, Xiao-Yi, Jin-Yang Chen, Xiang-Min Tong, Jian-Guo Mei, Yun-Fang Chen and Xiao-Zhou Mou. 2020. Recent advances in the use of microcarriers for cell cultures and their *ex vivo* and *in vivo* applications. Biotechnology Letters 42: 1–10.

Chui, Chih-Yao, Akinlolu Odeleye, Linh Nguyen, Naresh Kasoju, Erfan Soliman and Hua Ye. 2019. Electrosprayed genipin cross-linked alginate–chitosan microcarriers for *ex vivo* expansion of mesenchymal stem cells. Journal of Biomedical Materials Research Part A 107: 122–33.

Clegg, John R., Kolade Adebowale, Zongmin Zhao and Samir Mitragotri. Hydrogels in the clinic: An update. Bioengineering & Translational Medicine: e10680.

Datta, Sudipto. 2023. Advantage of alginate bioinks in biofabrication for various tissue engineering applications. International Journal of Polymer Science 2023: 6661452.

Deptuła, Milena, Małgorzata Zawrzykraj, Justyna Sawicka, Adrianna Banach-Kopeć, Robert Tylingo and Michał Pikuła. 2023. Application of 3D-printed hydrogels in wound healing and regenerative medicine. Biomedicine & Pharmacotherapy 167: 115416.

Derakhti, Sorour, Seyed Hamid Safiabadi-Tali, Ghassem Amoabediny and Mojgan Sheikhpour. 2019. Attachment and detachment strategies in microcarrier-based cell culture technology: A comprehensive review. Materials Science and Engineering: C 103: 109782.

Dias, A. D., J. M. Elicson and W. L. Murphy. 2017. Microcarriers with synthetic hydrogel surfaces for stem cell expansion. Adv Healthc Mater 6.

Dong, Ruonan and Baolin Guo. 2021. Smart wound dressings for wound healing. Nano Today 41: 101290.

Fang, Dawei, Yang Liu, Shan Jiang, Jun Nie and Guiping Ma. 2011. Effect of intermolecular interaction on electrospinning of sodium alginate. Carbohydrate Polymers 85: 276–79.

Farshidfar, Nima, Siavash Iravani and Rajender S. Varma. 2023. Alginate-based biomaterials in tissue engineering and regenerative medicine. Marine Drugs 21: 189.

Fitzpatrick, Emer, Celine Filippi, Barath Jagadisan, Dharshene Shivapatham, Hanish Anand, Mike Lyne et al. 2023. Intraperitoneal transplant of Hepatocytes co-Encapsulated with mesenchymal stromal cells in modified alginate microbeads for the treatment of acute Liver failure in Pediatric patients (HELP)—An open-label, single-arm Simon's two stage phase 1 study protocol. PloS One 18: e0288185.

Focaroli, Stefano, Gabriella Teti, Viviana Salvatore, Isabella Orienti and Mirella Falconi. 2016. Calcium/cobalt alginate beads as functional scaffolds for cartilage tissue engineering. Stem Cells International 2016: 2030478.

Gheorghita Puscaselu, Roxana, Andrei Lobiuc, Mihai Dimian and Mihai Covasa. 2020. Alginate: From food industry to biomedical applications and management of metabolic disorders. Polymers 12: 2417.

Gorshkova, Natalia, Olga Brovko, Irina Palamarchuk, Konstantin Bogolitsyn and Artem Ivakhnov. 2021. Preparation of bioactive aerogel material based on sodium alginate and chitosan for controlled release of levomycetin. Polymers for Advanced Technologies 32: 3474–82.

Gregurec, Danijela, Guocheng Wang, Ricardo H. Pires, Marija Kosutic, Tanja Lüdtke, Mihaela Delcea et al. 2016. Bioinspired titanium coatings: self-assembly of collagen–alginate films for enhanced osseointegration. Journal of Materials Chemistry B 4: 1978–86.

Grellier, Maritie, Pedro L. Granja, Jean-Christophe Fricain, Sílvia J. Bidarra, Martine Renard, Reine Bareille et al. 2009. The effect of the co-immobilization of human osteoprogenitors and endothelial cells within alginate microspheres on mineralization in a bone defect. Biomaterials 30: 3271–78.

Guo, Ning, Yu Xia, Weishen Zeng, Jia Chen, Quanxin Wu, Yaxin Shi et al. 2022. Alginate-based aerogels as wound dressings for efficient bacterial capture and enhanced antibacterial photodynamic therapy. Drug Delivery 29: 1086–99.

Hecht, Hadas and Simcha Srebnik. 2016. Structural characterization of sodium alginate and calcium alginate. Biomacromolecules 17: 2160–67.

Hsu, Shan-hui, Shu Wen Whu, Shu-Chih Hsieh, Ching-Lin Tsai, David Chanhen Chen and Tai-Sheng Tan. 2004. Evaluation of chitosan-alginate-hyaluronate complexes modified by an RGD-containing protein as tissue-engineering scaffolds for cartilage regeneration. Artificial Organs 28: 693–703.

Hurtado, Alejandro, Alaa A. A. Aljabali, Vijay Mishra, Murtaza M. Tambuwala and Ángel Serrano-Aroca. 2022. Alginate: Enhancement strategies for advanced applications. International Journal of Molecular Sciences 23: 4486.

Ismail, Sarosh, Faisal Akhlaque Ali Khan, Sadaf Gulzar, Waqas Sami, Bushra Zulfiqar, Suneel Kumar et al. 2023. A single centre, randomized control trial of donor site wound dressings after split-thickness skin grafting. Journal of Ayub Medical College Abbottabad-Pakistan 35.

Jabbar, Ali A., Dhia H. Hussain, Kamal H. Latif, Salim Albukhaty, Adel Kareem Jasim, Ghassan M. Sulaiman et al. 2024. Extremely efficient aerogels of graphene oxide/graphene oxide nanoribbons/ sodium alginate for uranium removal from wastewater solution. Scientific Reports 14: 1285.

Jadbabaei, Sorour, Majid Kolahdoozan, Farid Naeimi and Hassan Ebadi-Dehaghani. 2021. Preparation and characterization of sodium alginate–PVA polymeric scaffolds by electrospinning method for skin tissue engineering applications. Rsc Advances 11: 30674–88.

Jahani, Afsaneh, Mohammad Sadegh Nourbakhsh, Mohammad H. Ebrahimzadeh, Marzieh Mohammadi, Davood Yari and Ali Moradi. 2024. Biomolecules-loading of 3D-printed alginate-based scaffolds for cartilage tissue engineering applications: a review on current status and future prospective. Archives of Bone and Joint Surgery 12: 92.

Kistler, Samuel Stephens. 1931. Coherent expanded aerogels and jellies. Nature 127: 741–41.

Koehler, Julia, Ferdinand P. Brandl and Achim M. Goepferich. 2018. Hydrogel wound dressings for bioactive treatment of acute and chronic wounds. European Polymer Journal 100: 1–11.

Kothale, Dhalendra, Utsav Verma, Nagesh Dewangan, Partha Jana, Ankit Jain and Dharmendra Jain. 2020. Alginate as promising natural polymer for pharmaceutical, food, and biomedical applications. Current Drug Delivery 17: 755–75.

Kumar, S. Logesh, S. Anandhavelu and M. Swathy. 2019. Preparation and characterization of goat hoof keratin/gelatin/sodium alginate base biofilm for tissue engineering application. Integrated Ferroelectrics 202: 1–12.

Kundu, Joydip, Jin-Hyung Shim, Jinah Jang, Sung-Won Kim and Dong-Woo Cho. 2015. An additive manufacturing-based PCL–alginate–chondrocyte bioprinted scaffold for cartilage tissue engineering. Journal of Tissue Engineering and Regenerative Medicine 9: 1286–97.

Lázár, István, Ladislav Čelko and Melita Menelaou. 2023. Aerogel-based materials in bone and cartilage tissue engineering—a review with future implications. Gels 9: 746.

Lee, K. Y. and D. J. Mooney. 2012. Alginate: properties and biomedical applications. Prog Polym Sci. 2012 Jan; 37(1): 106–126. doi: 10.1016/j.progpolymsci.2011.06.003.

Lee, B.-B., Pogaku Ravindra and E.-S. Chan. 2013. Size and shape of calcium alginate beads produced by extrusion dripping. Chemical Engineering & Technology 36: 1627–42.

Lehnert, Sarah and Pawel Sikorski. 2022. Application of temporary, cell-containing alginate microcarriers to facilitate the fabrication of spatially defined cell pockets in 3D collagen hydrogels. Macromolecular Bioscience 22: 2100319.

Li, Chunge, Yufeng Qian, Shuang Zhao, Yuji Yin and Junjie Li. 2016. Alginate/PEG based microcarriers with cleavable crosslinkage for expansion and non-invasive harvest of human umbilical cord blood mesenchymal stem cells. Materials Science and Engineering: C 64: 43–53.

Li, Shuangyang, Qixuan Yu, Hongpeng Li, Meiqi Chen, Ye Jin and Da Liu. 2023. Self-assembled peptide hydrogels in regenerative medicine. Gels 9: 653.

Li, Xinda, Boxun Liu, Ben Pei, Jianwei Chen, Dezhi Zhou, Jiayi Peng et al. 2020. Inkjet bioprinting of biomaterials. Chemical Reviews 120: 10793–833.

Liu, Huanbao, Huixing Zhou, Haiming Lan, Tianyu Liu, Xiaolong Liu and Hejie Yu. 2017. 3D printing of artificial blood vessel: study on multi-parameter optimization design for vascular molding effect in alginate and gelatin. Micromachines 8: 237.

Liu, Jun, Hongzhi Zhou, Michael D. Weir, Hockin H. K. Xu, Qianming Chen and Carroll Ann Trotman. 2012. Fast-degradable microbeads encapsulating human umbilical cord stem cells in alginate for muscle tissue engineering. Tissue Engineering Part A 18: 2303–14.

Liu, Xiaoli, Huan Liu, Xue Qu, Miao Lei, Chuchu Zhang, Hua Hong et al. 2017. Electrical signals triggered controllable formation of calcium-alginate film for wound treatment. Journal of Materials Science: Materials in Medicine 28: 1–10.

López-Gutierrez, Jorge, Rosalío Ramos-Payán, Alfredo Ayala-Ham, José Geovanni Romero-Quintana, Hipolito Castillo-Ureta, Carlos Villegas-Mercado et al. 2023. Biofunctionalization of hydrogel-based scaffolds for vascular tissue regeneration. Frontiers in Materials 10: 1168616.

Majidi Ghatar, Jilla, Arian Ehterami, Simin Nazarnezhad, Maryam Sadat Hassani, Nariman Rezaei Kolarijani, Solmaz Mahami et al. 2023. A novel hydrogel containing 4-methylcatechol for skin regeneration: *In vitro* and *in vivo* study. Biomedical Engineering Letters 13: 429–39.

Mallakpour, Shadpour, Elham Azadi and Chaudhery Mustansar Hussain. 2021. State-of-the-art of 3D printing technology of alginate-based hydrogels—An emerging technique for industrial applications. Advances in Colloid and Interface Science 293: 102436.

Mitrovic, Jovana, Gabriella Richey, Sarah Kim and Mustafa O. Guler. 2023. Peptide hydrogels and nanostructures controlling biological machinery. Langmuir 39: 11935–45.

Mohandesnezhad, Sanam, Mahdieh Hajian Monfared, Saeed Samani, Ali Farzin, S. Ali Poursamar, Jafar Ai et al. 2023. 3D-printed bioactive Chitosan/Alginate/Hardystonite scaffold for bone tissue engineering: Synthesis and characterization. Journal of Non-Crystalline Solids 609: 122261.

Montes, Susan and Hajar Maleki. 2020. Aerogels and their applications. *In*: Colloidal Metal Oxide Nanoparticles (Elsevier).

Moshaverinia, Alireza, Sahar Ansari, Chider Chen, Xingtian Xu, Kentaro Akiyama, Malcolm L. Snead et al. 2013. Co-encapsulation of anti-BMP2 monoclonal antibody and mesenchymal stem cells in alginate microspheres for bone tissue engineering. Biomaterials 34: 6572–79.

Muñoz-Ruíz, Abraham, Diana M. Escobar-García, Mildred Quintana, Amaury Pozos-Guillén and Héctor Flores. 2019. Synthesis and characterization of a new collagen-alginate aerogel for tissue engineering. Journal of Nanomaterials 2019: 2875375.

Najafiasl, Maryam, Shahriar Osfouri, Reza Azin and Sasan Zaeri. 2020. Alginate-based electrospun core/shell nanofibers containing dexpanthenol: A good candidate for wound dressing. Journal of Drug Delivery Science and Technology 57: 101708.

Najdahmadi, Avid, Jonathan R. T. Lakey and Elliot Botvinick. 2018. Structural characteristics and diffusion coefficient of alginate hydrogels used for cell based drug delivery. MRS Advances 3: 2399–408.

Nazarnezhada, Simin, Ghasem Abbaszadeh-Goudarzi, Hadi Samadian, Mehdi Khaksari, Jila Majidi Ghatar, Hossein Khastar et al. 2020. Alginate hydrogel containing hydrogen sulfide as the functional wound dressing material: *In vitro* and *in vivo* study. International Journal of Biological Macromolecules 164: 3323–31.

Negahdari, Nazanin, Sanaz Alizadeh, Jila Majidi, Mahdi Saeed, Tayyeb Ghadimi, Kobra Tahermanesh et al. 2024. Heat-treated alginate-polycaprolactone core-shell nanofibers by emulsion electrospinning process for biomedical applications. International Journal of Biological Macromolecules 275: 133709.

Nguyen, Duong, Daniel A. Hägg, Alma Forsman, Josefine Ekholm, Puwapong Nimkingratana, Camilla Brantsing et al. 2017. Cartilage tissue engineering by the 3D bioprinting of iPS cells in a nanocellulose/alginate bioink. Scientific Reports 7: 658.

Nie, Huarong, Aihua He, Jianfen Zheng, Shanshan Xu, Junxing Li and Charles C. Han. 2008. Effects of chain conformation and entanglement on the electrospinning of pure alginate. Biomacromolecules 9: 1362–65.

Ojansivu, Miina, Ahmad Rashad, Astrid Ahlinder, Jonathan Massera, Ayush Mishra, Kristin Syverud et al. 2019. Wood-based nanocellulose and bioactive glass modified gelatin–alginate bioinks for 3D bioprinting of bone cells. Biofabrication 11: 035010.

Omo, J. O., J. E. Enabulele, M. A. Sede, J. Igunma, O. Akhigbemen and J. Atat. 2025. Effect of microwave and sodium hypochlorite disinfection of alginate impressions on dimensional accuracy of such impressions. J. Dent. Res. 10(1): 22–27. https://dx.doi.org/10.4314/njdr.v10i1.4.

Op't Veld, Roel C., X. Frank Walboomers, John A. Jansen and Frank A. D. T. G. Wagener. 2020. Design considerations for hydrogel wound dressings: strategic and molecular advances. Tissue Engineering Part B: Reviews 26: 230–48.

Ouwerx, C., N. Velings, M. M. Mestdagh and Ma A. V. Axelos. 1998. Physico-chemical properties and rheology of alginate gel beads formed with various divalent cations. Polymer Gels and Networks 6: 393–408.

Pahlevanzadeh, F., H. Mokhtari, H. R. Bakhsheshi-Rad, R. Emadi, M. Kharaziha, A. Valiani et al. 2020. Recent trends in three-dimensional bioinks based on alginate for biomedical applications. Materials (Basel) 2020 Sep 8; 13(18): 3980. doi: 10.3390/ma13183980.

Park, Ju An, Yunji Lee and Sungjune Jung. 2023. Inkjet-based bioprinting for tissue engineering. Organoid, 3.

Pawar, Siddhesh N. and Kevin J. Edgar. 2012. Alginate derivatization: A review of chemistry, properties and applications. Biomaterials 33: 3279–305.

Pereira, Rúben, Anabela Carvalho, Daniela C. Vaz, M. H. Gil, Ausenda Mendes and Paulo Bártolo. 2013. Development of novel alginate based hydrogel films for wound healing applications. International Journal of Biological Macromolecules 52: 221–30.

Perteghella, Sara, Elisa Martella, Laura De Girolamo, Carlotta Perucca Orfei, Michela Pierini, Valentina Fumagalli et al. 2017. Fabrication of innovative silk/alginate microcarriers for mesenchymal stem cell delivery and tissue regeneration. International Journal of Molecular Sciences 18: 1829.

Rastogi, Prasansha and Balasubramanian Kandasubramanian. 2019. Review of alginate-based hydrogel bioprinting for application in tissue engineering. Biofabrication 11: 042001.

Reig-Vano, Belen, Bartosz Tylkowski, Xavier Montané and Marta Giamberini. 2021. Alginate-based hydrogels for cancer therapy and research. International Journal of Biological Macromolecules 170: 424–36.

Rezvani Ghomi, Erfan, Shahla Khalili, Saied Nouri Khorasani, Rasoul Esmaeely Neisiany and Seeram Ramakrishna. 2019. Wound dressings: Current advances and future directions. Journal of Applied Polymer Science 136: 47738.

Rosellini, Elisabetta, Caterina Cristallini, Niccoletta Barbani, Giovanni Vozzi and Paolo Giusti. 2009. Preparation and characterization of alginate/gelatin blend films for cardiac tissue engineering. Journal of Biomedical Materials Research Part A: An Official Journal of The Society for Biomaterials, The Japanese Society for Biomaterials, and The Australian Society for Biomaterials and the Korean Society for Biomaterials 91: 447–53.

Rossi, Arianna, Teresa Pescara, Alberto Maria Gambelli, Francesco Gaggia, Amish Asthana, Quentin Perrier et al. 2024. Biomaterials for extrusion-based bioprinting and biomedical applications. Frontiers in Bioengineering and Biotechnology 12: 1393641.

Safonova, L. A., M. M. Bobrova, A. E. Efimov, O. I. Agapova, I. I. Agapov and S. V. Gautier. 2021. Surface Modification of Alginate Microcarriers for Improvement of Their Biological Properties. In Doklady Biological Sciences, 97–98. Springer.

Sahoo, Deepti Rekha and Trinath Biswal. 2021. Alginate and its application to tissue engineering. SN Applied Sciences 3: 30.

Saquing, Carl D., Christina Tang, Brinda Monian, Christopher A. Bonino, Joshua L. Manasco, Eben Alsberg et al. 2013. Alginate–polyethylene oxide blend nanofibers and the role of the carrier polymer in electrospinning. Industrial & Engineering Chemistry Research 52: 8692–704.

Satish, Aishwarya, R. Aswathi, J. Caroline Maria and Purna Sai Korrapati. 2019. Triiodothyronine impregnated alginate/gelatin/polyvinyl alcohol composite scaffold designed for exudate-intensive wound therapy. European Polymer Journal 110: 252–64.

Senturk Parreidt, Tugce, Kajetan Müller and Markus Schmid. 2018. Alginate-based edible films and coatings for food packaging applications. Foods 7: 170.

Soran, Zeliha, R. Seda Tığlı Aydın and Menemşe Gümüşderelioğlu. 2012. Chitosan scaffolds with BMP-6 loaded alginate microspheres for periodontal tissue engineering. Journal of Microencapsulation 29: 770–80.

Sudhakar, Kuncham, Sabata Jonas Moloi and Kummara Madhusudhana Rao. 2017. Graphene oxide/poly (N-isopropyl acrylamide)/sodium alginate-based dual responsive composite beads for controlled release characteristics of chemotherapeutic agent. Iranian Polymer Journal 26: 521–30.

Tarun, Agarwal, Kabiraj Prajna, Narayana Gautham Hari, Kulanthaivel Senthilguru, Kasiviswanathan Uvanesh, Pal Kunal et al. 2016. Alginate bead based hexagonal close packed 3D implant for bone tissue engineering.

Teo, Wee E. and Seeram Ramakrishna. 2006. A review on electrospinning design and nanofibre assemblies. Nanotechnology 17: R89.

Thapliyal, Prakash C. and Kirti Singh. 2014. Aerogels as promising thermal insulating materials: An overview. Journal of Materials 2014: 127049.

Tonsomboon, Khaow and Michelle L. Oyen. 2013. Composite electrospun gelatin fiber-alginate gel scaffolds for mechanically robust tissue engineered cornea. Journal of the Mechanical Behavior of Biomedical Materials 21: 185–94.

Tripathi, Swikriti, Subham Shekhar Mandal, Sudepta Bauri and Pralay Maiti. 2023. 3D bioprinting and its innovative approach for biomedical applications. MedComm 4: e194.

Uyen, Nguyen Thi Thanh, Zuratul Ain Abdul Hamid, Nguyen Xuan Thanh Tram and Nurazreena Ahmad. 2020. Fabrication of alginate microspheres for drug delivery: A review. International Journal of Biological Macromolecules 153: 1035–46.

Varaprasad, Kokkarachedu, Tippabattini Jayaramudu, Vimala Kanikireddy, Claudio Toro and Emmanuel Rotimi Sadiku. 2020. Alginate-based composite materials for wound dressing application: A mini review. Carbohydrate Polymers 236: 116025.

Vitamia, Cszahreyloren, Ghina Nadhifah Iftinan, Irma Rahayu Latarissa, Gofarana Wilar, Arief Cahyanto, Ahmed Fouad Abdelwahab Mohammed et al. 2024. α-Mangostin hydrogel film with chitosan alginate base for recurrent aphthous stomatitis (RAS) treatment: study protocol for double-blind randomized controlled trial. Frontiers in Pharmacology 15: 1353503.

Wang, Bo, Randall J. Lee and Ling Tao. 2023. First-in-human transcatheter endocardial alginate-hydrogel implantation for the treatment of heart failure. European Heart Journal 44: 326–26.

Wang, Yaquan and Yao Lu. 2023. Sodium alginate-based functional materials toward sustainable applications: water treatment and energy storage. Industrial & Engineering Chemistry Research 62: 11279–304.

Wei, Qinghua, Rongbin Yang, Daocen Sun, Jiayi Zhou, Mingyang Li, Yingfeng Zhang et al. 2022. Design and evaluation of sodium alginate/polyvinyl alcohol blend hydrogel for 3D bioprinting cartilage scaffold: Molecular dynamics simulation and experimental method. Journal of Materials Research and Technology 17: 66–78.

Wu, Yi-Fan, Ya-Ting Wen, Eisner Salamanca, Lwin Moe Aung, Yan-Qiao Chao, Chih-Yun Chen et al. 2024. 3D-bioprinted alginate-based bioink scaffolds with β-tricalcium phosphate for bone regeneration applications. Journal of Dental Sciences 19: 1116–25.

Yeo, MyungGu and GeunHyung Kim. 2015. Fabrication of cell-laden electrospun hybrid scaffolds of alginate-based bioink and PCL microstructures for tissue regeneration. Chemical Engineering Journal 275: 27–35.

Yu, Chia-Cherng, Jung-Jhih Chang, Yen-Hsien Lee, Yu-Cheng Lin, Meng-Hsiu Wu, Ming-Chien Yang et al. 2013. Electrospun scaffolds composing of alginate, chitosan, collagen and hydroxyapatite for applying in bone tissue engineering. Materials Letters 93: 133–36.

Yu, Jiashing, Kim T. Du, Qizhi Fang, Yiping Gu, Shirley S. Mihardja, Richard E. Sievers et al. 2010. The use of human mesenchymal stem cells encapsulated in RGD modified alginate microspheres in the repair of myocardial infarction in the rat. Biomaterials 31: 7012–20.

Zdiri, Khmais, Aurélie Cayla, Adel Elamri, Annaëlle Erard and Fabien Salaun. 2022. Alginate-based bio-composites and their potential applications. Journal of Functional Biomaterials 13: 117.

Zhai, Peng, X. B. Chen and David J. Schreyer. 2013. Preparation and characterization of alginate microspheres for sustained protein delivery within tissue scaffolds. Biofabrication 5: 015009.

Zhang, Miao and Xia Zhao. 2020. Alginate hydrogel dressings for advanced wound management. International Journal of Biological Macromolecules 162: 1414–28.

Zhang, Zhenkun, Zhe Li, Ya Li, Yingying Wang, Minghao Yao, Kun Zhang et al. 2021. Sodium alginate/collagen hydrogel loaded with human umbilical cord mesenchymal stem cells promotes wound healing and skin remodeling. Cell and Tissue Research 383: 809–21.

Zhou, Hongzhi and Hockin H. K. Xu. 2011. The fast release of stem cells from alginate-fibrin microbeads in injectable scaffolds for bone tissue engineering. Biomaterials 32: 7503–13.

Zhu, Jingjing, Shuangshuang Zheng, Hanbo Liu, Yiwei Wang, Zeren Jiao, Yi Nie et al. 2021. Evaluation of anti-tumor effects of crocin on a novel 3D tissue-engineered tumor model based on sodium alginate/gelatin microbead. International Journal of Biological Macromolecules 174: 339–51.

CHAPTER 9

Microbial Exopolysaccharide Based Composites

A Remedial Add-on in Biomedical Sciences

Asha Arora[1,*] and *Vinita Paliwal*[2]

1. Introduction

Wellbeing is one of the utmost pre-requisite of the human evolution and basically involves both physical and mental health. Various therapeutic systems deployed by different civilizations in prehistoric times of clinical expertise of contemporary era reside on some basic fundamentals as safety, time ensured action, target oriented cure, etc. Biomedicine involves two basic lines- drug based therapies and physical procedures including surgeries and implants. Both procedures used number of synthetic polymers but limitations and shortcomings as rejection of graft, inflammatory responses, restricted applications and high cost has made researchers to look for better option. Ethno-medicinal applications of various natural biopolymers have opened new vistas. Natural biopolymers include various biomolecules of different metabolic pathways among proteins and carbohydrates are preferred for their biocompatibility. Biopolymers from higher plants and animal sources are restricted for their bioavailability whereas culture methods applied for the same increase cost burden. In the same lineage, bacteria serve as better option as they serve as bio-factories. In medical field, exopolysaccharides obtained through bacteria find application as sutures, scaffold, implants, soft tissue regeneration, bio sealants,

[1] Department of Botany, B N University, Udaipur (Raj.) India.
[2] Department of Botany, B N Girls College, Rajsamand (Raj.) India.
* Corresponding author: araudr@bnuniversity.ac.in, araudr@gmail.com

DOI: 10.1201/9781003518792-9

hydrogels, drug carriers, etc. Exopoysaccharides has many advantages over synthetic biopolymers but their production varies with bacterial strain/s and growth/culture conditions and therefore faces some challenges and limitations in their commercial application in medical field. To overcome such limitations, additives are added to form composites which can provide enhanced applicability of exopolysaccharides. The current review presents overview of EPS, natural and synthetic polymer based EPS-composites and their applicability in medical sciences.

2. Exopolysaccharide

The term exopolysaccharide (EPS) was first introduced by Sutherland in 1972 (Osemwegie et al., 2020) for high molecular weight carbohydrate biopolymers produced by microorganisms, including bacteria, fungi, and microalgae. They show considerable diversity in their composition and structure as some of the polymers resemble cell-wall components while majority possess distinct chemical structures totally unrelated to cellular constituents (Wang et al., 2023). EPS have involved in the interaction of surviving cells with the environment and assist in nutrients uptake along with protects cells against aridity, chemical agents (antibiotics, heavy metals and oxidants), extreme temperature, osmotic stress, phagocytosis, salinity, unfavorable pH values and UV-rays.

Bacteria survive in almost all environmental biomes with variable temperature and pH due to which more adaptive behavior of EPS is observed in both pathogenic and non-pathogenic bacteria. Nature of medium also effects microbial colonial size and quorum sensing within population is also related to the physio-chemical properties of EPS as-

a. In liquid medium, EPS producing cultures become very viscous or, exceptionally, may solidify as a gel and also impart slime.
b. On solid surfaces exposed to aqueous environments, bacterial growth is seen as biofilms, in which the microbial cells are associated with large amounts of EPS.

Biofilm formation is an adaptive physiological process for survival in certain environments for the adjustment to external and internal demands. It is a transformation from unicellular conditions to multicellular integrity as biofilms are formed when biofilm producing bacteria in an aqueous environment adhere to solid surface and produce a network of extracellular polymeric substances adapting multicellular life. Within biofilms, EPS is distributed in a non-homogenous pattern. EPS interacts and forms a matrix that encompasses living cells, in both Gram-positive and Gram negative bacteria (Secor et al., 2022; Balducci et al., 2023). EPSs of the biofilm matrix promote horizontal gene transfer, and intercellular interactions prevent the dehydration of bacteria and provide protection against external agents including antibiotics. Therefore, biofilm formation is associated with higher antibiotic resistance. EPS also plays an important role in controlling virulence. The stability of the matrix depends upon the non-covalent interactions between EPS that involve physicochemical forces. This EPS network interacts with the bacterial aggregates which confers cohesion and viscoelasticity to the structure.

Functionally EPS have been categorized in seven categories as constructive or structural, sorptive, surface-active, active, informative, redox-active and nutritive exopolysaccharides for water retention, cell protection, biofilm formation, antibacterial or antifungal activities and cell-surface interactions (Flemming and Wingender, 2010) due to their physio-chemical properties.

2.1 Physio-Chemical Properties of EPS

Exopolysaccharide composition varies among bacterial strains and possess properties according to microbial needs and environment but some physio-chemical properties are generally shared by exopolysaccharides of all bacteria and include-

i. **Adhesion**- Adherence of microbial colonies on abiotic and biotic surfaces for long-term attachment
ii. **Cell aggregation**- Immobilization of bacterial population thereby development of high cell densities and cell–cell recognition.
iii. **Cohesion of biofilms**- Formation of hydrated polymer network mediating the mechanical stability of biofilms determining biofilm architecture, as well as allowing cell-cell communication
iv. **Nutrient source**- Source of carbon, nitrogen and phosphorus containing compounds
v. **Protective barrier**- Resistance to non-specific and specific host defenses, tolerance to various antimicrobial agents
vi. **Sink for excess energy**- Stores excess carbon under unbalanced carbon to nitrogen ratios
vii. **Sorption of organic compounds and inorganic ions**- Accumulation of nutrients from the environment, sorption of xenobiotic and recalcitrant materials. gel formation resulting in ion exchange, mineral formation and the accumulation of toxic metal ions
viii. **Water retention**- Formation of hydrated microenvironment

2.2 Biological Activities of EPS

Bacterial EPS are also known for their biological activities as antitumor and anti cancer (Angelin and Kavitha, 2020; Gargouch et al., 2021), antidiabetic (Al-Nabulsi et al., 2022), hypoglycemic/hypolipidemic (Sun et al., 2021; Qi et al., 2022), cholesterol-lowering (Zou et al., 2021), immuno-modulatory (Chen and Huang, 2018), anti-inflammatory (Hou et al., 2020; Stellavato et al., 2022), probiotic activities (Oerlemans et al., 2021; Xu et al., 2022), etc.. Due to extensive biological activities, their biochemical formulation for novel curative application becomes demanding (Wu et al., 2021).

2.3 Chemical Modifications of EPS

EPS are generally modified by group substitutions to alter the structure and enhance targeted biological activity. Principal modifications include acetylation, carboxymethylation, phosphorylation and sulfonation.

a. **Acetylation-** Acetylation alters the spatial structure of EPS and thereby affects their biological activities (Ren et al., 2019).
b. **Carboxymethylation-** It involves the etherifcation mediated introduction of carboxymethyl groups to EPS polysaccharides with acids or carboxylic acid derivatives to achieve changes in the spatial structure and water solubility of the EPS (Ren et al., 2019).
c. **Phosphorylation-** It involves phosphorylation of EPS (Zhang et al., 2017; Guo et al., 2013).
d. **Sulfonation-** EPS are sulfonated to achieve water solubility and desired chain length. It enhances antioxidant activity and protects cells by exchanging more hydrogen atoms (Jin et al., 2014; Li and Shah, 2014).

2.4 Advantages of Bacterial EPS

EPSs synthesized by bacteria have an advantage over those isolated from Algae (agar, fucoidan, carrageenans, and alginates), plants (cellulose, starch, and pectin) and animals (glycogen and chitin) as-

a. Can be obtained regardless of the season and weather conditions on an industrial scale.
b. They are extracellular slime that is easily released from cells into the environment. Therefore, the methods for their extraction from the cell-free supernatant are quite simple and cost-effective.
c. Bacteria multiply rapidly and are characterized by metabolic flexibility and a variety of physiological and biochemical properties. By means of the methods for optimizing cultivation conditions, genetic and metabolic engineering, it is possible to modulate the yield as well as the structural and functional properties of bacterial EPSs (Pereira et al., 2019; Xiao et al., 2022; Manan et al., 2022).
d. Bacterial EPS are characterized by the presence of a large number of functional groups (hydroxyl, carboxyl, carbonyl, acetate, etc.), that enable them to modify their molecules in order to impart new valuable properties to them (Aditya et al., 2022).

In addition, their peculiar characters includes- biocompatibility, biodegradability, higher purity, hydrophilic nature and non-toxicity. Their ability for gelation, high adhesive ability, viscoelasticity, pseudo-plasticity, and thixotropic nature makes new vistas for blending them with other natural and synthetic polymers to obtain composites and fabricate materials with improved functionality for biomedical applications (Aziz et al., 2022).

3. Exopolysaccharides Producing Bacteria

EPS is produced by various bacteria belong to different phylogenetic groups including both Gram negative and Gram-positive bacteria. Among Gram negative bacteria the major genera of different class, i.e., Alphaproteobacteria, Betaproteobacteria, Gammaproteobacteria include various diversified genera (Netrusov et al., 2023) as-

a. **Alphaproteobacteria-** *Acetobacter, Gluconobacter, Gluconacetobacter, Komagataeibacter, Kozakia, Neoasaia, Agrobacterium, Rhizobium* and *Zymomonas*
b. **Betaproteobacteria-** *Alcaligenes* and *Achromobacter*
c. **Gammaproteobacteria-** *Azotobacter, Pseudomonas, Enterobacter, Alteromonas, Pseudoalteromonas, Xanthomonas, Halomonas, Erwinia, Vibrio* and *Klebsiella*

While Gram-positive bacteria includes

a. **Bacilli** - *Bacillus, Paenibacillus, Lactobacillus, Leuconostoc* and *Streptococcus*
b. **Clostridia** - *Sarcina* Sp.
c. **Actinomycetia** - *Bifidobacterium* and *Rhodococcus*

Commercial EPS for specific application is obtained from distinct genus depending on the EPS composition as cellulose from *Komagataei bacter*, levan from *Bacillus* sp., *Paenibacillus* sp., Halomonas sp., and *Zymomonas* sp,, gellan from the *Sphingomonas* and *Pseudomonas* genera, dextran from the *Leuconostoc, Streptococcus*, and *Lactobacillus* genera, alginate from the *Azotobacter* and *Pseudomona*s genera, hyaluronan from *Streptococcus* sp., curdlan from *Alcaligenes faecalis*, *Rhizobium radiobacter*, and *Agrobacterium* sp., clavan by *Clavibacter michiganensis,* etc. EPS composition may vary with strain of same species and growth conditions (Barcelos et al., 2019; Osemwegie et al., 2020) (Table 1).

4. Microbial Bioploymers: Types

Microbial biopolymers apart from structural constituent serve different functions and are therefore classified as-

A. **Structural biopolymers** - They are cell structure components or integral parts of the cell wall as well as membranes.
B. **Functional biopolymers** - They assist microbial cell in many aspects and are therefore divided as-
    a. **Intracellular biopolymers** - They are storage polysaccharides which also serve as energy pool for microbe
    b. **Extracellular biopolymers** - They are generally produced in internal cell milieu and transported across cell for various functions or in some cases produced extracellular. Depending on strain they may form -

Table 1. EPS of microbial origin and their pros and cons for biomedical utility.

EPS (Molecular Composition)	Bacterial Producers	Biomedical utility properties	Limitations	Biological Activity and Biomedical Applications
Acetan Polysaccharide with repeat unit consists of a cellobiose unit solubilised by attachment of a charged pentasaccharide side chain to one of the glucose residues. The repeat unit contains two sites of acetylation	*Komagataeibacter* Sp., *Acetobacter* Sp.	Bio-thickeners, encapsulation of drugs or probiotics and controlled released of therapeutic proteins	Low yield, Poor stability	Induces production of interleukin-12 p40 and tumor necrosis factor by macrophages and *in vitro* effect by regulating Th2-mediated allergic response (Trcek et al., 2021; Schilling et al., 2024)
Alginate Binary, anionic linear copolymers of (1 → 4) linked ß-D-mannuronic acid and α-L-guluronic acid residues	*Azotobacter* Sp., *Pseudomonas* Sp.	Hydrocolloid, Gelling capacity, Film-formers	Poor stability, low mechanical and barrier properties, incompatible with heavy metals, heat treatment instability	Induces cytokine production (Wang et al., 2023)
Alternan Soluble-branched glucan with α-1,6 and α-1,3 glycosidic linkages	*Leuconostoc mesenteroides*	Resistance to enzymatic hydrolysis, High solubility in water and low viscosity, Microbial resistant	Low yield	Cell growth and differentiation (Yucel et al., 2021; Venkatachalam et al., 2021)
Cellulose (Bacterial) Linear, ribbon-shaped glucose ploymer. Glucose units are connected via glycosidic bonds between their C1 and C4 positions, and the anomeric C1 carbon adopts the β-configuration	*Komagataei bacter*	Fabrication of biopolymers, manufacturing scaffolds, implants, artificial blood vessels, and wound dressing materials for wound and burn cases.	Does not exhibit antimicrobial activity	Provides strength to bio-structures (Zanotti et al., 2023)

Table 1 contd. ...

...Table 1 contd.

EPS (Molecular Composition)	Bacterial Producers	Biomedical utility properties	Limitations	Biological Activity and Biomedical Applications
Clavan Tetrasaccharide with repeating units of glucose, galactose, fucose and pyruvic acid	*Clavibacter michiganensis*	Richest polymers with sugar D-fucose, thermostability	High viscosity and non-Newtonian, pseudoplastic flow	To treat rheumatoid arthritis, prevent the colonization of the lung by tumor cells, regulate the production of white blood cells (Xiao et al., 2022)
Curdlan Linear structure is composed exclusively of repeating (1,3)-β-glucosidic linkages and without any substituents or branching	*Alcaligenes faecalis, Rhizobium radiobacter, Agrobacterium* Sp.	Gel-forming ability Water insolubility, Non toxic,Thermal stability	Less water solubility	Immunomodulatory, Antimicrobial (Chaudhari et al., 2021; Suflet et al., 2024; Aquinas et al., 2024)
Dextran Linear chain of d-glucoses linked by α-(1→6) bonds, with possible branches of d-glucoses linked by α-(1→4), α-(1→3), or α-(1→2) bonds	*Lactic acid bacteria, Leuconostoc* Sp., *Streptococcus* Sp., *Lactobacillus* Sp.	Non-ionic, Good stability Newtonian fluid behavior; pH sensitivity, High encapsulation efficacy, Across membrane trafficking, tendency to expand in the presence of water, contributing to the volume expansion in the case of plasma volume compensation, Gelling agent, Drug delivery efficacy	Solubility and rheological properties of dextran are affected by its molecular weight and branching	Blood volume expander (Luanda and Badalamoole, 2023; Sharma et al., 2024)
Elsinan Extracellular α-D-glucan of maltotriose residues and maltotetraose residues (minor) joined by α-(1→3)-linkages	Elsinoe leuicospila, *E. fawcetti, E. ampelina*	High viscosity in aqueous, solution, stability over wide range of pH (3–11), temperature (30–70°), and salt concentrations, ability to form a strong, resilient and oxygen-impervious film	Enzyme susceptibility	Anti-hypercholesterolemic (Ullah et al., 2021)

Emulsan Extracellular polysaccharide of a sugar backbone decorated by fatty acids, i.e., α- and β-hydroxydodecanoic acid	*Acinetobacter venetianus* *Acinetobacter calcoaceticus*	Hydrocarbon stabilizer Amphipathic properties Tailorable for controlled release, Drug carrier	Structural constraints, where small change in side chain results either in increased toxicity or decreased efficacy	Antioxidant , Antimicrobial (Benjamin et al., 2024; D'Almeida et al., 2024)
Fucogel Anionic polysaccharide with L-fucose, D-galactose and galacturonic acid units	*Klebsiella pneumoniae*	Psychosensorial, High viscosity, Hydrating, Self-emulsifying	Elusive hydrolyzation	Affinity with fucose receptors in the skin, unique ability to modulate neuronal sensitivity (Guetta et al., 2003; Xiao et al., 2022)
Fucopol Polysaccharide composed of fucose, galactose, glucose and glucuronic acid with an acyl groups substituents (acetyl, pyruvyl and succinyl)	*Enterobacter* Sp.	Viscous shear thinning solutions in aqueous media Film-forming Emulsifying capacity, Flocculating capacity	Polyanionic	Antioxidant, cryo- and photoprotective (Guerreiro et al., 2021)
Galactopol Linear molecule of galactose and three other neutral sugars (mannose, glucose and rhamnose), substituted with three anionic groups (succinate, pyruvate, acetate)	*Pseudomonas oleovorans*	Viscous shear thinning solutions in aqueous media, Film-forming Emulsifying capacity, Flocculating capacity	Delayed Gelation	Protective against stressful environmental conditions, harsh materials, toxins, pathogenesis (Devi et al., 2024)
Gellan Linear tetrasaccharide with repeating unit of two β-D-glucose, β-D glucuronic acid, and α L-rhamnose subunits.	*Sphingomonas elodea*, *Pseudomonas* Sp.	Strong gelling agent with low-viscosity suspensions Drug delivery Clinincal gums and Thermo reversible hydrogels	Lacks stability against stress and shear tolerance for tissue engineering	Antifungal (Khare et al., 2022; Lalebeigi et al., 2024; Lin et al., 2024; Barroca-Ferreira et al., 2024)

Table 1 contd. ...

...Table 1 contd.

EPS (Molecular Composition)	Bacterial Producers	Biomedical utility properties	Limitations	Biological Activity and Biomedical Applications
Hyaluronic acid/Hyaluronan Anionic, non-sulfated glycosaminoglycan	*Streptococcus pyogenes*, *Streptococcus zooepidemicus*, *Bacillus cereus*	Highly hydrophilic and Biocompatible	Poor mechanical Stability	Cell proliferation and cell Migration (Di Mola et al., 2022; Guo et al., 2021)
Inulin Linear molecule composed mainly of fructose units (fructans) joined by a β(2→1) glycosidic bond, with terminal glucose	*Bacillus* stercoris, *Pseudomonas* Sp.	High water solubility, Resistant to enzymes, Potent substrate for the production of inulinases	High degree of polymerization (2–60)	Prebiotic Antioxidant, lipid metabolism regulator (Schropp et al., 2023)
Kefiran Glucogalactan with roughly equal amounts of D-glucose and D-galactose in a chain sequence	*Lactobacillus kefiranofaciens*, *L. kefirgranum*, *L. parakefir*, *L. kefir* and *L. delbrueckii* subsp. *bulgaricus*	Rheological behavior, emulsifier effect, stabilizing effect, resistance against hydrolysis, water vapor permeability	High permeability to water	Biocidal properties, wound healing, modulation of gut immune-system , protection of epithelial cells against microbial toxins (Moradi and Kalanpour, 2019; de Oliveira et al., 2023)
Levan Fructose subunits interlinked by β-(2→6) glycosidic linkages, along with side chains attachment via β-(2→1) linkage to the main backbone	*Aerobacter levanicum*, *Bacillus subtilis*, *B. polymyxa*, *Corynebacterium laevaniformans*, *Pseudomonas* Sp., *Streptococcus* Sp.	Thickening/gelling agent, Encapsulating material, Drug carrier	Low yield	Antitumor, Immunomodulatory, Antihyperlipidemic and Radioprotector activities (Hundschell et al., 2020; Domzał-Kędzia et al., 2023)
Mutan α-(1,3)- linked glucose residues in main chains and α- (1,6) bonds in side chains	*Streptococcus mutans* *S. sobrinus*	Physical integrity and matrix stability, Tight adherence and coherence of bacterial cells	Acidification over time	Antifungal (Rush et al., 2024)

Pullulan Polymer consisting of maltotriose units (three glucose units connected by α-1,4 glycosidic bonds) connected to each other by an α-1,6 glycosidic bond	*Aureobasidium pullulans*	Aqueous solubility and flexibility, biodegradable, nontoxic, biocompatible	Sensitive to modification by hydrophobic functional groups, high-water solubility and hygroscopic nature; poor support for cell adhesion and proliferation	Antibacterial, antifungal, antiviral, antitumor activities (de Souza et al., 2023; Elangwe et al., 2023; Roy et al., 2024)
Reuteran α-glucan with a large amount of alternating (α1 →4) and (α1 →6) linkages.	*Lactobacillus reuteri*	Viscosifying, Stabilising, Emulsifying	Low yield	Antiinflammatory Antimicrobial, Immuno-protective (Dailin et al., 2022)
Scleroglucan High molecular weight non-ionic branched glucan with a backbone of (1,3)-β-linked D-glucopyranosyl residues bearing a single (1,6)-β-linked D-glucopyranosyl unit every three sugar residues of the main chain	*Sclerotium rolfsii*	Stabilizers, Gelling agents, Thickeners, Drug carriers	Cost constraints	Wound Healing, Antimicrobial, Antitumor, Antiviral, Immune-stimulatory effects (Giri, 2024; Castro et al., 2024)
Succinoglycan Saccharide-based oligomers in which sugar molecules are derivatized with acyl, pyruvate, and succinic acid	*Rhizobium* Sp., *Alcaligenes* Sp., *Pseudomonas* Sp., *Agrobacterium* Sp.	Viscous shear thinning solutions in aqueous media, Acid stability	Low yield	Antihyperlipidemic, Antibacterial, Anticancer, Immune-stimulatory (Kim et al., 2024)

Table 1 contd. ...

...Table 1 contd.

EPS (Molecular Composition)	Bacterial Producers	Biomedical utility properties	Limitations	Biological Activity and Biomedical Applications
Welan Repeating glucose, rhamnose and glucuronic acid units with a single side chain of either an α-L-rhamnopyranosyl or an α-L-mannopyranosyl unit linked (1,3) to the 4-O-substituted β-D-glucopyranosyl unit in the backbone	*Sphingomonas* Sp.	Alkali stability, maintaining viscosity under extreme pH conditions (pH 2 to 12)	Non-gelling	Hydroxyl, ABTS, DPPH radical-scavenging activities (Wang et al., 2024)
Xanthan/Xanthan Gum Backbone has cellobiose as repeating unit and side-chains consisting of a trisaccharide composed of D-mannose (β-1,4), D-glucuronic acid (β-1,2) and D-mannose, which are attached to alternate glucose residues in the backbone by α-1,3 linkages	X*anthomonas* Sp. (*X. arboricola, X. axonopodis, X. campestris, X. citri, X. fragaria, X. gummisudans, X. juglandis, X. phaseoli, X. vasculorium*)	Hydrocolloid, high viscosity, emulsion stability, shear thinning activity	Prone to bacterial contamination, poor electrical conductivity and stability of heat	Antimicrobial, Antioxidant (Hussain et al., 2017; Patel et al., 2020; Dzionek et al., 2022)

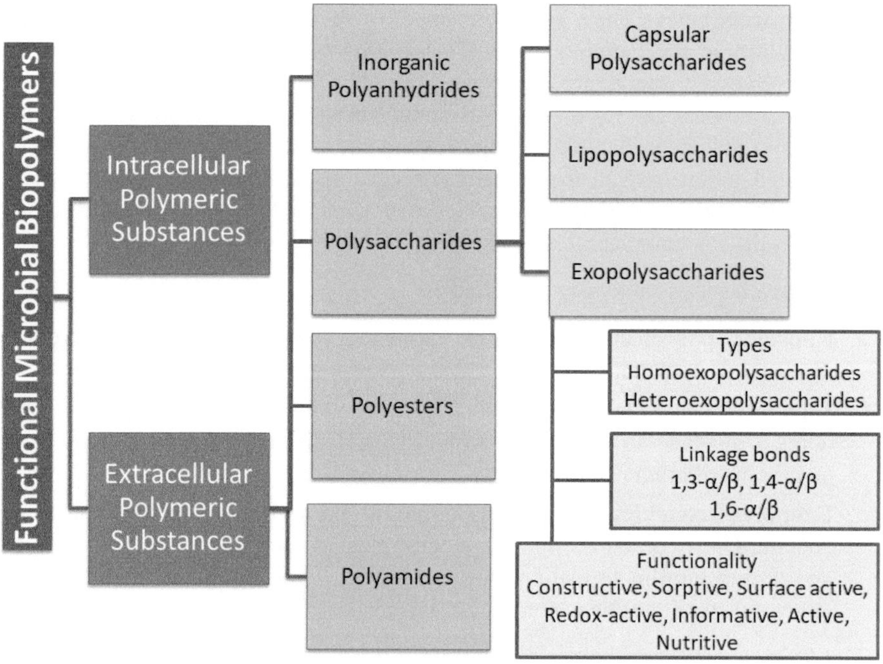

Figure 1. Classification of functional microbial biopolymers.

i. capsules outside the cell, thereby becoming an integral part of the cell wall, or
ii. loose slimes that collect outside the cell wall and gradually diffuse in the liquid phase during the fermentation.

On the basis of chemical composition, extracellular biopolymers are generally divided into four groups as inorganic polyanhydrides, polyamides, polyesters and polysaccharides (Figure 1)

a. **Inorganic polyanhydrides-** Three to several hundred inorganic phosphates condense to form highly negatively charged polyphosphate polymer rich in 'high-energy' anhydride bonds that is used by bacteria as energy storage molecules.
b. **Polyamides-** Poly (amino acid), polymers in general include secretory ε-poly-L-lysine and poly (γ-Dglutamic acid) along with the intracellular cyanophycin granule peptide with compositional and functional role in biofilm matrix.
c. **Polyesters-** Polyesters are generally linear molecules of Polyhydroxyalkanoates (PHAs)/poly (R)-3-hydroxybutyrate polyesters that are synthesized and pieced into hydrophobic spherical inclusions for carbon and energy storage.

d. **Polysaccharides-** They are polymers of sugar and sugar acids (glucuronic and iduronic acid), amino sugars (D-galactosamine, D-glucosamine) and their derivatives (N-acetylneuraminic acid and N-acetylmuramic acid) linked by glycosidic bonds as 1, 4-β- or 1, 3-β-linkage in strongly rigid polymers or 1, 2-α or 1,6-α linkage in flexible polymers (Miao and Guthmiller, 2022). They can be charged or noncharged, non-repeating or repeating, and branched or unbranched. In addition, they are diverse molecules associated with the structural (membrane and wall) and compositional (cytosolic polysaccharides) integrity of cells.

Exopolysaccharides are further classified as-

a. **Capsular polysaccharides-** High molecular polysaccharides that provide virulence to bacteria.
b. **Lipopolysaccharides-** Amphipathic glycoconjugates with hydrophobic lipid domain attached to core oligosaccharide and distal polysaccharide in outer membrane of gram negative bacteria.
c. **Exopolysaccharides-** Exopolysaccharides or extracellular polymeric substance contain DNA, proteins, lipids and enzymes in addition to polysaccharides and during biofilm formation they encapsulate cells along with nutrition and structural stability (Díaz-Montes, 2021).

EPS structures have the general formula of $(C_6H_{10}O_5)n$ where n is $\geq$40 and $\leq$3000 and are classified depending on different monosaccharide compositions, linkages, charges, side-chains, substitutions and functions (Zayed et al., 2022). Based on chemical composition, exopolysaccharides could be classified into two main classes:

A. **Homoexopolysaccharides (HoEPS)** - HoPS are polymers of single-type monosaccharides which are either unbranched or branched with approximate molecular mass of 10^7 Da. They are classified into four groups based on linkage bonds and monomeric unit as:

   a. Sialic Acid

   b. Polyglycans

   c. Fructans

   d. Glucans

Fructans include two important EPS as inulin and levan while Glucans include α-D-glucans and β-D-glucans. Bacterial Glucans include Curdlan, Cellulose, Scleroglucan, Dextran, Elsinan, Pullulan, Alternan, Mutan, Reuteran and Glycogen (Figure 2).

B. **Heteroexopolysaccharides (HeEPS)** - HeEPS contain two or more units of different monosaccharides (glucose, fructose, galactose, mannose, rhamnose, fucose, N-acetylglucosamine, and uronic acids) with a molecular mass is in the range of 10^4–10^6 Da. Acetan, Alginates, Gellan, Emulsan, Welan, Heteroglycans, Succinoglycans, Dermatan, Keratan, Hylauronic acid, Heparin, Xanthan and Kefiran possess many potential capabilities as emulsion, gelling agent, stability over longer temperature and pH bands along with biological activity which make them suitable candidate for therapeutic applications (Moradali and Rehm, 2020).

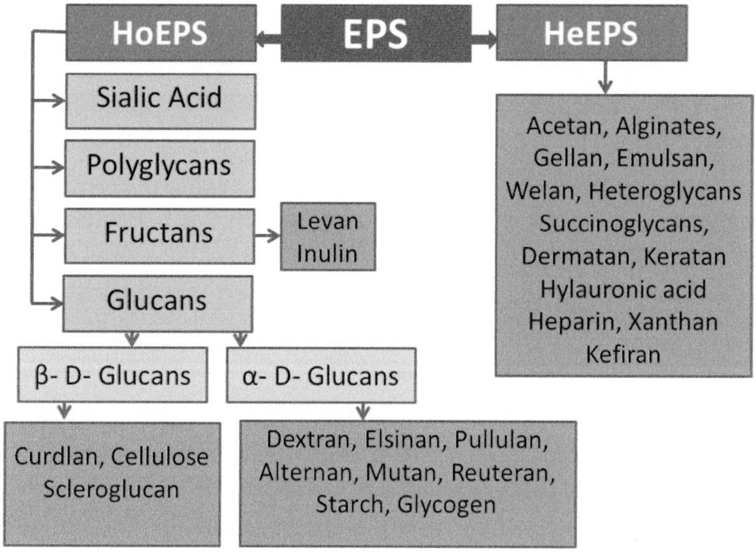

EPS-Exopolysaccharide; HoEPS-Homoexopolysccharides; HeEPS-Heteroexopolysccharides

Figure 2. Classification of exopolysaccharide, homoexopolysccharides and heteroexopolysccharides.

Most of the HoEPS are neutral, while the majority of HeEPS are polyanionic. These polysaccarhides not only differ in linkage bonds and chemical nature but also in their enzymatic machinery and site of synthesis.

On the basis of functionality, EPS are divided into following categorize– constructive, sorptive, surface active, redox-active, informative, active and nutritive. Sorptive are charged polymers that help in sorption to other charged molecules during cell-surface interactions. Structural EPS are utile in the matrix, as easing water retention and cell protection. Surface active molecules possess amphiphilic behavior and exhibit different chemical structures and surface properties and assist in biofilm formation or may exhibit antimicrobial activities. Redox-active properties are exhibited by electroactive bacterial species as *Escherichia coli*, *Pseudomonas putida* and some species of *Shewanella*. EPS from electroactive bacterial species also help in migration and biotransformation of redox-sensitive compounds (Santra and Banerjee, 2021; Kaur and Dey, 2023).

5. Biosynthesis of Bacterial Exopolysaccharides

Bacterial population size and environmental stress responses are critical triggers for EPS production. Depending on the microbial strain and microbe colonial environment, some EPS are synthesized throughout the bacterial life cycle while others are only formed in the late logarithmic or stationary phases (Schmid, 2018; Panchal et al., 2022). There is a linear relationship between EPS production rate and biomass growth rate, i.e., EPS production is more when the cells are growing with slow rate (Knirel and Kenyon, 2021). Generally EPS is produced in two ways:

a. Intracellular and
b. Extracellular

Homopolysaccharides are synthesized extracellularly whereas heteropolysaccharides are synthesized intracellularly and transferred to the extracellular environment (Sun and Zhang, 2021).

5.1 Intracellular EPS Biosynthesis

The general pathway for intracellular biosynthesis include four steps

I. Uptake of substrate
II. Central metabolite pathway (By Group 1 and Group 2 Enzymes)
III. Polymerization (By Group 3 Enzymes)
IV. Modification and Transportation (By Group 4 Enzymes)

A diverse set of enzymes play an important role in bacterial EPS synthesis. Based on their involvement in the last three of the four phases of bacterial EPS production, these enzymes are divided into four groups (Whitfield et al., 2020; Sun and Zhang, 2021) (Figure 3):

Group 1: This group includes intracellular hexokinase phosphorylase which converts glucose (Glu) molecule into glucose-6-phosphate (Glu-6-p) followed Phosphoglucomutase which transforms glucose-6-phosphate to glucose-1-phosphate.

Group 2: The second set of enzymes are responsible for the synthesis of sugar nucleotides as Uridine diphosphate-glucose pyrophosphorylase (UDP-Glu Pyrophosphorylase) converts glucose-1-phosphate to uridine diphosphate glucose (UDP-Glu). In the presence of UDP-Gal4-epimerase, this sugar nucleotide is also interconverted into other sugarsas UDP-Glu to UDP-Gal.

Group 3: Periplasmic membrane localized glycosyltransferases transport the sugar nucleotide UDP-Glu or UDP-Gal to the repeating unit linked to the glycosyl carrier lipid, isoprenoid alcohol and thereby fascinating polymerization.

Group 4: Polysaccharides modified by acetylation, acylation, sulphation, and methylation are exported through hydrophobic enzymes such as flippase, permease or ABC transporters to the extracellular surface.

Intracellular EPS biosynthesis begins with uptake of substrate, i.e., a specific sugar moiety by active transport, diffusion, or group translocation. The second step proceeds with the formation of phosphorylated and nucleotide sugars. Intracellular enzyme "hexokinase" phosphorylates glucose to glucose-6-phosphate which is further converted to glucose-1-phosphate by phosphoglucomutase. The phosphorylated glucose may be utilized for energy or for the formation of polysaccharides anabolism. Group II enzyme UDP-glucose pyrophosphorylase

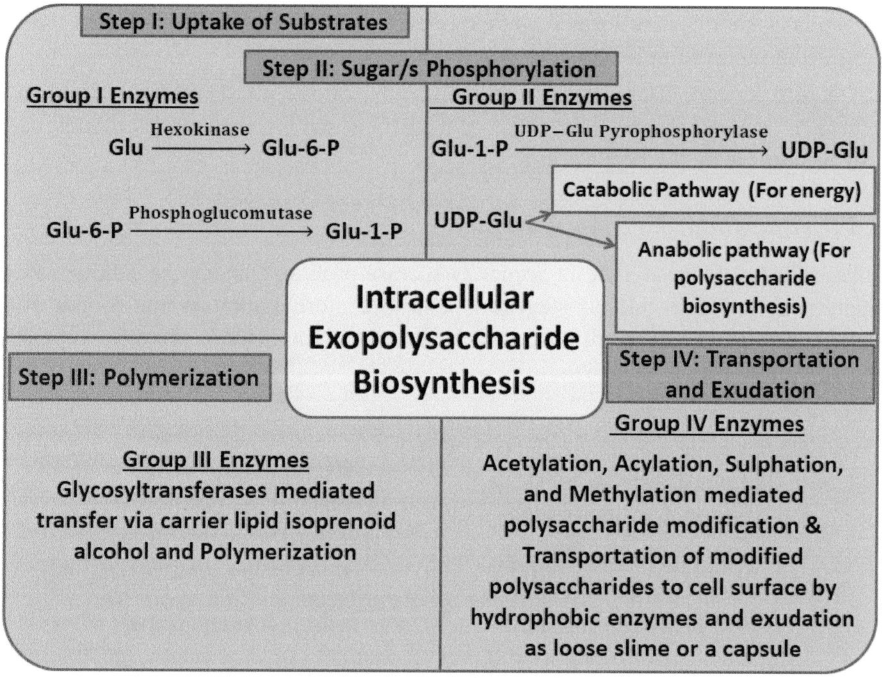

Figure 3. Biosynthesis of Intracellular Exopolysaccharide/s through Uptake of Substrates, their Phosphorylation, Polymerization and Transportation and Exudation across membrane.

converts glucose-1-phosphate to uridine diphosphate glucose. UDP-glucose is either interconverted to other sugars or enters in third step of polymerization (Fernandes et al., 2021). Polymerization step involves Glycosyltransferases of third group enzymes which transfers a sugar moiety to a repeating unit, attached to a glycosyl carrier lipid and recognized as isoprenoid alcohol. The terminal alcohol group of lipid carrier is linked to a monosaccharide unit by a pyrophosphate bridge. In this step, monomers are polymerized. In fourth step polysaccharides are modified by different enzymatic activities like acylation, acetylation, sulphation, and methylation. These modified polysaccharides are transported to the cell surface and finally exuded from the cell in the form of loose slime or a capsule by hydrophobic enzymes or transporters (Rana and Upadhyay, 2020; Das, 2022). Three pathways explain basic units' translocation through periplasmic space to the cell surface via cytoplasmic membrane:

1. **Wzx/Wzy dependent pathway**- Repeating units are transported across the cytoplasmic membrane by the Wzx flippase. Prior to polymerization, translocated oligosaccharide undergoes enzymatic changes by Wzy protein. ABC transporters deliver the formed polysaccharide to the cell surface.
2. **ABC transporters dependent pathway**- It involves three members of the ABC Transporter-Dependent Periplasmic protein, Polysaccharide co-polymerase and Outer Membrane Polysaccharide Export families. Glycosyltransferase mediated

polysaccharides are formed and exported from the cytoplasm through tripartite efflux pump complex.

3. **Synthase dependent pathway**- Single synthase protein and membrane-localized GTFs manage the polymerization of the EPS which is transported across the cell membrane by the flippase enzyme

5.2 Extracellular EPS Biosynthesis

Extracellular EPS biosynthesis occurs by sucrase protein. Outside the cellular outer membrane, the extracellular sucrase enzyme transforms sucrose into monomeric units followed by glycosyltransferase mediated polymerization (Juraskova et al., 2022) (Figure 4).

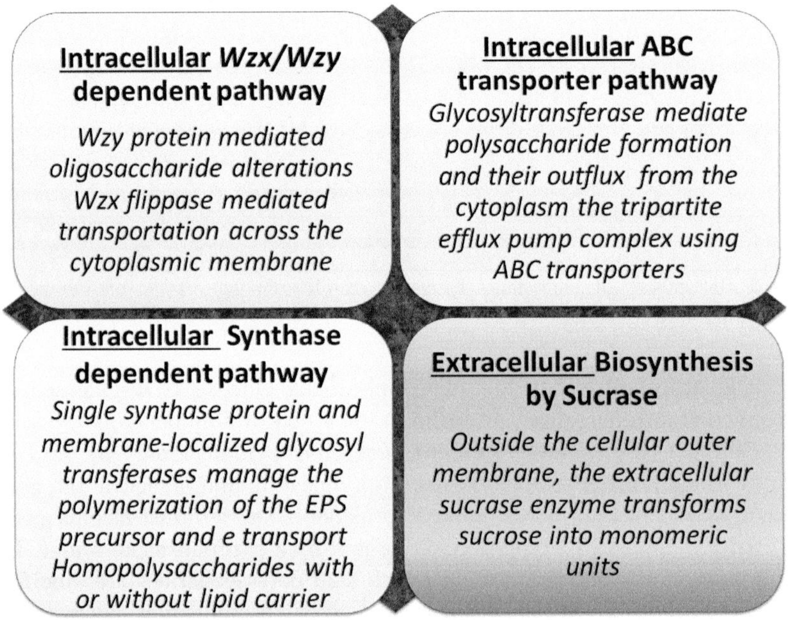

Figure 4. Bacterial strategies for intracellular and extracellular biosynthesis of EPS.

6. Production of Bacterial Exopolysaccharides

Bacteria differ in their habit and habitat as well their adaptability in varied region makes their diversified physiological nature for nutritional requirement as well as their response for physical and chemical triggers (Habibi and Khosravi-Darani, 2017). As bacteria differ in their bio-physio-chemical attributes, no single set of culture conditions ensure high exopolysaccharide yields. EPS production includes extreme fermentation rheology, product concentration; process mediated structural and conformational changes and the discernable effect of such changes on the product's application and performance (Sengupta et al., 2018). In general, EPS production essentially involves following steps (Figure 5).

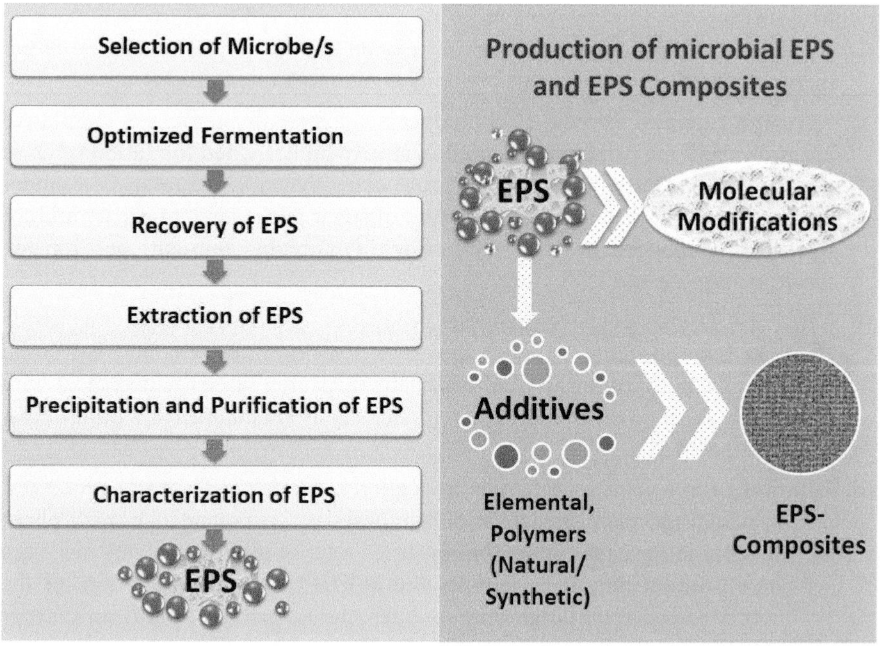

Figure 5. General plan of EPS and EPS composite production for biomedical applications.

6.1 Optimized Fermentation

Exoploysaccharides are mainly produced via fermentation process. The effectual designing for fermentation product largely depends on microbial strain, nutritional requirements, operating process, extraction and purification of the product. Handling the microbial strain is the most affective portion of fermentation as it affects the cost price (Das et al., 2023). Since the cost of culture media is about 30% of the cost of the entire fermentation process, media as well as factors controlling microbial growth has to be optimized (Barcelos et al., 2019; Efremenko et al., 2022). Various factors that need to be governed during fermentation include nutrients, temperature, pH, fluidity, osmolality, biological features of cell life shell, etc. (Puertas-Bartolome et al., 2021; Rabiya and Sen, 2022). Major factors include-

i. **Carbon source**- EPS size is affected by carbon source. Some of the carbon source utilized include glucose, lactose, maltose, mannitol, methanol, sucrose, starch, whey, etc. C_9 to C_{16} n-alkanes generally stimulate the yield of EPS.
ii. **Nitrogen source**- Organic as well as inorganic sources as sodium nitrate, urea, ammonium sulfate, peptone, yeast extract etc are used as nitrogen sources. In addition, organic nitrogen source in combination with additional growth factors in trace amount results into higher EPS production and growth rate.
iii. **Carbon/Nitrogen ratio (C: N ratio)**- High carbon/nitrogen ratio (10:1) favors high EPS production. Availability of nitrogen in culture pool marks EPS synthesis.

iv. **Mineral source**- Minerals impart micronutrient role and some as ion also effect pH of the media. Elemental role in enzyme activity effect metabolic entity thereby they affects the growth of microbe and production of EPS.

v. **Aeration rate and oxygen**- EPS producers are generally aerobes or facultative anaerobes and the production is comparatively higher when limitation of O_2 is not reenacted. During biosynthesis molecular oxygen is required for primary energy metabolism and for the oxidation of sugar to the alcohol carrier and for re-oxidizing reduced pyridine nucleotides. Optimizing nutrients and oxygen result in an increased EPS yield.

vi. **Viscosity**- During exopolysaccharide production quality of the polysaccharide becomes heterogeneous as broth develops non-newtonian characteristics and behaves as a pseudo-plastic fluid where the measured viscosity decreases with increasing shear rate. In such phases viscosity is maintained for a particular fermentative media.

vii. **Dilution rate**- Exopolysaccharide production is a batch or fed-batch processes, during which the exopolysaccharide synthesis and exponential growth phase do not occur at the same time. The residence time of microorganisms increases with low dilution rate and promotes the utilization of excess carbon for the production whereas with high dilution rates, the isoprenoid lipid insufficiently goes with simultaneous synthesis of surface polymers.

viii. **Osmolarity**- Osmolarity plays crucial role in both, growth and exopolysaccharide production, increasing osmolarity of the culture medium comparatively increase EPS yield

ix. **Incubation temperature**- In fermentative process for better yield, desired temperature is generally lower than the optimum growth temperature. With precursor availability decrease in temperature also results in an increase in EPS production due to decrease in cell wall polymer synthesis and growth rate.

x. **Incubation pH**- pH is an essential parameter in fermentation process. Neutral pH is required for the culturing of numerous microorganisms whereas some multiply in slightly acidic pH.

xi. **Age of the exopolysaccharide producing cells and growth**- Exopolysaccharide producing bacteria obtain their optimal growth within the initial 24 h of incubation whereas maximal EPS production occurs in the later stationary phase of growth. In this phase some polysaccharides are hydrolyzed thereby reducing viscosity of the medium. In addition, a competition between cell-wall polymer and EPS leads to stagnant phase. The culture can be optimized by reversal phase.

6.2 Recovery of Exopolysaccharides

Lower exopolysaccharide content in fermentation broth, presence of cell debris and solutes along with high viscosity of the broth increases production cost of exopolysaccharide (Das et al., 2023). Therefore, recovery process is deployed for-

a. Concentration of broth for stability, transportation and storage

b. Reduction of non-polymer solid/s to improve functional performance
c. Deactivation of non-essential enzymes

6.3 Extraction

Prior to Exopolysaccharide extraction, EPS is dissociated from the cells using centrifugation/ultracentrifugation. The duration and speed of centrifugation depends on the nature and viscosity of the polysaccharide. In cell-EPS strong conjunction, heat or alkaline treatments are used depending on the thermo sensitivity of the microbe (Majee et al., 2017).

6.4 Precipitation and Purification

Exopolysaccharides are generally obtained by solvent precipitation from culture broth. They are precipitated from the supernatant by the addition of polar organic solvent miscible with water. During precipitation, some proteins and salts of the medium also precipitate which are subsequently removed (Karadeniz et al., 2021). Depending on the operational and end product requisites the final EPS precipitate is harvested following centrifugation, filtration, pressing or settling. Drying is carried out under vacuum or with an inert gas and the dry exopolysaccharide is milled to required mesh size (Bagnol et al., 2022; Zanotti et al., 2023).

6.5 Characterization of EPS

EPS is characterized through instrumentation techniques or chemical methods. Among instrumentation, EPS is chiefly characterized through-

a. **Chromatography methods**- Among chromatographic methods thin Layer Chromatography is deployed by using mixture of acetic acid, chloroform, ethanol, water as the mobile phase or analytical gas chromatography technique is used for hydrolyzed EPS. High Performance Liquid Chromatography analysis is performed to check the purity of the sample in which the mobile phase and sample are degassed using a Sonicator (Bertsch et al., 2019).
b. **Spectroscopic methods**- The major functional groups of the purified EPS are determined by Fourier transform infrared spectroscopy. For sugars, strong absorption bands are obtained at 3500–3200 cm^{-1}, 1150–1000 cm^{-1} with moderately strong bands at 1000–750 cm^{-1} fingerprint zone. One dimensional NMR (Nuclear Magnetic Resonance) spectroscopy is used to determine the basic molecular confirmation of EPS (Shukla et al., 2019; Zaghloul and Ibrahim, 2022).

Chemical characterization of EPS include-

Chemical characterization is carried out by characterizing molecules for their chemical behavior, rheological features and antioxidant activities (Gunatillake and Adhikari, 2016; Gurav et al., 2023; Jusoh et al., 2021).

c. **Estimation of Biomolecules**- Biomolecules are estimated through conventional detection methods as-
   i. Folin-phenol method is used to estimate the protein Phenol
   ii. Sulfuric acid method is used to quantify the total carbohydrate content
   iii. Amount of EPS is determined from the calibration curve
   iv. DNA and RNA contents of the EPS are estimated by diphenylamine and orcinol methods respectively
   v. Estimation of uronic acid by m-hydroxy diphenyl

d. **Rheological study**- Among rheological parameters viscosity of EPS is measured through viscometer.

e. **Kaolin flocculating activity**- Kaolin flocculating activity is detected using kaolin clay suspension (0.4% w/v) as test material and flocculating efficacy is calculated as-

$$\text{Flocculating activity (in \%)} = \frac{A - B}{A} \times 100$$

A = optical density of control; B = optical density of sample

f. **Scavenging activity**- DPPH (2,2-Diphenyl-1-picrylhydrazyl) radical scavenging activity is measured through optical density of EPS, using vitamin C (ascorbic acid) as positive a control.

$$\text{Scavenging activity (in \%)} = \frac{A_1 - A_2}{A_0} \times 100$$

A_1 = O.D. of reaction mixture; A_2 = O.D. of reaction mixture without DPPH; A_0 = O.D. of reaction mixture with DPPH without sample

7. Exopolysaccharide Composites

In contemporary era emerging new techniques has enabled the clinicians and researchers to cope with maladies and to provide better health care systems. Microbes act as a small bio factories and have capabilities to produce various chemicals and metabolites, many of which are the important part of life saving therapeutic systems. Biopolymers have served the biotic components in all its evolutionary phases. Synthetic chemicals and molecules have made commendable changes in the society with polymers as a broad pan. Though, synthetic polymers hold immense therapeutic market, obstacles and issues such as non-degradability, toxicity, inadequate cell adhesion and allergic reactions cannot be ignored (Reddy et al., 2021; Wei et al., 2018). To overcome the constrictions and shortcomings, natural biopolymers as acetan, cellulose, chitin, chitosan, dextran, levan, etc., are preferred and catching eye due to their non-toxicity, environment and user friendly approach, supportive cell adhesion for surgeries and biodegradability (Puertas-Bartolome et al., 2021).

Among natural biopolymers exopolysaccharides are preferred for their biocompatibility, non-toxicity and biodegradation, structural diversity, purity, hydrophilic nature, strong water binding and retention ability along with species dependent biological activities. These natural polymers are commonly employed in high-value applications for their non-Newtonian behavior and high viscosity in aqueous media (Netrusov et al., 2023).

Microbial exopolysaccharides are biopolymers of high molecular weight and are composed of carbohydrates and non-carbohydrates moieties like acetate, carboxylate, glycerol, phosphates, pyruvate, sulfate and succinate besides proteins, phospholipids, and DNA (Angelin and Kavitha, 2020). They show diverse applications, variation in structure and adaptability along with presence of different functional groups defined by enantio- and stereo-selectivity, biostability and biocompatibility (Roychoudhury et al., 2024).

In nature EPS is generally produced by microbes in response to stress while some exopolysaccharides as alginate, cellulose, sphingan, and xanthan form a protective barrier on the bacterial cell surfaces (Angelin and Kavitha, 2020). These chemicals also protect cells against extreme environments like aridity, osmotic stress, phagocytosis, pH, salinity, temperature, UV-rays and certain incompatible chemical agents (Netrusov et al., 2023). At large scale, almost all bacterial species produce EPS but some groups of bacteria are preferred for their production as LAB group including *Bifidobacterium, Enterococcus, Lactobacillus, Lactococcus, Leuconostoc, Pediococcus, Streptococcus* and *Weissella* sp., for their non-pathogenicity (Angelin and Kavitha, 2020).

Some polysaccharides (e.g., alginate, FucoPol, gellan, GalactoPol, etc.) can establish physical and/or chemical intermolecular interactions and form cohesive polymeric matrix mediated film/s. Polymeric matrices with modified properties as transparency, barrier and mechanical properties or biocompatibility or bioactivity serve many applications for structured materials (nanoparticles, scaffolds or hydrogels) that are used for biomedical applications as drug delivery, imaging, tissue engineering, wound dressings, etc. (Guerreiro et al., 2021; Xiao et al., 2022; Devi et al., 2024).

Bacterial EPS bear many biomedical utility properties as gelling capacity of alginate, dextran, gellan, levan and scleroglucan. Acetan and scleroglucan are additionally bio-thickeners which make them suitable for topical applications. Prolonged vitality is the basic feature required in pharmaceutical products. Some EPS have good stability over time as dextran (Sharma et al., 2024), Elsinan (Ullah et al., 2021), emulsan (D'Almeida et al., 2024), Kefiran (de Oliveira et al., 2023), mutan (Rush et al., 2024), reuteran (Dailin et al., 2022), Scleroglucan (Castro et al., 2024), xanthan/xanthan gum (Dzionek et al., 2022). Though, xanthan gum stability varies with temperature indicating its sensitivity for temperature. Succinoglycan and welan points a unique stability in acidic and alkaline conditions while clavan and curdlan are thermo stable (Aquinas et al., 2024; Kim et al., 2024; Wang et al., 2024). EPS are synthesized to provide the optimum conditions for bacterial growth under stress conditions so many their acclimatizing properties do not cope with medical utility and forms limitation in applications as poor stability of acetan, alginate, gellan, hyaluronic acid/hyaluronan or non-gelling property of welan. Some bacteria yields low EPS as acetan, alternan, levan, reuteran and succinoglycan but in overall sum these biopolymers have great biological activity as promontory Cell growth, cell differentiation efficacy, immuno-modulatory, antimicrobial, anti-hypercholesterolemic, neuronal sensitivity, Prebiotic capacity, wound healing, antitumor, anticancer, etc. (Gunatillake and Adhikari, 2016; Zanotti et al., 2023) (Table 1).

Despite of advantages and limitations, exopolysaccharides have some bottlenecks like hydrophilic nature, rapid degradation, low tensile and mechanical strength, subsidized stress tolerance specifically for scaffold preparation which restricts their application. In addition, some EPS come with certain shortcomings as cellulose lacks antimicrobial properties while xanthan has poor electrical conductivity and poor stability against heat while pullulan has poor support for cell adhesion and proliferation (Ahuja et al., 2023). But the basic properties of EPS can be changed dramatically using mixtures with other biopolymers. Therefore, to overcome the mechanical and technical bottlenecks of EPS, composites are prepared to alter the lacunas and increase their biomedical application like drug delivery, scaffolding, surgical sealants, tissue engineering, and repair, etc. (Netrusov et al., 2023; Waoo et al., 2023).

Exopolysaccharide composites exhibit higher potential than native EPSs due to secondary polymer molecules that offer hybrid characteristics of both the polymers but face a challenge of cost in commercial production along with some operational challenges as stability and activity after prolonged storage (Table 1).

Exopolysaccharide composites exhibit higher potential than native EPSs due to secondary polymer molecules that offer hybrid characteristics of both the polymers but face a challenge of cost in commercial production along with some operational challenges as stability and activity after prolonged storage. In addition, they also help to overcome the limitations on solo EPS as some lack biological activity while other may have low yield thereby imposing high retrieval cost (Table 1). EPS composites are prepared with large number of additives including inorganic molecules, clays, natural and synthetic, polymers. Among additives polymers play significant role for composites for biomedical applications and their selection relies upon the type of application such as drug loading and release, tissue support and engineering, or *ex vivo* applications as a sealant (Tabernero and Cardea, 2020; Asgher et al., 2021).

7.1 Composites of Exopolysaccharide with Natural Biomolecules

EPS is a natural biopolymer so when it is blended with other natural biomolecules as proteins, enzymes, lipids, carbohydrates, etc., its increases its healthcare-associated applications as being both natural components the possibility of inflammation, rejection and allergenic responses decline (El-Naggar et al., 2020). Alongside such composites generally not have any toxic effect and sum up with bioactivity, biodegradability, and biocompatibility (Laubach et al., 2021). Large scale production of these composites also becomes cost effective with high recovery rates. Some vivid applications of such composites include medical sealants, scaffolds in tissue recovery and repair, conferred bioactivity as an antimicrobial, anti-inflammatory, antioxidant, etc. (He et al., 2023; Biswas et al., 2023; Ul-Islam et al., 2023).

7.2 Composites of Exopolysaccharide with Synthetic Biomolecules

Exopolysaccharides with synthetic biopolymers improves the binding ability and woven strength of any fiber. These fibers find great applications in wound closure, tissue fabrication and skin grafts. Porosity and scaffold play an important role in their redesigning (Kumar et al., 2023).

7.3 Therapeutic Applications of EPS Composites

Drug delivery systems (DDSs) are generally designed to check fluctuating blood drug concentrations and to reduce dosage frequency. Their design architecture needs complete drug noesis, system biocompatibility, metabolic synchronization and administration route (Ding and Li, 2017). Among physical parameters size of the delivery system makes a prominent role for the target organs as the transportation molecules and membrane permeability varies with organs/tissues (Danaei et al., 2018). The carrier selected is usually a biocompatible and biodegradable biopolymer. The selection of the carrier is key driver for controlling the subsequent release mechanism i.e. when the drug diffuses from the polymer and/or the polymer swells to improve the drug's release, or the drug can be released due to a dual mechanism erosion-diffusion. DDSs are generally designed as beads, particles, fibers, scaffolds, hydrogels, cryogels, aerogels, etc. (Guastaferro et al., 2021; Idumah et al., 2021).

Hydrogels are composed of a large amount of water (~ 70–99%) and a cross-linked polymer network with biocompatibility and efficacy to encapsulate hydrophilic drugs (Das et al., 2023). Hydrophilic groups confer high water binding capacity and allow intermolecular interaction and crosslinking like polymer-drug or polymer-host tissue. Hydrogel based drug delivery through EPS composites ensures controlled drug delivery as that of Alginate and Gelatin (Wang et al., 2019a), dextran and polydopamine (Zhang et al., 2021), oxidized dextran and gelatin, curcumin, cerium oxide (Andrabi et al., 2020), gellan gum (oxidized) and carboxy methyl chitosan (Zhang et al., 2020), etc. These hydrogels also prove their application in wound healing when deployed as wound healers as fiber of EPS composites, e.g., gellan gum and carboxy methyl chitosan (Zhang et al., 2020), levan and catechol (Osman et al., 2023), pullulan and dopamine (Baron et al., 2022), xanthan gum and alginate (Kumar et al., 2017), etc.

The physiological aspects of organs differ in their histo-architecture, modalities and functionalities. Many EPS-composites are designed for specific tissues, organs or systems as scaffold of cellulose and chitosan (Zhu et al., 2023), curdlan and chitosan (Toullec et al., 2021), cellulose-poly vinyl alcohol (Aki et al., 2020) and composites membranes of Gellan gum- Calcium–poly-glutamic acid and glycerol (Lin et al., 2021) for Bone tissue regeneration, bone growth and bone tissue engineering. Biosealant hydrogels of cellulose and epoxides (Orlando et al., 2020), pullulan and collagen (Chen et al., 2022), dextran with poly vinyl alcohol and citric acid (Kenawy et al., 2023), gellan gum and poly-acrylamide (Li et al., 2021) have been prepared for wound closure (Tables 2, 3) while composite of cellulose and chitosan along with alginates (Ahuja et al., 2023) for organ reconstructions and dental implants. Ocular rectification composites are made from pullulan and dexamethasone (Kicková

Table 2. Composite formation/s from bacterial exopolysaccharides and natural polymers for enhanced biomedical application.

EPS	Additives in Composites	Composite's Functionality	Modified Biomedical Application
Alginate	Gelatin	Hydrogel with increased water absorption and retention, controlled drug release	Wound dressing (Wang et al., 2019a)
Cellulose	Carboxymethylated diethylaminoethyl cellulose	Scaffold fabrication	Hydrogel with higher drug loading capacity and swelling ratio (Ojagh et al., 2021)
Cellulose	Epoxides	Biosealant Hydrogel with Bacteriostatic and bacteriolytic	Wound closure and antimicrobial properties (Orlando et al., 2020)
Cellulose	Methacrylated gelatin	Biosealant Hydrogel	Hydrogel with sustained cell viability for possible application in cardiovascular systems (Ma et al., 2022)
Cellulose	Chitosan	Improved mechanical strength, porosity, cytocompatibility, and drug release rate	Bone tissue regeneration (Zhu et al., 2023)
Cellulose	Chitosan + Alginates	3D scaffolds and Gelling system	Organ reconstructions; Dental implants (Ahuja et al., 2023)
Curdlan	Chitosan	Fibrillary scaffold	Bone growth (Toullec et al. 2021)
Dextran	Periodate-assisted oxidation + Chitosan hydrochloride	Biosealant and biological adhesive	To control bleeding during surgical procedures (Balakrishnan et al., 2017)
Dextran	Chitosan, Collagen, and Polydopamine	Procoagulation, and adhesion , Activation of Platelets	Cryogels of high coagulation potency; Reduced bleeding time and blood loss (Ma et al., 2022)
Dextran	Thyme magnesium-doped hydroxyapatite	Bacteriostatic and Bacteriolytic	Antimicrobial (Iconaru et al., 2020)

Dextran	Polydopamine	Carrier for antibiotics in macroporous Hydrogels	Increase in negative charge, surface area mechanical strength with decrease in pore size, making suitable for drug loading and release (Zhang et al., 2021)
Dextran (oxidized)	Gelatin, Curcumin and Cerium oxide	Drug delivery with increased surface area	Hydrogel with rapid tissue repair and wound healing along with anti-inflammatory and antioxidant properties (Andrabi et al., 2020)
Gellan	Alginate	3D Scaffold	Tissue engineering (Akkineni et al., 2016)
Gellan	Gelatin	3D Scaffold	Hydrogel for Skin regeneration with shear-thinning, and recovering abilities (Zheng et al., 2018)
Gellan	Silk sericin	Cell carriers	Hydrogels causing replacements in retinal pigment epithelium cells (Kim et al., 2022)
Gellan	Methacryl	3D Scaffold	Tissue engineering (D'Amora et al., 2023)
Gellan Gum	Alginate and Nisin-Enriched Lipid Nanoparticles	Cytocompatible Hydrogel	Enhanced Wound recovery (Reczynska-Kolman et al., 2022)
Gellan gum	Egg shell membranes	Hydrogel	Regeneration of retinal pigment epithelium no cytotoxicity (Choi et al., 2020)
Gellan gum	Alginate-Calcium chloride	Hydrogel with induced neovascularization	Osteochondral repair (Xing et al., 2021)
Gellan gum (Oxidized)	Carboxy methyl chitosan	Hydrogel with decreased gelation time	Drug delivery and wound dressing (Zhang et al., 2020)
Hyaluronic acid	Gelatin	Hydrogel with adjusted viscosity and giffusion coefficient	Adipose stem cells cultivation (Sakai et al., 2019)

Table 2 contd. ...

...Table 2 contd.

EPS	Additives in Composites	Composite's Functionality	Modified Biomedical Application
Hyaluronic acid	Methacrylated gelatin	Hydrogel scaffold with higher elastic modulus and porosity	Tissue engineering (Velasco-Rodriguez et al., 2021)
Hyaluronic acid	Gelatin-hydroxyphenyl propionic acid	Polymer network	Retinal ganglion cells replacement therapy (Dromel et al., 2021)
Hylauronic acid/ Hyaluronan	Poly(Ne-acryloyl L-lysine)	Scaffold fabrication	Breast cancer (Hussain et al., 2017; Xu et al., 2016)
Hylauronic acid/ Hyaluronan	Self-assembling peptide	Scaffold fabrication	Breast cancer (Sieni et al., 2020)
Hylauronic acid/ Hyaluronan	Silk fibroin–gelatin and heparan sulfate	Scaffold fabrication	Cholangiocarcinoma (Buhome et al., 2022)
Levan	Carboxymethyl cellulose	Biofiller hydrogel with improved cell proliferation and anti-wrinkle activity	Regeneration of soft tissue with collagen synthesis (Choi et al., 2018)
Levan	Catechol	Wound dressing sealants through improved adhesion	Rapid blood clotting and wound healing with lower endotoxin level (Osman et al., 2023)
Pullulan	Dextran with interfacial Polyelectrolyte fibers	Improved cell adhesion, scaffold fabrication	Hydrogels for bone grafts (Arora et al., 2015)
Pullulan	Dexamethasone	Drug delivery	Ocular rectifications (Kicková et al., 2021)
Pullulan	Collagen	Transformative cellular response	Rapid wound closure and Sealants (Chen et al., 2022)
Pullulan (Oxidized)	Dopamine	Drug delivery through hydrogels	Hemostatic wound dressing (Baron et al., 2022)

Pullulan	Gelatin	Nanofibers with increased viscosity, conductivity and surface tension	Tissue engineering (Wang et al., 2019b)
Pullulan (Dialdehyde)	Dopamine	Altered water retention capacity, drug delivery	Cryogels for wound healing (Baron et al., 2022)
Xanthan gum	Gelatin	Increased cell viability	Hydrogel for Skin cell growth (Alves et al., 2020)
Xanthan gum	Alginate	Hydrogels with increased cyto-compatability with osteoblastic cells and nanotube viability, Increased water retention	Increased cell adhesion and mechanical stability of wound dressing material (Kumar et al., 2017)
Xanthan gum	Soybean protein	Increased adhesion strength	Medical adhesives (Feng et al., 2018)
Xanthan gum	Chitosan and nanohydroxyapatite and polyelectrolyte complex	Osteo-conductive tri-composite scaffolds	Osteo-regeneration (Zia et al., 2020)
Xanthan gum	Chitosan, hydroxyapatite and brushite	Polymeric porous scaffolds	Higher cell viability and low inflammation (Neves et al., 2022)
Xanthan gum	Konjac glucomannan	Increased Adhesion	Wound healing (Alves et al., 2020)

Table 3. Composite formation/s from bacterial exopolysaccharides and synthetic polymers for enhanced biomedical application.

EPS	Additives in Composites	Composite's Functionality	Modified Biomedical Application/s
Cellulose	Electrospun poly vinyl alcohol and hexagonal boron nitride	3D scaffold with reduced pore size and increased viscosity	Enhanced cell adhesion surface and proliferation, Bone tissue engineering (Aki et al., 2020)
Cellulose (Acetate)	Poly(-caprolactone)	Tubular scaffold fabrication in nanofiber membranes through stainless-steel collectors	Preparation of artificial blood vessels with small diameters for thrombosis patients (Wang et al., 2023)
Curdlan (Phosphorylated)	Metronidazole	Hydrogel	Controlled drug release (Suflet et al., 2021)
Dextran	Electrospun poly vinyl alcohol and citric acid	Hydrogel with proliferation of fibroblasts and increased anti-microbial activity	Accelerated Wound gap closure (Kenawy et al., 2023)
Dextran (Aldehyde)	Poly vinyl alcohol	Hydrogel of increased fluid adsorption capacity and decreased tensile strength	Wound coverage area increased thereby promoted wound healing (Zheng et al., 2019)
Gellan gum	Calcium ions and synthetic clay Laponite	Beads loaded with theophylline and cyanocobalamin	Controlled drug entrapment efficiency, slower drug release into the gastric environment (Adrover et al., 2019)
Gellan Gum	Poly-Acrylamide	Hydrogel with controlled release	Fibroblast cells with better proliferation and complete wound closure without scare (Li et al., 2021)
Gellan gum	Calcium–poly--glutamic acid and glycerol	Composite membranes with uniform calcium aggregates for protein adsorption, and bone cell proliferation	Increased secretion of alkaline phosphatase and induced mineralization assisted in bone repair and regeneration; Improved tensile and mechanical strength and delayed degradation of membranes (Lin et al., 2021)
Levan (Hydrolyzed)/ Levan (Sulfated)	Polycaprolactone	Coaxial electrospinning of composite fiber for e tensile strength	Increased viability and biocompatibility of fibroblasts (Avsar et al., 2018)

Levan	Hydroxyapatite	Hydrogel with improved cell proliferation and long stability	Dermal filler improved collagen production and anti-wrinkle activity (Hwang et al., 2021)
Pullulan	Cholesterol succinate	Alteration in particle size and hydrophobicity based drug release	Amphiphilic composites with ensured drug release (Yuan et al., 2020)
Pullulan	Thiolactone	Hydrogels with biocompatible mesh size	Improved fiblroblast and hepatocyte proliferation (Mommer et al., 2021)
Pullulan	Pendant carboxy methyl groups and poloxamer	Hydrogel for scaffold fabrication and drug delivery with improved thermo sensitivity, increased flexibility and elasticity	Prolonged and sustained drug release (Constantin et al., 2021)
Xanthan gum	Poly(3,4-ethylene dioxythiophene)	Scaffolds with increased interconnected pores	Cell growth and biocompatibility in epithelial cells (del Agua et al., 2018)
Xanthan gum	Citric acid, gelatin and glutaraldehyde,	Hydrogels with >90% of water-holding capacity	Increased wound healing efficacy in deep second-degree skin burns (Shawan et al., 2019)
Xanthan gum	Chitosan, 2-acrylamido-2-methylpropane sulphonic acid and potassium persulphate	Higher thermal stability and porous structure encapsulated Acyclovir drug loading	Oral antiviral drug carrier (Malik et al., 2020)
Xanthan gum/ Hypermellose/ Sodium carboxy methyl cellulose	Poly vinyl alcohol	Hydrogel with bactericidal effect along with free radical scavenging behavior and reduction in the inflammatory response	Higher antimicrobial potential with inhibition of > 99% (Singh et al., 2022)

et al., 2021) while hyaluronic acid and gelatin-hydroxyphenyl propionic acid (Dromel et al., 2021) assist in retinal ganglion cells replacement therapy (Table 2).

Blood loss due to over bleeding during severe injuries toll lives. Biosealant and coagulants are designed using Dextran and Periodate-assisted oxidation with Chitosan hydrochloride (Balakrishnan et al., 2017) to prevent over bleeding. Patients suffering from thrombosis have therapeutic aid of EPS composite cellulose acetate-poly(-caprolactone) prepared artificial blood vessels with small diameters (Wang et al., 2023) (Table 3). On similar lineage cellulose and methacrylated gelatin (Ma et al., 2022) composed hydrogels function as biosealant for cardiovascular systems. Breast cancer is one of the most reported disease, threating all the women's around the world. EPS composites of hylauronic acid/hyaluronan with poly(Ne-acryloyl L-lysine) (Hussain et al., 2017; Xu et al., 2016) and self-assembling peptide (Sieni et al., 2020) are reported for their therapeutic assistance in breast cancer (Tables 2, 3).

Scaffolds with specificity as high porosity, pore interconnectivity, suitable pore size distribution, nanostructure, biocompatibility, biodegradability, lack of toxicity, etc. can be deployed for multi curative uses (Reddy et al., 2021; Santos et al., 2020, Unnikrishnan et al., 2021; Nikolova and Chavali, 2019; Eltom et al., 2019). Scaffold mediated drug aid is also made from cellulose- carboxymethylated diethylaminoethyl cellulose (Ojagh et al., 2021) cellulose- chitosan and alginates (Ahuja et al., 2023), curdlan- chitosan (Toullec et al., 2021) gellan - alginate (Akkineni et al., 2016), gellan- gelatin (Zheng et al., 2018), gellan- methacryl (D'Amora et al., 2023), hyaluronic acid- methacrylated gelatin (Velasco-Rodriguez et al., 2021) hyaluronic acid-poly(Ne-acryloyl L-lysine) (Hussain et al., 2017; Xu et al., 2016) hyaluronic acid-self-assembling peptide (Sieni et al., 2020), hyaluronic acid-silk fibroin–gelatin and heparan sulfate (Buhome et al., 2022) pullulan-dextran (Arora et al., 2015), xanthan gum-chitosan and nanohydroxyapatite and polyelectrolyte complex (Zia et al., 2020), xanthan gum-chitosan, hydroxyapatite and brushite (Neves et al., 2022) for various maladies (Table 2).

8. Challenges and Future Outlook

EPS and its composites application/s especially in medical field depends mainly on two factors. The first factor involves production of EPS on commercial scale and second formation of appropriate EPS composite for specific application. Despite of unique properties and great application/s, bacterial EPS are synthesized and produced in extremely limited amount. Low productivity goes with high cost of the resulting products. Therefore, for highly efficient production of EPSs and to reduce costs further research needs on some lineages as Central Composite Design (CCD) and Artificial Neural Network mediated optimization of culture conditions, improved downstream processing, comparatively cheap/low cost culture media has to be designed along with development of highly productive strains through genetic engineering. Formation of Composites adds or enhances structural, functional, and other physico-chemical properties to EPS. Currently, many composites face a problem of stability as ESPs are mainly negatively charged and therefore they exert electrostatic repulsion in many composites. Though production of EPS and their respective composites face many challenges and limitations but their applications especially for wound healing, skin and bone regeneration, cell proliferation and many more cannot be ignored. The futuristic image as projected through contemporary work indicates possibilities of organogenesis, more implants, scaffolds, imprinting, etc. for non-curable and hard tasking maladies.

References

Aditya, T., J. P. Allain, C. Jaramillo and A. M. Restrepo. 2022. Surface modification of bacterial cellulose for biomedical applications. Int. J. Mol. Sci. 23: 610–617. https://doi.org/10.3390/ijms23020610.

Adrover, A., P. Paolicelli, S. Petralito, L. Di Muzio, J. Trilli, S. Cesa et al. 2019. Gellan gum/laponite beads for the modified release of drugs: experimental and modeling study of gastrointestinal release. Pharmaceutics 11: 187–194. https://doi.org/10.3390/pharmaceutics11040187.

Ahuja, V., A. K. Bhatt, J. R. Banu, V. Kumar, G. Kumar, Y. H. Yang et al. 2023. Microbial exopolysaccharide composites in biomedicine and healthcare: trends and advances. Polymers 15(7): 1801. https://doi.org/10.3390/polym15071801.

Aki, D., S. Ulag, S. Unal, M. Sengor, N. Ekren, C. C. Lin et al. 2020. 3D printing of PVA/hexagonal boron nitride/bacterial cellulose composite scaffolds for bone tissue engineering. Mater. Des. 196: 109094. https://doi.org/10.1016/j.matdes.2020.109094.

Akkineni, A. R.,T. Ahlfeld, A. Funk, A. Waske, A. Lode and M. Gelinsky. 2016. Highly concentrated alginate-gellan gum composites for 3D plotting of complex tissue engineering scaffolds. Polymers 8: 170. https://doi.org/ 10.3390/polym8050170.

Al-Nabulsi, A. A., Z. W. Jaradat, F. R. Al Qudsi, L. Elsalem, T. M. Osaili and A. N. Olaimat. 2022. Characterization and bioactive properties of exopolysaccharides produced by *Streptococcus thermophilus* and *Lactobacillus bulgaricus* isolated from Labaneh. LWT-Food Sci Technol. 167: 113817. https://doi.org/10.1016/j.lwt.2022.113817.

Alves, A., S. P. Miguel, A. R. T. S. Araujo, M. J. de Jesús Valle, A. Sánchez Navarro, I. J. Correia et al. 2020. Xanthan gum–konjac glucomannan blend hydrogel for wound healing. Polymers 12: 99. https://doi.org/ 10.3390/polym12010099.

Andrabi, S. M., S. Majumder, K. C. Gupta and A. Kumar. 2020. Dextran based amphiphilic nano-hybrid hydrogel system incorporated with curcumin and cerium oxide nanoparticles for wound healing. Colloids Surf. B Biointerfaces 195: 111263. https://doi.org/10.1016/j.colsurfb. 2020.111263.

Angelin, J. and M. Kavitha. 2020. Exopolysaccharides from probiotic bacteria and their health potential. International Journal of Biological Macromolecules 162: 853–865. https://doi.org/10.1016/j.ijbiomac.2020.06.190.

Aquinas, N., R. Bhat and S. Selvaraj. 2024. Curdlan based hydrogels. pp. 203–213. *In*: Giri, T. K., B. Ghosh and H. Badwaik (eds.). Polysaccharide Hydrogels for Drug Delivery and Regenerative Medicine. Elsevier. https://doi.org/10.1016/B978-0-323-95351-1.00005-3.

Arora, A., P. Sharma and D. S. Katti. 2015. Pullulan-based composite scaffolds for bone tissue engineering: Improved osteoconductivity by pore wall mineralization. Carbohydr. Polym. 123: 180–189. https://doi.org/10.1016/j.carbpol.2015.01.038.

Asgher, M., S. A. Qamar and H. M. Iqbal. 2021. Microbial exopolysaccharide-based nano-carriers with unique multi-functionalities for biomedical sectors. Biologia 76(2): 673–685. https://doi.org/10.2478/s11756-020-00588-7.

Avsar, G., D. Agirbasli, M. A. Agirbasli, O. Gunduz and E. T. Oner. 2018. Levan based fibrous scaffolds electrospun via co-axial and single-needle techniques for tissue engineering applications. Carbohydr. Polym. 193: 316–325. https://doi.org/10.1016/ j.carbpol.2018.03.075.

Aziz, T., A. Farid, F. Haq, M. Kiran, A. Ullah, K. Zhang et al. 2022. A review on the modification of cellulose and its applications. Polymers 14: 3206. https://doi.org/10. 3390/polym14153206.

Bagnol, R., D. Grijpma, D. Eglin and T. F. Moriarty. 2022. The production and application of bacterial exopolysaccharides as biomaterials for bone regeneration. Carbohydrate Polymers 291: 119550. https://doi.org/10.1016/j.carbpol.2022.119550.

Balakrishnan, B., D. Soman, U. Payanam, A. Laurent, D. Labarre and A. A. Jayakrishnan. 2017. Novel injectable tissue adhesive based on oxidized dextran and chitosan. Acta Biomater. 53: 343–354. https://doi.org/10.1016/j.actbio.2017.01.065.

Balducci, E., F. Papi, D. E. Capialbi and L. Del Bino. 2023. Polysaccharides' structures and functions in biofilm architecture of antimicrobial-resistant (AMR) pathogens. Int. J. Mol. Sci. 24: 4030. https://doi.org/10 3390/ijms24044030.

Barcelos, M. C. S., K. A. C. Vespermann, F. M. Pelissari and G. Molina. 2019. Current status of biotechnological production and applications of microbial exopolysaccharides. Crit. Rev. Food Sci. Nutr. 60: 1475–1495. https://doi.org/10.1080/10408398.2019.1575791.

Baron, R. I., I. A. Duceac, S. Morariu, A. C. Bostănaru-Iliescu and S. Coseri. 2022. Hemostatic cryogels based on oxidized pullulan/dopamine with potential use as wound dressings. Gels 8: 726. https://doi.org/10.3390/gels8110726.

Barroca-Ferreira. J., D. Gomes, A. Sousa, L. A. Passarinha and D. Costa. 2024. Gellan gum–based biomedical gels. pp. 95–121. *In*: Nayak, A. K. and M. S. Hasnain (eds.). Application of Gellan Gum as a Biomedical Polymer. Academic Press. https://doi.org/10.1016/B978-0-323-91815-2.00012-0.

Benjamin, S. R. and E. J. Júnior. 2024. Biocomposites for biomedical applications. pp. 227–255. *In*: Karak, N. (ed.). Advances in Biocomposites and their Applications. Woodhead Publishing. https://doi.org/10.1016/B978-0-443-19074-2.00008-3.

Bertsch, A., D. Roy and G. LaPointe. 2019. Enhanced exopolysaccharide production by *Lactobacillus rhamnosus* in co-culture with *Saccharomyces cerevisiae*. Appl Sci. 9: 4026. https://doi.org/10.10.3390/app9194026.

Biswas, A., T. Ahmed, M. R. Rana, M. M. Hoque, M. F. Ahmed, M. Sharma et al. 2023. Fabrication and characterization of ZnO nanoparticles-based biocomposite films prepared using carboxymethyl cellulose, taro mucilage, and black cumin seed oil for evaluation of antioxidant and antimicrobial activities. Agronomy 13: 147. https://doi.org/10.3390/agronomy13010147.

Buhome, O., M. Wongwattanakul, J. Daduang and T. Limpaiboon. 2022. 3D silk fibroin-gelatin/hyaluronic acid/heparan sulfate scaffold enhances expression of stemness and EMT markers in cholangiocarcinoma. Vivo 36: 1155–1167. https://doi.org/10.21873/invivo.12815.

Castro, R. H., I. Burgos, L. M. Corredor, S. Llanos, C. A. Franco, F. B. Cortés et al. 2024. Carboxymethyl scleroglucan synthesized via O-alkylation reaction with different degrees of substitution: rheology and thermal stability. Polymers 16(2): 207. https://doi.org/10.3390/polym16020207.

Chaudhari, V., H. S. Buttar, S. Bagwe-Parab, H. S. Tuli, A. Vora and G. Kaur. 2021. Therapeutic and Industrial applications of Curdlan with Overview on its recent Patents. Front. Nutr. 8: 646988. https://doi.org/10.3389/fnut.2021.646988.

Chen, F. and G. Huang. 2018. Preparation and immunological activity of polysaccharides and their derivatives. Int J Biol Macromol. 112: 211–216. https://doi.org/10.1016/ j.ijbiomac.2018.01.169.

Chen, K., D. Sivaraj, M. F. Davitt, M. C. Leeolou, D. Henn, S. R. Steele et al. 2022. Pullulan-Collagen hydrogel wound dressing promotes dermal Remodelling and wound healing compared to commercially available collagen dressings. Wound Repair Regen. 30: 397–408. https://doi.org/10.1111/wrr.13012.

Choi, J., J. Lee, M. E. Shin, S. Been, D. H. Lee and G. Khang. 2020. Eggshell membrane/gellan gum composite hydrogels with increased degradability, biocompatibility, and anti-swelling properties for effective regeneration of retinal pigment epithelium. Polymers 12: 2941. https://doi.org/10.3390/polym12122941.

Choi, W. I., Y. Hwang, A. Sahu, K. Min, D. Sung, G. Tae et al. 2018. An injectable and physical Levan-based hydrogel as a Dermal filler for soft tissue augmentation. Biomater. Sci. 6: 2627–2638. https://doi.org/10.1039/C8BM00524A.

Constantin, M., B. Cosman, M. Bercea, G. L. Ailiesei and G. Fundueanu. 2021. Thermosensitive poloxamer-graft-carboxymethyl pullulan: a potential injectable hydrogel for drug delivery. Polymers 13: 3025. https://doi.org/10.3390/polym13183025.

D'Amora, U., A. Ronca, S. Scialla, A. Soriente, P. Manini, J. W. Phua et al. 2023. Bioactive composite methacrylated gellan gum for 3D-printed bone tissue-engineered scaffolds. Nanomaterials 13: 772. https://doi.org/10.3390/nano13040772.

Dailin, D. J., S. Selvamani, K. Michelle, Y. M. Jusoh, L. F. Chuah, A. Bokhari et al. 2022. Production of high-value added exopolysaccharide by biotherapeutic potential *Lactobacillus reuteri* strain. Biochemical Engineering Journal 188: 108691. https://doi.org/10.1016/j.bej.2022.108691.
D'Almeida, A. P., T. L. de Albuquerque and M. V. Rocha. 2024. Recent advances in Emulsan production, purification, and application: Exploring Bioemulsifiers unique potentials. International Journal of Biological Macromolecules 5: 133672. https://doi.org/10.1016/j.ijbiomac.2024.133672.
Danaei, M., S. Dehghankhold, S. Ataei, F. H. Davarani, R. Javanmard, A. Dokhani et al. 2018. Impact of particle size and polydispersity index on the clinical applications of Lipidic nanocarrier systems. Pharmaceutics 10: 57. https://doi.org/10.3390/pharmaceutics10020057.
Das, S. 2022. Genetic regulation, biosynthesis and applications of extracellular polysaccharides of the biofilm matrix of bacteria. Carbohydrate Polymers 291: 119536. https://doi.org/10.1016/j.carbpol.2022.119536.
Das, A., T. Ringu, S. Ghosh and N. Pramanik. 2023. A comprehensive review on recent advances in preparation, physicochemical characterization, and bioengineering applications of biopolymers. Polymer Bulletin 80(7): 7247–7312. https://doi.org/ 10.1007/s00289-022-04443-4.
de Oliveira Filho, J. G., C. de Oliveira Silva, M. B. Egea, H. M. de Azeredo and L. H. Mattoso. 2023. Employing alternative culture media in Kefiran exopolysaccharide production: Impact on microbial diversity, physicochemical properties, and bioactivities. International Journal of Biological Macromolecules 246: 125648. https://doi.org/10.1016/j.ijbiomac.2023.125648.
de Souza, C. K., T. Ghosh, N. Lukhmana, S. Tahiliani, R. Priyadarshi, T. G. Hoffmann et al. 2023. Pullulan as a sustainable biopolymer for versatile applications: A Review. Mat. Today Communications 36: 106477. https://doi.org/10.1016/j.mtcomm.2023.106477.
del Agua, I., S. Marina, C. Pitsalidis, D. Mantione, M. Ferro, D. Iandolo et al. 2018. Conducting polymer scaffolds based on poly (3,4-ethylenedioxythiophene) and xanthan gum for live-cell monitoring. ACS Omega 3: 7424–7431. https://doi.org/10.1021/acsomega.8b00458.
Devi, M., K. Thakur, D. Kumari, M. Behl, R. K. Rathour, A. K. Bhatt et al. 2024. Advances in fermentation technology for EPS production. pp. 37–76. *In*: Bhatia, S. K., P. S. Panesar and S. Mehariya (eds.). Microbial Exopolysaccharides. CRC Press. https://doi.org/10.1201/ 9781003342687.
Di Mola, A., M. R. Landi, A. Massa, U. D'Amora and V. Guarino. 2022. Hyaluronic acid in biomedical fields: new trends from chemistry to biomaterial applications. Int. J. Mol. Sci. 23: 14372. https://doi.org/10.3390/ijms232214372.
Díaz-Montes, E. 2021. Dextran: Sources, structures, and properties. Polysaccharides 2: 554–565. https://doi.org/10.3390/polysaccharides2030033.
Ding, C. and Z. Li. 2017. A review of drug release mechanisms from nanocarrier systems. Mat. Sci. Eng. 76: 1440–1453. https://doi.org/10.1016/j.msec.2017.03.130.
Domzał-Kędzia, M., M. Ostrowska, A. Lewińska and M. Łukaszewicz. 2023. Recent developments and applications of microbial levan, a versatile polysaccharide-based biopolymer. Molecules 28(14): 5407. https://doi.org/10.3390/molecules28145407.
Dromel, P. C., D. Singh, E. Andres, M. Likes, M. Kurisawa, A. Alexander-Katz et al. 2021. A bioinspired gelatinhyaluronic acid-based hybrid interpenetrating network for the enhancement of retinal ganglion cells replacement therapy. Npj Regen. Med. 6: 85. https://doi.org/10.1038/s41536-021-00195-3.
Dzionek, A., D. Wojcieszynska and U. Guzik. 2022. Use of xanthan gum for whole cell immobilization and its impact in bioremediation—A Review. Bioresour. Technol. 351: 126918. https://doi.org/10.1016/j.biortech.2022.126918.
Efremenko, E., O. Senko, O. Maslova, N. Stepanov, A. Aslanli and I. Lyagin. 2022. Biocatalysts in synthesis of microbial polysaccharides: properties and development trends. Catalysts 12: 1377. https://doi.org/10.3390/catal12111377.
Elangwe, C. N., S. N. Morozkina, R. O. Olekhnovich, V. O. Polyakova, A. Krasichkov, P. K. Yablonskiy et al. 2023. Pullulan-based hydrogels in wound healing and skin tissue engineering applications: a review. International Journal of Molecular Science 24: 4962. https://doi.org/10.3390/ijms24054962.
El-Naggar, M. E, S. I. Othman, A. A. Allam and O. M. Morsy. 2020. Synthesis, drying process and medical application of polysaccharide based aerogels. Int. J. Biol. Macromol. 145: 1115–1128. https://doi.org/10.1016/j.ijbiomac.2019.10.037.

Eltom, A., G. Zhong and A. Muhammad. 2019. Scaffold techniques and designs in tissue engineering functions and purposes: a review. Adv. Mater. Sci. Eng. 3429527. https://doi.org/10.1155/2019/3429527.

Feng, C., F. Wang, Z. Xu, H. Sui, Y. Fang, X. Tang et al. 2018. Characterization of soybean protein adhesives modified by xanthan gum. Coatings 8: 342. https://doi.org/10.3390/coatings8100342.

Fernandes, J., D. Deo and R. Kulkarni. 2021. Molecular engineering of bacterial exopolysaccharide for improved properties. pp. 85–103. *In*: Nadda, A. K., K. V. Sajna and S. Sharma (eds.). Microbial Exopolysaccharides as Novel and Significant Biomaterials. Springer, Cham. https://doi.org/10.1007/978-3-030-75289-7_4.

Flemming, H. C. and J. Wingender. 2010. The biofilm matrix. Nat. Rev. Microbiol. 8: 623–633. https://doi.org/10.1038/nrmicro2415.

Gargouch, N., F. Elleuch, I. Karkouch, O. Tabbene, C. Pichon, C. Gardarin et al. 2021. Potential of exopolysaccharide from *Porphyridium marinum* to contend with bacterial proliferation, bioflm formation, and breast cancer. Mar Drugs 19: 66. https://doi.org/10.3390/md19020066.

Giri, T. K. 2024. Hydrogels based on scleroglucan. pp. 139–150. *In*: Giri, T. K., B. Ghosh and H. Badwaik (eds.). Polysaccharide Hydrogels for Drug Delivery and Regenerative Medicine. Elsevier. https://doi.org/10.1016/B978-0-323-95351-1.00004-1.

Guastaferro, M., E. Reverchon and L. Baldino. 2021. Polysaccharide-based aerogel production for biomedical applications: a comparative review. Materials 14: 1631. https://doi.org/10.3390/ma14071631.

Guerreiro, B. M., J. C. Silva, J. C. Lima, M. A. M. Reis and F. Freitas. 2021. Antioxidant potential of the bio-based fucose-rich polysaccharide fucopol supports its use in oxidative stress-inducing systems. Polymers 13: 3020. https://doi.org/10.3390/polym 13183020.

Guetta, O., K. Mazeau, R. Auzely, M. Milas and M. Rinaudo. 2003. Structure and properties of a bacterial polysaccharide named Fucogel. Biomacromolecules 4(5): 1362–71. https://doi.org/10.1021/bm030033h.

Gunatillake, P. A. and R. Adhikari. 2016. Nondegradable synthetic polymers for medical devices and implants. pp. 33–62. *In*: Warren, L. P., P. Martens and R. Green (eds.). Biosynthetic Polymers for Medical Applications. Elsevier. https://doi.org/10.1016/B978-1-78242-105-4.00002-X.

Guo, Y., D. Pan, Y. Sun, L. Xin, H. Li and X. Zeng. 2013. Antioxidant activity of phosphorylated exopolysaccharide produced by *Lactococcus lactis* subsp. *lactis*. Carbohydr Polym. 97: 849–54. https://doi.org/10.1016/j.carbpol.2013.06.024.

Guo, L., Y. Chai, F. Zhou and P. Wang. 2021. Preparation and properties of hyaluronic acid hydrogel modified by L-cysteine hydrochloride. IOP Conf. Ser. Earth Environ. Sci. 651: 042022. https://doi.org/10.1088/1755-1315/651/4/042022.

Gurav, R., S. Hwang, S. K. Bhatia, A. D. Kshirsagar, S. Mandal and Y. H. Yang. 2023. Exopolysaccharide production using pinewood hydrolysate as a substrate for psychrotrophic bacterium isolated from Svalbard glacier soil. Biomass Conversion and Biorefinery 1: 1. https://doi.org/10.1007/s13399-023-05039-2.

Habibi. H. and K. Khosravi-Darani. 2017. Effective variables on production and structure of xanthan gum and its food applications: A review. Biocatal Agric Biotechnol 10: 130e40. https://doi.org/10.1016/j.bcab.2017.02.013.

Karadeniz, D. G., B. Kaskatepe, M. E. Kiymaci, K. C. Tok, M. Gumustas and C. Karaaslan. 2021. Microbial exopolysaccharide production of *Streptococcus thermophilus* and its antiquorum sensing activity. Arch Microbiol 203: 3331–3339. https://doi.org/10.1007/s00203-021-02313-7.

He, W., X. Wang, T. Hang, J. Chen, Z. Wang, D. A. Mosselhy et al. 2023. Fabrication of Cu^{2+}-loaded phase transited lysozyme nanofilm on bacterial cellulose: Antibacterial, anti-inflammatory, and pro-angiogenesis for bacteria-infected wound healing. Carbohydr. Polym. 309: 120681. https://doi.org/10.1016/j.carbpol.2023.120681.

Hou, C., L. Chen, L. Yang and X. Ji. 2020. An insight into anti-inflammatory effects of natural polysaccharides. Int J Biol Macromol. 153: 248–255. https://doi.org/10.1016/j.ijbiomac.2020.02.315.

Hundschell, C. S., F. Jakob and A. M. Wagemans. 2020. Molecular weight dependent structure of the exopolysaccharide Levan. Int J Biol Macromol. 161: 398–405. https://doi.org/10.1016/j.ijbiomac.2020.06.019.

Hussain, A., K. M. Zia, S. Tabasum, A. Noreen, M. Ali, R. Iqbal et al. 2017. Blends and composites of exopolysaccharides; properties and applications: A Review. Int. J. Biol. Macromol. 94: 10–27. https://doi.org/10.1016/j.ijbiomac.2016.09.104.

Hwang, Y., J. S. Lee, H. An, H. Oh, D. Sung, G. Tae et al. 2021. Hydroxyapatite-embedded levan composite hydrogel as an injectable dermal filler for considerable enhancement of biological efficacy. J. Ind. Eng. Chem. 104: 491–499. https://doi.org/10.1016/j.jiec.2021.08.040.

Iconaru, S. L., M. V. Predoi, M. Motelica-Heino, D. Predoi, N. Buton, C. Megier et al. 2020. Dextran-thyme magnesium-doped hydroxyapatite composite antimicrobial coatings. Coatings 10: 57. https://doi.org/10.3390/coatings10010057.

Idumah, C. I., A. C. Ezika and V. U. Okpechi. 2021. Emerging trends in polymer aerogel nanoarchitectures, surfaces, interfaces and applications. Surf. Interfaces 25: 101258. https://doi.org/10.1016/j.surfin.2021.101258.

Jin, M., Y. Wang, M. Huang, Z. Lu and Y. Wang. 2014. Sulphation can enhance the antioxidant activity of polysaccharides produced by *Enterobacter cloacae* Z0206. Carbohyd Polym. 99: 624–629. https://doi.org/10.1016/j.carbpol.2013.08.072.

Jusoh, N., N. Othman and M. Bukhari Rosly. 2021. Extraction and recovery of organic compounds from aqueous solution using emulsion liquid membrane process. Materials Today: Proceedings 47: 1301–1306. https://doi.org/10.1016/j. matpr.2021.02.800.

Kaur, N. and P. Dey. 2023. Bacterial exopolysaccharides as emerging bioactive macromolecules: from fundamentals to applications. Research in Microbiology 174(4): 104024. https://doi.org/10.1016/j.resmic.2022.104024.

Kenawy, E. R. S., E. A. Kamoun, M. S. Eldin, H. M. A. Soliman, S. H. El-Moslamy, E. M. El-Fakharany et al. 2023. Electrospun PVA–Dextran nanofibrous scaffolds for acceleration of topical wound healing: Nanofiber optimization, characterization and *in vitro* assessment. Arab. J. Sci. Eng. 48: 205–222. https://doi.org/10.1007/s13369-022-06856-9.

Khare, P., M. Chogale, P. Kakade and V. Patravale. 2022. Gellan gum-based in situ gelling ophthalmic nanosuspension of Posaconazole. Drug Deliv. Transl. Res. 12: 2920–2935. https://doi.org/10.1007/s13346-022-01155-0.

Kicková, E., S. Salmaso, F. Mastrotto, P. Caliceti and A. Urtti. 2021. Pullulan based bioconjugates for ocular dexamethasone delivery. Pharmaceutics 13: 791. https://doi.org/10.3390/pharmaceutics13060791.

Kim, J., J. P. Jeong, Y. Kim and S. Jung. 2024. Physicochemical and rheological properties of Succinoglycan overproduced by *Sinorhizobium meliloti* 1021 Mutant. Polymers 16(2): 244. https://doi.org/10.3390/polym16020244.

Kim, S., G. Y. Jeon, S. E. Kim, S. H. Choe, S. J. Kim, J. S. Seo et al. 2022. Injectable hydrogel based on gellan gum/silk sericin for application as a retinal pigment epithelium cell carrier. ACS Omega 7: 41331–41340. https://doi.org/10.1021/ acsomega.2c05113.

Knirel, Y. A. and J. J. Kenyon. 2021.Biosynthesis of bacterial polysaccharides. pp. 1–28. *In*: Oliveira, J. M., H. Radhouani and R. Reis (eds.). Polysaccharides of Microbial Origin: Biomedical Applications. Cham Springer International Publishing. https://doi.org/10.1007/978-3-030-35734-4_28-1.

Kumar, A., R. K. Mishra, K. Verma, S. M. Aldosari, C. K. Maity, S. Verma et al. 2023. A comprehensive review of various biopolymer composites and their applications: From biocompatibility to self-healing. Materials Today Sustainability 23: 100431. https://doi.org/10.1016/j.mtsust.2023.100431.

Kumar, A., K. M. Rao and S. S. Han. 2017. Development of sodium alginate-xanthan gum based nanocomposite scaffolds reinforced with cellulose nanocrystals and halloysite nanotubes. Polym. Test. 63: 214–225. https://doi.org/10.1016/j.polymertesting.2017.08.030.

Lalebeigi, F., A. Alimohamadi, S. Afarin, H. A. Aliabadi, M. Mahdavi, F. Farahbakhshpour et al. 2024. Recent advances on biomedical applications of gellan gum: A review. Carbohydrate Polymers 15(334): 122008. https://doi.org/10.1016/j.carbpol.2024.122008.

Laubach, J., M. Joseph, T. Brenza, V. Gadhamshetty and R. K. Sani. 2021. Exopolysaccharide and biopolymer-derived films as tools for transdermal drug delivery. Journal of Controlled Release 329: 971–87. https://doi.org/10.1016/j.jconrel.2020.10.027.

Li, S. and N. P. Shah. 2014. Antioxidant and antibacterial activities of sulphated poly-saccharides from *Pleurotus eryngii* and *Streptococcus thermophilus* ASCC 1275. Food Chem. 165: 262–270. https://doi.org/10.1016/j.foodchem.2014.05.110.

Li, W., X. Jian, Y. Zou, L. Wu, H. Huang, H. Li et al. 2021. The fabrication of a gellan gum-based hydrogel loaded with magnesium ions for the synergistic promotion of skin wound healing. Front. BioEng. Biotechnol. 9: 709679. https://doi.org/10.3389 /fbioe.2021.709679.

Lin, G. D., P. N. Smith and R. W. Li. 2024. Biomedical property of gellan gum. pp. 81–93. *In*: Nayak, A. K. and M. S. Hasnain (eds.). Application of Gellan Gum as a Biomedical Polymer. Academic Press. https://doi.org/10.1016/B978-0-323-91815-2.00006-5.

Lin, C. C. and J. Y. Chiu. 2021. A novel γ-PGA composite gellan membrane containing glycerol for guided bone regeneration. Mater. Sci. Eng. C 118: 111404. https://doi.org/10.1016/j.msec.2020.111404.

Luanda, A. and V. Badalamoole. 2023. Past, present and future of biomedical applications of dextran-based hydrogels: A Review. International Journal of Biological Macromolecules 228: 794–807. https://doi.org/10.1016/j.ijbiomac.2022.12.129.

Ma, N., D. Y. Cheung and J. T. Butcher. 2022. Incorporating nanocrystalline cellulose into a multifunctional hydrogel for heart valve tissue engineering applications. J. Biomed. Mater. Res. Part A. 110: 76–91. https://doi.org/10.1002/jbm.a.37267.

Majee, S. B., D. Avlani and G. R. Biswas. 2017. Rheological behavior and pharmaceutical applications of bacterial exopolysaccharides. J Appl Pharmaceut Sci. 7: 224e32. https://doi.org/10.7324/JAPS.2017.70931.

Malik, N. S., M. Ahmad, M. U. Minhas, R. Tulain, K. Barkat, I. Khalid et al. 2020. Chitosan/xanthan gum based hydrogels as potential carrier for an antiviral drug: fabrication, characterization, and safety evaluation. Front. Chem. 8: 50. https://doi.org/10.3389/fchem.2020.00050.

Manan, S., M. Wajid Ullah, M. Ul-Islam, Z. Shi, M. Gauthier and G. Yang. 2022. Bacterial cellulose: Molecular regulation of biosynthesis, supramolecular assembly, and tailored structural and functional properties. Prog. Mater. Sci. 129: 100972. https://doi.org/10.1016/j.pmatsci.2022.100972.

Miao, K. H. and K. B. Guthmiller. 2022. Dextran; StatPearls Publishing: Treasure Island, FL, USA.

Mommer, S., D. Gehlen, T. Akagi, M. Akashi, H. Keul and M. Moller. 2021. Thiolactone-functional pullulan for *in situ* forming biogels. Biomacromolecules 22: 4262–4273. https://doi.org/10.1021/acs.biomac.1c00807.

Moradali, M. F. and B. H. Rehm. 2020. Bacterial biopolymers: from pathogenesis to advanced materials. Nature Reviews Microbiology 18(4): 195–210. https://doi.org/10.1038/s41579-019-0313-3.

Moradi, Z. and N. Kalanpour. 2019. Kefiran, a branched polysaccharide: Preparation, properties and applications: A review. Carbohydrate Polymers 223: 115100. https://doi.org/10.1016/j.carbpol.2019.115100.

Netrusov, A. I., E. V. Liyaskina, I. V. Kurgaeva, A. U. Liyaskina, G. Yang and V. V. Revin. 2023. Exopolysaccharides producing bacteria: a review. Microorganisms 11(6): 1541. https://doi.org/10.3390/microorganisms11061541.

Neves, J. G., D. Navarro da Rocha, C. C. Lopes, R. M. Barbosa, L. F. Ferreira, C. B. Westin et al. 2022. Calcium phosphates Chitosan-Xanthan composite scaffolds associated with mesenchymal stem cells for regenerative dentistry application. Ceram. Int. 48: 23088–23095. https://doi.org/10.1016/j.ceramint.2022.04.288.

Nikolova, M. P. and M. S. Chavali. 2019. Recent advances in biomaterials for 3D scaffolds: A review. Bioact. Mater. 4: 271–292. https://doi.org/10.1016/j.bioactmat.2019.10.005.

Oerlemans, M. M. P., R. Akkerman, M. Ferrari, M. T. C. Walvoort and P. de Vos. 2021. Benefits of bacteria-derived exopolysaccharides on gastrointestinal microbiota, immunity and health. J Funct Food. 76: 104289. https://doi.org/10.1016/j.jff.2020.104289.

Ojagh, S. M. A., F. Vahabzadeh and A. Karimi. 2021. Synthesis and characterization of bacterial cellulose-based composites for drug delivery. Carbohydr. Polym. 273: 118587. https://doi.org/10.1016/j.carbpol.2021.118587.

Orlando, I., P. Basnett, R. Nigmatullin, W. Wang, J. C. Knowles and I. Roy. 2020. Chemical modification of bacterial cellulose for the development of an antibacterial wound dressing. Front. Bioeng. Biotechnol. 8: 557885. https://doi.org/10.3389/fbioe.2020.557885.

Osemwegie, O. O., C. O. Adetunji, E. A. Ayeni, O. I. Adejobi, R. O. Arise, C. O. Nwonuma et al. 2020. Exopolysaccharides from bacteria and fungi: Current status and perspectives in Africa. Heliyon 6: e04205. https://doi.org/10.1016/j.heliyon.2020. e04205.

Osman, A., E. Lin and D. S. Hwang. 2023. A sticky carbohydrate meets a mussel adhesive: Catechol-conjugated Levan for hemostatic and wound healing applications. Carbohydr. Polym. 299: 120172. https://doi.org/10.1016/j.carbpol.2022.120172.

Panchal, R., K. Prajapati, M. Prajapati, S. Sharma and M. S. Saraf. 2022. Bacterial exopolysaccharides: Types, its biosynthesis and their application in different fields. Acta Sci Biotechnol. 3(2): 3–11.

Patel, J., B. Maji, N. S. H. Narayana Moorthy and S. Maiti. 2020. Xanthan gum derivatives: Review of synthesis, properties and diverse applications. RSC Adv. 10: 27103–27136. https://doi.org/10.1039/D0RA04366D.

Pereira, S. B., A. Sousa, M. Santos, M. Araújo, F. Serôdio, P. Granja et al. 2019. Strategies to obtain designer polymers based on cyanobacterial extracellular polymeric substances (EPS). Int. J. Mol. Sci. 20: 5693. https://doi.org/10.3390/ijms20225693.

Puertas-Bartolome, M., A. Mora-Boza and L. García-Fernandez. 2021. Emerging biofabrication techniques: a review on natural polymers for biomedical applications. Polymers 13(8): 1209. https://doi.org/10.3390/polym13081209.

Qi, Y., D. Wang, L. Fang, X. Liu, C. Liu, F. Zhao et al. 2022. Hypoglycemic effect of exopolysaccharide from *Lactiplantibacillus plantarum* JLAU103 on streptozotocin and high-fat diet-induced type 2 diabetic mice. Foods. 11: 3571. https://doi.org/10.3390/foods 11223571.

Rabiya, R. and R. Sen. 2022. Artificial intelligence driven advanced optimization strategy vis-a-vis ` response surface optimization of production medium: Bacterial exopolysaccharide production as a case-study. Biochemical Engineering Journal 178: 108271. https://doi.org/10.1016/j.bej.2021.108271.

Rana, S. and L. S. Upadhyay. 2020. Microbial exopolysaccharides: Synthesis pathways, types and their commercial applications. International Journal of Biological Macromolecules 157: 577–583. https://doi.org/10.1016/j.ijbiomac.2020.04.084.

Reczynska-Kolman, K., K. Hartman, K. Kwiecien, M. Brzychczy-Włoch and E. Pamuła. 2022. Composites based on gellan gum, alginate and nisin-enriched lipid nanoparticles for the treatment of infected wounds. Int. J. Mol. Sci. 23: 321. https://doi.org/10.3390/ijms23010321.

Reddy, M. S. B., D. Ponnamma, R. Choudhary and K. K. Sadasivuni. 2021. A comparative review of natural and synthetic biopolymer composite scaffolds. Polymers 13(7): 1105. https://doi.org/10.3390/polym13071105.

Ren, Y. Y., P. P. Sun, Y. P. Ji, X. T. Wang, S. H. Dai and Z. Y. Zhu. 2019. Carboxymethylation and acetylation of the polysaccharide from Cordyceps militaris and their α-glucosidase inhibitory activities. Nat Prod Res. 34(3): 369–77. https://doi.org/ 10.1080/14786419.2018.1533830.

Roy, S., S. Biswas, S. Chakraborty, S. K. Das and R. Mondal. 2024. Biomedical applications of pullulan. Biopolymers for Biomedical Applications 25: 253–85. https://doi.org/10.1002/9781119865452.ch10.

Roychoudhury, A., S. Pahari and S. Banerjee. 2024. Biopolymer production by natural and engineered microbes. IJRES 12(3): 438–459.

Rush, J. S., S. Zamakhaeva, N. R. Murner, P. Deng, A. J. Morris, C. W. Kenner et al. 2024. Structure and mechanism of biosynthesis of *Streptococcus mutan*s cell wall polysaccharide. BioRxiv. 9: 2024.05.09.593426. https://doi.org/10.1101/2024.05.09. 593426.

Sakai, S., H. Ohi and M. Taya. 2019. Gelatin/hyaluronic acid content in hydrogels obtained through blue light-induced gelation affects hydrogel properties and adipose stem cell behaviors. Biomolecules 9: 342. https://doi.org/10.3390/biom9080342.

Santos, M. V. B., A. L. Oliveira, J. A. Osajima and E. C. Silva-Filho. 2020. Development of composites scaffolds with calcium and cerium-hydroxyapatite and gellan gum. Ceramics International 46(3): 3811–3817. https://doi.org/10.1016/j. ceramint.2019.10.104.

Santra, H. K. and D. Banerjee. 2021. Microbial exopolysaccharides: structure and therapeutic properties. pp. 375–420. *In*: Vaishnav, A. and D. K. Choudhary (eds.). Microbial Polymers: Applications and Ecological Perspectives. Springer, Singapore. https://doi.org/10.1007/978-981-16-0045-6_17.

Schilling, J. and J. Schmid. 2024. Comprehensive rheological analysis of structurally related acetan-like heteroexopolysaccharides from two *Kozakia baliensis* strains in surfactants and galactomannan blends. New Biotechnology 82: 75–84. https://doi.org/10.1016/j.nbt.2024.05.003.

Schmid, J. 2018. Recent insights in microbial exopolysaccharide biosynthesis and engineering strategies. Current Opinion in Biotechnology 53: 130–6. https://doi.org/10.1016/j.copbio. 2018.01.005.

Schropp, N., V. Stanislas, K. B. Michels and K. Thriene. 2023. How do prebiotics affect human intestinal bacteria?—assessment of bacterial growth with inulin and XOS *in vitro*. International Journal of Molecular Sciences 24(16): 12796. https://doi.org/10.3390/ ijms241612796.

Secor, P. R., L. A. Michaels, D. C. Bublitz, L. K. Jennings and P. K. Singh. 2022. The depletion mechanism actuates bacterial aggregation by exopolysaccharides and determines species distribution and composition in bacterial aggregates. Front. Cell Infect. Microbiol. 12: 869736. https://doi.org/10.3389/fcimb.2022.869736.

Sengupta, D., S. Datta and D. Biswas. 2018. Towards a better production of bacterial exopolysaccharides by controlling genetic as well as physico-chemical parameters. Applied Microbiology and Biotechnology 102: 1587–1598. https://doi.org/10.1007/s00253-018-8745-7.

Sharma, S. and K. Jagadeesh Chandra Bose. 2024. Biomedical applications of dextran. Biopolymers for Biomed. App. 25: 219–251. https://doi.org/10.1002/9781119865452.ch9.

Shawan, M. M. A. K., N. Islam, S. Aziz, N. Khatun, S. R. Sarker, M. Hossain et al. 2019. Fabrication of xanthan gum: gelatin (xnt:gel) hybrid composite hydrogels for evaluating skin wound healing efficacy. Mod. Appl. Sci. 13: 101–111. https://doi.org/10.5539/ mas.v13n3p101.

Shukla, A., K. Mehta, J. Parmar, J. Pandya and M. Saraf. 2019. Depicting the exemplary knowledge of microbial exopolysaccharides in a nutshell. Eur Polym J. 119: 298e310. https://doi.org/10.1016/j.eurpolymj.2019.07.044.

Sieni, E., B. Bazzolo, F. Pieretti, A. Zamuner, A. Tasso, M. Dettin et al. 2020. Breast cancer cells grown on hyaluronic acid-based scaffolds as 3D *in vitro* model for electroporation. Bioelectrochemistry 136: 107626. https://doi.org/10.1016/j.bioelechem. 2020.107626.

Singh, S., O. F. Nwabor, D. M. Sukri, S. Wunnoo, K. Dumjun, S. Lethongkam et al. 2022. Poly (vinyl alcohol) copolymerized with xanthan gum/hypromellose/sodium carboxymethyl cellulose dermal dressings functionalized with biogenic nanostructured materials for antibacterial and wound healing application. Int. J. Biol. Macromol. 216: 235–250. https://doi.org/10.1016/ j.ijbiomac.2022.06.172.

Stellavato, A., A. Dabous, S. D'Ambrosio, D. Cimini and C. Schiraldi. 2022. Anti-inflammatory effect of exopolysaccharides from *Lactobacillus brevis* on co-culture models of macrophage-Like and enterocyte. FEBS Open Bio. 12: 186.

Suflet, D. M., I. Popescu, M. C. Stanciu and C. M. Rimbu. 2024. Antimicrobial hydrogels based on cationic curdlan derivatives for biomedical applications. Gels 10(7): 424. https://doi.org/10.3390/gels10070424.

Suflet, D. M., I. Popescu, A. I. Prisacaru and I. M. Pelin. 2021. Synthesis and characterization of curdlan–phosphorylated curdlan based hydrogels for drug release. Int. J. Polym. Mater. Polym. Biomater. 70: 870–879. https://doi.org/10.1080/00914037. 2020.1765360.

Sun, H., X. Yu, T. Li and Z. Zhu. 2021. Structure and hypoglycemic activity of a novel exopolysaccharide of *Cordyceps militaris*. Int J Biol Macromol. 166: 496–508. https://doi.org/10.1016/j.ijbiomac.2020.10.207.

Sun, X. and J. Zhang. 2021. Bacterial exopolysaccharides: Chemical structures, gene clusters and genetic engineering. International Journal of Biological Macromolecules 173: 481–90. https://doi.org/10.1016/j.ijbiomac.2021.01.139.

Juraskova. D., S. C. Ribeiro and C. C. Silva. 2022. Exopolysaccharides produced by lactic acid bacteria: from biosynthesis to health-promoting properties. Foods 11(2): 156. https://doi.org/10.3390/foods11020156.

Tabernero, A. and S. Cardea. 2020. Microbial exopolysaccharides as drug carriers. Polymers 12(9): 2142. https://doi.org/10.3390/polym12092142.

Toullec, C., J. LeBideau, V. Geoffroy, B. Halgand, N. Buchtova, R. Molina-Peña et al. 2021. Curdlan–chitosan electrospun fibers as potential scaffolds for bone regeneration. Polymers 13: 526. https://doi.org/10.3390/polym13040526.

Trcek, J., I. Dogsa, T. Accetto and D. Stopar. Acetan and acetan-like polysaccharides: genetics, biosynthesis, structure, and viscoelasticity. Polymers (Basel) 13(5): 815. https://doi.org/10.3390/polym13050815.

Ul-Islam, M., W. Alhajaim, A. Fatima, S. Yasir, T. Kamal, Y. Abbas et al. 2023. Development of low-cost bacterial cellulose-pomegranate peel extract-based antibacterial composite for potential biomedical applications. Int. J. Biol. Macromol. 231: 123269. https://doi.org/10.1016/j.ijbiomac.2023.123269.

Ullah, M. W., M. Ul-Islam, T. Khan and J. K. Park. 2021. Recent developments in the synthesis, properties, and applications of various microbial polysaccharides. pp. 975–1015. *In*: Phillips, G. O. and P. A. Williams (eds.). Handbook of Hydrocolloids. Woodhead Publishing. https://doi.org/10.1016/B978-0-12-820104-6.00032-2.

Unnikrishnan, K., L. V. Thomas and R. M. Ram Kumar. 2021. Advancement of scaffold based 3D cellular models in cancer tissue engineering: An update. Frontiers in Oncology 11: 733652. https://www.frontiersin.org/articles/10.3389/fonc.2021.733652.

Velasco-Rodriguez, B., T. Diaz-Vidal, L. C. Rosales-Rivera, C. A. García-González, C. Alvarez-Lorenzo, A. Al-Modlej et al. 2021. Hybrid methacrylated gelatin and hyaluronic acid hydrogel scaffolds. Preparation and Systematic Characterization for Prospective Tissue Engineering Applications. Int. J. Mol. Sci. 22: 6758. https://doi.org/10.3390/ijms22136758.

Venkatachalam, G., S. Arumugam and M. Doble. 2021. Industrial production and applications of α/β linear and branched glucans. Indian Chemical Engineer. 63(5): 533–47. https://doi.org/10.1080/00194506.2020.1798820.

Wang, J., S. Liu, J. Huang, K. Ren, Y. Zhu and S. Yang. 2023. Alginate: Microbial production, functionalization, and biomedical applications. International Journal of Biological Macromolecules 242: 125048. https://doi.org/10.1016/j.ijbiomac.2023.125048.

Wang, W., J. Yuhao, L. Nan, S. Shengbo and H. Lujiang. 2023. Structural characteristics of microbial exopolysaccharides in association with their biological activities: A review. Chem. Biol. Technol. Agric. 10: 137. https://doi.org/10.1186/s40538-023-00515-3.

Wang, Y., T. Zhang, L. Zhu, R. Li, Y. Jiang, Z. Li et al. 2024. Optimization of welan gum extraction and purification using lysozyme and alkaline protease. Appl Microbiol Biotechnol 108: 70. https://doi.org/10.1007/s00253-023-12880-y.

Wang, H., H. Xia, Z. Xu, T. Natsuki and Q. Q. Ni. 2023. Effect of surface structure on the antithrombogenicity performance of poly(-caprolactone)-cellulose acetate small-diameter tubular scaffolds. Int. J. Biol. Macromol. 226: 132–142. https://doi.org/10.1016/j.ijbiomac.2022.11.315.

Wang, Q. Q., Y. Liu, C. J. Zhang, C. Zhang and P. Zhu. 2019a. Alginate/gelatin blended hydrogel fibers cross-linked by Ca^{2+} and oxidized starch: Preparation and properties. Mater. Sci. Eng. C. 99: 1469–1476. https://doi.org/10.1016/j.msec.2019.02.091.

Wang, Y., Z. Guo, Y. Qian, Z. Zhang, L. Lyu, Y. Wang et al. 2019b. Study on the electrospinning of gelatin/pullulan composite nanofibers. Polymers 11: 1424. https://doi.org/10.3390/polym11091424.

Waoo, A. A., S. Singh, A. Pandey, G. Kant, K. Choure, K. K. T. Amesho et al. 2023. Microbial exopolysaccharides in the biomedical and pharmaceutical industries. Heliyon. 9(8): e18613. https://doi.org/10.1016/j.heliyon.2023.e18613.

Wei, Q., N. N. Deng, J. Guo and J. Deng. 2018. Synthetic polymers for biomedical applications. International Journal of Biomaterials e7158621. https:// doi.org/10.1155/2018/7158621.

Whitfield, C., S. S. Wear and C. Sande. 2020. Assembly of bacterial capsular polysaccharides and exopolysaccharides. Annual Review of Microbiology 74(1): 521–43. https://doi.org/10.1146/annurev-micro-011420-075607.

Wu, J. Y., K. C. Siu and P. Geng. 2021. Bioactive ingredients and medicinal values of *Grifola frondosa* (Maitake). Foods. 10: 95. https://doi.org/10.3390/foods10010095.

Xiao, M., X. Ren, Y. Yu, W. Gao, C. Zhu, H. Sun et al. 2022. Fucose-containing bacterial exopolysaccharides: Sources, biological activities, and food applications. Food Chem. X 13: 100233. https://doi.org/10.1016/j.fochx.2022.100233.

Xing, J., X. Peng, A. Li, M. Chen, Y. Ding, X. Xu et al. 2021. Gellan gum/alginate-based Ca-enriched acellular bilayer hydrogel with robust interface bonding for effective osteochondral repair. Carbohydr. Polym. 270: 118382. https://doi.org/10.1016/j.carbpol.2021.118382.

Xiong, Z. Q., Y. Z. Fan, X. Song, X. X. Liu, Y. J. Xia and L. Z. Ai. 2020. The second messenger c-di-AMP mediates bacterial exopolysaccharide biosynthesis: A review. Molecular Biology Reports 47: 9149–9157. https://doi.org/10.1007/s11033-020-05930-5.

Xu, M., Z. Li, X. Zhao and W. Li. 2022. Prebiotic properties of exopolysaccharides from Lactobacillus helveticus LZ-R-5 and L. pentosus LZ-R-17 evaluated by *in vitro* simulated digestion and fermentation. Foods 11: 2501. https://doi.org/10.3390/foods11162501.

Xu, W., J. Qian, Y. Zhang, A. Suo, N. Cui, J. Wang et al. 2016. A double-network poly(Nε-acryloyl L-lysine)/hyaluronic acid hydrogel as a mimic of the breast tumor microenvironment. Acta Biomater. 33: 131–141. https://doi.org/10.1016/j.actbio.2016 .01.027.

Yuan, H., W. Zhong, R. Wang, P. Zhou, Y. Nie, W. Hu et al. 2020. Preparation of cholesteryl-modified aminated pullulan nanoparticles to evaluate nanoparticle of hydrophobic degree on drug release and cytotoxicity. J. Nanomater. e717120. https://doi.org/10.1155/2020/7171209.

Yucel, N., H. Ispirli, E. Mercan, U. Erdogan and E. Dertli. 2021. Synthesis of alternan-stabilized zinc nanoparticles: morphological, thermal, antioxidant and antimicrobial characterization. Preparative Biochemistry and Biotechnology 51(4): 331–9. https://doi.org/10.1080/ 10826068.2020.1815059.

Zaghloul, E. H. and M. I. A. Ibrahim. 2022. Production and characterization of exopolysaccharide from newly isolated Marine Probiotic *Lactiplantibacillus plantarum* EI6 With *in vitro* wound healing activity. Frontiers in Micro. 13: 903363. https://www.frontiersin.org/articles/10.3389/ fmicb.2022.903363.

Zanotti, A., L. Baldino and E. Reverchon. 2023. Production of exopolysaccharide-based porous structures for biomedical applications: a review. Nanomaterials 13: 2920. https://doi.org/10.3390/ nano13222292.

Zayed, A., M. K. Mansour, M. S. Sedeek, M. H. Habib, R. Ulber and M. A. Farag. 2022. Rediscovering bacterial exopolysaccharides of terrestrial and marine origins: novel insights on their distribution, biosynthesis, biotechnological production, and future perspectives. Critical Reviews in Biotechnology 42(4): 597–617. https://doi.org/10.1080/07388551 .2021.1942779.

Zhang, M., N. Su, Q. Huang, Q. Zhang, Y. Wang, J. Li et al. 2017. Phosphorylation and antiaging activity of polysaccharide from *Trichosanthes* peel. J Food Drug Anal. 25: 976–83. https://doi.org/10.1016/j. jfda.2016.12.013.

Zhang, M., Y. Huang, W. Pan, X. Tong, Q. Zeng, T. Su et al. 2021. Polydopamine-incorporated dextran hydrogel drug carrier with tailorable structure for wound healing. Carbohydr. Polym. 253: 117213. https://doi.org/10.1016/j.carbpol.2020.117213.

Zhang, X., Y. Pan, S. Li, L. Xing, S. Du, G. Yuan et al. 2020. Doubly crosslinked biodegradable hydrogels based on gellan gum and chitosan for drug delivery and wound dressing. Int. J. Biol. Macromol. 164: 2204–2214. https://doi.org/10.1016/j.ijbiomac. 2020.08.093.

Zheng, C., C. Liu, H. Chen, N. Wang, X. Liu, G. Sun et al. 2019. Effective wound dressing based on Poly (vinyl alcohol)/Dextranaldehyde composite hydrogel. Int. J. Biol. Macromol. 132: 1098–1105. https://doi.org/10.1016/j.ijbiomac.2019.04.038.

Zheng, Y., Y. Liang, D. Zhang, X. Sun, L. Liang, J. Li et al. 2018. Gelatin-based hydrogels blended with gellan as an injectable wound dressing. ACS Omega 3: 4766–4775. https://doi.org/10.1021/ acsomega.8b00308.

Zhu, Q., X. Chen, Z. Liu, Z. Li, D. Li, H. Yan et al. 2023. Development of alginate-chitosan composite scaffold incorporation of bacterial cellulose for bone tissue engineering. Int. J. Polym. Mater. Polym. Biomater. 72: 296–307. https://doi.org/10.1080/00914037.2021. 2007384.

Zia, I., R. Jolly, S. Mirza, M. S. Umar, M. Owais and M. Shakir. 2020. Hydroxyapatite nanoparticles fortified xanthan gum–chitosan based polyelectrolyte complex scaffolds for supporting the osteo-friendly environment. ACS Appl. Bio. Mater. 3: 7133–7146. https://doi.org/10.1021/acsabm.0c00948.

Zou, Y., W. Xue, X. Lin, T. Hu, S. W. Liu, C. H. Sun et al. 2021. Taxonomic description and genome sequence of *Christensenella intestinihominis* sp. Nov., a novel cholesterol-lowering bacterium isolated from human gut. Front Microbiol. 12: 632361. https://doi.org/10.3389/fmicb.2021.632361.

CHAPTER 10

Polyhydroxyalkanoates Based Biomaterials in Tissue Repair and Regeneration

Sunita Jain

1. Introduction

Petroleum-based plastics put environment and health at risk due to their low degradation rate, the emergence of carcinogenic substances when burned, and the formation of toxic by-products during its processing (Roy et al., 2021; Evode et al., 2021). To reduce the harmful effects of synthetic plastics, eco-friendly bio-polymers are needed (Naser et al., 2021). These biopolymers offer many beneficial properties compared to synthetic polymers. They can be derived from agricultural products, such as starch; produced through microbial activities such as polyhydroxyalkanoates (PHA) or synthesized through biotechnological methods such as polylactic acid (PLA). Among these, PHAs present a sustainable alternative to non-degradable plastics due to their comparable physicochemical properties, as well as their biodegradability, biocompatibility, eco-friendly production processes, non-carcinogenicity, renewable in nature and desirable mechanical properties. Consequently, PHAs have gained increasing importance globally (Zhou et al., 2023; Sridharan et al., 2022) and have wide-ranging applications across various sections including packaging materials, biomedical, agriculture, disposable items, bio-fuels, etc. contributing to a circular economy (Shahid et al., 2021; Yousefi and Wnek, 2024). In addition to their excellent biodegradibility, the low oxygen and CO_2 permeability of PHA make them suitable biomaterial for manufacturing packing films and other valuable substitutes (Roohi et al., 2024; Naser et al., 2021b; Bairwan et al., 2024). These linear polyesters

Department of Botany, Faculty of Science, Bhupal Nobles' University, Udaipur, Rajasthan, India.
Email: sunitajain130799@gmail.com

DOI: 10.1201/9781003518792-10

synthesized and stored by bacteria in the cytoplasm (Surendran et al., 2020) when essential nutrients such as P, N, S are lacking and there is a sufficient carbon source available (Kumar et al., 2020; Behera et al., 2022). According to the American National Institute of Health, a biomaterial is defined as "Any active substance or combination of active substances isolated naturally or synthetically and that could be used for the partial or complete replacement of tissue or organ to manage or improve quality of life". Farag (2023) categorized biomaterials into three generations.

In tissue repair and regeneration, the selection of a proper bio-material is critical as it serves as an extracellular matrix (ECM) that supports cell adhesion, multiplication, and differentiation. The selected bio-material should promote regeneration, have sufficient mechanical and other properties identical to the host tissue, and should degenerate at the implantation site at a rate comparable to the growth of new tissue (Li et al., 2023c). Several polymers have been reported as bio-materials for scaffold preparation (Khalid et al., 2020). Recently, PHA based biomaterials have gained great interest in the medical sector due to their excellent properties, which are not exhibited by synthetic polymers. Compared to conventional synthetic polymers, PHAs offer several major advantages, including their biodegradability, biocompatibility, and non-toxicity. Their biodegradable by-products such as monomers and oligomers, have medicinal value. Additionally, their large-scale production is cost-effective because they are produced by a variety of microorganisms and production process is relatively simple and efficient. The pH value during the degradation of PHAs makes them well-accepted by the host tissue and immune system. These unique properties make PHAs an ideal choice for medical devices and implants. Variation in structure and properties of members of PHA family depends on species from which they are extracted, their biosynthetic pathways, provided growth conditions and carbon rich substrates in their biosynthesis process and the polymer extraction method (Pryadko et al., 2021). PHA can be categorized into short chain, medium chain and large chain length on the basis of number of hydroxyalkanoate (HA) monomers present in the polymer chain. Tissue engineering is evolving as a novel strategy for repairing and regenerating tissue defects and restoring their functions. Short chain length PHAs are generally brittle and stiff as they have 55–80% crystallinity (Khatami et al., 2021), have slow degradation rate and are hydrophobic. These properties can make them less desirable for some biomedical applications. To address these drawbacks various methods have been proposed such as: (a) Developing PHA co-polymers (b) Blending with other biopolymers, and (c) Incorporating additives or fillers. Tissue- engineered scaffolds are the key materials and the fabrication methods commonly used for processing biomaterials into 3D scaffolds for tissue repair and regeneration are (i) conventional techniques and (ii) rapid prototyping techniques. Under conventional technique freeze-drying, solvent casting, thermal induced phase separation (TIPS), gas foaming and electrospinning are included whereas stereolithography (SLA), fused deposition molding (FDM), selective laser sintering (SLS) and 3D bioprinting are rapid prototyping technologies, where cytotoxic solvents are not required. Among 150 identified monomers of PHA, only a few such as PHB, PHV, P4HB, PHBHHX and P3HB4HB have been extensively used in tissue engineering and the biomedical field. They have been used in tissue repair, regenerative medicine, wound repairing, bone and cartilage scaffolds, medical devices like surgical sutures, cardiac patches, blood vessels, wound-healing patches, and implantable devices such as heart valve, stent and nerve guidance conduit (NGC) (Kalia et al., 2023; Pulingam et al., 2022; Gregory et al., 2022). For controlled liberation of bioactive compounds like growth hormones, anticancer agents, and antibiotics, PHA-based nanoparticles have also been used. However, challenges related to the PHA production cost and the functionalities of engineered scaffolds need to be solved.

2. PHA and PHA-Based Biomaterials

The development of sustainable biomaterial from natural resources is of great importance. The first generation biomaterials developed in 1970s, aimed to replace the defected or lost tissue. These bio-materials were bioinert and did not interact with the host biological system. In the 1990s, second-generation biomaterials were introduced, characterized by significant interactions with the host tissue system. Third generation biomaterials were emerged at the end of 2000s by integrating the beneficial effects of bioactive compounds with the advantages of bio-materials to stimulate cellular responses at the molecular level and promote tissue repair and regeneration. According to standard ENI3432, polymers obtained from renewable resources are categorized as biopolymers (Awasthi et al., 2022). These materials are not only accepted by the body but also exhibit functional activities (Liu et al., 2023b). They are compatible with living tissues after *in vivo* administration and stimulate the formation of new tissues. Natural polymers include proteins (myosin, collagen, actin, fibrinogen, gelatin, keratin, silk), polysaccharides (chitin, glycosaminoglycans, bacterial cellulose, amylose, dextran), and inorganic compounds such as hydroxyapatite, bioactive glasses (Schuhladen et al., 2020). Unlike other synthetic materials, their hydrophobicity and bioactive surface can be utilized to create various biological tools. The widespread application of PHAs across diverse fields are food packaging, biofuels, in protein purification, etc. A type of PHA first discovered in 1920 was PHB (Polyhydroxybutyrate). Since then, many types of PHAs have been explored for their medicinal value. Name and structure of some common PHAs are given in Table 1.

PHAs exhibit a wide range of physicochemical properties: glass transition temperatures from –52°C to 4°C, hydrophobocity, melting temperature from 60 to 177°C, stiffness, brittleness, crystallinity from 0% to 70%, tensile strength of 17–104 MPa and thermodegradation temperatures from 227–256°C. These variations in parameters can be adjusted or modified to manufacture desired biomaterial (Miu et al., 2022; Reyna-Urrutia et al., 2024). They have been gained a significant interest as biomaterials for applications in resorbable medical devices like surgical sutures, patches, meshes, implants and tissue engineering scaffolds. Compare to other polymers, PHA based devices are well-tolerated by surrounding tissues and the immune system (Zhang et al., 2018). Due to their diverse structural and mechanical properties, PHAs can be blended with natural polymers (e.g., chitosan, collagen, alginate) and synthetic polymers (e.g., PLA, PCL, PLGA) or combined with organic fillers (e.g., nanoparticles, glass) to achieve desirable biological properties and broaden their applications. PHAs have the capacity to mimic the extracellular matrix

Table 1. Name and structure of some common PHAs.

IUPAC Nomenclature (Name of the PHA)	**Abbreviation**	**n**	**R (Radical)**	**Structure of Polymer**
Poly(3-hydroxybutyrate)	P3HB	1	$-CH_3$	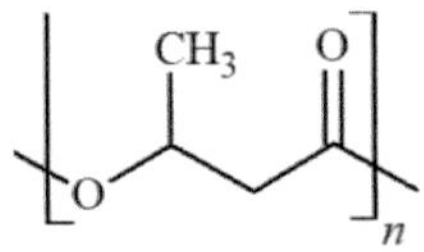
Poly(4-hydroxybutyrate)	P4HB	2	H	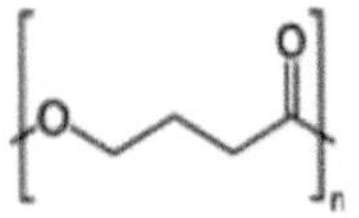
Poly(3-hydroxyval-erate)	P3HV	1	$-CH_2CH_3$	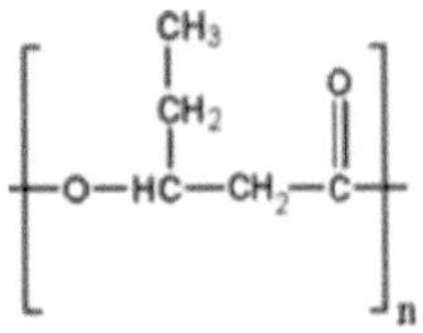
Poly(3-hydroxyhexanoate)	P3HH	1	$-(CH_2)_2CH_3$	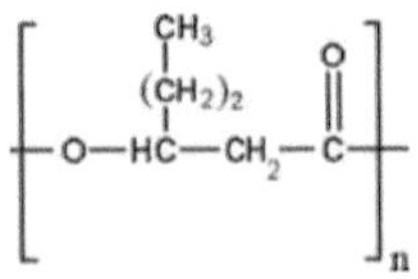
Poly(3-hydroxyoc-tanoate)	P3HO	1	$-(CH_2)_4CH_3$	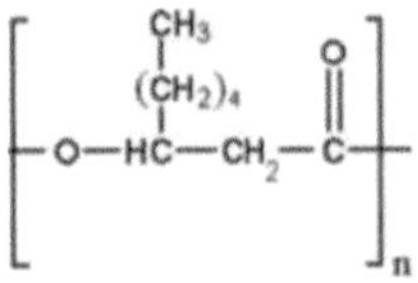
Poly(3-hydroxydecanoate)	P3HD	1	$-(CH_2)_6CH_3$	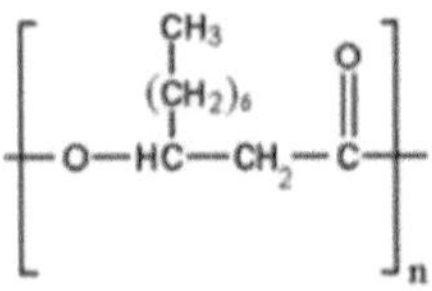

and support biological activities such as biodegradability, biocompatibility and other essential for better host-biomaterial interactions and the stimulation of various cellular processes in tissue repair and regeneration. Additionally their encapsulation capability and degradation kinetics enhance their biological activity making them suitable for the development of controlled drug delivery systems.

2.1 Structure and Properties

PHAs are a class of adaptable compounds with over 100 polymers studied to date. They are polyesters of 3-hydroxyalkanoic acid differing by the number of carbon atoms in the main chain or the radical (R). Structure and properties of PHAs depend mainly on the microbial species, bio-synthetic conditions and the substrate used in the production process. In addition, the physical properties of PHAs are closely related to the chemical structure and crystallization morphology (Wang et al., 2022b). To date various PHAs and their copolymers have been developed. Each HA monomer had a side chain attached to the main chain, which can be substituted by various functional groups, ranging from methyl (C1) to tridecyl (C13), or aromatic groups. This variation allows PHAs to be easily modified to synthesize polymers with desired properties (Liang et al., 2024).

2.2 PHA Categories

The PHA family exhibits a wide range of properties, from fragile to flexible, with melting points ranging from 40°C to 180°C, and crystallization up to 70%. Three categories of PHAs, based on the number of carbon atoms present in the monomer are given in Table 2.

1. **Short chain length PHA (scl-PHA):** These are coposed of monomers with 3 to 5 carbon atoms such as poly(3-hydroxy valerate) (PHV), poly(3-hydroxy butyrate) (PHB).
2. **Medium chain length PHA (mcl-PHA):** These contain 6–13 carbon atoms, such as poly(3-hydroxy hexanoate) (PHHx) and poly(3-hydroxy octanoate) (PHO).
3. **Long chain length PHA (lcl-PHA):** These have 14 or more carbon atoms, such as 3-hydroxyoctadecanoic acid (3HOD) and 3-hydroxy hexadecenoic acid (3HHD).

Scl-PHAs include homopolymers like P3HB, P4HB, P3HV and the co-polymer, PHBV. They are generally more stiff and brittle due to their crystalline structure (typically 55–80% crystallinity). In addition high melting temperature, and lower elongation at break, and high glass transition temperature (Tg) make them ideal scaffolding biomaterials in hard tissue repair and regeneration. In contrast, mcl-PHAs include homopolymers like PHHx and PHO and copolymers like PHH3HO. Their high elastomeric nature and low melting temperature make them a better choice for medical applications, particularly in soft tissue repair and regeneration (Asare et al., 2020; Kaniuk et al., 2021; Hahn et al., 2024). Among them, PHB and PHBV are most extensively used as biomaterials (Cesarino et al., 2024). PHB is best known and is extensively studied due to its highly desirable properties. Various PHA based homopolymers and copolymer, have been studied and compared (Yousefi and Wnek, 2024). Reyna-Urrutia and colleagues have detailed the advantages of PHB over other polymers such as its longer degradation time compared to PLA and PLGA, the fabrication methods for PHB and its derivatives, its applications in tissue engineering through studies *in vitro* and in animal models (Reyna-Urrutia et al., 2024). Continued research suggests a promising future for mcl-PHAs in expanding PHA applications (Azizi et al., 2024).

Table 2. PHA categories and their thermal and mechanical properties.

Carbon Source	Poly (3-hydroxyalkanoate) (PHA)			Thermal and Mechanical Properties		
	Type	**Side chain (R)**	**Name***	**Tg (°C)**	**Tm (°C)**	**Elongation (%)**
Sugar, glucose, acetic acid	scl-PHAs	Methyl, ethyl	PHB PHV	15	170	5
Alkanoic acids, alkanes and alkanols	mcl-PHAs	Propyl, butyl, pentyl hexyl, heptyl	PHHx PHHp PHO PHN PHD	–40	60	800
Plant oily acids	lcl-PHAs	More than 14 Carbon per repeating unit	3HOD 3HHD	–50	40	Soft, sticky

* B: butyrate, V: valerate, Hx: hexanoate, Hp: heptanoate, O: octanoate, N: nonanoate, D: decanoate, Tg: glass temperature, Tm: melting temperatures.

2.3 PHA Production

The PHA production process consists of several key stages: fermentation, recovery, and purification. Fermentation is again divided into two stages. Stage 1. Growth phase which occurs under balanced growth conditions in a bioreactor to produce a suitable amount of bacterial biomass. Stage 2. PHA accumulation phase which occurs under nutrient-limited growth conditions. The limitation of essential nutrients shifts the metabolic pathways from bacterial cell growth (biomass accumulation), toward PHA production by shrinking the TCA cycle, which is a competing pathway (Cavaille et al., 2013). Following fermentation, the recovery step involves techniques such as solvent extraction or digestion methods to break open the cells . After cell disruption, PHA is extracted, purified and processed for specific applications (Kessler et al., 2001). To reduce the production cost various affordable and sustainable substrates have been explored, including whey (Israni et al., 2020), lignin (Xu et al., 2021; Zhao et al., 2023; Li et al., 2023b), glycerol (de Meneses et al., 2020; Kim et al., 2023), waste cooking oil (Sangkharak et al., 2021; Elazzazy et al., 2024), waste polypropylene (Johnston et al., 2022), agricultural waste (Razzaq et al., 2024), food waste (Chandra et al., 2023), marine waste (Ranganadhareddy and Chandraskhar, 2021) and other carbon rich waste (Wang et al., 2023a). The use of hydrocarbon-degrading bacteria to produce PHAs is an excellent approach combine pollutants degradation and PHA production (Corti Monzon et al., 2024).

Three main methods are used to obtain PHAs: (i) Enzymatic catalysis (ii) Bacterial fermentation and (iii) Synthesis in genetically modified plants (Mochane et al., 2020; Li et al., 2020; Sharma et al., 2022). Among these bacterial fermentation is the most common method. *Pseudomonas* species are considered to be the best for PHA production using various carbon sources. Strategies to enhance PHA production include: Metabolic engineering to manipulate PHA synthesis pathways and down-regulating undesirable by-products (Zhang et al., 2020); Bacterial strain improvement through genetic engineering or random mutagenesis (adaptive evolution) (Gonzalez-Villanueva et al., 2019) and genetic engineering approach to modify the cell size and morphology (Kalia et al., 2024). Metabolic strategies to enhance the substrate utilization, eliminating competitive pathways and increase precursor availability are also employed to improve the production rate.

2.3.1 Microbial Cell Factories for PHA Production

Microorganisms act as PHAs biosynthetic machinery under both aerobic and anaerobic conditions (Ma et al., 2024). Various bacterial species, fungi and plants as well as diverse carbon sources, have been used to synthesize PHAs, making large-scale production cost-effective (Behera et al., 2022). Bacterial species such as *Alcaligenes latus, Baccilus subtilis, Cupriavadus necator* (formerly *Ralstonia eutropha*), *Pseudomonas mendocina* and *P. putida, Aeromonas hydrophila* are known for PHA production through bacterial fermentation of sugars and fatty acids under nutrient-limiting conditions (Kanbe et al., 2023). Highly studied species for PHAs production include, *B. megaterium, P. aeruginosa, P. oleovorans, P. stutzeri* and *Cupriavidus nector* (Muneer et al., 2020). PHAs can also be synthesized from various resources to minimize the production cost such as agro-food waste (Nath et al., 2024), industrial waste (Rana et al., 2023), household waste (Pratap and Krishnan, 2023), agro-waste (Surendra et al., 2022) and agricultural waste (Razzaq et al., 2024). They have been extracted from both gram positive as well as gram negative bacteria (Anjum et al., 2016) and archaea (Koller, 2021; Salgaonkar and Braganca, 2019; Tripathi et al., 2021). Some cynobacteria such as *Nostoc muscorium* (Bhati et al., 2010; Yogeswar et al., 2023) and extremophiles (Shakhmuradian et al., 2024) are known to produce PHB. However, genetic modifications are often needed to increase production levels. PHAs are stored within the prokaryotic cells as water insoluble granules ranging in size from 0.2 to 0.5 μm (Raza et al., 2018). Halophilic archaea capable of producing and accumulating PHAs have also been discovered (Nagarajan et al., 2021). For example the halophilic yeast strain, *Pichia* species TSLS24 shows promise for PHA synthesis due to its saline requirement, which helps prevent microbial contamination, high intracellular osmotic pressure that facilitates PHA recovery, and economic benefits from using low-cost substrates (Thu et al., 2023; Mitra et al., 2020). Genetic modifications in recombinant *E. coli* and *Pseudomonas* species can produce various homo- or copolymers of PHAs (Wang et al., 2017). The carbon-to-nitrogen (C/N) ratio and pH significantly influence PHA accumulation. They alter the PHA production as well as composition of the product. A low C/N ratio is favourable for cell growth, while higher PHA accumulation occurs under low nitrogen concentration (Rodriguez-Perez et al., 2018). Under nitrogen limitation microbial protein synthesis is inhibited leading to PHA being synthesized as the main product (Wen et al., 2010). A pH range of 5.5–9.0 is generally suitable for the growth of most of non-extremophilic bacteria (Padan et al., 2005), whereas a pH range of 7.5–8.5 favour PHA accumulation (Villano et al., 2010). Complex carbon sources like starch need to be converted into simpler forms, whereas glucose can be rapidly metabolized. An ideal PHA producer should have a high growth rate which directly correlates with carbon source consumption. Metabolic strategies, such as evolving recombinant strains using CRISPRi have been applied to enhance the substrate utilization. For instance, over-expressing *sul*A gene has resulted in a

100% increase in PHB accumulation (Wang et al., 2014). Utilizing carbon-rich waste enriched with highly biodegradable organic compounds is a promising approach to reducing production cost (Wang et al., 2023a).

2.3.2 Metabolic Pathways of PHA biosynthesis

PHA biosynthesis involves several metabolic pathways. Three major naturally occurring pathways prominently featuring the enzyme acetyl CoA (Ma et al., 2024; Khatami et al., 2021) are depicted in Figure 1:

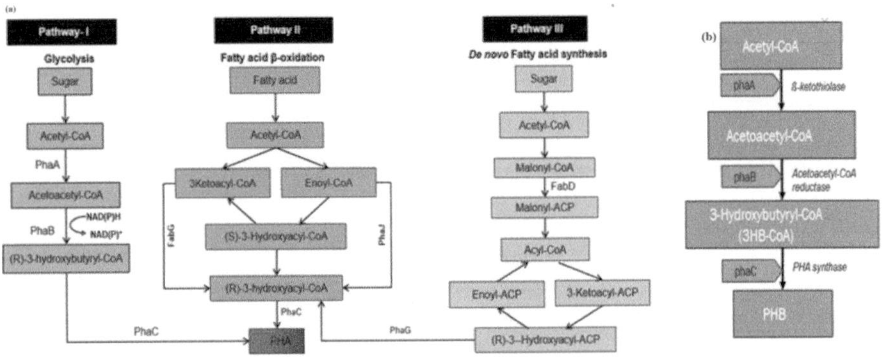

Figure 1. (a) Metabolic pathways involved in PHA biosynthesis (b) PHB biosynthetic pathway.

Main enzymes involved in PHA biosynthesis are-- PhaA: beta-ketothiolase; PhaB: Acetoacetyl-CoA reductase; PhaC: PHA synthase; PhaG: 3-Hydroxyacyl-CoA transferase; FabG: 3-ketoacyl carrier protein (ACP) reductase; PhaJ: (R)-enoyl-CoA hydratase: FabD: Malonyl-CoA-ACP transferase (Umesh et al., 2022).

Pathway I (Sugar based pathway): This is the most common pathway and utilizes sugar as substrate. Complete process includes—1. Acetoacetyl-coA formation: Two molecules of acetyl Co-A from the TCA cycle are condensed into a molecule of acetoacetyl-CoA with the help of enzyme beta-ketothiolase. 2. Reduction: Acetoacetyl-CoA is then converted into 3-hydroxybutyryl-CoA by the enzyme acetoacetyl-CoA reductase, which requires NADPH as a co-factor. 3. Polymerization: finally PHA synthase catalyses the formation of ester bond in 3-hydroxybutyryl-CoA leading to PHA production.

Pathway II (Fatty acids based pathway): This pathway uses fatty acids as the carbon source and involves—1. Fatty acid oxidation: fatty acids undergo beta-oxidation to generate different hydroxyalkanoate (HA) monomers. Key enzymes include (R)-enoyl-CoA hydratase, acyl-CoA oxidase and 3-ketoacyl-CoA reductase. 2. Polymerization: The enzyme PHA synthase catalyzes the polymerization of HA monomers into PHA.

Pathway III (Glucose and other carbon sources based pathway): Glucose and other carbon sources are converted into central intermediates, which are then transformed into HA-CoA thio-esters. These intermediates enable the synthesis of different types of PHAs (Miu et al., 2022). All the metabolic pathways have the same final step catalyzed by the enzyme PHA synthase. PHA production can be enhanced by different strategies such as—1. Enhancing Acetyl-CoA Availability: This can be achieved by overexpressing genes involved in its formation or by designing new pathways like the Synthetic Acetyl-CoA (SACA) pathway. SACA is a carbon conserving, ATP-independent pathway that provides acety CoA (Khatami et al., 2021). 2. PHA depolymerase inhibition: Blocking the enzyme, PHA depolymerase is a remarkable engineering strategy to increase PHA accumulation. This prevents the breakdown of PHA, allowing for greater accumulation within the cell (Li et al., 2020). 3. Inhibition of byproduct formation: By inhibiting pathways that produce byproducts, the carbon flux can be redirected towards PHA synthesis. For example, deleting genes related o beta-oxidation has been shown to promote the production of medium chain length (mcl) PHAs in *Psedomonas* strains (Zhao et al., 2020). Various homopolymers and heteropolymers of PHA can be formed due to the hydroxyl groups at 4, 5, and 6 positions of fatty acids (Ranganadharreddy and Chandrasekhar, 2021).

2.3.3 PHA Extraction Methods

Extraction process depends on the bacterial culture (single or mixed microbial culture) (Pagliano et al., 2021) and the physico-chemical properties of the polymer. Two main methods are: 1. Solvent Extraction Method: Most extensively used method for PHA recovery (Raza et al., 2018). It involves dissolving PHA granules using suitable solvents and then precipitating the polymer to obtain it in crystal form. Chlorinated solvents are commonly used for this purpose due to their effectiveness in dissolving PHAs. For mcl-PHAs extraction acetone is preferred. After dissolution, a precipitation agent, such as methanol or ethanol is added to the solution. This causes the PHAs to precipitate out of the solvent as crystals, which can be collected. 2. Digestion Method: This method involves disrupting the cell wall to release the intracellular PHA while conserving the polymer. This approach can be chemical or enzymatic.

2.4 Advantages of PHA Over Other Polymers

Most petrochemical based polymers are non-biodegradable and accumulate in the environment, leading to green house gases emission, the release of microplastics, and pollution of soil, water and air, which contributes to climate change. In contrast, PHAs offer several advantages over these synthetic plastics. They are biodegradable, decompose naturally and reduce environmental pollution. They do not accumulate in the environment and do not contribute to microplastic pollution and greenhouse gas emission. Excellent polymeric properties of PHAs such as crystallinity, melting temperature (Tm), glass transition temperature (T_g), young's modulus and stress-to-break, make them suitable replacement for synthetic plastics in various applications. PHAs are produced by a variety of microorganisms, making large scale production relatively simple and cost effective. Their desired structure and molecular weight can be achieved by changing the growth conditions of the microorganisms. PHAs can be sterilized without damaging their properties. These properties make

PHAs an ideal biomaterial of choice for medical devices and implants (Gregory et al., 2022). PHAs maintain pH stability during degradation, which enhances their biocompatibility compared to other biodegradable polymers such as PLA and PLGA. PLA, for instance, can lead to the accumulation of lactic acid in the host, while PHAs do not. PHAs degrade more slowly *in vivo* compared to PLA and PGLA. This slower degradation is advantageous for applications requiring extended support or controlled release of therapeutic agents (Naser et al., 2021b). The degradation products of PHAs, including monomers and oligomers are non-toxic and have therapeutic values, further enhancing their suitability for medical applications. PHAs can be blended with other polymers to adjust their mechanical properties, expanding their applicability beyond that of other biodegradable polymers like PLA, and polycaprolactone (PCL). Overall, PHAs present a more environmentally friendly and versatile alternative to synthetic and some biodegradable polymers, making them an ideal choice for a range of applications, particularly in the medical field (Liang et al., 2024; Gregory et al., 2022).

2.5 Contributing Properties of PHA for Their Application as Tissue Engineering Biomaterial

PHAs have emerged as versatile biomaterial due to their following unique properties:

1. **Biodegradability:** One of the outstanding properties of PHA is their biodegradability various conditions (Grigore et al., 2019). Under aerobic conditions, PHA degradation results in CO_2 and water, while under anaerobic conditions, it produces CO_2 and methane. PHAs can be broken down into low molecular weight monomers or oligomers by microbial depolymerases (hydrolases), which are then used up by microbial cells as nutrients. The rate of biodegradation is influenced by factors such as surface area, monomer type and composition, molecular weight, crystallinity, and environmental conditions (Tang et al., 2020). The biodegradability of polymer is inversely proportional to their crystallinity, meaning that mcl-PHA (e.g., P3HB4HB and PHBV) degrade faster than scl-PHAs (e.g., PHB). Additionally an increase in side chain monomer content accelerates degradation (Wu et al., 2009). These properties make PHAs ideal for manufacturing devices like films, nanoparticles, and prototypes (Gregory et al., 2022).
2. **Biocompatibility:** PHAs are highly biocompatible, primarily due to their short-chain degradation products, such as 3HB and PHBV. 3HB, a natural metabolite in human body, supports the excellent biocompatibility of PHB (Li et al., 2021), and implantable devices made from PHAs do not cause inflammatory reactions. PHAs do not induce thrombosis or antigenic response even being in contact with blood for longer time or even not trigger any adverse responses even when in prolonged contact with blood, and do not triger adverse responses like calcification, malignization, or inflammatory activity. Thus, the degradation products of PHA-based implants and surgical sutures are biocompatible and causing no adverse effects (Pullingam et al., 2022).

3. **Biological Properties:** Some of the biological properties that make it suitable and efficient biopolymer for medical applications are—
   - *Cell growth and tissue regeneration*: PHAs show diverse biological activities due to the presence of active groups. PHA and their derivatives support the tissue adhesion and proliferation (Ang et al., 2020). They promote the growth of various tissue cells in animal models, including epithelial cells, chondrocytes, and fibroblast (Xiao et al., 2007; Voinova et al., 2019).
   - *Non-carcinogenicity*: Rat fibroblast cultured with PHAs and their derivatives did not exhibit carcinogenicity. Even when rat osteosarcoma cells were grown on PHA, they did not transform into cancer cells, indicating that PHA's cell proliferating activity is regulated by the normal cell cycle (Peng et al., 2011).
   - *Antioxidant Properties*: PHAs can counteract oxidative stress, which is a common cause of inflammation due to an imbalance between reactive oxygen species (ROS) and the antioxidant system. The antioxidant potential of PHAs is an important functional property for tissue repair and regeneration Increased generation of free radicals and their poor quenching results in oxidative stress (Mirpoor et al., 2023). Electrospun PHA meshes incorporating olive leaf extract (OLE), rich in polyphenols, have shown immunomodulatory properties (De la Ossa et al., 2022).
   - *Anti-inflammatory effects*: Increased regeneration of ROS by neutrophils at the inflammation site causes tissue injury and endothelial dysfunction (Alsayari, 2021). Free radicals are the key signaling molecules in the progression of inflammatory disorders. PHAs can reduce inflammation by inhibiting inflammatory proteins like TNF-alpha (tumor necrosis factor), COX-2 (Cyclooxygenase-2), IL6 (Interleukin-6) and by modulating the proliferation of T- and B-lymphocyte cells (Fangkrathok et al., 2014).
4. **Mechanical Properties:** The mechanical properties depend on specific type of PHA, molecular weight, or any other modifications. Tensile strength of the material is the property to resist when forces are applied in tension whereas the Youngs' modulus tells how easily it can stretch and deform and is defined as the ratio of tensile stress to tensile strain. The mechanical properties of PHA like the tensile strength (40 MPa) and Youngs' modulus (3.5 GPA) are comparable to those of petroleum-based plastics. They have moderate tensile and flexural strength and have moderate to high tensile modulus. They are prone to cracking due to low elongation at a break of 5–7% and has a narrow thermal processing window (170–200°C). Generally a high crystalline material is stiffer and more brittle. This indicates that pure PHAs are not a material of choice for bearing tension. Comparatively PHB is the most rigid member of PHA family with tensile strength greater than 45 MPa (Singh et al., 2019).
5. **Thermal Properties:** The thermal properties are typically defined by two temperatures: 1. Glass transition temperature (Tg) for amorphous phase and 2. Melting temperature (Tm) for the crystalline phase. The literature value of Tm and Tg for PHB is 169°C and 1.1°C respectively. Tm decreases with increasing in side-chain length, while Tg increases. Scl-PHA and mcl-PHA exhibit different

thermodynamic properties. Mcl-PHAs have low crystallinity (about 40%), with a Tg between –52°C–25°C and a Tm between 38°C and 80°C making them elastic at room temperature. In contrast, scl-PHA, have higher crystallinity (about 60–80%) and exhibit poorer thermal stability (Chuah et al., 2013). A scl-PHA, P4HB is a highly thermoplastic and most elastic PHA homopolymer, and was the first PHA-based biomaterial approved by the FDA to manufacture surgical sutures (Williams et al., 2013). Thermal processing stability of PHAs can be modified and improved by combining, altering the surface, or blending it with other polymers, inorganic components, and enzymes to form reinforced composites.

6. **Biological Activities of PHA Degradation Products:** PHAs degrade into natural metabolites such as 3-hydroxybutyrate and hyroxyacyl-CoAs (Mukherjee and Koller, 2023). This contributes to their biocompatibility, bioresorbable and non-immunogenicity, facilitating their incorporation into biological systems (Lizarraga-Valderrama et al., 2019). Different PHA monomers and derivatives have expanded applications. For example, 3-HB, a monomer in the PHA polymerisation chain, is a natural metabolite produced in the human liver through fatty acid oxidation. It is found in blood, and performs some vital therapeutic roles, supports tissue regeneration, reduces inflammation (Williams et al., 2013), and acts as an energy source for the body. Additionally, 3HA, a medium chain length monomer, has shown excellent antibacterial potential against *E. coli, Salmonella typhimurium,* and *Listeria monocytogenes* (Kalia et al., 2021). Monomers like 3HB, 4HB and 3HHx are compatible with cells in the body, with 3HB specifically promoting Ca^{2+} influx, tissue regeneration, and osteoblasts differentiation (Zhao et al., 2007). Moreover, 3HB is recognized as a potential functional molecule for cancer therapy and immune regulation (Ren et al., 2023).

2.6 Limitations of PHA as Biomaterial

Despite the desirable properties, several drawbacks are associated with PHAs. Lengthy synthetic processes, expensive manufacturing methods and low productivity restrict its wide use and commercial viability. The production of highly pure and toxic-free PHAs for creating medical implants as well as micro/nano-carriers for targeted drug delivery through fermentation and downstream recovery processes has been challenging (Pramanik, 2023). Additionally, the low degradation rate, presence of contaminants like pyrogenic compounds, hydrophobicity, poor mechanical properties, high crystallinity, brittle nature, low thermal stability and low processability limit its use in manufacturing engineered tissue scaffolds (Foroughi et al., 2017; Eesaee et al., 2022). Thus, the use of unmodified PHAs as a replacement of conventional plastics still remains a major challenge.

3. PHA Modifications for Improved Properties and Bio-functionality

Although PHAs are often used in biodegradable applications to ease environmental problems, some undesirable properties restrict their applications. Polyhydroxybutyrate (PHB) is a crystalline, stiff, thermally unstable polymer with highly hydrophobic nature. To overcome the unsatisfactory mechanical properties and unfavourable biological responses, most PHAs need modifications by blending with other polymers or with additives.

3.1 PHA Based Blends

To overcome the limitations of PHAs, such as PHB and other scl-PHAs, several methods have been used to develop PHA based blends with improved physico-chemical and piezoelectric properties and controlled biodegradation rate, to achieve different and improved functionalities: (i) Developing PHA-heteropolymers (ii) Integrating with other absorbable polymers. (iii) Blending with additives or nanoparticles (Chuah et al., 2013; Nigmatullin et al., 2023). Mohamed and his coworkers categorized PHA based blends into three categories: (i) PHA/Inorganic composite (ii) PHA/Organic composite (iii) PHA/Other polymer (Mohamed et al., 2021). The blending of PHA with other polymers increased the range of their applications. Fabricated PHAs have enhanced cell viability and proliferation (Dhania et al., 2023). Desired mechanical strength of PHB/Polylactic acid (PLA) and remarkable biomedical application of PHA/PHB/PHBV-based composites have been studied (Cesarino et al., 2024). Films made using PHB, PHBV, and PHBHHx have been used for the attachment and multiplication of osteoblasts (Azimi et al., 2020). For cell adhesion and cell proliferation, hydrophilicity is a required property. Fabricating PHA with fillers has improved hydrophilicity. The lower degradation rate of PHAs can be changed by blending them with materials like poly-lactic acid (PLA), cellulose acetate (CA), chitosans (CTs), etc. which have a higher rate of degradation. Their poor processing performance due to high crystallinity, stiffness, high degradation rate, and low thermal stability can be improved by blending with monomers such as 5-hydroxyvaleric acid (5HV), 3-hydroxypropionic acid (3HP), and PLCL, to enhance their suitability for manufacturing into implanted devices (Gopi et al., 2018). Remarkable enhancement in degradation rate, mechanical properties, and toughness was observed by blending poly-l-lactic acid (PLLA) and PHBHHx into a film. Efforts have been made to produce P(4HB) homopolymer and P(3HB-co-4HB) copolymers, as well as their blends and composites, with improved biocompatible properties for a wide range of therapeutic applications (Guo et al., 2022; Gregory et al., 2022).

Enhanced processability of LA/PHB blends due to improved mechanical properties was observed (Olejnik et al., 2021). Raziyan and his team constructed three different composites using a combination of PLA, PHA, PHB and HA. The composite matrices, PHA/PHB/PLA/HA, showed enhanced bioactivity together with biocompatibility (Raziyan et al., 2024).

3.2 *Ideal Properties of Scaffolds Essential for Tissue Repair and Regeneration Applications*

3.2.1 Principal Elements and Major Steps of Tissue Engineering

The three basic components in tissue engineering are the scaffold, the cells, and growth regulating mechanism, called the tissue engineering triad (Krishnani et al., 2023).

Scaffolds + Growth factors + Cell = Neo-tissue (Sharma et al., 2021)

Neo-tissue is made from the cell grown on a bio-absorbable scaffold supplemented with growth factors. For tissue engineering application stem cells are the promising cell source. They can be categorized into: Embryonic Stem Cells (pluripotent cells that can be differentiate into any cell line), Adult Stem Cells (multipotent cells considered more advantageous for tissue engineering applications due to their lower risk of implant rejection), and induced pluripotent stem cells (somatic cells in a pluripotent state with autologous characteristics). Growth-regulating signals include mainly the growth factors, a subset of diffusible signaling proteins, and cytokines which regulate cell activity.

3.2.2 Scaffolds and their Fabrication

Scaffolds are biomaterials that serve as 3D framework to allow cell adhesion, multiplication, migration and the optimal supply of various bioactive compounds essential for tissue regeneration. The prerequisite characteristics of scaffolds include biodegradability, biocompatibility, 3D structure, wettability, mechanical strength, durability, and piezoelectric property. Due to the diverse characteristics of PHAs, ideal scaffolds required for tissue engineering can be prepared. Scaffolds can be divided on the basis of their geometry: 1. Porous Scaffolds, having high level porosity and a high surface-to-volume ratio, allowing cell planting and making possible the required nutrients through the interrelated pores. They are accessible in different types such as sponge, mesh, foam and nano- or micro-scale fibers; 2. Fibrous Scaffolds, consisting of nanofibers obtained through the process of electrospinning; 3. Acellular Scaffolds, lacking cellular components of tissue are considered to be most ideal for engineering different tissues because they allow the host cells to develop their own ECM; 4. Hydrogel Scaffolds; 5. Microsphere Scaffolds, suitable for encapsulation and controlled delivery of bioactive molecules to the target tissue; and 6. Composite Scaffolds, made by combining two or more components, which can offer better responses due to their synergistic/cooperative effects.

Fabrication Techniques to Produce Scaffolds: Several polymers have been described as biomaterials for scaffolds synthesis (Khalid et al., 2020). The prerequisites of polymer scaffold for tissue engineering applications are manifold and highly specific. PHA bio-fabrication is an emerging field for producing smart and functional scaffolds. They have possibilities of customization through physical, chemical and biological modifications. Manufacturing PHA scaffolds with ideal properties in various forms is crucial for applications in tissue repair and regeneration. Multiple fabrication techniques are being used to design and develop PHA-based scaffolds, microspheres, membranes, and nanoparticles. The commonly used technologies include solvent casting, particle leaching, gas foaming, fibre spinning, freeze-drying, melt forming, and specialized leaching injection molding (Mi et al., 2023).

Conventional Techniques

- **Solvent casting and particulate leaching-** Method is money saving and simple to perform. Mainly used for cartilage and bone tissue engineering (Prasad et al., 2017) as the porosity of scaffolds obtained by this method was reported to be 50–90% (Li et al., 2021). This technique can only prepare simple structured scaffolds, like flat sheets and tubes.
- **Freeze drying/Lyophilization-** High temperature, which is responsible for the loss of bioactivity of growth factors, is not required in this technique. This technique is mainly used for development of osteoconductive scaffolds. They showed remarkable water absorption and porosity.
- **Thermally induced Phase Separation-** Scaffold having a nano-sized fibrous structure and high porosity (greater than 95%) can be developed using this technique. Suitable for the integration of bioactive compounds.
- **Gas Foaming-** Spong-like structures having pore size range from 30–700 micrometers and porosity reaching up to 80% can be obtained by this method.
- **Electrospining-** It has gained interest in recent years due to the potential applicability of eletrospun fibers in tissue engineering, wound repair, and in drug delivery. Formation of ultrafine fibers (diameter ranging from microns to nanometers) can be achieved by this technique. In addition to high porosity, it has several other properties like biodegradability, biocompatibility, anti-toxicity, tunable geometry, high surface area-to-volume ratio, low density, and flexibility. This method is mainly used for skin tissue engineering. Electrospun nanofibers embeded with nanoparticles can be employed as drug delivery carriers at the wound site.

A. PHA biopolymers as Nanofibers: Electrospun nanofibers based on PHAs can mimic the struture and functionn of native extra cellular matrix (ECM) of cells. Their fibers have an average diameter of 50–500 nm, which is close to that of collagen (fundamental component of the ECM). Many researchers have summerized the potential applications of nanofibres in wound dressing, regenerative therapies including nerve repair, bone and cartilage regeneration, organ transplantation, cell encapsulation, and as a therapeutic carrier (Zhao et al., 2022; Datta et al., 2019; Olkhov et al., 2024). An innovative method combining electrospining and Wet Vertical Magnetic Rod Electrospining (WVMRE) has been developed to create a tri-layered scaffold mimicking the ECM of vascular tissues (Ma et al., 2024).

B. Rapid prototyping (RP)/Additive Manufacturing Techniques: These techniques are used for making customized and precisely structured scaffolds. 3D object is created layer by layer from a digital file. Their shape and dimensions can be tailored according to the patient's requirements. Various RP technologies have distinctive features and have overcome some of the shortcomings of conventional methods of scaffold fabrication in terms of speed, accuracy, and the range of materials utilized.

- Stereography (SLA): suitable for fabrication of scaffolds used for load-bearing applications.
- Fused Deposition Modelling (FDM)
- Selective Laser Sintering (SLS)
- Bioprinting: Any process that involves the transfer of living cells, bioactive molecules, and ECM in a defined pattern to obtain any specialized biological activity is considered bioprinting. 3D printing is a developing approach for fabricating complex-shaped biobased items such as biomedical devices, for drug screening, and for regenerative therapy (Yu et al., 2020; Gao et al., 2024; Singh et al., 2024). Acellular (cell-free) and cellular are two bioprinting types. Microextrusion, inkjet printing and laser-assisted are some forms of 3D printing which are being used for patterning biological material. 3D printing can be combined with electrospinning to fabricate scaffolds with hierarchical structure (Wang et al., 2021a; Vyas et al., 2020).

C. PHAs for 3D printing in tissue repair and regeneration: PHAs possess ideal properties for different 3D printing techniques, such as low melting temperature, low energy consumption, easy processability, and ease of customisation. The 3D implants made of P3HB/HA facilitated osteogenesis. Additionally, low degradation and enhanced growth of new bone tissues were observed (Shumilova et al., 2017). 3D printing shows more perfection in constructing implants for medical use (Culbreath et al., 2020). Using 3D printing technique, many tissue constructs have been constructed like cartilage, heart valves, and for cardiac tissue regeneration (Chingale et al., 2022).

3.2.3 An Ideal Scaffold for Tissue Repair and Regeneration

An ideal scaffold should have all the characteristics that bring the desired biological activities (Agoscsova et al., 2023; Yadav et al., 2024). The scaffold should mimic the natural extracellular matrix (ECM) of the host tissue. Its surface should facilitate cell adhesion, proliferation, differentiation, and most importantly, the secretion of extracellular matrix. They also act as a delivery vehicle for bioactive compounds, like drugs, growth hormones, and cytokines, to the target site (Dhania et al., 2022). Studies have revealed that cells likely to grow better on hydrophilic surfaces of scaffolds (Wei et al., 2009). The scaffold should be compatible with the cellular components in the host tissue and should not collapse during surgical implantation or during the patient's regular activities. The pore size and porosity play a versatile role in cell in-growth and neo-vascularization. The pore size should be more than 100 micrometers in diameter (optimum, 200–350 micrometers) to facilitate diffusion of oxygen and essential nutrients as well as the removal of metabolic waste. The scaffold should be easily degradable and must be properly sterilized to prevent infection. The degradation products should not be toxic and should not cause inflammation to the surrounding tissues. It should provide stability and versatility to the neotissues. According to the functionality, different scaffolds are required as mentioned in Table 3.

Table 3. Properties of scaffold for specific application.

S.N.	Application	Properties and Design Requirements	Technique Used to Design Scaffold	Reference
1.	Soft Tissue Repair and Regeneration (Mainly skin tissue)	➢ Skin-like surface topography and microstructure to promote cellular adhesion, proliferation and differentiation. Satisfied porosity (5–200 micrometer) ➢ Adequate and similar mechanical properties as the native skin. Tensile strength should range 5–30 MPa with elongation at break in the range of 35–115% and young's modulus between 4.6 and 20 MPa to support angiogenesis and development of nerves, lymphatic system, etc.	Electrospining, freeze-drying and 3D printing (Chen et al., 2023a).	Chen et al., 2023a; Janouskova, 2018; Zhang et al., 2022
2.	Wound Healing	➢ Anti-inflammatory; Prevent bacterial infection. Absorbing excess exudate from the wounds ➢ Allowing exchange of gases and adequate nutrients. Blood coagulation ability. ➢ Accelerating wound re-epithelialization and promoting tissue regeneration. ➢ Adhesion and proliferation of fibroblast and keratinocyte. Maintain a moist environment near the wound (Hydrophilicity). ➢ High porosity to allow cell ingrowth and neovascularization. Large surface area to volume ratio. ➢ Degradation profile concomitant to the wound healing period.	Electrospining (Negut et al., 2020).	Negut et al., 2023
3.	Dressing Material	➢ Non-immunogenic, bactericidal, antioxidant, non-toxic, simple to apply and anti-inflammatory properties. ➢ Excellent tissue adhesion and allows gaseous exchange ➢ Functionalized with bioactive compounds to enhance cellular response	Electrospining (Alven et al., 2021)	Priya et al., 2024
4.	Organ Tissue Repair and Regeneration	➢ Highly mimic the ECM of the native tissue ➢ Proper vascularization and formation of new tissues ➢ High mechanical strength to provide stability and versatility. Ability of remodeling for the integration of host tissue after implantation ➢ Non-immunogenic and non-toxic	3D Bioprinting (Shanmugam et al., 2023)	Suamte et al., 2023
	Liver tissue	➢ Proliferation and differentiation of hepatocytes, connective tissue and vessels ➢ ECM of hepatocyte: Type I collagen and fibronectin	Bioprinting (Inject-based and Extrusion-based) (Lv et al., 2022)	Lv et al., 2022; Wang et al., 2022a

Table 3 contd. ...

...Table 3 contd.

S.N.	Application	Properties and Design Requirements	Technique Used to Design Scaffold	Reference
5.	Cardiovascular Tissue Repair and Regeneration	➢ Biomimetic surface. Higher hydrophilicity ➢ Modifiable mechanical properties (match autologous tissue and resist the blood pressure); mechanical robustness (high tensile strength and good stiffness). Good flexibility ➢ Anticoagulation and antithombogenic (inhibit blood clot formation). Promote angiogenesis. Lack of chronic inflammatory response ➢ Facilitate the migration and growth of SMCs as well as inducing rapid adhesion of endothelial cells.	Hydrogel, electrospiningand 3D bioprinting (Rogozinski et al., 2022); Mould casting and 3D printing (Montero et al., 2020)	Smiga-Maluszowicz et al., 2023
6.	Surgical Sutures	➢ High tensile strength; absorbable; Scar prevention ➢ Microbe resistance to reduce risk of infection.Function as drug carrier ➢ Minimum propensity to tear through tissue (Low tissue drag). Good elasticity and flexibility. Knot retention (Good knot strength)	Spinning technologies like solution spinning, melt spinning and microfluidic spinning (Li et al., 2023c; Yadav et al., 2024)	Reddy et al., 2021; Chu, 2013; Xu et al., 2022; Robinson, 2021; Borchiellini et al., 2023
7.	Bone Tissue Repair and Regeneration	➢ Mimicking the natural bone matrix (composed of apatite). Should act as a temporary ECM by performing physico-chemical activities ➢ Guided immune response. Must promote the differentiation of monocytes into the anti-inflammatory M2 macrophage, responsible for synthesis of anti-inflammatory cytokines. ➢ Should promote vasculature throughout the defect space to deliver nutrients and cells. Induce neovascularization(allow exchange of gases, nutrients and metabolic waste through the newly formed blood vessels surrounding the implants). Prevent infection and biofilm formation ➢ Desirable mechanical properties (Compressive strength similar to host bone ~ 2–12 MPa; Young's Modulus ~ 0.1–5 GPa) ➢ Average porosity and pore size must be 80–88% and 200–300 micrometer resp. Interconnected porous network to allow cells to move in and grow in a damaged part. ➢ Osteoconductivity (recruit progenitor cells to healing site and to promote their osteogenic differentiation) ➢ Osteoinductivity (ability to allow the formation of new bone tissues on its surface and within its pores) ➢ Should degrade at a controlled resorption rate (Degradation rate has to coincide with the rate of tissue formation).	Electrospining, solvent casting, freeze drying, sponge method and via other methodologies to obtain porous, hydrogels, fibrous and microsphere (Abdelaziz et al., 2023)	Abdelaziz et al., 2023; Gundu et al., 2022; Feng et al., 2023; Li et al., 2023a

8.	Cartilage Tissue Repair and Regeneration	➢ Closely mimic the natural ECM. Mediate cell-cell signaling and interaction. Chondrocyte cells proliferation and production of type II collagen ➢ Mechanical integrity. High tensile strength. High hydrophilicity. Sterilizability. Capacity to degrade slowly ➢ High porosity (more than 80%); In favour of adhesion and proliferation of chondrocytes and their penetration into the electrospun fibers. Osteoconductive, non-cytotoxic, non-antigenic	Solvent casting, particulate leaching, electrospining methods (Deng et al., 2002; Wasyleczko et al., 2022)	Cao et al., 2014
8.	Heart Valve (Implantable Device)	➢ Withstand the haemodynamic pressure and flows of cardiac environment. Resistant to calcification and thrombosis. Mitigate the need for anticoagulant ➢ Mechanical properties with bipolar stiffness. Should mimic the physicochemical properties of native heart valve. ➢ Capable to grow and remodel. Post implantation viability and functionability. Drug delivery capabilities	Electrospinning (Wang et al., 2023b; Broadwin et al., 2024)	Pedersen et al., 2023; Jana et al., 2014
9.	Stunt (Implantable Device)	➢ Promote rapid endothelial recovery, Haemocompatibility, Anti-inflammatory ➢ High tensile strength. Appropriate temporal radial strength to resist recoil. ➢ Bioabsorbable (resist restenosis) .	3D printing tech., called stereolithography to fabricate stents (Seunarine et al., 2006)	Sakamoto et al., 2018; Dhania et al., 2022
10	NGCs (Implantable Device)	➢ Designed for use as conduits. Function as a nerve guide channel ➢ Promote development of mesenchymal stem cells ➢ Enhance regeneration of axons and schwann cells (SCs) ➢ Increased porosity. Decreased angle of contact with water	Electrospining (Wan et al., 2023)	Khorasani et al., 2011
11.	Drug Delivery	➢ Biocompatible. Biologically favourable environment for tissue formation ➢ Appropriate drug release profile. Can organize active compounds into different structural families ➢ Molecular medicinal properties like-- pharmacodynamics, physicochemical and pharm ➢ acokinetic properties	*In-slico* supported nano-emultion methods (Procopio et al., 2022)	Calori et al., 2020

4. Applications of PHA-Based Biomaterials in Tissue Repair and Regeneration

Biodegradable polymers have been used as a key component in tissue engineering and their uses are still expanding. Due to their high biological suitability, desirable mechanical properties, antibacterial properties and good processability, the PHA-based biomaterials are satisfying a wide range of biomedical requirements (Pandey et al., 2022). PHB and its copolymers, such as PHBV or PHB4HB, are the most extensively studied PHAs in biomedical applications (Cesarino et al., 2024). The objective of tissue repair and regeneration is to reconstruct the defective or diseased tissue and to improve their function using appropriate scaffolds. There are mainly four components of the repair process: angiogenesis/neovascularization, migration and multiplication of cells, deposition of ECM, and maturation and reformation of the tissue. A literature survey of PHA-based biomaterials with customised properties and execution in multiple tissue repair, regeneration and in drug delivery has been carried out. The published *in vivo* work in Table 4, claimed that PHA-based composites have better performance than pure PHA. The Biomedical applications can be divided into five sub-headings: Soft tissue repair and regeneration, biomedical devices, hard tissue repair and regeneration, medical implants and drug delivery device-related applications (Figure 2).

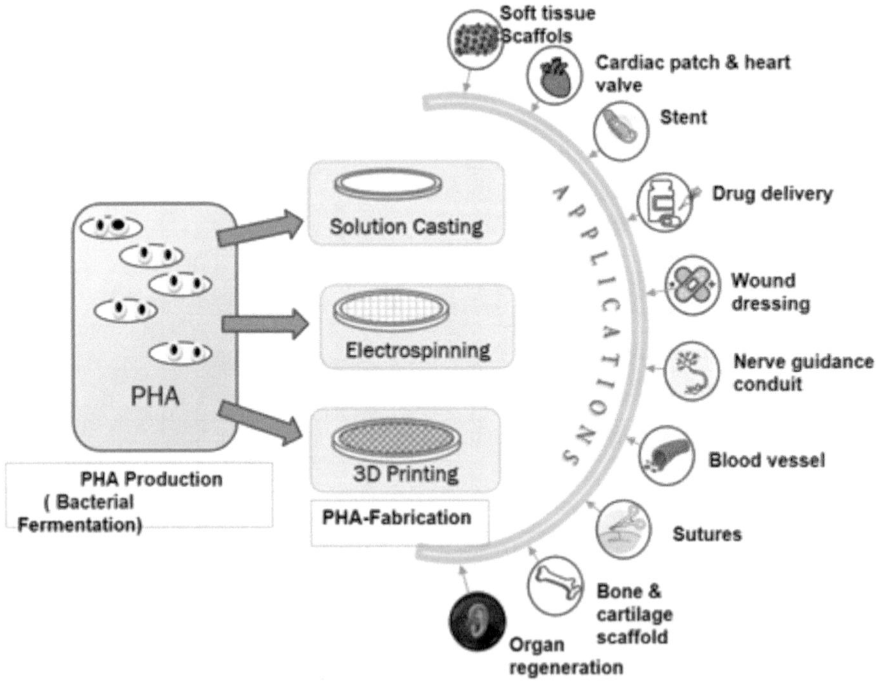

Figure 2. Applications of PHA-based biomaterials for multiple tissue repair and drug release.

Table 4. Applications of PHAs (*in vivo* studies).

4(A). PHAs in soft tissue repair and regeneration (Specially in Wound Healing)			
Type of PHA	**Combination**	**Application**	**Reference**
P3HB4HB	Bacterial Cellulose	Bacterial cellulose and degradable copolymer P3HB/4HB in combination with actovegin and fibroblast are more efficacious than a commercial wound dressing material	Volova et al., 2019
PHBV	$nCeO_2$/Growth Factor	The drug loaded hydrogel patches, absorb excess exudates, promoted angiogenesis and migration and proliferation of keratinocytes, fibroblast and endothelial cells. Promoted diabetic wound healing	Augustine et al., 2020
PHBV	–	By regulating myofibroblast formation using nano fibrous meshes made of biodegradable elastiomere PHBV, excessive scar formation was reduced. Accelerated healing and replacing of lost and damaged skin	Kim et al., 2020
PHBV	HMK/AgNPs	The electrospun nanofibrous mat provided suitable biocompatible and antibacterial properties that result in improved wound healing ability	Ye et al., 2020
PHB	Gelatin	Electrospun scaffold of gelatin/PHB promoted the healing of diabetic wound, increased amount of fibroblast (responsible for collagen and elastic synthesis and organization of ECM). Enhanced skin regeneration	Sanhueza et al., 2021
PHA	3HB, 4HB	Showed better biocompatibility and antibacterial activity against oral specific bacteria. Significantly improved healing process.	Chen et al., 2023a
PHB	Ag Nps	Ag Nps-incorporated electrospun membranes showed potent antibiotic and antibiofilm activity and prevented microbial infection associated with peri-implantitis-related complications.	Sanhueza et al., 2024
4(B). PHAs in organ tissue repair and regeneration			
PHB	PLA	Blending of PLA & PHB provided stable support conditions for urethral substitution.	Balogova et al., 2018
PHB	Chitosan and nanobioglass	Expression of type I collagen and dentin sialic acid increased 6-fold with increased cell multiplication and increased rates of Ca & P ion uptake	Khoroushi et al., 2018

Table 4 contd. ...

...Table 4 contd.

Type of PHA	Combination	Application	Reference
PHB	CNTs	Improved vascularization with a mild foreign body reaction.	Zarei et al., 2020
PHBV	Insulin producing cells (iPSCs)	Increased survival rate of insulin producing cells. Electrospun nanofibre could be used for pancreatic tissue repair and regeneration	Abazari et al., 2021
4(C). PHAs in cardio-vascular tissue repair and regeneration			
PHBV	PVA/elastin	Tri-layered vascular scaffold behaved as a native artery and was compatible to blood. Promising for small diameter vascular grafting	Deepthi et al., 2018
PHBV	Polycaprolactone (PCL)/RGD peptides	The PHBV based patches significantly construct the artery walls and promoted the formation of neointima due to proliferation in the vessels. Inhibited calcification.	Sevostianova et al., 2020
PHBV	PCL/VEGF/ bFGF/SDF	PHBV based electrospun patches showed high biocompatibility and calcification resistance. Cartid artery grafts afforded free passage and promoted vascular tissue regeneration.	Antonova et al., 2021
PHB	PLGA and PE	Enhanced cardiomyogenic differentiation of mesenchymal stem cells. Damaged cardiac tissue repair accelerated.	Mokhames et al., 2023
4(D). PHAs in bone tissue repair and regeneration.			
PHB	HA/ALG/MSCs	Scaffold seeded with mesenchyma stem cells stimulated bone regeneration rate in a critical sized parietal bone defected rats	Volkov et al., 2020
PHBV	*aloe vera*-derived gel	Stimulated the transcription of TGF-Beta1 and BMP-7 genes. Promoted activity of ALP (bone related genes and proteins)	Tahmasebi et al., 2020
PHBHHx	rhBMP-2	Stimulated bone regeneration and facilitated vascularisation.	Liu et al., 2020b
PHB	BC	Facilitated preadipocytes multiplication and osteoblast specialization to induce new bone formation	Codreanu et al., 2020
PHB	Carbon nanotubes	The PHB and nanotubes scaffold imicked fibrous connective tissue of the PDL, suitable for ligament tissue engineering	Zarei et al., 2020

Table 4 contd. ...

...Table 4 contd.

Type of PHA	Combination	Application	Reference
PHBV	MBGN and CIN	Enhanced proliferation and attachment of human osteosacoma cell and rapid formation of hydroxyapatite on the composite microspheres	Chotchindakun et al., 2021
PHA	Beta- TCP	Improved adhesion, proliferation and migration of MC3TB-E1 cells. Enhanced osteogenesis related gene expression	Ye et al., 2022
PHB	PDA	Enhanced osteogenic activity and bone regeneration	Zhang et al., 2022a
PHBHV	nHA and C60	Increased flexural elastic modulus (29% compared to neat PHBHV) was achieved in cortical bone filling applications	Ribeiro et al., 2023
PHB	Lignin/cellulose	Enhanced biological responses and promoted osteogenesis	Mohammadalipour et al., 2023
PHA	PLA and nHAp	Showed improvement in tensile strength, osteoconductivity, porosity and thermal stability	Injorhor, 2023
PHA	HA and/ or pDA	Enhanced cell proliferation, differentiation and viability	Kim et al., 2024
PHB	BaTiO3	Promoted osteogenesis and creation of new blood vessels	Strangis et al., 2024
4(E). PHAs in cartilage tissue repair and regeneration.			
PHB/PHBHHx	SVFCs/hASCs	Enhanced chondrogenic celll proliferation and migration for rapid cartilage repair	Ba et al., 2019
PHBV	Quercetin	Reduced oxidative stress. Inhibited chondrocyte apoptosis. Enhanced cartilage regeneration by regulating relevant gene transcription	Chen et al., 2021
PHB	Chitosan	Supported osteochondral regeneration in knee articular cartilage damage. Reconstructed CCL in damaged cartilage.	Petrovova et al., 2021
PHB	CS/GAS	Enhanced viability and proliferation of adipose stem cell and their differentiation to chondrocyte cells	Golshayan et al., 2022
PHB	DWJM	Increased chondrocyte cell adhesion and promoted articular cartilage tissue regeneration.	Soleymani et al., 2024
PHB	Cs and ECM nanoparticles	Enhanced viability, attachment and proliferation of chondrocytes.	Mohammadi et al., 2024

Table 4 contd. ...

...Table 4 contd.

Type of PHA	Combination	Application	Reference
4(F). PHA in nerve tissue repair and regeneration.			
PHBHHx	Neural stem cells (NSCs)	Facilitated the long term survival and growth of transplanted neuronal stem cells in traumatic brain injured rats.	Wang et al., 2019
PHB	CS (chitosan)	Increased % porosity and decreased the angle of contact with water	Karimi et al., 2020
P(3HB-co-3HHx)	Graphene/gold	Supported Schwinn cell attachment, proliferation and migration and support peripheral nerve regeneration	Liu et al., 2020a
P(3HB/P(3HO)	1% bioactive glass	Significantly promoted NG108-15 neuronal cell adhesion and differentiation	Lizariaga-Valderrama et al., 2020
mcl-PHA	PCL	High neuro-regenerative potential and biodegradable rates	Mendibil et al., 2021
PHB	PHO	Superior biocompatibility for NG108-15 and promoted peripheral nerve regeneration.	Nigmatullin et al., 2023

4.1 In Soft Tissue Repair and Regeneration

Under soft tissue, cartilage, ligament and tendon, skin, neurons, and vasculature are included. Collagen, a polymer of amino acids, is abundantly found in soft tissues and is a main ECM element of connective tissues within the human body. The use of mcl-PHAs, like P(3HO) and P(3HO-Co-3HD), is preferred in soft tissue repair and regeneration due to their elastic properties. These include biomedical devices like surgical sutures, cardiac patches, blood vessels, wound dressings, auricular reconstructions, and medical implants like heart valves, stents and nerve guidance conduits (NGCs).

a. Skin Tissues (Wound Healing and Skin Regeneration)

Skin is the largest organ in the human body and is most exposed to dangers, including chemicals, microbial, thermal, mechanical, diabetic ulcers, injuries, burns, and trauma. The primary functions of skin include protection against pathogens, fluid homeostasis, thermoregulation, sensory detection, and self-healing. In the last few decades, PHA-based biomaterials, including membranes, nanofibers, and hydrogels are being used extensively in skin tissue repair and regeneration. For healing of wound and skin recovery, collagen I and fibrinogen are essential ECM proteins (Sun et al., 2021). Fabricated nanofibers possess suitable tensile strength and perform dual functions, antibacterial as well as wound healing (Zou et al., 2020). PHA scaffolds as carriers for alkanins and skikonins have also been used as wound healing agents for wound management (Arampatzis et al., 2021).

Phases of wound healing: Wound healing is a complex process and can be divided into four overlapping phases (Zarrintaj et al., 2023). Three processes involved in the scar formation are; (i) migration and multiplication of fibroblasts in the injury site, (ii) deposition of ECM and (iii) tissue remodelling.

Wound Repair Mechanism: The wound healing process includes—1. Hemostasis (clot formation) via platelets, 2. Differentiation of mesenchymal cells into various cell types such as fibroblast, endothelial cells and keratinocytes, 3. Proliferation and migration of keratinocytes to cover the wound surface (Re-epithelialization), 4. Proliferation of fibroblasts and production of collagen, elastin and ECM), 5. Matrix/collagen deposition and tissue remodelling, 6. Angiogenesis via endothelial cells to restore blood supply to the wound area, 7. Scar formation and restoration (Pena and Martin, 2024). They are controlled by different growth factors and cytokines. Growth factors involved in wound repair are:--Epidermal Growth Factor (EGF), Transforming Growth Factor-beta (TGF-β), Fibroblast Growth Factor (FGF), Platelete-Derived Growth Factor (PDGF) and Vascular Endothelial Growth Factor (VEGF) (Zandi et al., 2021). Enhanced level of the growth factor, TGF-beta accelerate collagen deposition via accelerating proliferation of fibroblast. VEGF plays an important role in angiogenesis whereas FGF participates in formation of new epidermis. **Cytokines**, are signaling proteins that affect the immune system. Several pro-inflammatory cytokines such as Interleukin-1, interleukin-6 and tumor necrosis factor-alpha are upregulated during inflammatory phase of wound healing. They activate inflammation and contribute to delayed wound healing whereas anti-inflammatory cytokines, like Interleukin-1 receptor antagonist, helps in preventing excessive inflammation (Schafers et al., 2003).

PHA based Antibacterial Material: Microbial contamination is a major issue in the application of PHAs in medical devices, bio-implants and tissue engineering. Rise in multi-drug resistant pathogens has now become a global health risk. Biodegradable polymers can be used to produce various materials such as nano-sized particles, films and coatings to combat bacterial infections (Prajapati et al., 2023). The feasibility of incorporating antimicrobial activity into PHA-based biomaterials can broaden its application in several medical sectors associated with wound repair, coatings against biofilm formation, adhesive strips, implantable material (Ladhari et al., 2023). PHAs in wound healing as scaffolding, hydrogels and films are useful.

***In vivo* studies:** Basnett and her coworkers loaded lime oil (known for its antimicrobial activity) into a copolymer of 3-hydroxyoctanoate and 3-hydroxydecanoate, i.e., P(3HO-co-3HD) and developed a natural antimicrobial biomaterial film which exhibited remarkable activity against the Gram-positive bacteria, *Staphylococcus aureus* (Basnett et al., 2020). Porous PHA scaffolds containing 3HB and 4HB restrained bacterial infection and promoted the healing rate of soft tissue grafts (Chen et al., 2023a). Volova et al. prepared P3HB4HB fibrous membranes by eletrospining techniques and found that they were more suitable to stem cell proliferation and differentiation, and promoted wound healing by enhancing angiogenesis and by reducing inflammation (Volova et al., 2019). Augustine and co-workers blended cerium oxide nanoparticles into PHBV. The electrospun wound healing patch based on PHBV, Gelatin-methacryloyl (GelMA) hydrogel, and cerium oxide showed free radical scavenging activity along with promoted migration and proliferation of various cells like keratinocytes, fibroblasts, and endothelial cells, which support angiogenesis and boost diabetic wound healing. The resulting scaffold healed the diabetic wound rapidly, as they exhibit potential for cell proliferation and vascularization (Augustine et al., 2020). In order to resist excessive scare formation, Kim and co-workers prepared PHBV electrospun scaffolds of three different types using three different contents of 3HV. It was observed that with an increase in 3HV content, the crystallinity of the PHBV polymer decreased. The results suggested that the higher the 3HV number in PHV, the more elastic the scaffold, and it can better maintain minimum dermal fibroblast count to reduce excessive scar proliferation (Kim et al., 2020). High molecular keratin (HMK) was electrospun with PHBV and beaded with silver nitrate nanoparticles. The prepared PHBV/HMK/AgNPs nanofibrous mats showed antibacterial activity, good biocompatibility and improved wound healing ability (Ye et al., 2020). Electrospun scaffolds of dual-sized gelatin/ poly 3HB nanofibers prepared by Sanhueza et al. promoted fibroblast for skin regeneration and spread the force required to contract wound. PHB facilitated the cross-linking of the gelatin, Ge nanofibers, preventing the collapse of the 3D structure of the scaffold (Sanhueza et al., 2021). PHBV nanonofibers promoted insulin-producing cells differentiation. Rapid wound healing was observed as it enhanced the hydrophilicity and wound bacteria inhibition efficiency (Abazari et al., 2021a). PHB membranes loaded with silver nanoparticles, AgNPs showed a combination of antibacterial and antibiofilm activity and controlled microbial infection associated with peri-implantitis complications (Sanhueza et al., 2024).

b. Organ Tissues

Whole organ engineering approach is a clinically viable substitution for organ transplantation. Key technologies to support are--1. Decellularisation and recellularization of whole organ, 2. 3D bioprinting, 3. Stem cell technologies, 4. Genetic engineering (Sohn et al., 2020). Polymeric biomaterials have been used to develop viable and functional whole organ construct ready to implant. They are used to develop soft tissue organ construct with the structural and biochemica properties similar to native organ and have potential to minimize the immune response. ECM and decellularization are promising techniques. To overcome the cell availability challenge, stem cells have been used. Recently, 3D bio-printing has great ability to provide fully functional organs. Bio-printing blood vessel, cardiac patches, etc. are currently possible for treatments. Researchers got a little success in engineering a viable and functional whole heart, liver, urinary bladder, lung, kidney and pancreas (Sohn et al., 2020). Organoid technique can also be used to whole tissue engineering as it has potential to replicate native organ function (Kim et al., 2016).

***In vivo* Studies:** Balogova et al. succesfully utilized a 3D printed PLA/PHB scaffold to provide stable support conditions for urethral repair (Balogova et al., 2020). The effect of PHB/CS/nano-bioactive glass, nBG composite fiber scaffolds were investigated by Khoroushi et al., and the results showed increased cell multiplication rates and increased uptake rates of Ca and phosphorous. Type I collagen and dentin sialic acid phosphoprotein (DSPP) also increased six-folds, indicating its utility in dental disease treatment (Khoroushi et al., 2018). A PHB/carbon nanotube scaffold prepared using electrospinning technique showed improved vascularization and a mild foreign body reaction (Zarei et al., 2020). Electrospun nanofibers using PHBV and induced pluripotent stem cells (iPSCs), promoted the survival rate of insulin-producing cells and could be used for pancreatic repair and regeneration (Abazari et al., 2021b).

Liver tissue regeneration: Liver is the Largest gland in the human body. It is mainly composed of hepatocytes (parenchyma), connective tissue and endothelial cells. Type I collagen and fibronectin are mainly found around hepatocytes and form the ECM. Recent advancements in using biomaterials for liver tissue engineering include the fabrication of integrated liver cells and biomaterials (Kaur et al., 2020; Kim et al., 2021). A vascularized liver model was developed using 3D bioprinting techniques (Lv et al., 2022).

c. Cardiovascular Tissues

Cardiovascular tissues are the building blocks of blood vessels, heart chambers and valves. Advantages like modifications in strength, elasticity and the ability to promote the formation of elastin make P3HB4HB a potentially efficient biomaterial for artificial vascular engineering. Similarly, P4HB has been used for manufacturing vascular scaffolds to strengthen the pulmonary arteries *in vivo*. Ma and coworkers created a tri-layered scaffold to mimic the ECM of vascular tissues. The ECM has potential roles in maintaining vascular homeostasis and in preventing pathological remodeling (Fu et al., 2024; Ma et al., 2020).

***In vivo* studies:** Using electrostatic spinning technique, Deepthi et al. developed a 3-layered vascular scaffold of elastin nanofibers, PVA and PHBV to mimic the multilayer structure of natural vessels (Deepthi et al., 2018). Sevostianova and co-workers synthesized biodegradable PCL/PHBV patches modified with Arg-Gly-Asp (RGD) peptides for arterial reconstruction. The composite scaffolds could significantly promote the formation of new vascular tissues *in vivo* (Sevostianova et al., 2020). Antonova and colleagues compared PHBV/poly-caprolactone, PCL vascular grafts with expanded polytetrafluoroethylene (ePTFE) artificial vessels in a sheep carotid artery interposition model. They incorporated fibroblast growth factor (FGF) and vascular endothelial growth factor (VEGF) into the vascular graft and observed that the grafts were replaced by newly generated blood vessels. The regenerated arteries were resistant to calcification (Antonovo et al., 2021). In an another experiment, a functional scaffold was prepared by electrospinning technique using PHB, human placenta extract (PE) and polylactic glycolic acid (PLGA) to repair damaged cardiac muscle tissues. The results showed that the PE-added PLGA-PHB construct exhibited improved bioactivity and enhanced cardiomyogenic differentiation of mesenchymal stem cells (Mokhames et al., 2023).

4.2 Biomedical Devices

In medical applications, there is ongoing research and development effort for the improvement of devices for more efficient and successful treatment of diseases.

a. **Surgical Sutures:** According to the depth and severity of the wound, sutures, wound dressings and skin grafts are advised. They are used for various purposes such as ligation, tissue attachment, hemostasis, and wound repair. They are of two types—Non-absorbable and absorbable. Absorbable sutures have many desirable properties such as high tensile strength, biocompatibility, antibacterial properties, ease of sterilization, easy tying with a good grip (Manavitehrani et al., 2016), and ultimate absorption and elimination. Baptist et al. in 1965 described a method for making absorbable surgical sutures from polyesters. Polymers have been used as adhesives and surgical sutures (Ironpour et al., 2024). PHAs fulfilled the mentioned properties, and two PHA-based sutures manufactured using P(4HB) were approved in 2007 by FDA, named Phanton Fiber™ and MonoMax (Manavitehrani et al., 2016). Members of PHA family like P(3HB), P(4HB), P(3HB-co-HV0), and P(3HB-co-3HHX) have all been investigated. For absorbable sutures PHB is noted to be a promising biomaterial when compared to PCL and PGA because its degradation products are less acidic. Additionally, its degradation rate is higher than PCL, PLLA and PHB (Gregory et al., 2022). Sutures synthesised by blending PHB and PHBV had good strength needed for healing myofascial wounds (Guo et al., 2022).
b. **Cardiac Patches:** PHA-based patches have ideal biomaterial properties for cardiovascular applications and the capability to bear heart muscle stresses. During cardiac surgery, PHB has been applied as an anti-adhesive pericardial patch. For the first time, PHB was used as cardiac patch in human patients in 1995. When a multifunctional PHO heart patch was compared to myocardial tissue, it was reported that the biocompatibility of the P(3HO) patches was equivalent to collagen in cell survivability, proliferation and adhesion. Thus, the fabricated cardiac patch was close to myocardial tissue. The porous patch of PHO is highly elastomeric at body temperature, with an elongation at break of about 447%, which can cope with the repeated contraction and relaxation of the heart muscle, myocardium (Pulingam et al., 2022).
c. **Blood Vessels:** To treat blood vessel disorder, tissue engineering is a versatile therapy. PHA, PCL, PLA, and PLGA have been used in vascular tissue engineering to repair and deliver bioactive ingredients to the damaged vessels (Mi et al., 2023). Their flexible nature allows the expansion of blood vessels, showing magnificent benefits for those suffering from cardiovascular blockage. Tissue engineered blood vessels combining with mcl-PHA-based scaffolds are the best choice for this purpose as they have the required qualities like highly adjustable elasticity and strength, their degradation into non-toxic components after the regeneration of blood vessels, potential for cell growth and remodeling and their capability to induce elastin formation. Additionally, no encouragement of infection, stenosis (narrowing of the blood vessels passage), thrombosis (formation of blood clots), or calcification (deposition of Ca and

tissue hardening) was observed (Opitz et al., 2004). PHB and P3 4HB have been evaluated for their applications as blood vessel. P3HB was found to dilute antherosclerosis in mice model by reducing M1 macrophages and enhancing cholesterol efflux through its receptor G protein (Zhang et al., 2021).

d. **Wound Dressings:** Bioactive wound dressings are easy to use, non-toxic and effective in preventing bacterial infection. They promote acute as well as chronic wound healing by covering and protecting it (Yang et al., 2023a). The choice of material for wound dressing must have properties like non-adherent, easily usable, sterilization, non-cytotoxic, hypoallergenic, and can prevent infection. Additionally, the biomaterial is chosen based on its inherent medical properties like promoting angiogenesis, maintaining a moist environment, permitting gaseous exchange, allowing keratinocyte, fibroblast, and epidermal cell proliferation and migration (Gregory et al., 2022). Numerous studies have reported that the polysaccharides present in dressings enhanced healing by accelerating wound re-epithelialization (Hu and Xu, 2020; Mistry et al., 2024). PHA-based bio-adhesive films as wound dressings have been used. Studies revealed that the adhesion of keratinocytes and fibroblasts was better on PHA-based biomaterials when compared to PLLA, a synthetic polymer. *In vitro* and *in vivo* studies showed that P(3HB-co-4HB)/collagen nanofibers are a promising biomaterial and perform better in promoting wound contraction. mcl-PHAs showed good elastic and adhesive properties, suggesting their potential application in the development of natural adhesives for wound dressings and as transdermal patches (Pereira et al., 2019).

4.3 *In Hard Tissue Repair and Regeneration*

a. **Bone Tissue Regeneration:** Bone tissue is made up of collagen protein and hydroxyapatite (HAp) mineral. Autograft or allograft is the first choice for bone tissue repair, but their limited availability restricts this option. To overcome this limitation, there has been an interest in biomaterials composed of apatite, a material mimicking the natural bone matrix. Adipose-derived mesenchymal stem cells are an excellent source of cells in bone tissue engineering (Gao et al., 2024). Biomimetic scaffolds overcome the limits and the adverse effects of inert implants (Rodrigues and Raz, 2022). Bone tissue engineering has been developed as a promising alternative to bone grafting and a solution exhibiting the three required processes of bone healing: osteogenesis, osteoinduction and osteoconduction. The four types of cells in bone tissue are osteogenic cells, osteoblast (bone-building cells that form bone matrix), osteocytes (mature and principal cells of bone tissue) and osteoclasts (derived from monocytes).

Both osteoblasts and osteocytes lack mitosis, and the osteogenic cell is the only cell that can divide and differentiate into osteoblast cells. In the bone regeneration process, the improvement of osteogenesis and vascularization can be achieved by the incorporation of growth factors like bone morphogenetic proteins and vascular endothelial growth factors (Kim et al., 2022; Yang et al., 2023b). The main problem associated with biomaterials that use ceramics as a component of the polymeric matrix is their crystallinity and difficulty in processing (Soleymani et al., 2020). PHA-based scaffolds and composites are important in bone tissue engineering (Bonartsev et al., 2022; Vach Agocsova et al., 2023). Here, polymers of exceptionally high mechanical strength and stiffness are required. It was reported that piezoelectric scaffolds could enhance both osteogenesis and vasculogenesis *in vivo* (D' Alessandro et al., 2021). To improve the osteogenic properties, the incorporation of some organic or inorganic materials such as hydroxyaptite (HA), alginate (ALG), bioactive glass (BG), etc. could improve the osteoinductivity and osteoconductivity of PHA composites. The mineral HA is a key component of natural bone and has good osteoinductive properties.

***In vivo* studies:** PHBV, a commonly used PHA polymer when mixed with bioactive compounds exhibited comparable mechanical strength to human bone and improved the bioactivity of the PHA scaffold. Volkov et al. fabricated ALG hydrogel-based PHB/ HA composite scaffold loaded with MSCs and planted in a parietal bone defective rat. A high regeneration rate (94%) was achieved as the scaffold induced the osteogenic differentiation of mesenchymal cells (Volkov et al., 2020). The *Aloe vera*-derived gel-blended PHBV fibrous scaffold showed promising osteo-inductive potential and complete biodegradation without injurious products (Tahmasebi et al., 2020). Recombinant human bone morphogenetic protein-2 loaded PHA-based scaffold, PHBHHx/rhBMP-2, facilitated angiogenesis and vasculogenesis, and therefore, accelerated bone regeneration was observed (Liu et al., 2020b). In another experiment, PHA scaffolds promoted bone formation in calvarial bone defects in mice. They supported 3T3-L1 preadipocyte viability and proliferation without toxicity (Codreanu et al., 2020). PHB/1% carbon nanotubes (CNTs) electrspun scaffold showed more attachment and proliferation of periodontal ligament stem cells (PDLSCs) on the scaffolds in rat model (Zarei et al., 2020). PHBV with mesoporous bioactive glass nanoparticles (MBGN) and cinnamaldehyde (CIN) fabricated by emulsion solvent extraction method showed enhanced proliferation and attachment of human osteosarcoma cells. Additionally, the rapid formation of hydroxyapatite (HA) on the composite microspheres revealed its bioactivity in bone tissue engineering (Chotchindakun et al., 2021). A 3D printed PHA/β-tricalcium phosphate scaffold (PHA/beta-TCP) synthesized using fused deposition modeling (FDM) exhibited desirable mechanical properties and enhanced osteogenesis (Ye et al., 2022). A designed material, PHB/P3 4HB-BMP2/PDA, showed improved adhesion, proliferation, and osteogenic activity of stem cells in bone tissue engineering (Zhang et al., 2022a). By combining PHBHV matrix and nanoparticles, PHBHV/nHA/C60 was formed with a compression elastic modulus of 0.022 GPa, a promising material to trabecular bone replacement (Ribeiro et al., 2023). When lignin and cellulose were added to a PHB-based nanocomposite, enhanced biological responses and promoted osteogenenic potential of the PHB-lignin/cellulose construct compared to PHB alone was observed due to their synergistic effect (Mohammadalipour et al., 2023). PLA-PHA based nanocomposites with nanohydroxyapatite (nHAp) were prepared and investigated by Injorhor and team. PLA is a polymer with significant porosity, being moldable in different structures, and is of great value in bone tissue engineering. Medical properties such as osteoconductivity from nHAp, biocompatibility, and biodegradability from PHA and PLA make the scaffold suitable for bone tissue engineering (Injorhor et al., 2023). Promoted cell viability, proliferation and differentiation were observed when PHA scaffold fabricated by fused deposition modeling technique was coated with polydopamine (pDA) and hydroxyapatite (HA) (Kim et al., 2024). BaTiO3/PHB nanocomposites made of PHB as a matrix and barium titanate nanoparticles as a filler have shown the best biocompatibility and demonstrated promising features for bone replacement and reconstruction (Strangis et al., 2024).

b. **Cartilage Tissue Regeneration:** Cartilage is a soft, flexible connective tissue present in ribs, joints, nose, trachea, etc., made of chondrocytes that produce large amount of ECM composed of collagen fibers, elastin fibers and proteoglycans. Unlike bone tissue, cartilage is poorly self-repairing because cartilage tissue is avascular and lacks regeneration capabilities. Therefore, providing chondro-inductive growth factors with scaffolds is an effective method to repair and restore the cartilage defects (Wang et al., 2021c; Haghighi and Shamloo, 2021; Heirani-Tabasi et al., 2021). The stiffness of the scl-PHAs can be enhanced by blending or by making copolymers, e.g., P(3HB)/P(3HB-co-3HHx) and P(3HB)/P(3HO), making them suitable for use in cartilage tissue engineering (Liu et al., 2020b). Other polymers applied for cartilage tissue regeneration are chitosan, collagen, gelatin, fibrin, and alginate.

***In vivo* studies:** PHAs have shown promising results in cartilage tissue repair and regeneration. Improved hydrophilicity and increased regeneration of cartilage matrix were noted when PHBV-BG was designed by introducing bioglass in PHBV (Zhou and Yu, 2014). PHB with PHBHHx copolymer, PHB/PHBHHx scaffolds showed good viability, chondrogenic differentiation, and glycosaminoglycan synthesis. Enhanced production of chondrocytes and collagen accelerated the rate of cartilage healing in a cartilage defect in rabbit knee joints (Ba et al., 2019). Quercetin-modified electrospun PHBV fibrous scaffold accelerated chondrocyte adhesion and multiplication, inhibited their apoptosis, managed oxidative stress in them, regulated the cartilage ECM related genes, and showed enhanced cartilage regeneration (Chen et al., 2021). PHB/Chitosan based accelular scaffold when implanted in a sheep knee cartilage defect repair model, acted as a template for repair tissue formation. Promoted integration between repair tissue and native articular cartilage resulted in stimulated chondrogenesis. Chitosan prevent cartilage degradation and promoted healing process (Petrovova et al., 2021). Glucosamine sulphate (GAS) is a major components of ECM of cartilage tissue and plays a positive role in hydrophilicity and in the initiation of chondrogenic differentiation (Golshayan et al., 2022). Electrospun fibrous scaffold based on PHB and decellularised Wharton's jelly extracellular matrix (DWJM), showed excellent biocompatibility, improved hydrophilicity, degradation rate and chondrocyte adhesion, suitable for articular cartilage tissue engineering (Soleymani et al., 2024). ECM nanoparticles were synthesized and incorporated into the PHB-Cs eletrospun scaffold containing PHB, chitosan (CS), and ECM nanoparticles showed promoted proliferation of human costal chondrocytes (Mohammadi et al., 2024).

4.4 PHA as Medical Implants

PHAs are been used as implants to replace, repair and regenerate damaged and/or diseased body parts. *In vivo* implants of PHA-based biomaterials have been made possible due to their bio-compatibility, non-toxic degradation products, and a wide range of physico-chemical properties, desired surface modifications, cellular growth support, and attachment without carcinogenic effects.

a. **Heart Valves:** The bio-mimetic approach involves an integrated use of cells, biomaterials, scaffolds, and bioreactors to engineer compact, multi-meter thick, synchronously contracting cardiac-tissue contracts. Recent studies have shown that PHA based absorbable polymers are excellent substrates for regeneration of functional heart valves due to some promising characteristics. Heart valve tissue scaffolds should be elastic and conductive. PHAs have been used for tissue engineering of heart valves because they are thermoplastic polymers and can be molded into the desired shape. Sodian et al. synthesised a P3HB4HB-based scaffold using salt leaching technique. The trileaflet heart valve construct showed desired elasticity and satisfactory hemodynamic function (Sodian et al., 2000). PHA polymers can also be used as heart valve coatings or patches. Coating with PHBHHx showed better resistance and tensile strength along with a reduction in calcification *in vivo.* For proper regeneration, hydrogels were loaded with drugs, growth factors, or cells (Bagheri et al., 2019).

b. **Stent:** One of the common difficulties with the use of metal stents is their association with the risk of restenosis, which occurs when the artery again narrows around an implanted stent. To overcome this, synthesis of biodegradable polymer stent, which open a blocked artery and then degrade before restenosis occurs, is an effective option (Gregory et al., 2022). Lesser restenosis in implanted stents proved them to be a promising biomaterial for their application as cardiac stents.

c. **Nerve guidance Conduit (NGC):** There is no effective therapy to treat nerve injuries. Due to the required second surgery as well as donar site morbidity associated with auto grafts, researchers have focused on the biosynthesis of artificial nerve guidance conduits as an alternative option for critical gap injury use (Zarrintaj et al., 2023). The peripheral nervous system has the capability of axonal regeneration, whereas the axons in the central nervous system (CNS) do not regenerate due to presence of growth inhibitors. For nerve regeneration, the basic components required are: A 3D scaffold, support cells, neurotrophic factors and cellular matrix. NGCs are artificial nerve grafts for guiding axonal regrowth. They are used to protect and bridge the small gap (10–30 mm). PHA scaffolds enhanced Schwann cells (SCs) activity and supported axonal regeneration (Mendibil et al., 2021). PHA strips seeded with regenerative cells accelerate the peripheral regeneration process. Adding CS (chitosan) to the PHB scaffolds increased the percent porosity and decreased the angle of contact with water (Karimi et al., 2020). Mcl-PHAs are blended with P(3HB) to produce P(3HO-co-3HB) as a potential matrix for peripheral nerve regeneration using NG108-15 neuronal cells (Nigmatullin et al., 2023). Liu and co-workers added graphene/gold into P(3HB-co-3HHx) fibers. The synthesized electrospun fibrous PHA/RGO/Au hybrid supported Schwann cells multiplication and peripheral nerve regeneration (Liu et al., 2020a). In an another study, scaffold based on PHA and bioactive glass (BG) showed enhanced neuronal growth. Is 1% addition into 25:75 PHO/PHB blend, promoted peripheral nerve regeneration (Lizariaga-Valderrama et al., 2020). Such techniques are also being tried for nerve repair in the CNS, but here, nerve regeneration is difficult because its axons do not regenerate appreciably in their native environment.

4.5 PHA-Based Micro and Nano-Carriers for Drug Delivery

Conventional drug delivery systems suffer from poor bioavailability, insolubility in water, fluctuations in plasma drug level, low absorption by the intestine and an inability to achieve controlled and sustained drug release to the appropriate site of action (Adepu and Ramakrishna, 2021). One of the major aims of drug delivery systems (DDS) is to improve the bioavailability of active compounds and to improve the therapeutic profile of drugs (Li et al., 2022). An ideal DDS should be inert, biocompatible, stable, provide controlled and prolonged release of pharmaceutical substances, be easy to manufacture, and be compatible with wide range of drugs. Modern drug delivery systems have been made possible by advances in polymeric sciences (Duran-Lara et al., 2024; Elmowafy et al., 2019). PHAs are excellent candidates for controlled and targeted drug delivery systems (Nigmatullin et al., 2015; Liu et al., 2023a; Murueva et al., 2023). In addition, they are economic, environmentally friendly, and without any side effects (Ashwini and Mahalingam, 2020).

***In vivo* Studies:** PHAs have been potentially applied as drug carriers for delivery systems against a range of diseases (Diniz et al., 2023; Lopes, 2022; Paul et al., 2024). PHA micro- and nano-sized carriers can efficiently deliver the bioactive compounds such as antibiotics (Pulingam et al., 2022; Rakkan et al., 2024), anticancer drugs (Babos et al., 2020), and growth factors (Mohlashami et al., 2020; Cendal et al., 2024) to the target tissue as shown in Table 5. PHA-based biodegradable implants themselves can release the drug in prescribed dosages and transport them by stimulating the nervous system and dissolve in the body after a specific time frame. Nanoparticles may be synthesized by matrixing or encapsulating with different types of polymers (Singh et al., 2024). The encapsulation of bioactive pharmacological compounds in a polymeric matrix helps in protecting and enhancing thermal and oxidative stability, bioavailability, and bio-accessibility of the bioactive compounds, resulting in a significant improvement in the efficacy of treatment (Prakash et al., 2022). An alginate-based encapsulation system has been used to maintain cell viability and release of therapeutic molecules. It is anionic polymer having desirable characteristics such as excellent chemical stability, non-toxic, hydrophilic, and high biocompatibility. In addition, alginate do not interfere with the survival, differentiation and maturation of cells. Micro and nano particles are being developed using copolymers with 3-hydroxybutyric-co-3-hydroxyvaleric acid P(3HB-co-3HV), 3-hydroxybutyric-co-3-hydroxyhexanoic acid P(3HB-co-

Table 5. Drug release studies in animal models (In Last Five Years).

***In vivo* study**	**Reference**
Lime oil loaded P(3HO-co-3HD) film showed antimicrobial activity against *Staphylococcus aureus* and *E. coli.* Reduction in antimicrobial activity was noted with decreased lime oil content with time.	Basnett et al., 2020
Controlled and prolonged release of anti-inflammatory agent, diclofenac using PHA scaffolds of tricalcium phosphate (TCP) and PHO after bone surgery	Agarwal et al., 2020
CS/PHB composite membranes coated with kaemferol has powerful antibacterial potential against multidrug resistant bacteria and supported wound healing	Rofeal et al., 2021
Incorporation of polyenes in 3-hydroxydecanoic acid added mcl-PHA film showed excellent inhibitory activity against fungi and formation of biofilm	Pekmezovic et al., 2021
Cinnamaldehyde-loaded mesosporous bioactive glass nanoparticles/PHBV-based composite, PHBV/MBGN/CIN significantly exhibited antibacterial activity.	Chotchindakun et al., 2021
Encapsulated Nys and AmB in mcl PHA microshpheres reduced the systemic drug toxicity and showed improved effect against fungal infection	Pavic et al., 2022
Hybrid nanocomposites of PHBHV with nanohydroxyapatite, Fullerene and vancomycin, PHBHV/nHA/C60/VC showed good ability in adhering macrophage cell and antibacterial effects in osteomyelitis, a bacterial disease.	Ribeiro et al., 2023
PHAs based composite, P(3HB)/PEG and P(3HB)/PLGA, loaded with 5-fluorouracil or ceftriaxone showed antibacterial and antitumor activity in *E. coli* and HeLa cultures	Murueva et al., 2023
PHA based scaffold containing (3HB) and (4HB) (P34HB) and was seeded with zinc oxide nanoparticles showed significant improvement in healing process	Chen et al., 2023a
PHA nanoparticles for controlled and sustained release of transforming growth factor, beta-3(TGF-beta3) in osteoarthritis therapy.	Cendal et al., 2024
Promoted antibacterial activity by enhanced drug release and reduced cytotoxicity activity of fabricated PHA film incorporated with antibacterial drug, OCT	Rakkan et al., 2024

3HHx), and also composites with other biocompatible materials (Pramanik, 2023). The biocompatibility of these materials, their degradation into components already present in the blood, non-immunogenicity, low crystallinity, and low melting point are some excellent properties that make them suitable for the development of drug carriers to provide controlled release of both hydroplilic and hydrophobic therapeutic agents in constant doses over long periods (Carvalho et al., 2021). Nanoparticles, having unique physical properties, can be designed to protect the loaded drug against molecular degradation and denaturation within the body by the natural degradation process and reduce the initial drug concentrations and the number of administrations (Scicluna and Vella-Zarb, 2020). The selected fabrication methodology influences the size, stability, entrapment efficiency, and polymeric microparticles' release kinetics (Lagreca et al., 2020). PHB and PHBV have been evaluated for the development of effective drug delivery systems as they have desirable properties like the capacity to connect with other polymers and no adverse effect on platelete reactivity (Ansari et al., 2021). Entrapping natural antimicrobial compounds (NACs) within PHAs matrices has recently been used to deliver antibiotics, growth factors, vaccines, nucleic acids, steroids, etc. (Romero-Castelan et al., 2023). The higher activity of lime oil loaded P(3HO-co-3HD) film against *Staphylococcus aureus* and *E. coli* was prepared as an effective material for a drug delivery system (Basnett et al., 2020). Marian and colleagues prepared a chitosan/PHB blend loaded with kaempferol, which demonstrated powerful killing effects against multi-drug resistant bacteria at low concentrations and helped in infected wound healing (Rofeal et al., 2021). A PHA monomer, C10-OH was added to mcl-PHA film to achieve an enhanced synergistic effect with polyenes against fungal growth (Pekmezovic et al., 2021). Mesoporous bioactive glass nanoparticles (MBGN) were incorporated in PHBV to enhance its bioactivity, and cinnamaldehyde (CIN) was loaded to introduce antibacterial activity (Chotchindakun et al., 2021). The use of the broad-spectrum antifungal drug, amphotericin B (AmB) has been limited due to its side effects and toxicity. Its poor stability, low membrane permeability and low aqueous solubility were improved by encapsulating nystatin (Nys) and AmB in mcl-PHA microspheres. In another study, reduced systemic drug toxicity and improved effect against fungal infection were also observed (Pavic et al., 2022). The incorporation of 3-hydroxy-pentadecanoic acid (3OHC15) improved the antibacterial activity of an antibiotic (OCT, octenidine dihydrochloride) loaded PHA drug carrier (Rakkan et al., 2024). It enhanced the antibacterial activity and reduced the cytotoxic activity.

5. Conclusion

The search for biodegradable and sustainable plastics for packing and biomedical applications is currently encouraging. Microbial PHAs are the novel candidates due to their promising capabilities in mitigating pollution issues. They can be extracted from a wide range of bacteria, along with fungi, plants, and carbon sources. Additionally, they are materials of specific applications, such as being biocompatible, having non-toxic degradable products, ease of customization, and piezoelectric properties. Among others biological orgin, thermosoftening polymers with physico-chemical properties suitable for the various strategies have been used to modify their properties, with blending being one of the most attractive options due to its ease in processability. PHAs, along with their blends and variants, have established themselves as a versatile biomaterial for multiple biological applications in the biomedical field, such as medical devices, wound healing and skin regeneration, cardiovascular tissue, organ tissue and other soft tissue repair and regeneration, engineered scaffolds for hard tissue (bone and cartilage), implantable devices, and as PHA-based micro- and nano-carriers for controlled drug liberation. The potential of multiple tissue repair and restoration of the functions using PHAs-based constructs are discussed with suitable examples. In conclusion, this chapter serves as a valuable resource to understand PHA-based biomaterials with improved properties and their wide range of applications in tissue repair, regeneration, and as controlled drug delivery system. Rapid advancement have solve certain drawbacks, however, further work needed to improve PHA production methods, assess the longstanding ecological impacts, and evaluate the workability and feasibility of PHAs in various sectors.

6. Challenges and Future Perspectives

PHAs have made remarkable contributions in the medical sector with a wide range of applications, ranging from tissue-engineered scaffolds and 3D printing medical implants to drug delivery carriers. However, their extensive applications in tissue repair and regeneration are limited due to their brittleness, lack of elasticity, low fabricability, and unsuccessful re-usability (Zhou et al., 2023). Along with limitations in mechanical properties, customizability and processing performance, their higher production costs associated with fermentation and purification are the main challenge. Cost-effective strategies to decline the production cost and enhance the process efficiency need to be addressed. PHA extraction and purification methods should be replaced by more environmentally friendly methods in the future. Further study to improve diversity and functionalities of scaffolds is also needed. Metabolic engineering approaches, including genetic modification of the bacterial genetic constitution using recombinant DNA technology and the enzyme engineering can be employed. Collaborative research work is needed to get the enormous benefits of PHAs for well-being. In the near future, PHAs and their blends can replace non-biodegrable polymers in various applications. The diversity in the physico-chemical properties of PHAs will be used more effectively in the biomedical sector, as surgical sutures, implants, tissue-engineered scaffolds; in the pharmacological sector as controlled-release carriers for drug delivery; in the agriculture sector as mulching films, greenhouse films, biocontrol agents, and carriers for the gradual release of fertilizers, herbicides and insecticides for sustainable agriculture, as biofuel, antimicrobial coated films, eco-friendly carry bags, and many more (Saravanan et al., 2022). Controlled and sustained release of anticancer drugs is a significant approach of drug delivery system and can be achieved by PHA nanocarriers. The potential of this strategy and other underutilized customized properties of PHAs are valuable, with promising ongoing research indicating that this polymer can provide an excellent biomaterial for various fields, marking towards a more sustainable and adaptive future.

References

Abazari, M. F., S. Gholizadeh, S. Z. Karizi, N. H. Birgani, D. Abazari, S. Paknia et al. 2021. Recent advances in cellulose-based structures as the wound-healing biomaterials: A clinically oriented review. Applied Sciences 11(17): 7769.

Abazari, M. F., S. Z. Karizi, N. Hajati-Birgani, S. Norouzi, Z. Khazeni, J. Hashemi et al. 2021. PHBV nanofibers promotes insulin-producing cells differentiation of human induced pluripotent stem cells. Gene 768: 145333.

Abdelaziz, A. G., H. Nageh, S. M. Abdo, M. S. Abdalla, A. A. Amer, A. Abdal-Hay et al. 2023. A review of 3D polymeric scaffolds for bone tissue engineering: principles, fabrication techniques, immunomodulatory roles, and challenges. Bioengineering 10(2): 204.

Adak, S., K. Ramalingam, M. Bishai, S. Jacob and D. Kundu. 2024. Advancements in microbial production of polyhydroxyalkanoates (PHA) from wastes for sustainable active food packaging: an eclectic review. Biocatalysis and Agricultural Biotechnology 103288.

Adepu, S. and S. Ramakrishna. 2021. Controlled drug delivery systems: Current status and future directions. Molecules 26(19): 5905.

Agarwal, S. 2020. Biodegradable polymers: Present opportunities and challenges in providing a microplastic-free environment. Macromolecular Chemistry and Physics 221(6): 2000017.

Alsayari, A., A. B. Muhsinah, D. Almaghaslah, S. Annadurai and S. Wahab. 2021. Pharmacological efficacy of ginseng against respiratory tract infections. Molecules 26(13): 4095.

Alven, S., B. Buyana, Z. Feketshane and B. A. Aderibigbe. 2021. Electrospun nanofibers/nanofibrous scaffolds loaded with silver nanoparticles as effective antibacterial wound dressing materials. Pharmaceutics 13(7): 964.

Ang, S. L., R. Sivashankari, B. Shaharuddin, J. A. Chuah, T. Tsuge, H. Abe et al. 2020. Potential applications of polyhydroxyalkanoates as a biomaterial for the aging population. Polymer Degradation and Stability 181: 109371.

Anjum, A., M. Zuber, K. M. Zia, A. Noreen, M. N. Anjum and S. Tabasum. 2016. Microbial production of polyhydroxyalkanoates (PHAs) and its copolymers: A review of recent advancements. International Journal of Biological Macromolecules 89: 161–174.

Ansari, A. I. and N. A. Sheikh. 2023. A review on materials application in scaffold design by fused deposition method. Journal of The Institution of Engineers (India): Series C 104(6): 1247–1265.

Ansari, S., N. Sami, D. Yasin, N. Ahmad and T. Fatma. 2021. Biomedical applications of environmental friendly poly-hydroxyalkanoates. International Journal of Biological Macromolecules 183: 549–563.

Antonova, L. V., E. O. Krivkina, V. V. Sevostianova, A. V. Mironov, M. A. Rezvova, A. R. Shabaev et al. 2021. Tissue-engineered carotid artery interposition grafts demonstrate high primary patency and promote vascular tissue regeneration in the ovine model. Polymers 13(16): 2637.

Arampatzis, A. S., K. Giannakoula, K. N. Kontogiannopoulos, K. Theodoridis, E. Aggelidou, A. Rat et al. 2021. Novel electrospun poly-hydroxybutyrate scaffolds as carriers for the wound healing agents alkannins and shikonins. Regenerative Biomaterials 8(3): rbab011.

Asare, E., D. A. Gregory, A. Fricker, E. Marcello, A. Paxinou, C. S. Taylor et al. 2020. Polyhydroxyalkanoates, their processing and biomedical applications. pp. 255–284. In The Handbook of Polyhydroxyalkanoates. CRC Press.

Ashwini, D. and G. Mahalingam. 2020. Green synthesized metal nanoparticles, characterization and its antidiabetic activities—a review. Research Journal of Pharmacy and Technology 13(1): 468–474.

Augustine, R., A. A. Zahid, A. Hasan, Y. B. Dalvi and J. Jacob. 2020. Cerium oxide nanoparticle-loaded gelatin methacryloyl hydrogel wound-healing patch with free radical scavenging activity. ACS Biomaterials Science & Engineering 7(1): 279–290.

Awasthi, S. K., M. Kumar, V. Kumar, S. Sarsaiya, P. Anerao, P. Ghosh et al. 2022. A comprehensive review on recent advancements in biodegradation and sustainable management of biopolymers. Environmental Pollution 307: 119600.

Azimi, B., L. Thomas, A. Fusco, O. I. Kalaoglu-Altan, P. Basnett, P. Cinelli et al. 2020. Electrosprayed chitin nanofibril/electrospun polyhydroxyalkanoate fiber mesh as functional nonwoven for skin application. Journal of Functional Biomaterials 11(3): 62.

Azizi, N., R. Eslami, S. Goudarzi, H. Younesi and H. Zarrin. 2024. A review of current achievements and recent challenges in bacterial medium-chain-length polyhydroxyalkanoates: production and potential applications. Biomacromolecules 25(5): 2679–2700.

Ba, K., X. Wei, D. Ni, N. Li, T. Du, X. Wang et al. 2019. Chondrocyte co-cultures with the stromal vascular fraction of adipose tissue in polyhydroxybutyrate/poly-(hydroxybutyrate-co-hydroxyhexanoate) scaffolds: evaluation of cartilage repair in rabbit. Cell Transplantation 28(11): 1432–1438.

Babos, G., J. Rydz, M. Kawalec, M. Klim, A. Fodor-Kardos, L. Trif et al. 2020. Poly (3-hydroxybutyrate)-based nanoparticles for sorafenib and doxorubicin anticancer drug delivery. International Journal of Molecular Sciences 21(19): 7312.

Bagheri, B., P. Zarrintaj, S. S. Surwase, N. Baheiraei, M. R. Saeb, M. Mozafari et al. 2019. Self-gelling electroactive hydrogels based on chitosan–aniline oligomers/agarose for neural tissue engineering with on-demand drug release. Colloids and Surfaces B: Biointerfaces 184: 110549.

Bairwan, R. D., E. B. Yahya, D. Gopakumar and H. A. Khalil. 2024. Recent advances in poly (3-Hydroxybutyrate-co-3-Hydroxyvalerate) biocomposites in sustainable packaging applications. Adv Mater Lett 24011739.

Balogová, A. F., R. Hudák, T. Tóth, M. Schnitzer, J. Feranc, D. Bakoš et al. 2018. Determination of geometrical and viscoelastic properties of PLA/PHB samples made by additive manufacturing for urethral substitution. Journal of Biotechnology 284: 123–130.

Balogova, A., M. Trebuňová, R. Hudák and J. Zivcák. 2020. Possibilities of an artificial urethra reconstruction. IJISET-International Journal of Innovative Science 7: 125–135.

Basnett, P., E. Marcello, B. Lukasiewicz, R. Nigmatullin, A. Paxinou, M. H. Ahmad et al. 2020. Antimicrobial materials with lime oil and a poly (3-hydroxyalkanoate) produced via valorisation of sugar cane molasses. Journal of Functional Biomaterials 11(2): 24.
Behera, S., M. Priyadarshanee and S. Das. 2022. Polyhydroxyalkanoates, the bioplastics of microbial origin: Properties, biochemical synthesis, and their applications. Chemosphere 294: 133723.
Bhati, R., S. Samantaray, L. Sharma and N. Mallick. 2010. Poly-β-hydroxybutyrate accumulation in cyanobacteria under photoautotrophy. Biotechnology Journal 5(11): 1181–1185.
Bonartsev, A. P., V. V. Voinova, A. V. Volkov, A. A. Muraev, E. M. Boyko, A. A. Venediktov et al. 2022. Scaffolds based on poly (3-hydroxybutyrate) and its copolymers for bone tissue engineering. Современные технологии в медицине 14(5 (eng)): 78–90.
Borchiellini, P., A. Rames, F. Roubertie, N. L'heureux and F. Kawecki. 2023. Development and characterization of biological sutures made of cell-assembled extracellular matrix. Biofabrication 15(4): 045018.
Broadwin, M., F. Imarhia, A. Oh, S C. R. tone, F. W. Sellke, S. Bhowmick and M. R. Abid. 2024. Exploring electrospun scaffold innovations in cardiovascular therapy: a review of electrospinning in cardiovascular disease. Bioengineering 11(3): 218.
Calori, I. R., G. Braga, P. D. C. C. de Jesus, H. Bi and A. C. Tedesco. 2020. Polymer scaffolds as drug delivery systems. European Polymer Journal 129: 109621.
Cao, Z., C. Dou and S. Dong. 2014. Scaffolding biomaterials for cartilage regeneration. Journal of Nanomaterials 2014(1): 489128.
Carvalho, L. T., T. A. Vieira, Y. Zhao, A. Celli, S. F. Medeiros and T. M. Lacerda. 2021. Recent advances in the production of biomedical systems based on polyhydroxyalkanoates and exopolysaccharides. International Journal of Biological Macromolecules 183: 1514–1539.
Cavaillé, L., E. Grousseau, M. Pocquet, A. S. Lepeuple, J. L. Uribelarrea, G. Hernandez-Raquet et al. 2013. Polyhydroxybutyrate production by direct use of waste activated sludge in phosphorus-limited fed-batch culture. Bioresource Technology 149: 301–309.
Cendal, A. I. R., I. G. Seoane, J. S. Rodríguez, F. J. D. T. Santos, I. M. F. Boquete, L. C. Mouzo et al. 2024. Polyhydroxyalkanoate nanoparticles for strategic drug delivery in cartilage tissue engineering. Osteoarthritis and Cartilage 32: S290–S291.
Cesarino, I., I. Moroz, A. F. Carbonera, G. de Oliveira Martins, L. G. L. Fernandes and A. L. Leao. 2024. PHA/PHB/PHBV-based composites: development and biomedical applications. pp. 107–123. In Bioresorbable Polymers and their Composites. Woodhead Publishing.
Chandra, R., A. Thakor, T. H. Mekonnen, T. C. Charles and H. S. Lee. 2023. Production of polyhydroxyalkanoate (PHA) copolymer from food waste using mixed culture for carboxylate production and Pseudomonas putida for PHA synthesis. Journal of Environmental Management 336: 117650.
Chen, F., X. Liu, X. Ge, Y. Wang, Z. Zhao, X. Zhang et al. 2023a. Porous polydroxyalkanoates (PHA) scaffolds with antibacterial property for oral soft tissue regeneration. Chemical Engineering Journal 451: 138899.
Chen, L., L. Wei, X. Su, L. Qin, Z. Xu, X. Huang et al. 2023b. Preparation and characterization of biomimetic functional scaffold with gradient structure for osteochondral defect repair. Bioengineering 10(2): 213.
Chen, S., W. Chen, Y. Chen, X. Mo and C. Fan. 2021. Chondroitin sulfate modified 3D porous electrospun nanofiber scaffolds promote cartilage regeneration. Materials Science and Engineering: C 118: 111312.
Chingale, M., K. Cheng and K. Huang. 2022. 3D Bioprinting technology–one step closer towards cardiac tissue regeneration. Frontiers in Materials 8: 804134.
Chotchindakun, K., J. Pekkoh, J. Ruangsuriya, K. Zheng, I. Unalan and A. R. Boccaccini. 2021. Fabrication and characterization of cinnamaldehyde-loaded mesoporous bioactive glass nanoparticles/PHBV-based microspheres for preventing bacterial infection and promoting bone tissue regeneration. Polymers 13(11): 1794.
Chu, C. C. 2013. Types and properties of surgical sutures. pp. 231–273. In Biotextiles as Medical Implants. Woodhead Publishing.
Chuah, J. A., M. Yamada, S. Taguchi, K. Sudesh, Y. Doi and K. Numata. 2013. Biosynthesis and characterization of polyhydroxyalkanoate containing 5-hydroxyvalerate units: effects of 5HV units

on biodegradability, cytotoxicity, mechanical and thermal properties. Polymer Degradation and Stability 98(1): 331–338.

Codreanu, A., C. Balta, H. Herman, C. Cotoraci, C. V. Mihali, N. Zurbau et al. 2020. Bacterial cellulose-modified polyhydroxyalkanoates scaffolds promotes bone formation in critical size calvarial defects in mice. Materials 13(6): 1433.

Corti Monzon, G., G. Bertola, M. K. Herrera Seitz and S. E. Murialdo. 2024. Exploring polyhydroxyalkanoates biosynthesis using hydrocarbons as carbon source: a comprehensive review. Biodegradation 1–20.

Culbreath, C. J., B. Gaerke, M. S. Taylor, S. D. McCullen and O. T. Mefford. 2020. Effect of infill on resulting mechanical properties of additive manufactured bioresorbable polymers for medical devices. Materialia 12: 100732.

D'Alessandro, D., C. Ricci, M. Milazzo, G. Strangis, F. Forli, G. Buda et al. 2021. Piezoelectric signals in vascularized bone regeneration. Biomolecules 11(11): 1731.

Datta, S. and G. Menon. 2019. Nanofibers from polyhydroxyalkanoates and their applications in tissue engineering. Biotechnological Applications of Polyhydroxyalkanoates 409–420.

De la Ossa, J. G., S. Danti, J. Esposito Salsano, B. Azimi, V. Tempesti, N. Barbani et al. 2022. Electrospun poly(3-hydroxybutyrate-co-3-hydroxyvalerate)/olive leaf extract fiber mesh as prospective bio-based scaffold for wound healing. Molecules 27(19): 6208.

de Meneses, L., J. R. Pereira, C. Sevrin, C. Grandfils, A. Paiva, M. A. Reis and F. Freitas. 2020. Pseudomonas chlororaphis as a multiproduct platform: Conversion of glycerol into high-value biopolymers and phenazines. New Biotechnology 55: 84–90.

Deepthi, S., M. N. Sundaram, P. Vijayan, S. V. Nair and R. Jayakumar. 2018. Engineering poly (hydroxy butyrate-co-hydroxy valerate) based vascular scaffolds to mimic native artery. International Journal of Biological Macromolecules 109: 85–98.

Deng, Y., K. Zhao, X. F. Zhang, P. Hu and G. Q. Chen. 2002. Study on the three-dimensional proliferation of rabbit articular cartilage-derived chondrocytes on polyhydroxyalkanoate scaffolds. Biomaterials 23(20): 4049–4056.

Dhania, S., M. Bernela, R. Rani, M. Parsad, S. Grewal, S. Kumari et al. 2022. Scaffolds the backbone of tissue engineering: Advancements in use of polyhydroxyalkanoates (PHA). International Journal of Biological Macromolecules 208: 243–259.

Diniz, M. S. D. F., M. M. Mourão, L. P. Xavier and A. V. Santos. 2023. Recent biotechnological applications of polyhydroxyalkanoates (PHA) in the biomedical sector—a review. Polymers 15(22): 4405.

Durán-Lara, E. F., D. Rafael, F. Andrade and S. Vijayakumar. 2024. Bacterial polyhydroxyalkanoates-based therapeutics-delivery nanosystems. Current Medicinal Chemistry.

Eesaee, M., P. Ghassemi, D. D. Nguyen, S. Thomas, S. Elkoun and P. Nguyen-Tri. 2022. Morphology and crystallization behaviour of polyhydroxyalkanoates-based blends and composites: A review. Biochemical Engineering Journal 187: 108588.

Elazzazy, A. M., K. Ali Abd, N. M. Bataweel, M. M. Mahmoud and A. M. Baghdadi. 2024. Microbial biosynthesis of medium-chain-length polyhydroxyalkanoate (mcl-PHA) from waste cooking oil. Polymers 16(15): 2150.

Evode, N., S. A. Qamar, M. Bilal, D. Barceló and H. M. Iqbal. 2021. Plastic waste and its management strategies for environmental sustainability. Case Studies in Chemical and Environmental Engineering 4: 100142.

Fangkrathok, N., J. Junlatat, K. Umehara, H. Noguchi and B. Sripanidkulchai. 2014. Cytotoxic and immunomodulatory effects of polyhydroxyoctane isolated from Lentinus polychrous mycelia. Journal of Natural Medicines 68: 302–309.

Farag, M. M. 2023. Recent trends on biomaterials for tissue regeneration applications. Journal of Materials Science 58(2): 527–558.

Feng, P., R. Zhao, W. Tang, F. Yang, H. Tian, S. Peng et al. 2023. Structural and functional adaptive artificial bone: materials, fabrications, and properties. Advanced Functional Materials 33(23): 2214726.

Foroughi, M. R., S. Karbasi, M. Khoroushi and A. A. Khademi. 2017. Polyhydroxybutyrate/chitosan/bioglass nanocomposite as a novel electrospun scaffold: Fabrication and characterization. Journal of Porous Materials 24: 1447–1460.

Fu, E. and Z. Li. 2024. Extracellular vesicles: A new frontier in the theranostics of cardiovascular diseases. iRADIOLOGY.

Gao, C., W. Yuan, D. Wang, X. Zhang, T. Zhang and Z. Zhou. 2024. Adipose-derived mesenchymal stem cell-incorporated PLLA porous microspheres for cartilage regeneration. Animal Models and Experimental Medicine.

Golshayan, N. S., S. Karbasi, E. Masaeli, E. Bahremandi-Toloue, M. H. Nasr-Esfahani and M. Rafienia. 2022. Evaluation of the effects of glucosamine sulfate on poly (3-hydroxybutyrate)-chitosan/carbon nanotubes electrospun scaffold for cartilage tissue engineering applications. Polymer-Plastics Technology and Materials 61(11): 1244–1264.

González-Villanueva, M., H. Galaiya, P. Staniland, J. Staniland, I. Savill, T. S. Wong et al. 2019. Adaptive laboratory evolution of Cupriavidus necator H16 for carbon co-utilization with glycerol. International Journal of Molecular Sciences 20(22): 5737.

Gopi, S., M. Kontopoulou, B. A. Ramsay and J. A. Ramsay. 2018. Manipulating the structure of medium-chain-length polyhydroxyalkanoate (MCL-PHA) to enhance thermal properties and crystallization kinetics. International Journal of Biological Macromolecules 119: 1248–1255.

Gregory, D. A., C. S. Taylor, A. T. Fricker, E. Asare, S. S. Tetali, J. W. Haycock et al. 2022. Polyhydroxyalkanoates and their advances for biomedical applications. Trends in Molecular Medicine 28(4): 331–342.

Grigore, M. E., R. M. Grigorescu, L. Iancu, R. M. Ion, C. Zaharia and E. R. Andrei. 2019. Methods of synthesis, properties and biomedical applications of polyhydroxyalkanoates: a review. Journal of Biomaterials Science, Polymer Edition 30(9): 695–712.

Gundu, S., N. Varshney, A. K. Sahi and S. K. Mahto. 2022. Recent developments of biomaterial scaffolds and regenerative approaches for craniomaxillofacial bone tissue engineering. Journal of Polymer Research 29(3): 73.

Guo, W., K. Yang, X. Qin, R. Luo, H. Wang and R. Huang. 2022. Polyhydroxyalkanoates in tissue repair and regeneration. Engineered Regeneration 3(1): 24–40.

Haghighi, P. and A. Shamloo. 2021. Fabrication of a novel 3D scaffold for cartilage tissue repair: *In-vitro* and *in-vivo* study. Materials Science and Engineering: C 128: 112285.

Hahn, T., M. O. Alzate, S. Leonhardt, P. Tamang and S. Zibek. 2024. Current trends in medium-chain-length polyhydroxyalkanoates: Microbial production, purification, and characterization. Engineering in Life Sciences 24(6): 2300211.

Heirani-Tabasi, A., S. Hosseinzadeh, S. Rabbani, S. H. A. Tafti, K. Jamshidi, M. Soufizomorrod et al. 2021. Cartilage tissue engineering by co-transplantation of chondrocyte extracellular vesicles and mesenchymal stem cells, entrapped in chitosan–hyaluronic acid hydrogel. Biomedical Materials 16(5): 055003.

Hu, H. and F. J. Xu. 2020. Rational design and latest advances of polysaccharide-based hydrogels for wound healing. Biomaterials Science 8(8): 2084–2101.

Injorhor, P., T. Trongsatitkul, J. Wittayakun, C. Ruksakulpiwat and Y. Ruksakulpiwat. 2023. Biodegradable polylactic acid-polyhydroxyalkanoate-based nanocomposites with bio-hydroxyapatite: preparation and characterization. Polymers 15(5): 1261.

Iranpour, H., A. Alipour, S. Zhand and H. Shahsavarani. 2024. Polymers as medical adhesives and sutures. pp. 191–205. In Polymeric Materials for Biomedical Implants. Woodhead Publishing.

Israni, N. and S. Shivakumar. 2020. Polyhydroxyalkanoate (PHA) biosynthesis from directly valorized ragi husk and sesame oil cake by Bacillus megaterium strain Ti3: Statistical optimization and characterization. International Journal of Biological Macromolecules 148: 20–30.

Jana, S., B. J. Tefft, D. B. Spoon and R. D. Simari. 2014. Scaffolds for tissue engineering of cardiac valves. Acta Biomaterialia 10(7): 2877–2893.

Janoušková, O. 2018. Synthetic polymer scaffolds for soft tissue engineering. Physiological Research 67.

Johnston, B., G. Adamus, A. I. Ekere, M. Kowalczuk, F. Tchuenbou-Magaia and I. Radecka. 2022. Bioconversion of plastic waste based on mass full carbon backbone polymeric materials to value-added polyhydroxyalkanoates (PHAs). Bioengineering 9(9): 432.

Kalia, V. C., S. K. S. Patel, R. Shanmugam and J. K. Lee. 2021. Polyhydroxyalkanoates: Trends and advances toward biotechnological applications. Bioresource Technology 326: 124737.

Kalia, V. C., S. K. Patel and J. K. Lee. 2023. Exploiting polyhydroxyalkanoates for biomedical applications. Polymers 15(8): 1937.

Kalia, V. C., S. K. Patel, K. K. Karthikeyan, M. Jeya, I. W. Kim and J. K. Lee. 2024. Manipulating microbial cell morphology for the sustainable production of biopolymers. Polymers 16(3): 410.

Kanbe, M. T., L. F. da Silva and J. G. C. Gomez. 2024. Cross-sectoral citrus sugarcane biorefinery: polyhydroxyalkanoates production in Brazil. Biofuels, Bioproducts and Biorefining 18(1): 322–339.

Kaniuk, Ł. and U. Stachewicz. 2021. Development and advantages of biodegradable PHA polymers based on electrospun PHBV fibers for tissue engineering and other biomedical applications. ACS Biomaterials Science & Engineering 7(12): 5339–5362.

Karimi Tar, A., S. Karbasi, E. Naghashzargar and H. Salehi. 2020. Biodegradation and cellular evaluation of aligned and random poly (3-hydroxybutyrate)/chitosan electrospun scaffold for nerve tissue engineering applications. Materials Technology 35(2): 92–101.

Kaur, S., D. M. Tripathi, J. R. Venugopal and S. Ramakrishna. 2020. Advances in biomaterials for hepatic tissue engineering. Current Opinion in Biomedical Engineering 13: 190–196.

Kessler, B., R. Weusthuis, B. Witholt and G. Eggink. 2001. Production of microbial polyesters: fermentation and downstream processes. Biopolyesters 159–182.

Khalid, N., M. Aqeel and A. Noman. 2020. Microplastics could be a threat to plants in terrestrial systems directly or indirectly. Environmental Pollution 267: 115653.

Khatami, K., M. Perez-Zabaleta, I. Owusu-Agyeman and Z. Cetecioglu. 2021. Waste to bioplastics: How close are we to sustainable polyhydroxyalkanoates production? Waste Management 119: 374–388.

Khorasani, M. T., S. A. Mirmohammadi and S. Irani. 2011. Polyhydroxybutyrate (PHB) scaffolds as a model for nerve tissue engineering application: fabrication and *in vitro* assay. International Journal of Polymeric Materials 60(8): 562–575.

Khoroushi, M., M. R. Foroughi, S. Karbasi, B. Hashemibeni and A. A. Khademi. 2018. Effect of Polyhydroxybutyrate/Chitosan/Bioglass nanofiber scaffold on proliferation and differentiation of stem cells from human exfoliated deciduous teeth into odontoblast-like cells. Materials Science and Engineering: C 89: 128–139.

Kim, B., S. J. Oh, J. H. Hwang, H. J. Kim, N. Shin, S. K. Bhatia et al. 2023. Polyhydroxybutyrate production from crude glycerol using a highly robust bacterial strain Halomonas sp. YLGW01. International Journal of Biological.

Kim, C., J. W. Lee, J. H. Heo, C. Park, D. H. Kim, G. S. Yi et al. 2022. Natural bone-mimicking nanopore-incorporated hydroxyapatite scaffolds for enhanced bone tissue regeneration. Biomaterials Research 26(1): 7.

Kim, D., S. J. Lee, D. Lee, J. M. Seok, S. J. Yeo, H. Lim et al. 2024. Biomimetic mineralization of 3D-printed polyhydroxyalkanoate-based microbial scaffolds for bone tissue engineering. International Journal of Bioprinting 10(2): 1806.

Kim, G. A., J. R. Spence and S. Takayama. 2017. Bioengineering for intestinal organoid cultures. Current Opinion in Biotechnology 47: 51–58.

Kim, H. S., J. Chen, L. P. Wu, J. Wu, H. Xiang, K. W. Leong et al. 2020. Prevention of excessive scar formation using nanofibrous meshes made of biodegradable elastomer poly (3-hydroxybutyrate-co-3-hydroxyvalerate). Journal of Tissue Engineering 11: 2041731420949332.

Koller, M. 2021. Production, properties, and processing of microbial polyhydroxyalkanoate (PHA) biopolyesters. pp. 3–55. In Microbial and Natural Macromolecules. Academic Press.

Krishnani, S., N. Rishi, R. Yadav and R. Singh. 2023. Valorization of agro-industrial waste through microbes for production of industrial protease. pp. 259–277. In Microbial Bioprocessing of Agri-food Wastes. CRC Press.

Kumar, V., S. Kumar and D. Singh. 2020. Microbial polyhydroxyalkanoates from extreme niches: Bioprospection status, opportunities and challenges. International Journal of Biological Macromolecules 147: 1255–1267.

Ladhari, S., N. N. Vu, C. Boisvert, A. Saidi and P. Nguyen-Tri. 2023. Recent development of polyhydroxyalkanoates (PHA)-based materials for antibacterial applications: a review. ACS Applied Bio Materials 6(4): 1398–1430.

Lagreca, E., V. Onesto, C. Di Natale, S. La Manna, P. A. Netti and R. Vecchione. 2020. Recent advances in the formulation of PLGA microparticles for controlled drug delivery. Progress in Biomaterials 9: 153–174.

Li, D., L. Lv, J. C. Chen and G. Q. Chen. 2017. Controlling microbial PHB synthesis via CRISPRi. Applied Microbiology and Biotechnology 101: 5861–5867.

Li, F., J. Li, X. Song, T. Sun, L. Mi, J. Liu et al. 2022. Alginate/Gelatin hydrogel scaffold containing nCeO2 as a potential osteogenic nanomaterial for bone tissue engineering. International Journal of Nanomedicine 17: 6561.

Li, J., X. Zhang, A. Udduttula, Z. S. Fan, J. H. Chen, A. R. Sun et al. 2021. Microbial-derived polyhydroxyalkanoate-based scaffolds for bone tissue engineering: biosynthesis, properties, and perspectives. Frontiers in Bioengineering and Biotechnology 9: 763031.

Li, M. and M. R. Wilkins. 2020. Recent advances in polyhydroxyalkanoate production: Feedstocks, strains and process developments. International Journal of Biological Macromolecules 156: 691–703.

Li, W., F. Xu, F. Dai, T. Deng, Y. Ai, Z. Xu et al. 2023a. Hydrophilic surface-modified 3D printed flexible scaffolds with high ceramic particle concentrations for immunopolarization-regulation and bone regeneration. Biomaterials Science 11(11): 3976–3997.

Li, Y. Q., M. J. Wang and C. B. Luo. 2023b. Highly efficient polyhydroxyalkanoate production from lignin using genetically engineered Halomonas sp. Y3. Bioresource Technology 370: 128526.

Li, Y., Q. Meng, S. Chen, P. Ling, M. A. Kuss, B. Duan et al. 2023c. Advances, challenges, and prospects for surgical suture materials. Acta Biomaterialia.

Liang, X., D. K. Cha and Q. Xie. 2024. Properties, production, and modification of polyhydroxyalkanoates. Resources, Conservation & Recycling Advances 200206.

Liu, Q., G. Liu, X. Liu, M. Yang, S. Xing, Y. Du et al. 2020a. Synthesis of an electrospun PHA/RGO/Au scaffold for peripheral nerve regeneration: An *in vitro* study. Applied Nanoscience 10(3): 687–694.

Liu, R., C. Luo, Z. Pang, J. Zhang, S. Ruan, M. Wu et al. 2023a. Advances of nanoparticles as drug delivery systems for disease diagnosis and treatment. Chinese Chemical Letters 34(2): 107518.

Liu, S., J. M. Yu, Y. C. Gan, X. Z. Qiu, Z. C. Gao, H. Wang et al. 2023b. Biomimetic natural biomaterials for tissue engineering and regenerative medicine: new biosynthesis methods, recent advances, and emerging applications. Military Medical Research 10(1): 16.

Liu, Y., S. Yang, L. Cao, X. Zhang, J. Wang and C. Liu. 2020b. Facilitated vascularization and enhanced bone regeneration by manipulation hierarchical pore structure of scaffolds. Materials Science and Engineering: C 110: 110622.

Lizarraga-Valderrama, L. R., C. S. Taylor, F. Claeyssens, J. W. Haycock, J. C. Knowles and I. Roy. 2019. Unidirectional neuronal cell growth and differentiation on aligned polyhydroxyalkanoate blend microfibres with varying diameters. Journal of Tissue Engineering and Regenerative Medicine 13(9): 1581–1594.

Loo, C. Y. and K. Sudesh. 2007. Biosynthesis and native granule characteristics of poly (3-hydroxybutyrate-co-3-hydroxyvalerate) in Delftia acidovorans. International Journal of Biological Macromolecules 40(5): 466–471.

Lopes, L. B., A. C. Apolinário, G. C. Salata, I. D. Malagó and J. S. Passos. 2022. Lipid nanocarriers for breast cancer treatment. pp. 3–44. In Cancer Nanotechnology. Cham: Springer International Publishing.

Lv, W., H. Zhou, A. Aazmi, M. Yu, X. Xu, H. Yang et al. 2022. Constructing biomimetic liver models through biomaterials and vasculature engineering. Regenerative Biomaterials 9: rbac079.

Ma, J., G. Sanchez-Duffhues, M. J. Goumans and P. Ten Dijke. 2020. TGF-β-induced endothelial to mesenchymal transition in disease and tissue engineering. Frontiers in Cell and Developmental Biology 8: 260.

Ma, R., J. Li, R. D. Tyagi and X. Zhang. 2024. A newly isolated strain for poly (3-hydroxybutyrate) production under anaerobic conditions and the key enzyme analysis. Chemical Engineering Journal 496: 154200. Macromolecules 236: 123997.

Manavitehrani, I., A. Fathi, H. Badr, S. Daly, A. Negahi Shirazi and F. Dehghani. 2016. Biomedical applications of biodegradable polyesters. Polymers 8(1): 20.

Mendibil, X., F. González-Pérez, X. Bazan, R. Díez-Ahedo, I. Quintana, F. J. Rodríguez et al. 2021. Bioresorbable and mechanically optimized nerve guidance conduit based on a naturally derived medium chain length polyhydroxyalkanoate and poly (ε-caprolactone) blend. ACS Biomaterials Science & Engineering 7(2): 672–689.

Mi, C. H., X. Y. Qi, Y. W. Ding, J. Zhou, J. W. Dao and D. X. Wei. 2023. Recent advances of medical polyhydroxyalkanoates in musculoskeletal system. Biomaterials Translational 4(4): 234.

Mirpoor, S. F., I. Corrado, R. Di Girolamo, G. Dal Poggetto, L. Panzella, E. Borselleca et al. 2023. Manufacture of active multilayer films made of functionalized pectin coated by polyhydroxyalkanoates: A fully renewable approach to active food packaging. Polymer 281: 126136.

Mistry, P., S. Kumar, R. S. Schloss, F. Berthiaume and N. A. Langrana. 2024. Chitosan-polygalacturonic acid complex dressing improves diabetic wound healing and hair growth in diabetic mice. Biochemical and Biophysical Research Communications 696: 149502.

Mitra, R., T. Xu, H. Xiang and J. Han. 2020. Current developments on polyhydroxyalkanoates synthesis by using halophiles as a promising cell factory. Microbial Cell Factories 19: 1–30.

Miu, D. M., M. C. Eremia and M. Moscovici. 2022. Polyhydroxyalkanoates (PHAs) as biomaterials in tissue engineering: production, isolation, characterization. Materials 15(4): 1410.

Mochane, M. J., J. S. Sefadi, T. S. Motsoeneng, T. E. Mokoena, T. G. Mofokeng and T. C. Mokhena. 2020. The effect of filler localization on the properties of biopolymer blends, recent advances: A review. Polymer Composites 41(7): 2958–2979.

Mohammadalipour, M., T. Behzad, S. Karbasi, M. B. Khorzoghi and Z. Mohammadalipour. 2023. Osteogenic potential of PHB-lignin/cellulose nanofiber electrospun scaffold as a novel bone regeneration construct. International Journal of Biological Macromolecules 250: 126076.

Mohammadi, N., Y. A. Amnieh, S. Ghasemi, S. Karbasi and S. Vaezifar. 2024. Evaluation of the effects of decellularized extracellular matrix nanoparticles incorporation on the polyhydroxybutyrate/nano chitosan electrospun scaffold for cartilage tissue engineering. International Journal of Biological Macromolecules 133217.

Mohtashami, Z., H. A. Javar, M. R. Tehrani, M. R. Esfahani, R. Roohipour, L. Aghajanpour et al. 2020. Fabrication, optimization, and *in vitro* and *in vivo* characterization of intra-vitreal implant of budesonide generally made of PHBV. AAPS PharmSciTech 21: 1–12.

Mokhames, Z., E. Seyedjafari, S. M. Seifati and M. D. Ashkezari. 2023. Human placental extract incorporated PLGA-PHB nanofibers enhances cardiomyocyte differentiation of mesenchymal stem cells.

Montero, P., M. Flandes-Iparraguirre, S. Musquiz, M. Pérez Araluce, D. Plano, C. Sanmartín et al. 2020. Cells, materials, and fabrication processes for cardiac tissue engineering. Frontiers in Bioengineering and Biotechnology 8: 955.

Mukherjee, A. and M. Koller. 2023. Microbial polyHydroxyAlkanoate (PHA) biopolymers—Intrinsically natural. Bioengineering 10(7): 855.

Muneer, F., I. Rasul, F. Azeem, M. H. Siddique, M. Zubair and H. Nadeem. 2020. Microbial polyhydroxyalkanoates (PHAs): efficient replacement of synthetic polymers. Journal of Polymers and the Environment 28: 2301–2323.

Murueva, A. V., A. M. Shershneva, E. I. Shishatskaya and T. G. Volova. 2023. Characteristics of microparticles based on resorbable polyhydroxyalkanoates loaded with antibacterial and cytostatic drugs. International Journal of Molecular Sciences 24(19): 14983.

Nagarajan, D., G. R. Aristya, Y. J. Lin, J. J. Chang, H. W. Yen and J. S. Chang. 2021. Microbial cell factories for the production of polyhydroxyalkanoates. Essays in Biochemistry 65(2): 337–353.

Naser, A. Z., I. Deiab and B. M. Darras. 2021a. Poly (lactic acid)(PLA) and polyhydroxyalkanoates (PHAs), green alternatives to petroleum-based plastics: a review. RSC Advances 11(28): 17151–17196.

Naser, A. Z., I. Deiab, F. Defersha and S. Yang. 2021b. Expanding poly lactic acid (PLA) and polyhydroxyalkanoates (PHAs) applications: A review on modifications and effects. Polymers 13(23): 4271.

Nath, P. C., R. Sharma, S. Debnath, P. K. Nayak, R. Roy, M. Sharma et al. 2024. Recent advances in production of sustainable and biodegradable polymers from agro-food waste: Applications in tissue engineering and regenerative medicines. International Journal of Biological Macromolecules 129129.

Negut, I. and C. Ristoscu. 2023. Bioactive glasses for soft and hard tissue healing applications—A short review. Applied Sciences 13(10): 6151.

Negut, I., B. Bita and A. Groza. 2022. Polymeric coatings and antimicrobial peptides as efficient systems for treating implantable medical devices associated-infections. Polymers 14(8): 1611.

Negut, I., G. Dorcioman and V. Grumezescu. 2020. Scaffolds for wound healing applications. Polymers 12(9): 2010.

Nigmatullin, R., C. S. Taylor, P. Basnett, B. Lukasiewicz, A. Paxinou, L. R. Lizarraga-Valderrama et al. 2023. Medium chain length polyhydroxyalkanoates as potential matrix materials for peripheral nerve regeneration. Regenerative Biomaterials 10: rbad063.

Nigmatullin, R., P. Thomas, B. Lukasiewicz, H. Puthussery and I. Roy. 2015. Polyhydroxyalkanoates, a family of natural polymers, and their applications in drug delivery. Journal of Chemical Technology & Biotechnology 90(7): 1209–1221.

Olejnik, O., A. Masek and J. Zawadziłło. 2021. Processability and mechanical properties of thermoplastic polylactide/polyhydroxybutyrate (PLA/PHB) bioblends. Materials 14(4): 898.

Olkhov, A. A., E. L. Kucherenko, Y. N. Zernova, V. S. Markin, R. Y. Kosenko, A. G. Filatova et al. 2024. Production methods and biomedical applications of materials based on poly-3-hydroxybutyrate and its compositions. Inorganic Materials: Applied Research 15(4): 1064–1076.

Opitz, C. A., M. C. Leake, I. Makarenko, V. Benes and W. A. Linke. 2004. Developmentally regulated switching of titin size alters myofibrillar stiffness in the perinatal heart. Circulation Research 94(7): 967–975.

Padan, E., E. Bibi, M. Ito and T. A. Krulwich. 2005. Alkaline pH homeostasis in bacteria: new insights. Biochimica et biophysica acta (BBA)-biomembranes 1717(2): 67–88.

Pagliano, G., P. Galletti, C. Samorì, A. Zaghini and C. Torri. 2021. Recovery of polyhydroxyalkanoates from single and mixed microbial cultures: A review. Frontiers in Bioengineering and Biotechnology 9: 624021.

Pandey, A., A. Ali, N. R. Patra and Y. S. Negi. 2022. Functionalized 3D bioactive polymeric materials in tissue engineering and regenerative medicine. pp. 257–277. In Nanoscale Engineering of Biomaterials: Properties and Applications. Singapore: Springer Nature Singapore.

Paul, V., S. Pandhi, D. K. Mahato, A. Agarwal and A. D. Tripathi. 2024. Polyhydroxyalkanoates (PHAs) and its copolymer nanocarrier application in cancer treatment: An overview and challenges. International Journal of Biological Macromolecules 134201.

Pavic, A., Z. Stojanovic, M. Pekmezovic, Đ. Veljović, K. O'connor, I. Malagurski et al. 2022. Polyenes in medium chain length polyhydroxyalkanoate (mcl-PHA) biopolymer microspheres with reduced toxicity and improved therapeutic effect against Candida infection in zebrafish model. Pharmaceutics 14(4): 696.

Pedersen, D. D., S. Kim, A. D'Amore and W. R. Wagner. 2023. Cardiac valve scaffold design: Implications of material properties and geometric configuration on performance and mechanics. Journal of the Mechanical Behavior of Biomedical Materials 146: 106043.

Pekmezovic, M., M. Kalagasidis Krusic, I. Malagurski, J. Milovanovic, K. Stępień, M. Guzik et al. 2021. Polyhydroxyalkanoate/antifungal polyene formulations with monomeric hydroxyalkanoic acids for improved antifungal efficiency. Antibiotics 10(6): 737.

Peng, S. W., X. Y. Guo, G. G. Shang, J. Li, X. Y. Xu, M. L. You et al. 2011. An assessment of the risks of carcinogenicity associated with polyhydroxyalkanoates through an analysis of DNA aneuploid and telomerase activity. Biomaterials 32(10): 2546–2555.

Pereira dos Santos, E., P. H. M. Nicácio, F. Coêlho Barbosa, H. Nunes da Silva, A. L. S. Andrade, M. V. Lia Fook et al. 2019. Chitosan/essential oils formulations for potential use as wound dressing: Physical and antimicrobial properties. Materials 12(14): 2223.

Petrovova, E., M. Tomco, K. Holovska, J. Danko, L. Kresakova, K. Vdoviakova et al. 2021. PHB/CHIT scaffold as a promising biopolymer in the treatment of osteochondral defects—An experimental animal study. Polymers 13(8): 1232.

Prajapati, S. K., A. Jain, D. Jain, N. Mody and M. Bajpai. 2023. Antibacterial and antiviral materials based on biodegradable polymers. pp. 111–138. In Antibacterial and Antiviral Functional Materials, Volume 1. American Chemical Society.

Prakash, P., W. H. Lee, C. Y. Loo, H. S. J. Wong and T. Parumasivam. 2022. Advances in polyhydroxyalkanoate nanocarriers for effective drug delivery: An overview and challenges. Nanomaterials 12(1): 175.

Pramanik, N. 2023. A tool for biomedical application: Synthesis and modification of polyhydroxyalkanoates. Sustainable Chemistry and Pharmacy 32: 101041.

Prasad, A., M. R. Sankar and V. Katiyar. 2017. State of art on solvent casting particulate leaching method for orthopedic scaffolds fabrication. Materials Today: Proceedings 4(2): 898–907.

Pratap, J. K. and K. Krishnan. 2023. Microbial production of Polyhydroxyalkonates (Bioplastic) using cheap household waste resources and their biomedical applications: a systematic review. Lett Appl NanoBioScience 12: 174.

Priya, S., M. Choudhari, Y. Tomar, V. M. Desai, S. Innani, S. K. Dubey et al. 2024. Exploring polysaccharide-based bio-adhesive topical film as a potential platform for wound dressing application: A review. Carbohydrate Polymers 121655.

Procopio, A., E. Lagreca, R. Jamaledin, S. La Manna, B. Corrado, C. Di Natale et al. 2022. Recent fabrication methods to produce polymer-based drug delivery matrices (Experimental and *In Silico* Approaches). Pharmaceutics 14(4): 872.

Pryadko, A., M. A. Surmeneva and R. A. Surmenev. 2021. Review of hybrid materials based on polyhydroxyalkanoates for tissue engineering applications. Polymers 13(11): 1738.

Pulingam, T., J. N. Appaturi, T. Parumasivam, A. Ahmad and K. Sudesh. 2022. Biomedical applications of polyhydroxyalkanoate in tissue engineering. Polymers 14(11): 2141.

Rakkan, T., S. Zhang, S. Lehner, R. Hufenus, K. Sangkharak and Q. Ren. 2024. Bio-based modification of polyhydroxyalkanoates (PHA) towards increased antimicrobial activities and reduced cytotoxicity. International Journal of Biological Macromolecules 133132.

Rana, A., V. Kumar, T. Dhewa and N. K. Taneja. 2023. Bioplastic production using whey (polyhydroxyalkanoates and polyhydroxybutyrates). pp. 103–113. In Whey Valorization: Innovations, Technological Advancements and Sustainable Exploitation. Singapore: Springer Nature Singapore.

Ranganadhareddy, A. and C. Chandrasekhar. 2021. Production of polyhydroxtbutyratye from marine source: a review.

Raza, Z. A., S. Abid and I. M. Banat. 2018. Polyhydroxyalkanoates: Characteristics, production, recent developments and applications. International Biodeterioration & Biodegradation 126: 45–56.

Raziyan, M. S., A. Palevicius, D. Perkowski, S. Urbaite and G. Janusas. 2024. Development and evaluation of 3D-printed PLA/PHA/PHB/HA composite scaffolds for enhanced tissue-engineering applications. Journal of Composites Science 8(6): 226.

Razzaq, S., S. Shahid, R. Farooq, S. Noreen, S. Perveen and M. Bilal. 2024. Sustainable bioconversion of agricultural waste substrates into poly (3-hydroxyhexanoate)(mcl-PHA) by Cupriavidus necator DSM 428. Biomass Conversion and Biorefinery 14(8): 9429–9439.

Reddy, M. S. B., D. Ponnamma, R. Choudhary and K. K. Sadasivuni. 2021. A comparative review of natural and synthetic biopolymer composite scaffolds. Polymers 13(7): 1105.

Ren, Z. W., Z. Y. Wang, Y. W. Ding, J. W. Dao, H. R. Li, X. Ma et al. 2023. Polyhydroxyalkanoates: the natural biopolyester for future medical innovations. Biomaterials Science 11(18): 6013–6034.

Reyna-Urrutia, V. A., G. Leyva-Gómez, C. Ríos and M. González-Torres. 2024. The role of poly (3-hydroxybutyrate)-based derivatives in skin, nerve, cartilage, and bone tissue engineering applications—An updated review. Polymers for Advanced Technologies 35(3): e6340.

Ribeiro, M. E. A., N. R. C. Huaman, J. G. C. Gomez and R. J. S. Rodríguez. 2023. Poly (3-hydroxybutyrate-co-3-hydroxyvalerate) and amino-functionalized nanodiamond bionanocomposites for bone tissue defect repair. International Journal of Biological Macromolecules 226: 1041–1053.

Robinson, J. 2021. Mechanically Tunable Extracellular Matrix Using Genipin Crosslinked Collagen and Its Effect on Endothelial Function. North Carolina Agricultural and Technical State University.

Rodriguez-Perez, S., A. Serrano, A. A. Pantión and B. Alonso-Fariñas. 2018. Challenges of scaling-up PHA production from waste streams. A review. Journal of Environmental Management 205: 215–230.

Rofeal, M., F. Abd El-Malek and X. Qi. 2021. *In vitro* assessment of green polyhydroxybutyrate/chitosan blend loaded with kaempferol nanocrystals as a potential dressing for infected wounds. Nanotechnology 32(37): 375102.

Rogozinski, N., A. Yanez, R. Bhoi, M. Y. Lee and H. Yang. 2022. Current methods for fabricating 3D cardiac engineered constructs. Iscience 25(5).

Roohi, Bano, N., A. Gupta, M. H. Siddiqui and M. R. Zaheer. 2024. Potential of PHA (Polyhydroxyalkanoates) polymers as packaging materials: from concept to commercialization. pp. 67–100. In Biobased Packaging Materials: Sustainable Alternative to Conventional Packaging Materials. Singapore: Springer Nature Singapore.

Roy, P. S., G. Garnier, F. Allais and K. Saito. 2021. Strategic approach towards plastic waste valorization: challenges and promising chemical upcycling possibilities. ChemSusChem 14(19): 4007–4027.

Sakamoto, A., H. Jinnouchi, S. Torii, R. Virmani and A. V. Finn. 2018. Understanding the impact of stent and scaffold material and strut design on coronary artery thrombosis from the basic and clinical points of view. Bioengineering 5(3): 71.

Salgaonkar, B. B. and J. M. Bragança. 2019. Production of polyhydroxyalkanoates by extremophilic microorganisms through valorization of waste materials. pp. 419–443. In Advances in Biological Science Research. Academic Press.

Sangkharak, K., P. Khaithongkaeo, T. Chuaikhunupakarn, A. Choonut and P. Prasertsan. 2021. The production of polyhydroxyalkanoate from waste cooking oil and its application in biofuel production. Biomass Conversion and Biorefinery 11: 1651–1664.

Sanhueza, C., J. Hermosilla, A. Bugallo-Casal, A. Da Silva-Candal, C. Taboada, R. Millán et al. 2021. One-step electrospun scaffold of dual-sized gelatin/poly-3-hydroxybutyrate nano/microfibers for skin regeneration in diabetic wound. Materials Science and Engineering: C 119: 111602.

Sanhueza, C., M. Pavéz, J. Hermosilla, S. Rocha, I. Valdivia-Gandur, M. C. Manzanares et al. 2024. Poly-3-hydroxybutyrate-silver nanoparticles membranes as advanced antibiofilm strategies for combatting peri-implantitis. International Journal of Biological Macromolecules 269: 131974.

Saravanan, K., M. Umesh and P. Kathirvel. 2022. Microbial polyhydroxyalkanoates (PHAs): a review on biosynthesis, properties, fermentation strategies and its prospective applications for sustainable future. Journal of Polymers and the Environment 30(12): 4903–4935.

Schaefer, M., M. Schwaiger, M. Pich, K. Lieb and A. Heinz. 2003. Neurotransmitter changes by interferon-alpha and therapeutic implications. Pharmacopsychiatry 36(S 3): 203–206.

Schuhladen, K., B. Lukasiewicz, P. Basnett, I. Roy and A. R. Boccaccini. 2020. Comparison of the influence of 45S5 and Cu-containing 45S5 bioactive glass (BG) on the biological properties of novel polyhydroxyalkanoate (PHA)/BG composites. Materials 13(11): 2607.

Scicluna, M. C. and L. Vella-Zarb. 2020. Evolution of nanocarrier drug-delivery systems and recent advancements in covalent organic framework–drug systems. ACS Applied Nano Materials 3(4): 3097–3115.

Seunarine, K., N. Gadegaard, M. Tormen, D. O. Meredith, M. O. Riehle and C. D. W. Wilkinson. 2006. 3D polymer scaffolds for tissue engineering. Nanomedicine 1(3): 281–296.

Sevostianova, V. V., L. V. Antonova, A. V. Mironov, A. E. Yuzhalin, V. N. Silnikov, T. V. Glushkova et al. 2020. Biodegradable patches for arterial reconstruction modified with RGD peptides: results of an experimental study. ACS Omega 5(34): 21700–21711.

Shahid, S., S. Razzaq and R. Farooq. 2021. Polyhydroxyalkanoates: Next generation natural biomolecules and a solution for the world's future economy. International Journal of Biological Macromolecules 166: 297–321.

Shakhmuradian, R., D. Ghevondyan, A. Margaryan and H. Panosyan. 2024. Polyhydroxyalkanoates produced by extremophiles: main structure, synthesis, and application. pp. 243–263. In Microbial Essentialism. Academic Press.

Shanmugam, D. K., S. C. Anitha, V. Souresh, Y. Madhavan, S. Sampath, D. Catakapatri Venugopal et al. 2023. Current advancements in the development of bionic organs using regenerative medicine and 3D tissue engineering. Materials Technology 38(1): 2242732.

Sharma, M. K., S. Singh, N. Kapoor and R. S. Tomar. 2022. Polyhydroxyalkanoate production in transgenic plants: green plastics for better future and environmental sustainability. pp. 287–301. In Agro-biodiversity and Agri-ecosystem Management. Singapore: Springer Nature Singapore.

Sharma, V., R. Sehgal and R. Gupta. 2021. Polyhydroxyalkanoate (PHA): properties and modifications. Polymer 212: 123161.

Shumilova, A. A., M. P. Myltygashev, A. K. Kirichenko, E. D. Nikolaeva, T. G. Volova and E. I. Shishatskaya. 2017. Porous 3D implants of degradable poly-3-hydroxybutyrate used to enhance regeneration of rat cranial defect. Journal of Biomedical Materials Research Part A 105(2): 566–577.

Singh, A. K., J. K. Srivastava, A. K. Chandel, L. Sharma, N. Mallick and S. P. Singh. 2019. Biomedical applications of microbially engineered polyhydroxyalkanoates: An insight into recent advances, bottlenecks, and solutions. Applied Microbiology and Biotechnology 103: 2007–2032.

Singh, A., S. Parikh, N. Sethi, S. Patel, N. Modi and K. Patel. 2024. Nanoparticle formulations: a sustainable approach to biodegradable and non-biodegradable products. Nanocarrier Vaccines: Biopharmaceutics-Based Fast Track Development 95–151.

Śmiga-Matuszowicz, M., J. Włodarczyk, M. Skorupa, D. Czerwińska-Główka, K. Fołta, M. Pastusiak et al. 2023. Biodegradable scaffolds for vascular regeneration based on electrospun poly (L-lactide-co-glycolide)/poly (isosorbide sebacate) fibers. International Journal of Molecular Sciences 24(2): 1190.

Sodian, R., S. P. Hoerstrup, J. S. Sperling, D. P. Martin, S. Daebritz, J. E. Mayer Jr and J. P. Vacanti 2000. Evaluation of biodegradable, three-dimensional matrices for tissue engineering of heart valves. Asaio Journal 46(1): 107–110.

Sohn, S., M. V. Buskirk, M. J. Buckenmeyer, R. Londono and D. Faulk. 2020. Whole organ engineering: approaches, challenges, and future directions. Applied Sciences 10(12): 4277.

Soleymani, M., E. S. Motiee, S. Karbasi and A. Basiri. 2024. Evaluation of the effects of decellularized umbilical cord Wharton's Jelly ECM on polyhydroxy butyrate electrospun scaffolds: A new strategy for cartilage tissue engineering. Materials Today Chemistry 39: 102145.

Sridharan, S., M. Kumar, M. Saha, M. B. Kirkham, L. Singh and N. S. Bolan. 2022. The polymers and their additives in particulate plastics: what makes them hazardous to the fauna? Science of The Total Environment 824: 153828.

Strangis, G., M. Labardi, G. Gallone, M. Milazzo, S. Capaccioli, F. Forli et al. 2024. 3D printed piezoelectric BaTiO3/polyhydroxybutyrate nanocomposite scaffolds for bone tissue engineering. Bioengineering 11(2): 193.

Suamte, L., A. Tirkey, J. Barman and P. J. Babu. 2023. Various manufacturing methods and ideal properties of scaffolds for tissue engineering applications. Smart Materials in Manufacturing 1: 100011.

Sun, L., J. Li, W. Gao, M. Shi, F. Tang, X. Fu and X. Chen. 2021. Coaxial nanofibrous scaffolds mimicking the extracellular matrix transition in the wound healing process promoting skin regeneration through enhancing immunomodulation. Journal of Materials Chemistry B 9(5): 1395–1405.

Surendra, K. C., I. Angelidaki and S. K. Khanal. 2022. Bioconversion of waste-to-resources (BWR-2021): Valorization of industrial and agro-wastes to fuel, feed, fertilizer, and biobased products. Bioresource Technology 347: 126739.

Surendran, A., M. Lakshmanan, J. Y. Chee, A. M. Sulaiman, D. V. Thuoc and K. Sudesh. 2020. Can polyhydroxyalkanoates be produced efficiently from waste plant and animal oils? Frontiers in Bioengineering and Biotechnology 8: 169.

Suriyamongkol, P., R. Weselake, S. Narine, M. Moloney and S. Shah. 2007. Biotechnological approaches for the production of polyhydroxyalkanoates in microorganisms and plants—a review. Biotechnology Advances 25(2): 148–175.

Syed Mohamed, S. M. D., N. F. Ansari, N. Md Iqbal and S. N. S. Anis. 2022. Polyhydroxyalkanoates (PHA)-based responsive polymers. International Journal of Polymeric Materials and Polymeric Biomaterials 71(17): 1283–1302.

Tahmasebi, A., A. S. Moghadam, S. E. Enderami, M. Islami, M. Kaabi, E. Saburi et al. 2020. Aloe vera–derived gel-blended PHBV nanofibrous scaffold for bone tissue engineering. Asaio Journal 66(8): 966–973.

Tang, X., A. H. Westlie, L. Caporaso, L. Cavallo, L. Falivene and E. Y. X. Chen. 2020. Biodegradable polyhydroxyalkanoates by stereoselective copolymerization of racemic diolides: stereocontrol and polyolefin-like properties. Angewandte Chemie 132(20): 7955–7964.

Taylor, C. S., M. Behbehani, A. Glen, P. Basnett, D. A. Gregory, B. B. Lukasiewicz et al. 2023. Aligned polyhydroxyalkanoate blend electrospun fibers as intraluminal guidance scaffolds for peripheral nerve repair. ACS Biomaterials Science & Engineering 9(3): 1472–1485.

Thu, N. T. T., L. H. Hoang, P. K. Cuong, N. Viet-Linh, T. T. H. Nga, D. D. Kim et al. 2023. Evaluation of polyhydroxyalkanoate (PHA) synthesis by Pichia sp. TSLS24 yeast isolated in Vietnam. Scientific Reports 13(1): 3137.

Tripathi, A. D., V. Paul, A. Agarwal, R. Sharma, F. Hashempour-Baltork, L. Rashidi et al. 2021. Production of polyhydroxyalkanoates using dairy processing waste—a review. Bioresource Technology 326: 124735.

Umesh, M., S. Shanmugam, T. Kikas, N. T. L. Chi and A. Pugazhendhi. 2022. Progress in bio-based biodegradable polymer as the effective replacement for the engineering applicators. Journal of Cleaner Production 362: 132267.

Vach Agocsova, S., M. Culenova, I. Birova, L. Omanikova, B. Moncmanova, L. Danisovic et al. 2023. Resorbable biomaterials used for 3D scaffolds in tissue engineering: A review. Materials 16(12): 4267.

Villano, M., M. Beccari, D. Dionisi, S. Lampis, A. Miccheli, G. Vallini et al. 2010. Effect of pH on the production of bacterial polyhydroxyalkanoates by mixed cultures enriched under periodic feeding. Process Biochemistry 45(5): 714–723.

Voinova, V., G. Bonartseva and A. Bonartsev. 2019. Effect of poly (3-hydroxyalkanoates) as natural polymers on mesenchymal stem cells. World Journal of Stem Cells 11(10): 764.

Volkov, A. V., A. A. Muraev, I. I. Zharkova, V. V. Voinova, E. A. Akoulina, V. A. Zhuikov et al. 2020. Poly (3-hydroxybutyrate)/hydroxyapatite/alginate scaffolds seeded with mesenchymal stem cells enhance the regeneration of critical-sized bone defect. Materials Science and Engineering: C 114: 110991.

Volova, T. G., A. A. Shumilova, E. D. Nikolaeva, A. K. Kirichenko and E. I. Shishatskaya. 2019. Biotechnological wound dressings based on bacterial cellulose and degradable copolymer P (3HB/4HB). International Journal of Biological Macromolecules 131: 230–240.

Vyas, C., H. Mishbak, G. Cooper, C. Peach, R. F. Pereira and P. Bartolo. 2020. Biological perspectives and current biofabrication strategies in osteochondral tissue engineering. Biomanufacturing Reviews 5: 1–24.

Wan, T., Y. L. Wang, F. S. Zhang, X. M. Zhang, Y. C. Zhang, H. R. Jiang et al. 2023. The porous structure of peripheral nerve guidance conduits: features, fabrication, and implications for peripheral nerve regeneration. International Journal of Molecular Sciences 24(18): 14132.

Wang, C., Y. Xu, J. Xia, Z. Zhou, Y. Fang, L. Zhang et al. 2021a. Multi-scale hierarchical scaffolds with aligned micro-fibers for promoting cell alignment. Biomedical Materials 16(4): 045047.

Wang, J., D. Huang, H. Yu, Y. Cheng, H. Ren and Y. Zhao. 2022a. Developing tissue engineering strategies for liver regeneration. Engineered Regeneration 3(1): 80–91.

Wang, J., S. Liu, J. Huang, R. Cui, Y. Xu and Z. Song. 2023a. Genetic engineering strategies for sustainable polyhydroxyalkanoate (PHA) production from carbon-rich wastes. Environmental Technology & Innovation 30: 103069.

Wang, L., Z. Yang, F. Fan, S. Sun, X. Wu, H. Lu et al. 2019. PHBHHx facilitated the residence, survival and stemness maintain of transplanted neural stem cells in traumatic brain injury rats. Biomacromolecules 20(9): 3294–3302.

Wang, Q., Y. Xu, P. Xu, W. Yang, M. Chen, W. Dong et al. 2022b. Crystallization of microbial polyhydroxyalkanoates: A review. International Journal of Biological Macromolecules 209: 330–343.

Wang, X., H. Jiang, W. Zhang, Y. Kong, D. Kong, J. Liu et al. 2023b. Biomimetic polymeric transcatheter heart valve leaflets with low calcification and good regenerative ability. Journal of Materials Chemistry B 11(25): 5805–5816.

Wang, X., Z. Li, C. Wang, H. Bai, Z. Wang, Y. Liu et al. 2021b. Enlightenment of growth plate regeneration based on cartilage repair theory: a review. Frontiers in Bioengineering and Biotechnology 9: 654087.

Wang, Y., A. Chung and G. Q. Chen. 2017. Synthesis of medium-chain-length polyhydroxyalkanoate homopolymers, random copolymers, and block copolymers by an engineered strain of Pseudomonas entomophila. Advanced Healthcare Materials 6(7): 1601017.

Wang, Y., Y. Xu, G. Zhou, Y. Liu and Y. Cao. 2021c. Biological evaluation of acellular cartilaginous and dermal matrixes as tissue engineering scaffolds for cartilage regeneration. Frontiers in Cell and Developmental Biology 8: 624337.

Wasyłeczko, M., Z. J. Krysiak, E. Łukowska, M. Gruba, W. Sikorska, A. Kruk et al. 2022. Three-dimensional scaffolds for bioengineering of cartilage tissue. Biocybernetics and Biomedical Engineering 42(2): 494–511.

Wei, J., F. Chen, J. W. Shin, H. Hong, C. Dai, J. Su et al. 2009. Preparation and characterization of bioactive mesoporous wollastonite–polycaprolactone composite scaffold. Biomaterials 30(6): 1080–1088.

Wen, Q., Z. Chen, T. Tian and W. Chen. 2010. Effects of phosphorus and nitrogen limitation on PHA production in activated sludge. Journal of Environmental Sciences 22(10): 1602–1607.

Williams, S. F., S. Rizk and D. P. Martin. 2013. Poly-4-hydroxybutyrate (P4HB): a new generation of resorbable medical devices for tissue repair and regeneration. Biomedizinische Technik/Biomedical Engineering 58(5): 439–452.

Wu, Q., Y. Wang and G. Q. Chen. 2009. Medical application of microbial biopolyesters polyhydroxyalkanoates. Artificial Cells, Blood Substitutes, and Biotechnology 37(1): 1–12.

Xiao, B., R. Heath, P. Saiu, F. C. Leiper, P. Leone, C. Jing et al. 2007. Structural basis for AMP binding to mammalian AMP-activated protein kinase. Nature 449(7161): 496–500.

Xu, L., Y. Liu, W. Zhou and D. Yu. 2022. Electrospun medical sutures for wound healing: A review. Polymers 14(9): 1637.

Xu, Z., M. Xu, C. Cai, S. Chen and M. Jin. 2021. Microbial polyhydroxyalkanoate production from lignin by Pseudomonas putida NX-1. Bioresource Technology 319: 124210.

Yadav, S., J. Khan and A. Yadav. 2024. Applications of scaffolds in tissue engineering: current utilization and future prospective. Current Gene Therapy 24(2): 94–109.

Yang, Y., Y. Ma, J. Wang, L. You, R. Zhang, Q. Meng et al. 2023a. Chitosan-based mussel-inspired hydrogel for rapid self-healing and high adhesion of tissue adhesion and wound dressings. Carbohydrate Polymers 316: 121083.

Yang, Z., H. Li, J. Lin, D. Xing, J. J. Li, E. M. Cribbin et al. 2023b. Research landscape of 3D printing in bone regeneration and bone repair: A bibliometric and visualized analysis from 2012 to 2022. International Journal of Bioprinting 9(4).

Ye, J. P., J. S. Gong, C. Su, Y. G. Liu, M. Jiang, H. Pan et al. 2020. Fabrication and characterization of high molecular keratin based nanofibrous membranes for wound healing. Colloids and Surfaces B: Biointerfaces 194: 111158.

Ye, X., Y. Zhang, T. Liu, Z. Chen, W. Chen, Z. Wu et al. 2022. Beta-tricalcium phosphate enhanced mechanical and biological properties of 3D-printed polyhydroxyalkanoates scaffold for bone tissue engineering. International Journal of Biological Macromolecules 209: 1553–1561.

Yeo, J. C. C., J. K. Muiruri, X. Fei, T. Wang, X. Zhang, Y. Xiao et al. 2024. Innovative biomaterials for food packaging: Unlocking the potential of polyhydroxyalkanoate (PHA) biopolymers. Biomaterials Advances 213929.

Yogeswar, M., N. Valle and A. Nagarajan. 2023. Cyanobacteria as a source of biodegradable plastics. In Cyanobacteria-Recent Advances and New Perspectives. IntechOpen.

Yousefi, A. M. and G. E. Wnek. 2024. Poly (hydroxyalkanoates): Emerging biopolymers in biomedical fields and packaging industries for a circular economy. Biomedical Materials & Devices 1–26.

Yu, J., S. A. Park, W. D. Kim, T. Ha, Y. Z. Xin, J. Lee et al. 2020. Current advances in 3D bioprinting technology and its applications for tissue engineering. Polymers 12(12): 2958.

Zandi, N., B. Dolatyar, R. Lotfi, Y. Shallageh, M. A. Shokrgozar, E. Tamjid et al. 2021. Biomimetic nanoengineered scaffold for enhanced full-thickness cutaneous wound healing. Acta Biomaterialia 124: 191–204.

Zarei, M., S. Karbasi, F. S. Aslani, S. Zare, O. Koohi-Hosseinabad and N. Tanideh. 2020. *In vitro* and *in vivo* evaluation of poly (3-hydroxybutyrate)/carbon nanotubes electrospun scaffolds for periodontal ligament tissue engineering. Journal of Dentistry 21(1): 18.

Zarrintaj, P., F. Seidi, M. Y. Azarfam, M. K. Yazdi, A. Erfani, M. Barani et al. 2023. Biopolymer-based composites for tissue engineering applications: A basis for future opportunities. Composites Part B: Engineering 258: 110701.

Zhang, B., Y. Yang, J. Yi, Z. Zhao and R. Ye. 2021. Hyperglycemia modulates M1/M2 macrophage polarization via reactive oxygen species overproduction in ligature-induced periodontitis. Journal of Periodontal Research 56(5): 991–1005.

Zhang, J., E. I. Shishatskaya, T. G. Volova, L. F. da Silva and G. Q. Chen. 2018. Polyhydroxyalkanoates (PHA) for therapeutic applications. Materials Science and Engineering: C 86: 144–150.

Zhang, R., J. Zong, Y. Peng, J. Shi, X. Du, H. Liu et al. 2021. GPR30 knockdown weakens the capacity of CAF in promoting prostate cancer cell invasion via reducing macrophage infiltration and M2 polarization. Journal of Cellular Biochemistry 122(9): 1173–1191.

Zhang, X., J. Li, J. Chen, Z. X. Peng, J. N. Chen, X. Liu et al. 2022a. Enhanced bone regeneration via PHA scaffolds coated with polydopamine-captured BMP2. Journal of Materials Chemistry B 10(32): 6214–6227.

Zhang, X., Y. Lin, Q. Wu, Y. Wang and G. Q. Chen. 2020. Synthetic biology and genome-editing tools for improving PHA metabolic engineering. Trends in Biotechnology 38(7): 689–700.

Zhang, X., X. Y. Liu, H. Yang, J. N. Chen, Y. Lin, S. Y. Han et al. 2022b. A polyhydroxyalkanoates-based carrier platform of bioactive substances for therapeutic applications. Frontiers in Bioengineering and Biotechnology 9: 798724.

Zhang, X., Y. Yang, Z. Yang, R. Ma, M. Aimaijiang, J. Xu et al. 2023. Four-dimensional printing and shape memory materials in bone tissue engineering. International Journal of Molecular Sciences 24(1): 814.

Zhao, T., J. Zhang, X. Gao, D. Yuan, Z. Gu and Y. Xu. 2022. Electrospun nanofibers for bone regeneration: from biomimetic composition, structure to function. Journal of Materials Chemistry B 10(32): 6078–6106.

Zhao, W. and G. Q. Chen. 2007. Production and characterization of terpolyester poly (3-hydroxybutyrate-co-3-hydroxyvalerate-co-3-hydroxyhexanoate) by recombinant Aeromonas hydrophila 4AK4 harboring genes phaAB. Process Biochemistry 42(9): 1342–1347.

Zhao, X. H., Y. N. Niu, C. H. Mi, H. L. Gong, X. Y. Yang, J. S. Y. Cheng et al. 2021. Electrospinning nanofibers of microbial polyhydroxyalkanoates for applications in medical tissue engineering. Journal of Polymer Science 59(18): 1994–2013.

Zhao, Z. M., Z. H. Liu, Y. Pu, X. Meng, J. Xu, J. S. Yuan et al. 2020. Emerging strategies for modifying lignin chemistry to enhance biological lignin valorization. ChemSusChem 13(20): 5423–5432.

Zhao, Z. M., X. Meng, Y. Pu, M. Li, Y. Li, Y. Zhang et al. 2023. Bioconversion of homogeneous linear C-lignin to polyhydroxyalkanoates. Biomacromolecules 24(9): 3996–4004.

Zhou, L., Z. Zhang, C. Shi, M. Scoti, D. K. Barange, R. R. Gowda et al. 2023. Chemically circular, mechanically tough, and melt-processable polyhydroxyalkanoates. Science 380(6640): 64–69.

Zhou, M. and D. Yu. 2014. Cartilage tissue engineering using PHBV and PHBV/Bioglass scaffolds. Molecular Medicine Reports 10(1): 508–514.

Zhu, M., W. Zhong, W. Cao, Q. Zhang and G. Wu. 2022. Chondroinductive/chondroconductive peptides and their-functionalized biomaterials for cartilage tissue engineering. Bioactive Materials 9: 221–238.

Zou, P., W. H. Lee, Z. Gao, D. Qin, Y. Wang, J. Liu et al. 2020. Wound dressing from polyvinyl alcohol/chitosan electrospun fiber membrane loaded with OH-CATH30 nanoparticles. Carbohydrate Polymers 232: 115786.

CHAPTER 11

Silk Fibroin

Bridging Biological and Biomaterials Science in Regeneration

Jila Majidi,[1,2] *Mozhgan Jahani*,[1,2] *Fatemeh Aboutalebi*,[3] *Saeed kariminasab*,[4] *Sanaz Alizadeh*,[5,*] *Narendra Pal Singh Chauhan*[6,*] and *Mazaher Gholipourmalekabadi*[1,7,8,*]

1. Introduction

Regenerative medicine is a pioneering field that aims to repair, replace, or regenerate damaged tissues and organs, offering potential solutions for irreparable injuries and degenerative diseases. With its interdisciplinary approach, this area combines insights from biology, materials science, and engineering to devise innovative therapies that harness the body's healing processes. One key focus of regenerative medicine is the development of biomaterials that support cellular growth and function. Silk fibroin

[1] Cellular and Molecular Research Center, Iran University of Medical Sciences, Tehran, Iran.
[2] Department of Tissue Engineering & Regenerative Medicine, Faculty of Advanced Technologies in Medicine, Tehran, Iran.
[3] Department of Molecular Medicine, Faculty of Advanced Technologies in Medicine, Iran University of Medical Sciences, Tehran, Iran.
[4] Student Research Committee, Alborz University of Medical Sciences, Karaj, Iran.
[5] Burn Research Center, Iran University of Medical Science, Tehran, Iran.
[6] Department of Chemistry, Faculty of Science, Bhupal Nobles' University, Udaipur-313002, Rajasthan, India.
[7] Department of Medical Biotechnology, Faculty of Allied Medicine, Iran University of Medical Sciences, Tehran, Iran.
[8] NanoBiotechnology and Regenerative Medicine Innovation Group, Noavarn Salamat ZHINO (PHC), Tehran, Iran.
* Corresponding authors: s.alizadeh@rsct.ir; alizadehsannaz@gmail.com, narendrapalsingh14@gmail.com, mazaher.gholipour@iums.ac.ir; mazaher.gholipour@gmail.com

DOI: 10.1201/9781003518792-11

has emerged as a remarkable candidate due to its unique mechanical properties, biocompatibility, and biodegradability. Derived from silkworms, silk fibroin provides a scaffold miming the natural extracellular matrix, promoting cellular attachment, proliferation, and differentiation. This natural polymer's versatility makes it suitable for various applications, from tissue engineering to drug delivery systems.

As regenerative medicine continues to evolve, integrating advanced biomaterials like silk fibroin will play a crucial role in pioneering therapeutic strategies that can restore function and enhance the quality of life for patients facing chronic conditions (Sorg et al., 2017). The wound healing process is fairly uniform across numerous tissue types, including skin tissue. In recent years, regenerative tissue engineering has emerged as a top-tier platform for the creation of synthetic tissues and the regeneration of damaged organs, to resolve major health issues that are associated with adults. The blossoming advancements in tissue engineering have been made possible by several different fields, including cell biology, biomaterial research, bioengineering, and others (Strodtbeck, 2001; Majidi Ghatar et al., 2023). Tissue engineering is based on the fundamental premise of restoring and improving the function of the tissues. This can be accomplished through the generation of innovative or biocompatible alternatives or the rebuilding of the tissues themselves. The three most important pioneering approaches in tissue regenerative medicine are the use of cells or cell implants, the administration of tissue development-promoting substances, and the utilization of diverse matrices, such as scaffolds, to produce three-dimensional (3D) cellular structures. Furthermore, artificial skin substitutes pave the way for skin regrowth in injured regions (Menon, 2002; Alizadeh et al., 2024). Although there are a variety of skin substitutes available, both synthetic and natural, their clinical efficacy is sometimes inadequate when it comes to treating skin wounds (Li et al., 2016). Allografts, Xenografts, and natural macromolecule-derived scaffolds (e.g., hyaluronic acid, silk, collagen, cellulose, and others) are all instances of skin substitutes obtained from natural sources. Among other positive attributes, these replacements' biocompatibility and biodegradability promote wound healing. Natural skin substitutes are scarce, expensive, and prone to cross-contamination, which limits their applications in medicine.

2. Overview of Silk Fibroin

2.1 Sericin and Fibroin

Silk consists of two proteins, a fibrous and globular protein called SF and sericin, respectively The cocoons of the B. mori silkworm contain 75–83.3% SF and 16.7–25% of sericin (Liu et al., 2015). SF is a hydrophobic protein with a semi-crystalline structure that provides the mechanical characteristic of silk fibers but sericin is a glue-like protein with an amorphous structure that acts as a gumming agent to preserve the integrity of the structure (Malay et al., 2016).

However, silk's low biocompatibility has been reported due to the sericin unit in its structure (Nguyen et al., 2019b). Studies have shown immune responses to the sericin moiety (Aramwit et al., 2009). Therefore, the creation of allergic reactions by sericin limits its biomedical application, so this protein component is usually separated from the fibroin unit (Sahu et al., 2016). Sericin isolation from fibroin affects its mechanical properties and it has been indicated that fibroin mechanical properties increased by up to %50 after sericin elimination. So, due to the better biocompatibility as well as biomechanical properties of the sericin-free fibroin fibers, different sericin isolation methods have been developed (Sun et al., 2021b).

The hydrophilicity of the sericin makes it easy to separate from fibroin. The degumming process for sericin isolation is normally done by boiling the B. mori cocoons in an alkaline solution such as sodium carbonate (Na_2CO_3) for a short time (usually 30 min). This method has a low cost and can be done in a short time (Chopra and Gulrajani, 1994; Dou and Zuo, 2015).

Nevertheless, working on the isolation methods for improving them to have a pure SF with high quality is still being done (Sun et al., 2021b).

2.2 Molecular Structure of Silk Fibroin

The SF is a macromolecule with three different subunits including, a light chain (L-chain), a heavy chain (H-chain), and a glycoprotein (P25, L-chain and H-chain are connected by a disulfide bond, while glycoprotein 25 is noncovalently connected to them (L-chain and H-chain) (Patil et al., 2020) (Figure 1).

As indicated, the H-chain consists of different domains including the amorphous domain, N-terminal, and C-terminal. Both L-chain and H-chain are composed of

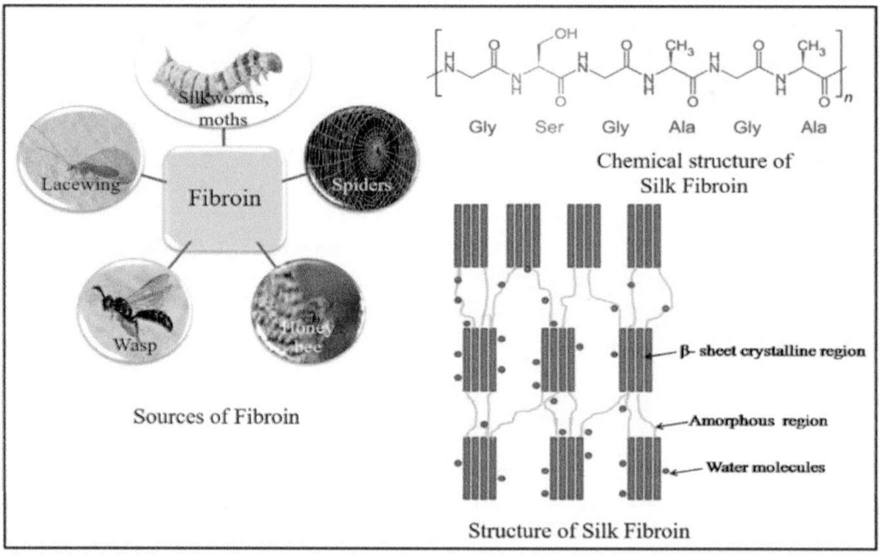

Figure 1. The fibroin derived from silk is primarily composed of light chains and heavy chains connected by disulfide bonds; the heavy chain is made up of crystalline and amorphous regions; the crystalline regions are made up of glycine-alanine repeats connected with serine and tyrosine amino acids, while the amorphous regions are made up of amino acids like aspartic acid. SF proteins can be found in silkworm moths, honey bees, spiders, wasps, and lacewings (Patil et al., 2020).

various amino acid residues (Rastogi et al., 2020; Hao et al., 2021). The L-chain contained the non-repetitive hydrophilic polypeptide segments consisting of the alanine (Ala) and aspartic acid (Asp) residues. The crystalline domain of the heavy chain consists of repetitive hydrophobic amino acids with two hexapeptide segments of (Gly-Ala-Gly-Ala-Gly-Ser (GAGAGS) and Gly-Ala-Gly-Ala-Gly-Tyr (GAGAGY)) which created the 12 crystalline domains. Furthermore, amorphous domain composed of hydrophobic domains which connected the hydrophilic domains in the secondary structure of the SF (Zhou et al., 2001; Fedic et al., 2003; Sun et al., 2021b).

Due to the high glycine (Gly) content of the repetitive polypeptides in the H-chain, the antiparallel β-sheet nanocrystals are formed by a strong hydrogen bond between the peptide segments. Good mechanical characteristics of the SF such as high tensile strength and stiffness related to its β-sheet conformation (Sun et al., 2021b; Qi et al., 2017; Patil et al., 2020).

2.3 Properties of Silk Fibroin

2.3.1 Biocompatibility

The materials used in tissue regeneration must be biocompatible. SF was recognized as a biomaterial by the US Food and Drug Administration (FDA) for the first time in 1993, and nowadays it is widely used as a suture material (Melke et al., 2016b). The low inflammatory reaction to the SF makes it a highly compatible material, and many FDA-approved medical devices are made based on the SF, such as Seriscaffold® (bioresorbable surgical mesh) and (SeriACL™) a ligament graft based on the SF (Melke et al., 2016b).

Different studies revealed the biocompatibility of the SF. For example, subcutaneously implantation of the SF scaffold in rats caused a deficient immune response after one year (Wang et al., 2008). In another study accumulation of the immune cells without any immune reaction was seen in the stained tissue of the rat after 8 weeks (Zhou et al., 2010). Furthermore, it has been shown that SF degradation by-products including the polypeptide chains and small peptides don't give rise to the hypersensitive response. Moreover, the cytocompatibility of the SF products was shown to be higher than the other synthetic biomaterials, like PLGA and PLA (Zhang et al., 2010).

2.3.2 Mechanical Strength

Natural SF fibers have high mechanical properties such as high tensile strength and elongation at breaks, etc. The mechanical behavior of the SF depends on its fiber's molecular structure (Luo et al., 2014). SF in its β-crystals form can bear up against the mechanical stress due to the protection of the connected sections together. So, the applied force is spread throughout the SF structure and the construct will not be damaged by the applied force. Changes in the SF hierarchical conformation and structure can affect its mechanical behavior (Xu et al., 2014).

The mechanical behaviors of the SF are highly regulated by its cross-linked arrangements (Cheng et al., 2014; Koh et al., 2015). Both amorphous and crystalline domains are involved in maintaining the SF structure under mechanical stress. The tension created by the mechanical force applied to the SF is initially supported by its amorphous domains. Thereafter, these non-crystalline domains are unfolded which results in the retrievable deformation. With the continuation of the loading force the amorphous domain is fully unraveled, and β-crystals support the tension. Deformation of SF may occur by breaking these β crystals, but the cross-linked arrangements allow to loaded force to be redistributed to surrounding β-crystals. Finally, the loading force can lead to the destruction of the remaining β-crystals and the final breakdown of the SF structure (Koh et al., 2015).

The mechanical characteristics of the SF are controlled and regulated by the size, orientation, and arrangements of the β-crystals domain (Koh et al., 2015). So, the mechanical strength of the β-crystals in the parallel arrangement is higher than the anti-parallel ones. Therefore, each processing way that improved the β-sheets parallel form in the SF, increase the stability of the whole SF structure (Koh et al., 2015). About the β-crystallites size effects on SF mechanical properties, studies indicated that a few nanometers size of caused more toughness, strength, and stiffness than the larger one. The optimal mechanical behaviors of the SF can be achieved in the β-crystallites dimension of 2 and 4 nm along the chain direction and 1–2 nm along the strand axis (Koh et al., 2015).

Due to the good mechanical properties of the SF, developing different products from it is not strange. For example, the good tensile strength of the SF is appropriate for suture production, however, its flexibility is suitable for making scaffolds with high mechanical resistance (Nguyen et al., 2019b). It is worth noting that the mechanical properties of the SF are affected by its processing methods. So, in most of these methods, native fibers possess better mechanical strength than the processed ones, and regenerated materials missing a suitable secondary structure than the native ones. Thus using the appropriate way to prepare the SF with excellent mechanical properties is important (Chopra and Gulrajani, 1994; Sun et al., 2021b).

2.3.3 Biodegradability

Two particular characteristics of TE scaffolds are the controlled degradation rate proportional to the speed of new tissue formation and possessing biocompatible degradation byproducts (Armentano et al., 2010).

Material biodegradation was defined by Vert et al. (Vert et al., 1992; Schneiderman and Hillmyer, 2016) as the degradability potential of the polymer material after implantation in the biological condition and the elimination of by-products from the implant site but not from the body, however, the complete elimination of the biomaterials is called bio-absorption. SF, as a protein, is sensitive to proteolytic degradation through various enzymes such as metalloproteinase. Previous report demonstrated that the SF degradation rate depends on various factors such as the extraction method, silk source, secondary structure, and type of scaffold (Sahoo et al., 2010). About the extraction method, it has been reported that the molecular weight (Mw) and degradation rate of the silk fibers are significantly affected by degumming methods. Accordingly, in the comparison study on the degumming of the silk with three different compounds of the bromelain, specific to sericin, urea- and Na_2CO_3. SF extracted with the bromelain-degumming method showed a higher Mw and lower degradation rate than those extracted with other methods (Wang et al., 2023). SF with higher β-sheet content needs more time for degradation than that with a lower β-sheet structure. According to Wang et al. (Wang et al., 2008) study water-based porous SF scaffolds with high content of the β-sheet, disintegrated after one year of transplantation in the rats. They also showed the bio-absorbability of SF in addition to its degradability in their study, which can indicate the effect of the immune system on SF degradation. Overall, in comparison to synthetic biomaterials such as PLGA and PLA, with high degradation rates, low mechanical properties, and residual acidic by-products (Gunatillake et al., 2003), SF has a low degradation rate and higher mechanical strength with safe and bio-absorbable by-products which make it more suitable for TE.

3. Silk Fibroin Application in Regenerative Medicine

3.1 Fabrication Technique

Silk fibroin derived from the Bombyx mori silkworm is a fibrous protein utilized for decades in the textile sector and for medical sutures. The common attributes of biocompatibility, environmental stability, and mechanical stability render all silk fibroins appropriate for tissue engineering applications. Silk fibroin has other advantageous features, including oxygen and nutrient permeability, gradual degradability, and little immunogenicity. It facilitates electrical conduction, demonstrated by its consistently reproducible shift to a higher oxidation state under an applied magnetic field, functioning as a carrier trap (Li and Sun, 2022). Moreover, silk fibroin has been utilized in flexible electronic devices, including energy harvesters, sensors, filters, and actuators. Silk fibroin may create tissue engineering scaffolds with highly repeatable topologies, such as foams, films, mats, fabrics, spongiform structures, hydrogels, and particles (Figure 2) (Bhattacharjee et al., 2017). Silk fibroin, like any material, might enhance or acquire features by the incorporation of additional chemicals, such as cell adhesiveness, biostability, immunomodulation, antibacterial activity, and superparamagnetic (Zuluaga-Vélez et al., 2021; Wani et al., 2020).

3.1.1 Electrospinning

Electrospinning is an up-and-coming technique employed in tissue engineering applications. This approach has several benefits, including the capability to produce micro- to nanometer-size fibers with extensive surface area and efficient therapeutic drug delivery (Sell et al., 2009). Consequently, interest in this study domain has increased, and several studies are currently being conducted to develop electrospun nanofibrous scaffolds for various tissue engineering applications (Rahmati et al., 2021; Negahdari et al., 2024b).

The electrospinning technology is classified into two types according to the electrospinning apparatus: needle-based electrospinning and needleless electrospinning. Moreover, needle-based electrospinning is categorized as single nozzle, blend, coaxial, and emulsion electrospinning (Figure 3) (Sell et al., 2009).

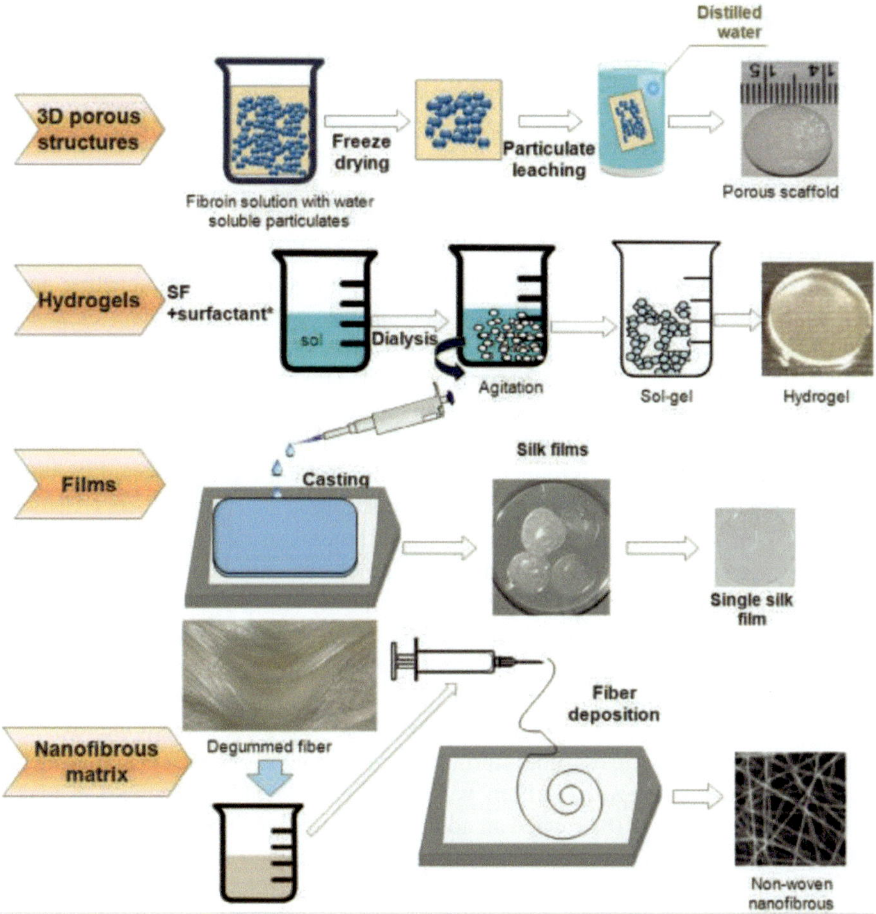

Figure 2. Graphical illustration of the manufacturing of four common SF scaffold types (Bhattacharjee et al., 2017).

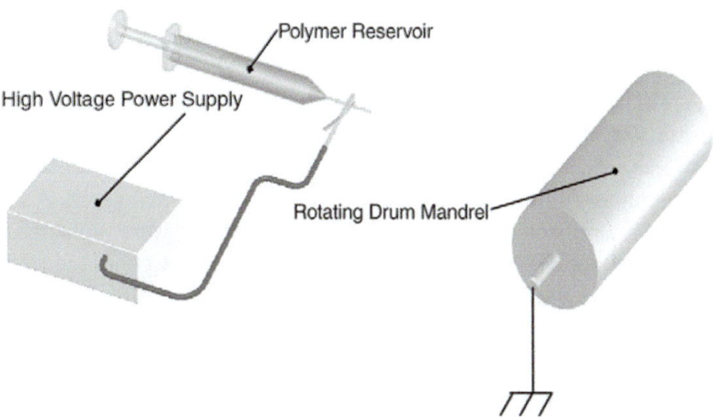

Figure 3. Schematic representation of electrospinning (Sell et al., 2009).

3.1.1.1 Needle-based Electrospinning

In needle electrospinning, a needle-shaped spinneret is employed, resulting in a pronounced "cone shape" at the needle tip due to the electric field's impact.

3.1.1.1.1 Blend Electrospinning

Electrospinning using blends is a technique employed in tissue engineering to replicate the extracellular microenvironment (Subbiah and Guldberg, 2019). In this procedure, bioactive compounds, like drugs, are dissolved or dispersed in the solution through blending electrospinning. The distribution of the bioactive agent within the fibers depends on the physicochemical properties of the solution and its interaction with the agent. However, this method has some limitations. For instance, delicate bioactive substances such as proteins and cytokines can lose their bioactivity and become denatured in solvents. Moreover, since most bioactive molecules carry a charge, they tend to accumulate on the surface of the jet during electrospinning due to charge repulsion. Consequently, the molecules are typically found in higher surface concentrations rather than uniformly distributed within the fibers (Alizadeh et al., 2024). SF mixes in the form of nanofibrous can help in wound healing. Recently, we created a two-layer scaffold using electrospun nanofibrous SF (ESF) and decellularized human amniotic membrane (AM) (Gholipourmalekabadi et al., 2018). They showed that as compared to AM alone, the mechanical characteristics of AM/ESF improved. When tested against adipose tissue-derived mesenchymal stem cells (AT-MSCs), AM and AM/ESF both demonstrated cell attachment without cytotoxicity. The AT-MSCs also exhibit stronger expression of the two main pro-angiogenesis factors, essential fibroblast growth factor, and vascular endothelial growth factor. For usage as a burn wound skin regeneration template, Kadakia et al. (Kadakia et al., 2018) created an additional SF scaffold blend with Poloxamers 407 (P407) and Manuka honey additives. According to their findings, the inclusion of P407 enhanced human dermal fibroblast cell adhesion and infiltration. Manuka honey-containing scaffolds demonstrated increased water absorption and a higher cell density after 28 days. In a different investigation, Migue et al. (Miguel et al., 2019) used electrospinning to create wound dressings, creating an asymmetric membrane. In the top layer, silk fibroin (SF) and poly (caprolactone) were used to mimic the thick structure of the epidermis. In contrast, the dermis-like bottom layer was made using SF and hyaluronic acid that had been combined with the herbal drug thymol (THY). According to the results, the electrospun asymmetric membrane possessed the wettability, porosity, and mechanical characteristics needed for the healing process. Additionally, the constructed scaffold's biocompatibility with adherence to human fibroblast cells was proven by the *in vitro* result. Additionally, the membrane's antibacterial and antioxidant qualities were enhanced by the THY-incorporated bottom layer. The Miguel et al. study (Miguel et al., 2019) used a combination of SF, PVA, Type I collagen Peptide (CP), a nitric oxide donor, and S-Nitrosoglutathione (GSNO) to treat non-healing diabetic ulcers. They created three distinct nanofibrous scaffold mats: SF-PVA, CP-SF-PVA, and CP-GSNO-SF-PVA. The *in vitro* biocompatibility experiment showed that the CP-GSNO-SF-PVA nanofibrous scaffold was cytocompatible and had strong cell adhesion qualities that increased cell proliferation.

An electrospun scaffold comprising silk fibroin/chitosan and collagen/chondroitin sulfate was developed for the healing of skin defects. The findings indicated that the CL/CS and SF/CS scaffolds exhibited biocompatibility and expedited skin wound healing in an *in vivo* rat model compared to the control group (Yu, Guo et al., 2017). Multiple SFP-based electrospun materials, including SFP/polyethylene glycol (PEG), SFP/polyvinyl alcohol (PVA), and SFP/polyethylene oxide (PEO), were produced to assess their efficacy in skin wound healing and compared to pure SFP. The SFP and PVA/SFP-treated groups exhibited enhanced wound healing and collagen regeneration *in vivo*. Furthermore, the anti-adhesion capabilities of the mix are as follows: PVA/ESF is superior to pure ESF, which is superior to ESF/PEG, and finally, ESF/PEO (Wang et al., 2018). SF was utilized to accelerate wound healing because of its capacity to enhance collagen regeneration. SF was combined with hydrophilic PVA and subjected to methanol treatment (PFSM) to create a wound-healing dressing. Furthermore, the epidermal stem cells derived from human exfoliated deciduous teeth and epidermal stem cells (ESCs) were integrated into the PFSM. The findings indicated that the PFSM/ESC facilitated the regeneration of granulation tissue, with a wound closure percentage of 53.49%, which is 1.18 times more than that of the PFSM/ESC (Huang et al., 2019). A biocompatible nanofibrous composite comprising soy protein isolate and silk (SPI/SF) was fabricated using electrospinning. Nanofibers were subjected to ethanol vapor treatment to enhance their water stability. The efficacy of constructed scaffolds for skin tissue regeneration was assessed *in vitro* utilizing common cell lines, including fibroblast cells (L929-RFP and NIH-3T3) and melanocytes (B16F10). The results indicated that none of the produced nanofibrous scaffolds exhibited toxicity to healthy mammalian cells. The efficacy of a nanofibrous scaffold in facilitating the healing of a full-thickness wound in rats within 14 days of therapy was evidenced (Varshney et al., 2020).

3.1.1.1.2 Coaxial Electrospinning

Coaxial electrospinning, or co-electrospinning, is an altered version of the traditional electrospinning technique (Negahdari et al., 2024a). It involves the use of two nozzles arranged in a straight line and connected to a high-voltage source. Using the core-sheath fiber shape is achieved by pumping two distinct solutions (core and shell materials) through nozzles. Coaxial electrospinning can be used to manufacture several types of fibers, such as fibers from different solution pairs, core-shell fibers, hollow fibers, and functional fibers. These fibers may also contain particles. Coaxial electrospinning is a commonly employed technique for fabricating hollow fibers, in which the material serves as the outer layer and a transitory substance serves as the inner core. Oil is commonly employed as a convenient ingredient in the post-spinning process due to its easier removal compared to other materials with larger molecular weights (Subbiah and Guldberg, 2019). The coaxial electrospinning process enables the production of multichannel nanotubes and nanofluidics, as well as the encapsulation of physiologically active compounds and controlled drug release. Nevertheless, there are other disadvantages associated with the utilization of this technology, such as the intricacy of the design and the requirement for meticulous management of process variables, including the interfacial tension and viscoelasticity of the two polymers.

This construction typically employs a high molecular weight polymer as the shell and pharmaceuticals as the cores. The characteristics of the shell (polymer solution) and core solution often affect the ultimate production of fibers, as seen in the manufacturing process of poly(glycerol sebacate) (PGS) fibrous membranes (You et al., 2016). The experiment revealed that augmenting PGS (core) concentration and incorporating PEO into the shell solution enhanced the mechanical strength of the resultant nanofibers. In contrast, the concentration of the polymer solution significantly affected the ultimate fiber production. The traditional coaxial electrospinning method is incapable of producing polymer fibers from a highly concentrated solution. A novel process method has successfully produced polymer fibers from a high concentration (35% w/v) of polyvinyl pyrrolidone (PVP) utilizing a concentric spinneret (Yu et al., 2011). Alongside the parameters of core and shell solutions, fabrication process variables significantly influence the manufacture of nanofibers using co-axial electrospinning. Navaporn Kaerkitcha et al. (Kaerkitcha et al., 2016) investigated the influence of applied voltage (10–25 kV) on the physical properties of polymethyl methacrylate (PMMA)/polyacrylonitrile (PAN) composite nanofibers produced from three distinct nozzle-end configurations (normal coaxial nozzle, inward coaxial nozzle, outward coaxial nozzle). The findings indicated that nanofibers with optimal morphologies were achieved at 15 kV across all three configurations.

E. Fallahiarezoudar et al. (Fallahiarezoudar et al., 2015) investigated the influence of various process parameters on the diameter of core (γFe_2O_3, a magneto-rheological fluid) and shell (polyvinyl alcohol (PVA)) nanofibers employing a complete factorial design. The investigations revealed that the primary determining elements were the applied voltage and flow rate. At a flow rate of 0.5 mL/h and a voltage of 35 kV, while maintaining a spacing of 5 cm and a rotational speed of 2000 rpm, nanofibers with a slender diameter and aligned configuration were produced. Additionally, the flow rate ratio between the core and shell channels was a critical metric that must be maintained between 4.25 and 6. If the value was very high or low, the production of electrospun nanofibers would be unsuccessful. Currently, the properties of manufactured fibers may be altered by encapsulating them in a single liquid stage during the co-axial electrospinning process, and innovative materials have been produced by coaxial electrospinning. Phase-change materials were encapsulated in nanofibers to enhance thermal stability and durability through coaxial electrospinning, utilizing structures such as polyethylene glycol (PEG)/cellulose acetate (CA) (core/shell) and poly(vinyl alcohol) (PVOH)/phase change material (PCM). The latter was synthesized by a mixture of emulsion electrospinning and coaxial electrospinning to prevent exposure to elevated relative humidity and achieve a high thermal storage capacity. Additionally, non-thermochromic dyes were encapsulated into core-shell fibers to provide reversible thermochromic properties with a restricted transition temperature range of 14–15°C (Malherbe et al., 2010).

3.1.1.1.3 Emulsion Electrospinning

Emulsion electrospinning is a process where two solutions that do not mix well are spun together at the same time. It uses the same setup as blend electrospinning. An emulsion is a mixture of two or more immiscible substances, such as oil and water, in which one substance is dispersed in the form of small Nanofibers that can be generated by the utilization of electrospinning technology, which allows the production of fibers with diverse compositions and structures. Fibers with core-shell structures are produced using a blend of two or more hydrophobic and hydrophilic polymers. Emulsion electrospun fibers consist of several polymer phases, regardless of whether they have a continuous core-shell form or not. Emulsion electrospinning regulates the core, sheath, and diameter of the fibers by altering the composition of the emulsion and the parameters of emulsification. This technique can be employed to transfer a pharmaceutical compound or a biologically active substance from an organic media to one that is only soluble in an inorganic solvent. Emulsion electrospinning is advantageous in tissue engineering applications due to its ability to achieve precise and controlled release of bioactive chemicals or drugs, guided by the degradation profiles of nanofibers. Emulsion electrospinning offers the benefit of utilizing water-based solvent systems instead of potentially dangerous solvents, resulting in simpler and safer manufacturing and application processes (Zhang et al., 2018).

A study by Trina Roy and Priti Prasanna Maity at the Indian Institute ofTechnology Kharagpur has developed a core-shell nanofibrous scaffold using polycaprolactone and silk fibroin emulsion electrospinning for tissue engineering applications. The researchers found that the stability of the water-in-oil emulsions depends on the concentration of silk fibroin in the PCL solution. The spinnability of the emulsions is more dependent on the viscosity of the solution, dominated by the molecular weight of PCL and its concentration. The nanofibers exhibited a distinct core-shell structure, better cytocompatibility, and cellular growth with the incorporation of the silk fibroin biopolymer (Roy et al., 2018). The study by Heydari Foroushani et al. explores the use of electrospun nanofibers for drug loading in wound dressings. The researchers fabricate silk fibroin (SF)-chitosan (CS)-silver (Ag)-curcumin (CUR) nanofibers using the electrospinning method, which facilitates the pH-responsive release of CUR, accelerates wound healing, and improves mechanical properties. The nanofibers showed improved drug loading and entrapment efficiencies after the incorporation of Ag nanoparticles. The study also found that nanofibers with lower CS concentrations had stronger inhibitory effects on bacterial growth. The nanofibers do not have any toxic effect on cell culture, as demonstrated by *in vitro* wound healing test on NIH 3T3 fibroblasts (Heydari Foroushani et al., 2022).

3.1.1.2 Needleless Electrospinning

Needleless electrospinning is a technique that seeks to enhance the effectiveness of the electrospinning process by generating several jets of fibers from a polymer solution that is placed on a compatible substrate. Utilizing this technology has been discovered to enable the mass production of nanowebs. While discussing needleless electrospinning, two main types need to be considered: stationary and rotational. The spinning needle-less arrangement utilizes mechanical force to facilitate the initiation of a jet from the surface of the polymer. Disturbance contributes to the production of Taylor cones, which in turn leads to the initiation, elongation, and formation of fibers. Needleless electrospinning utilizes external factors, such as magnetic fields, gas movement, and gravity, to create jets from the liquid surface (Partheniadis et al., 2020; Yu, Dong et al., 2017).

3.1.2 Film

In 2017, the SF film was developed by Zhang and his team and evaluated for its application potential in skin repair. The film showed outstanding properties in terms of light transmittance, moisture and vapour permeability, fluid transport capacity, waterproofness, biocompatibility, and antibacterial features. Indeed, *in-vivo* experiments on a rabbit with a full-thickness skin defect revealed an effective reduced healing time and noticeable skin regeneration in comparison with commercial wound dressings (Zhang et al., 2017).

In 2011, the efficiency of silk films for human corneal healing following injury was evaluated. Recently, Beena and her team proved that using different manufacturing methods and post-fabrication approaches may affect the properties and behavior of silk fibroin films in the context of corneal tissue engineering (Beena et al., 2022). In 2023, Sun reported the development of stretchable films with nanopores and tunable pore size through a nonsolvent-induced phase separation process. These fabricated films were able to protect the proliferation of human skin fibroblasts *in vitro* and accelerate the full-thickness skin wound healing process in rat models (Sun et al., 2023). Recently, a new regenerated silk fibroin (RSF) film with 2.5% pluronic F127 is developed which has acceptable mechanical and hydrophilicity properties. To address the lack of antibacterial activity, Duan and colleagues, incorporated different antibacterial agents into RSFPF127. *In-vitro* studies showed that these modified films could release drugs over time and effectively kill bacteria, promote fibroblast adhesion, and influence macrophage behavior. Regarding the *in vivo* studies, it was found that, the medicated RSFPF127 films eliminated antibiotic-resistant bacteria in wounds, reduced inflammation, and promoted blood vessel growth and skin regeneration (Zhang et al., 2024).

3.1.3 Hydrogel

There are a lot of reported studies with promising outcomes about using SF in wound repair and skin tissue engineering (Mukherjee et al., 2022). Since SF can form connections between the target tissue and the biomaterial being created (Mehrabani et al., 2018), is non-allergic (Liu et al., 2019a), can promote neovascularization (Li, Li, et al., 2017), can combine with other polymers to form composite structures (Akdag et al., 2023), and is simple to produce and use (Huang et al., 2017), it is a good option for use in the wound healing process. Jung Min Lee et al. (Lee et al., 2017) utilized freezing and thawing to physically crosslink SF and PVA and synthesized a high-strength silk hydrogel. The produced hydrogel exhibited comparable mechanical strength to ear cartilage and maintained its original form during the culture period. Interestingly, the final structure could provide a supportive environment for cartilage cell proliferation. In a different study, Wenqian Xiao used sonication and UV light to make a silk fibroin/methacrylate hyaluronic acid (SF-HAMA) double-network hydrogel. Since the produced hydrogel had high mechanical strength and biocompatibility, it was favorable for load-bearing soft tissue and contributed to osteogenic differentiation (Xiao et al., 2019).

In 2020, researchers have developed a natural green hydrogel called BVSF by incorporating biliverdin into silk fibroin. This hydrogel demonstrates remarkable versatility in glioma treatment. Under near-infrared light exposure, it rapidly heats up beyond 45°C, proving effective in eliminating cancer cells during *in-vitro* experiments and suppressing tumor growth *in-vivo* conditions. Simultaneously, the BVSF hydrogel exhibited impressive regenerative properties. It enhanced cell proliferation, migration, and adhesion while also possessing anti-inflammatory features. These properties significantly accelerate the processes of wound healing and tissue regeneration. This dual-function BVSF hydrogel offered a promising solution for both cancer treatment and post-treatment care, addressing current challenges in wound healing after photothermal therapy (Yao et al., 2020). Silk fibroin hydrogels have been extensively researched for their application in wound healing and skin tissue engineering. The hydrogels provide a moist environment conducive to wound healing and are capable of promoting cell proliferation and migration.

A study by Bogadi et al. (2024) demonstrated that silk fibroin hydrogels, combined with growth factors, could significantly accelerate wound healing and improve scar formation in animal models (Bogadi et al., 2024). Additionally, these hydrogels serve as carriers for drug delivery systems in chronic wounds, allowing for sustained release of antimicrobial agents (Withanage et al., 2021). Silk fibroin hydrogels have also shown significant potential in cartilage and bone tissue engineering. Due to their tunable mechanical properties and the ability to mimic the ECM of cartilage, silk fibroin hydrogels have been used as scaffolds for cartilage regeneration, promoting chondrocyte proliferation and matrix production (Lyu et al., 2023). For bone regeneration, composite hydrogels combining silk fibroin with bioceramics such as hydroxyapatite have demonstrated enhanced osteogenic potential, supporting bone growth in both *in vitro* and *in vivo* studies (Jin et al., 2023). The use of silk fibroin hydrogels in neural tissue engineering has gained attention due to their non-cytotoxic nature and ability to support neural cell differentiation. Modified silk fibroin hydrogels, incorporating peptides or growth factors, have been shown to promote the differentiation of neural stem cells and support axonal growth *in vitro* (Sun et al., 2017). These hydrogels also serve as vehicles for delivering neurotrophic factors to injured sites, making them promising for applications in spinal cord injury repair (Madappura and Madduri, 2023).

Finally, it should be pointed out that silk fibroin hydrogels have many applications in various aspects of tissue engineering, and it is expected that by changing their mechanical properties, a clear perspective can be envisioned for them (Zheng and Zuo, 2021).

3.1.4 Sponge

Silk sponge production usually involves degumming natural silk, followed by its dissolution in a solution of regenerated silk fibroin (RSF) (Gu et al., 2024). Various techniques, such as casting, spin coating, electrospinning, and 3D printing, are used to produce different forms of silk products, including sponges (Wen et al., 2021). Silk fibroin is one of the best scaffolds to be developed for bone engineering; it is highly biocompatible and degrades by proteases (Wu et al., 2022). The biocompatibility of these structures is dependent on the inorganic phase (Liang et al., 2024). Silk sponges have been used in the fields of blood vessels, cartilage, ligaments, fat, and cartilage. Porous sponge scaffolds were suitable for the formation of bone tissue by promoting cell attachment, proliferation, and migration. Furthermore, the high porosity simplifies the transportation of nutrients and waste into and out of the scaffolds. Porous silk sponges can be manufactured using porogens, gas foaming, or lyophilization approaches (Correia et al., 2012). In 2020, ZHU HuiMin could develop a 3D functional neural tissue based on an MEF-derived ECM-modified SF sponge and observed enhanced neuronal growth and axonal extension in comparison with a naked SF sponge (Zhu et al., 2020). The applicability of SF nanocomposites in tissue engineering has been recently extended by Afjeh-Dana and her colleagues. They developed a sponge-like structure targeting peripheral nervous tissue and demonstrated the expression of neural-specific proteins (nestin and neuron-specific enolase) by amplification of silk fibroin scaffolds with gold nanorods (Afjeh-Dana et al., 2019).

Recent studies show significant promise for SFSs in bone tissue engineering. Silk fibroin's osteoconductive properties can be enhanced by incorporating calcium phosphate or bioactive ceramics. Researchers have found that SFSs, when combined with hydroxyapatite or bone morphogenetic proteins (BMPs), support mesenchymal stem cell (MSC) differentiation and bone regeneration (Ai et al., 2021). A 2019 study by Li et al. demonstrated that SFSs functionalized with BMP-2 enhanced bone healing in rat femoral defects, showing higher bone regeneration compared to control groups (Nguyen et al., 2019a). Cartilage repair has seen advancements with silk fibroin sponges in combination with other bioactive molecules. Due to the avascular nature of cartilage, silk fibroin's porous structure aids in chondrocyte adhesion and ECM deposition. Recent innovations include blending SF with hyaluronic acid or chondroitin sulfate to mimic the natural cartilage matrix and enhance tissue regeneration (Yan et al., 2018). Jaipaew et al. (2016) demonstrated that a silk fibroin sponge-hyaluronic acid composite scaffold showed superior cartilage formation in osteochondral defect models in rabbits (Jaipaew et al., 2016).

Cardiac tissue engineering has also benefited from silk fibroin sponges. Researchers have developed SFSs that can deliver bioactive molecules and stem cells to repair myocardial tissue post-infarction. Dong lin et al. (2012) demonstrated that silk fibroin scaffolds, combined with vascular endothelial growth factor (VEGF), promoted angiogenesis and reduced scar formation in rat and pig myocardial infarction models (Lin et al., 2012). Silk fibroin sponges are a promising biomaterial for cardiac tissue engineering to support the integration of cardiomyocytes and improve cardiac function after injury (Song et al., 2020). Silk fibroin sponges are also utilized in skin regeneration, particularly in wound healing. Their biocompatibility and ability to support keratinocyte and fibroblast growth make them excellent scaffolds for skin tissue engineering. Recent innovations include incorporating antimicrobial agents or growth factors into SFSs to promote faster wound healing and reduce infection risks (Tran et al., 2023). Other studies showed that an SF sponge loaded with insulin like growth factor (EGF) significantly accelerated wound closure in diabetic mice (Lin et al., 2021).

3.1.5 3D Bioprinting

One of the main challenges in the field of tissue engineering is to generate soft artificial structures that are mechanically stable and biologically active (Mu et al., 2020). SF and new compounds derived from it can provide a suitable platform for the production of functional soft tissue structures in the future by creating desirable systems (Schneider et al., 2023). It has been established that 3D printing technology utilization can be effective in constructing complex multi-material structures for tissue engineering functions (Derakhshanfar et al., 2018). In the bioprinting method, biocompatible hydrogels are routinely used to allow cell encapsulation in an appropriate supportive 3D environment, similar to their native tissue (Dababneh and Ozbolat, 2014). Despite the 3D bioprinting applications in tissue engineering, there are two mains problems in this regard: limited material collection and cell types (Sun et al., 2021a). Due to its polymorphic nature and biocompatible properties, its stablished that rimarily use silk as dissolved silk fibroin to produce hydrogel bioinks (Mu et al., 2020). However, only a limited number of studies have used SF as a material for bioprinting processes (Melke et al., 2016a). In 2014, Suntivich et al. demonstrated that by utilizing inkjet printing procedures, SF could be fabricated into "nest" form with 70–100 μm diameters. The modified structure was able to serve as an anchored nest for *E. coli* cell incubation and proliferation (Suntivich et al., 2014). Das and her colleagues in 2015 described RSF gelatin scaffolds made with 3D bioprinting that were acceptable for nasal inferior turbinate tissue-derived mesenchymal progenitor cell culture (Das et al., 2015). Furthermore, the use of 3D bioprinting can modify the stiffness of the material by altering its sheet contents, ultimately influencing the amount of degradation. Interestingly, the produced scaffolds that were processed by sonication had a higher β sheet structure than the group of scaffolds that were treated with the tyrosinase enzyme and provided suitable platforms for osteogenic differentiation (Melke et al., 2016a; Sun et al., 2021a). According to this study, there is a tight relationship between the structural parameters of SF scaffolds (secondary conformation and matrix stiffness), which is involved in determining the differentiation fate of a cell population towards a specific lineage (Das et al., 2015; Sun et al., 2021a). Additionally, Rodriguez et al. (2018) could make scaffolds by using RSF, laponite, and PEG via the extrusion-based 3D printing method (34). In this method, because the gelation of the scaffolds occurs during the printing process, there is no need for additional processes such as chemical or photochemical cross-linking, and this method can be used to prepare complex structures of biomaterials on a microscale. Commonly, 3D printed SF scaffolds have structures in macroscopic dimensions that can be converted to micro and nanoscales by applying mechanical stresses and dopants (Jiang et al., 2009; Zhao et al., 2009; Wang et al., 2014). Another study used photocurable 3D printing technology to construct a nanohydroxyapatite (nHA)/methylacrylylated silk fibroin (MASF) scaffold to treat bone defects. To evaluate scaffold efficiency, an *in vivo* experiment was done to repair skull defects in rats. The team used micro-CT scans and two tissue staining methods to examine implant-site bone development. The results showed that after 4 and 8 weeks, the osteogenic effect of scaffolds was detected and an appropriate microenvironment was provided for cell adhesion, migration, and proliferation (Huiwen et al., 2024). Bone tissue engineering has greatly benefited from 3D bioprinting due to the ability to fabricate scaffolds that mimic the hierarchical structure of bone. Bioprinting allows the incorporation of osteoconductive materials such as hydroxyapatite (HA) and tricalcium phosphate into bio inks to enhance bone regeneration. Recent studies have demonstrated that 3D bio-printed scaffolds can promote osteogenesis and vascularization in bone defects. For example, Daly et al. (2016) developed a 3D bioprinted scaffold using a composite bioink of gelatin and HA, which supported osteoblast proliferation and enhanced bone regeneration *in vitro* and *in vivo* (Daly et al., 2016). 3D bioprinting offers significant potential in skin tissue engineering, especially for wound healing and skin graft applications. The ability to print multilayered structures mimicking the epidermal and dermal layers of skin is a key advantage. Advances in bioink development have led to the creation of cell-laden bioink capable of promoting keratinocyte and fibroblast proliferation, essential for skin regeneration. Kim et al. (2017) developed a 3D bioprinted skin model using a bioink composed of collagen and fibrin, which demonstrated enhanced wound healing and integration in preclinical studies (Kim et al., 2017). Additionally, the use of 3D bioprinting to fabricate skin substitutes with vascular networks has shown promise in accelerating the wound healing process (Lee, Singh et al., 2014). Cardiac tissue engineering aims to restore heart function following myocardial infarction or heart failure. 3D bioprinting has enabled the creation of cardiac patches and vascularized heart tissues that replicate the complex architecture of native myocardium. The use of bioink containing cardiomyocytes and endothelial cells allows for the development of functional, contractile cardiac tissues (Kato et al., 2021). Hwang et al. (2024) demonstrated the remarkable capacity of the 3D bioprinted vascular patch to enhance endogenous cardiac regeneration while providing exogenous cardioprotective functions. The dual mechanism of action, combining structural support with biological signaling, represents a significant advancement in cardiac tissue engineering. The successful integration of endothelial cells with FSTL1 peptide suggests potential clinical applications for acute MI treatment (Hwang et al., 2024). Additionally, recent advancements in stem cell-based bioink are paving the way for the development of patient-specific cardiac tissue constructs (Das et al., 2021).

The liver is a complex organ with a highly vascularized structure, making it one of the most challenging tissues to engineer. 3D bioprinting has been used to create liver constructs that replicate hepatic architecture and function. Bioink containing hepatocytes, along with endothelial cells, have been employed to create bio-printed liver tissues that exhibit key liver functions, including albumin production and drug metabolism (Wang et al., 2024). Moreover, the ability to create vascularized liver constructs holds promise for addressing liver failure and advancing drug testing platforms. Recent research has demonstrated significant progress in bioprinting liver tissue constructs with enhanced cell viability, liver-specific functions, and vascularization. Three-dimensional (3D) bioprinting techniques have been used to create biomimetic hepatic tissues with potential applications in regenerative medicine and drug discovery (Sanyal and Ghosh, 2024).

All in all, 3D bioprinting has revolutionized the field of tissue engineering by enabling the fabrication of complex, biomimetic tissue constructs. From bone and cartilage regeneration to skin and cardiac tissue engineering, 3D bioprinting offers a versatile platform for developing functional tissue substitutes. Despite existing challenges, ongoing advancements in bioink development, bioprinting technologies, and tissue vascularization are pushing the boundaries of what is possible in regenerative medicine. With continued progress, 3D bioprinting holds the potential to significantly impact clinical tissue engineering applications and organ transplantation in the future.

4. Application in Tissue Engineering

4.1 Skin Regeneration

Silk fibroin-based wound dressing has different applications on different kinds of skin wounds like full thickness, burn, and diabetic wounds. In the following topic, we discuss these applications. SF scaffolds activate the NF-kB signaling pathway which results in lesser reactive oxygen agents and lesser inflammatory response. In the SF-treated wounds, an increase in tumor necrosis factor receptors and toll-like receptors (Vidya and Rajagopal, 2021; Mazurek et al., 2022). Based on Park et al. studies on SF, SF solutions from BM worm can boost the skin wound healing process both *in vitro* and *in vivo* (Park et al., 2018).

The three-dimensional structure obtained by electrospinning, gives an appropriate ECM like scaffold that provides a setting for cell adhesion and future differentiation (Lee, Ju et al., 2014; Gholipourmalekabadi et al., 2018; Park et al., 2016). This technique gives special antibacterial properties to the scaffold which is based on ionic transmission from fibers to the environment (Ige et al., 2012; Rodríguez-Tobías et al., 2019). The special morphology can activate the tension-sensitive genes which is very important in preventing pathological structure in the cutaneous wounds (Krishna et al., 2020; Meka et al., 2018; Meka et al., 2019).

Also using coaxial electrospinning can build a structure as a promising method for drug delivery (Ma and Zhang, 1999). SF hydrogels are naturally accruing hydrogels that provide a suitable environment for cell attachment, proliferation, migration, and differentiation besides improving mechanical strength which distinct it from other natural-accruing hydrogels like chitosan (Murakami et al., 2010), alginate (Bouhadir et al., 2001), collagen, hyaluronic acid (Burdick and Prestwich, 2011). SF sponges are bio scaffolds linked porous with a setting suitable for tissue growth and cell adhesion especially in full-thickness skin wounds, According to Lee et al. (Lee, Ju et al., 2014) SF collagen sponges showed improvement in mechanical properties. Liu et al. (Liu et al., 2019b) engineered a freeze-dried SF sponge loaded with neurotensin-loaded gelatin microspheres in diabetic ulcers. A posttreatment 28-day follow-up with SF film loaded with growth factors showed faster and better Deep wound healing and less inflammatory response in the Rat model (Shen et al., 2010; Liu et al., 2008).

Clinical administrations of SF in skin healing tasted in various of wounds like full-thickness, burn-induced, and diabetic wounds. In a study by GUO et al. (Guo et al., 2021), using random SF scaffold (RSS) and aligned SF scaffold (ASS) in full-thickness wound model results in promoted epithelization and neovascularization in both scaffolds. In comparison to these two scaffolds, however, ASS had better evidence of healing. Hodgin et al. (Hodgkinson et al., 2014) designed as a study to compare the effect of different diameters of SF fibers in wound healing effect *in vitro* on primary human dermal fibroblast and *Ex vivo* on human full-thickness skin model resulting in better results in both re-epithelization and cell migration and hosting in scaffolds with thinner fibers (~ 205–300 nm).

Another study by Millan-Rivero et al. (Millán-Rivero et al., 2019) compared SF scaffolds loaded with human Warton jelly mesenchymal stem cell (WJ-MSC-SF) and the commercial dressing Linitul® as the control group. The WJ-MSC-SF scaffold showed longer healing time but less inflammatory (lesser PMN and macrophage M1) in compare to the control group.

The addition of chitosan into the SF solution to achieve a new dressing is also a way to achieve antibacterial (Bano et al., 2017) characteristics and hemostatic activity. Guang et al. (Guang et al., 2015) used this scaffold in both *in vitro* and *in vivo* studies. Disk diffusion tests antibacterial effect on *E. coli*, *S. aureus*, *P. aeruginosa* and monilia albicans. The *in vivo* study on rat models showed up promoted wound healing process with less inflammatory response.

In treating partial thickness burn wounds which do not require surgical intervention, SF films with or without additions showed promising results as Hyung Woo Ju et al. (Ju et al., 2016), Guan et al. (Guan et al., 2020) and Chouhan et al. (Chouhan et al., 2018) designed studies to prove these properties.

Wounds in the setting of DM can convert into chronic unhealing wounds that may accompany infections which can be a life-threatening phenomenon. Poor blood supply in the wound environment due to microangiopathies leads to insufficient nutrition and oxygen delivery and disruption of the healing process (Greenhalgh, 2003; Anderson and Hamm, 2012; Burgess et al., 2021; Blakytny and Jude, 2009).

Two study by Yang et al. (Yang et al., 2020) and li et al. (Li, Liu et al., 2017) Compare SF and insulin-loaded SF in DM wound healing. In comparison to the solo wound model, both scaffolds showed better improvement; however, the insulin-loaded scaffold even induced faster repair with lesser inflammatory response.

4.2 Bone Regeneration

Biomaterial used to regenerate hard tissue application in bone regeneration should have high mechanical strength besides biocompatibility and bioactivity (Paladini and Pollini, 2022; Xu et al., 2020). Among natural biopolymers, SF has gained attention due to its specific characteristics, such as high biocompatibility, proper degradation rate, and unique mechanical properties, along with various methods for obtaining it. Studies demonstrate that it can increase bone regeneration in both *in vitro* and *in vivo*.

Silk composite micro-patterning can provide an ECM-like environment allowing cells to migrate, attach, and secrete related ECM matrix. The stiffens beside its active sites for mineralization can promote bone regeneration. On the other hand, as mentioned earlier, SF is an appropriate drug delivery setting so the addition of bioactive molecules, drugs, and even stem cells can boost its function (Sun et al., 2021a; Bhattacharjee et al., 2017; Long et al., 2021).

Silk molecules pattern (beta-sheets) activates the pathways related to osteo-differentiation, as Long et al. showed that beta-sheet structures of SF can influence bone tissue evaluation through protein-focal adhesion-cytoskeleton links and intra-cellular pathways (Long et al., 2021).

SF can also promote active biomolecule aggregation niche with ionic concentration (e.g., Calcium) and increase bone structure mineralization (Zhang et al., 2020) its construction promotes MSC (hAMSC) differentiation into osteoblasts and further ECM secretion and biomineral deposition (Huang et al., 2019; Li et al., 2020; Luo et al., 2021; Jo and Oh, 2018).

Combining SF with other biopolymers favors new features such as increasing its biocompatibility, more adjustable morphology, better degradation, and conformation. For example, SF in combination with hydroxy appetite (HAP) end in promising results with excellent advantages for bone formation as HAP is one of the major components of the bone structure while presenting SF increases its mechanical strength and neovascularization as proven in both *in vivo* and *in vitro* animal models (Bhattacharjee et al., 2017; Ding et al., 2021; Farokhi et al., 2018; Wu et al., 2019).

Ting et al. used a 3-D printed porous composite using HAP/SF and sodium alginate as cross-linkers with hBMSC, the study showed expansion and differentiation of the cells all over the scaffold (Huang et al., 2019). Wu et al. add MgO into the HAP/SF composite to induce a local alkaline microenvironment which results in the promotion of hBMSC differentiation and proliferation (Wu et al., 2020). In two studies by Ding et al. and Wu et al. (Patil and Singh, 2019; Paladini et al., 2016) addition of graphene/oxide into scaffolds demonstrates enhancement of bone formation in both 3-D printed scaffold (+ 0.5 graphene) and electro spined scaffolds, as well as *in vitro* osteogenesis promotion and calvaria bone defect in Rat models.

4.3 Cartilage Regeneration

There are various methods for cartilage repair, some approaches like joint preservation operation, arthropathy and non-operative are among it. However, lack of nerve and blood supply slow this regeneration process and providing an appropriate microenvironment for cells to proliferate and secrete ECM can be a revolutional approach in cartilage regeneration. Among all of the bio polymers used in tissue engineering, silk fibroin is the point of interest due to its special characteristics like notable mechanical strength, elasticity, degradability and biocompatible. There are different sites of cartilage with different structure and each require an exact setting.

Intervertebral disc (IVD) degradation is a disease that can cause lots of complaints and lower the quality of life many people annually undergo surgery or endure its pain and complications. IVD anatomy consists of two parts, a central nucleus pulposus (NP) made of type 2 Collagen and an outer comprised layered segment called annulus fibrosus (AF) which is comprised of aligned collagen fibers (Coventry et al., 1945; Bruehlmann et al., 2002). Bio-engineered *in vitro* models of IVD (Richardson et al., 2008) eventuate good results in animal models in comparison to arthroplasty and prosthetic discs (Park et al., 2012). Recently studies focused on SF based on the results showing promising IVD repair.

Tissue engineering approaches need to focus on both NP and AF regeneration. In some studies, lamellar SF scaffolds were applied to mimic AF. Seeding hMSC on this AF-like tissue and differentiation into two types of cells based on their location in the composite, as the strength of the lamellar layer increased from inner to outer layer, the cells in the outer layer consist fibroblast-like phenotype and the inner cells had chondrocyte-like morphology. Also, the ECM secreted in these layers was different and the inner layer consists of more cartilage markers like aggrecan and type 2 collagen than the outer layer (Best et al., 1994; Bhunia and Mandal, 2018). As mentioned, the addition of bioactive molecules can favor new features to the scaffold. One of the materials is chondronectin sulfate (CS) which studies demonstrate that its able to promote the regeneration of the AF (Bhattacharjee et al., 2016; Hayes et al., 2011; Frauchiger et al., 2018; Chang et al., 2007).

Regenerating the NP part of the disc is also important. Approaches were trying to use a combination of SF and polyurethane or polyvinyl alcohol as porous injectable composites to reproduce the inner part of the VD. Another study used a biphasic composite made of SF and fibrin/HA to mimic both AF and NP structures. The results demonstrate that cells seeded on the biphasic composites had better proliferation in comparison to porous scaffold (Hu et al., 2017; Neo et al., 2014; See et al., 2012).

Anterior cruciate ligament (ACL) dysfunction is a common disabling problem that occurs every year. The main part of the ligaments is collagen but regeneration using solo collagen fibers didn't provide satisfactory results as it has low mechanical strength studies considered various combinations of silk and other bioactive molecules with or without stem cells to achieve an applicable approach. They include a combination of polyethylene terephthalate (PET) (Rising et al., 2005), exogenous vascular cells or angiogenic factors (Miyasaka, 1991) and poly-lactic-co-glycolic acid (PLGA) that all led to collagen deposition and promoting ACL reconstruction (Ran et al., 2017; Rising et al., 2005). Some approaches were obtained to repair meniscal ruptures. Also, SF-based scaffolds in various forms like composites, injectable hydrogels, and films in co-culture with MSC used to repair osteochondral defects (Chen et al., 2011; Verdonk et al., 2008; Leroy et al., 2017).

However, among all of these, the application of SF in cartilage and osteochondral regeneration needs more assessment to be ready for market.

4.4 Cornea Regeneration

The cornea has a simple structure consisting of three distinct layers: an outer corneal epithelium with 3–5 layers of stratified squamous cells that come into an acellular

Bowman's layer (BL) the corneal stroma is avascular populated with keratinocytes and the innermost layer known as Descent's membrane (DM). The endothelial layer of the cornea plays a role in supplying and homeostasis of the corneal environment. Then is corneoscleral limbus separates the clear cornea from the surrounding with sclera and consists of the limbal epithelial stem cells (Stepp and Zieske, 2005; KA, 1993; Tsai et al., 2000).

Lots of irritations can damage the corneal and limbal stem cells, the common therapy is limbal tissue transplant which unfortunately has a high risk of rejection because of lack of vasculature and the presence of immune cells (Zakaria et al., 2010). Using hdAM is a common approach to increase the stem cell life span, however, there are currently no reliable techniques for long-term limbal stromal repair. A suitable biomaterial for limbal repair should support sufficient homeostasis and regulatory molecule exchange. It has to support both epithelial and stromal cells (both fibroblast and vascular). The next priority or appropriate biomaterial is transparency. Besides all, like any other bio scaffold, thickness, permeability, cell interaction, degradation and biocompatibility are among the crucial characteristics (Schwab, 1999; Rama et al., 2001).

Infection, trauma, and degenerative disease are among the irritations that lead to scaring of the corneal stroma and the common treatment method is penetrating keratoplasty (PK) that aims to change the stromal full-thickness. However, presenting the grafted tissue is still challenging, the goal is to preserve the life span of the limbal stem cell and provide a microenvironment to lower contamination and promote supply (Harkin et al., 2004; Duan et al., 2007).

Regeneration of the cornea is complicated due to its special physical and cellular formation. SF is hypothesized to be used for corneoscleral, limbus, corneal stroma, corneal endothelium, and Ruysch's complex (Harkin et al., 2011; Calonge et al., 2021). 3D scaffolds could be a suitable bed for epithelial cell growth. Among two methods used for fabricating SF scaffolds for corneal regeneration (centrifugal casting and dry casting), researchers demonstrate that centrifugal casting is a more promising technique (Lee et al., 2016). Evolution of biocompatibility of SF scaffolds for corneal epithelial cells showed increased expression of DNp63a (Pellegrini et al., 2001) and keratin 3/12 marker (Schermer et al., 1986) which are related to corneal cell deviation. Cells cultured on these scaffolds proliferate and form a 2–3 layer of cells with compact columnar cells in the base and squamous cells on the apical surface and no cytotoxicity effect was proven (Liu et al., 2012; Lujerdean et al., 2022). Farasatkia et al. (Farasatkia et al., 2021; Farasatkia and Kharaziha, 2021) in two different studies, provide a combination of SF and gelatin and alginate which results in no scar formation, better stromal repair, and corneal regeneration.

SF films and matts fabricated through electrospinning methods form an architectural pattern mutual to corneal stroma with collagen fibers while their porous pattern allows nutrition and waste diffusion and promotes intercellular interactions (Lawrence et al., 2009).

5. Challenges and Limitations

Besides the attractive features of SF in biomedical applications, considerable limits exist that claim more attention. Obtaining raw SF for medical grade is the first problem because its quality is highly affected by the worm, temperature, humidity, and nutrition. To overcome this problem attaining a standard line for SF production is crucial. On the other hand, the degumming process followed by further cross-linking may also impact the final result and change its biocompatibility and toxicity state. Unlike other biopolymers, SF is not stored in powder, and storing its aqueous solution for a long time is impossible. More studies on the process and the relationship between variable factors should be done to overcome this challenge. Developing standard protocols to optimize production, extraction, and storage should be taken into consideration.

6. The Future of Silk Fibroin in Regenerative Medicine

Regenerative medicine is a pioneering field that aims to repair, replace, or regenerate damaged tissues and organs, offering potential solutions for otherwise irreparable injuries and degenerative diseases. With its interdisciplinary approach, this area combines insights from biology, materials science, and engineering to devise innovative therapies that harness the body's healing processes. One key focus of regenerative medicine is the development of biomaterials that support cellular growth and function. Among these, silk fibroin has emerged as a remarkable candidate due to its unique mechanical properties, biocompatibility, and biodegradability. Derived from silkworms, silk fibroin provides a scaffold that mimics the natural extracellular matrix, promoting cellular attachment, proliferation, and differentiation. This natural polymer's versatility makes it suitable for various applications, from tissue engineering to drug delivery systems. As regenerative medicine continues to evolve, the integration of advanced biomaterials like silk fibroin will play a crucial role in pioneering therapeutic strategies that can restore function and enhance the quality of life for patients facing chronic conditions.

References

Afjeh-Dana, E., P. Naserzadeh, H. Nazari, F. Mottaghitalab, R. Shabani, N. Aminii et al. 2019. Gold nanorods reinforced silk fibroin nanocomposite for peripheral nerve tissue engineering applications. Int J Biol Macromol 129: 1034–39.

Ai, Chengchong, Ling Liu and James Cho-Hong Goh. 2021. Pore size modulates *in vitro* osteogenesis of bone marrow mesenchymal stem cells in fibronectin/gelatin coated silk fibroin scaffolds. Materials Science and Engineering: C 124: 112088.

Akdag, Zekiye, Songul Ulag, Deepak M. Kalaskar, Liviu Duta and Oguzhan Gunduz. 2023. Advanced applications of silk-based hydrogels for tissue engineering: a short review. Biomimetics 8: 612.

Alizadeh, Sanaz, Jila Majidi, Mozhgan Jahani, Zahra Esmaeili, Zeinab Nokhbedehghan, Zahra Aliakbar Ahovan et al. 2024. Engineering of a decellularized bovine skin coated with antibiotics-loaded electrospun fibers with synergistic antibacterial activity for the treatment of infectious wounds. Biotechnology and Bioengineering 121: 1452–63.

Anderson, Kristin and Rose L. Hamm. 2012. Factors that impair wound healing. Journal of the American College of Clinical Wound Specialists 4: 84–91.

Aramwit, Pornanong, Sorada Kanokpanont, Wanchai De-Eknamkul and Teerapol %J. Srichana. 2009. Monitoring of inflammatory mediators induced by silk sericin. Journal of Bioscience and Bioengineering 107: 556–61.

Armentano, I., M. Dottori, E. Fortunati, S. Mattioli and J. M. J. P. d. Kenny. 2010. Biodegradable polymer matrix nanocomposites for tissue engineering: A review. Polymer Degradation and Stability 95: 2126–46.

Bano, Ijaz, Muhammad Arshad, Tariq Yasin, Muhammad Afzal Ghauri and Muhammad Younus. 2017. Chitosan: A potential biopolymer for wound management. International Journal of Biological Macromolecules 102: 380–83.

Beena, Maya, Jimna Mohamed Ameer and Naresh Kasoju. 2022. Optically clear silk fibroin films with tunable properties for potential corneal tissue engineering applications: a process–property–function relationship study. ACS Omega 7: 29634–46.

Best, Barbara A., Farhid Guilak, Lori A. Setton, Wenbo Zhu, Fatemeh Saed-Nejad, Anthony Ratcliffe et al. 1994. Compressive mechanical properties of the human anulus fibrosus and their relationship to biochemical composition. Spine 19: 212–21.

Bhattacharjee, Maumita, Shikha Chawla, Shibu Chameettachal, Sumit Murab, Neel Sarovar Bhavesh and Sourabh Ghosh. 2016. Role of chondroitin sulphate tethered silk scaffold in cartilaginous disc tissue regeneration. Biomedical Materials 11: 025014.

Bhattacharjee, Promita, Banani Kundu, Deboki Naskar, Hae-Won Kim, Tapas K. Maiti, Debasis Bhattacharya et al. 2017. Silk scaffolds in bone tissue engineering: An overview. Acta Biomaterialia 63: 1–17.

Bhunia, Bibhas K. and Biman B. Mandal. 2018. Modulation of extracellular matrix by annulus fibrosus cells on tailored silk based angle-ply intervertebral disc construct. Materials & Design 158: 74–87.

Blakytny, Robert and Edward B. Jude. 2009. Altered molecular mechanisms of diabetic foot ulcers. The International Journal of Lower Extremity Wounds 8: 95–104.

Bogadi, Subhasri, Rajkumar Malayandi, P. Vasanth Raj, P. Suresh Kumar, Madhukiran Parvathaneni, Milton Kumar Kundu et al. 2024. Silk fibroin and sericin: Multifunctional formulations for treating diabetic wound healing. European Polymer Journal: 113465.

Bouhadir, Kamal H., Kuen Yong Lee, Eben Alsberg, Kelly L. Damm, Kenneth W. Anderson and David J. Mooney. 2001. Degradation of partially oxidized alginate and its potential application for tissue engineering. Biotechnology Progress 17: 945–50.

Bruehlmann, Sabina B., Jerome B. Rattner, John R. Matyas and Neil A. Duncan. 2002. Regional variations in the cellular matrix of the annulus fibrosus of the intervertebral disc. Journal of Anatomy 201: 159–71.

Burdick, Jason A. and Glenn D. Prestwich. 2011. Hyaluronic acid hydrogels for biomedical applications. Advanced Materials 23: H41–H56.

Burgess, Jamie L., W. Austin Wyant, Beatriz Abdo Abujamra, Robert S. Kirsner and Ivan Jozic. 2021. Diabetic wound-healing science. Medicina 57: 1072.

Calonge, Margarita, Teresa Nieto-Miguel, Ana de la Mata, Sara Galindo, José M. Herreras and Marina López-Paniagua. 2021. Goals and challenges of stem cell-based therapy for corneal blindness due to limbal deficiency. Pharmaceutics 13: 1483.

Chang, G., H.-J. Kim, D. Kaplan, G. Vunjak-Novakovic and R. A. Kandel. 2007. Porous silk scaffolds can be used for tissue engineering annulus fibrosus. European Spine Journal 16: 1848–57.

Chen, Fa-Ming, Li-An Wu, Min Zhang, Rong Zhang and Hai-Hua Sun. 2011. Homing of endogenous stem/progenitor cells for *in situ* tissue regeneration: promises, strategies, and translational perspectives. Biomaterials 32: 3189–209.

Cheng, Y., L.-D. Koh, D. Li, B. Ji, M.-Y. Han and Y.-W. J. J. o. t. R. S. I. Zhang. 2014. On the strength of β-sheet crystallites of Bombyx mori silk fibroin. Journal of the Royal Society Interface 11: 20140305.

Chopra, Sheetal and M. L. Gulrajani. 1994. Comparative evaluation of the various methods of degumming silk.

Chouhan, Dimple, Tshewuzo-u Lohe, Pavan Kumar Samudrala and Biman B. Mandal. 2018. *In situ* forming injectable silk fibroin hydrogel promotes skin regeneration in full thickness burn wounds. Advanced Healthcare Materials 7: 1801092.

Correia, C., S. Bhumiratana, L. P. Yan, A. L. Oliveira, J. M. Gimble, D. Rockwood et al. 2012. Development of silk-based scaffolds for tissue engineering of bone from human adipose-derived stem cells. Acta Biomater 8: 2483–92.

Coventry, Mark B., Ralph K. Ghormley and James W. Kernohan. 1945. The intervertebral disc: Its microscopic anatomy and pathology: Part I. Anatomy, development, and physiology. JBJS 27: 105–12.

Dababneh, Amer B. and Ibrahim T. Ozbolat. 2014. Bioprinting technology: a current state-of-the-art review. Journal of Manufacturing Science and Engineering 136: 061016.

Daly, A. C., S. E. Critchley, E. M. Rencsok and D. J. Kelly. 2016. A comparison of different bioinks for 3D bioprinting of fibrocartilage and hyaline cartilage. Biofabrication 8: 045002.

Das, S., H. Nam and J. Jang. 2021. 3D bioprinting of stem cell-laden cardiac patch: A promising alternative for myocardial repair. APL Bioeng, 5: 031508.

Das, Sanskrita, Falguni Pati, Yeong-Jin Choi, Girdhari Rijal, Jin-Hyung Shim, Sung Won Kim et al. 2015. Bioprintable, cell-laden silk fibroin–gelatin hydrogel supporting multilineage differentiation of stem cells for fabrication of three-dimensional tissue constructs. Acta Biomaterialia 11: 233–46.

Derakhshanfar, Soroosh, Rene Mbeleck, Kaige Xu, Xingying Zhang, Wen Zhong and Malcolm Xing. 2018. 3D bioprinting for biomedical devices and tissue engineering: A review of recent trends and advances. Bioactive Materials 3: 144–56.

Ding, Zhaozhao, Weinan Cheng, Md Shipan Mia and Qiang Lu. 2021. Silk biomaterials for bone tissue engineering. Macromolecular Bioscience 21: 2100153.

Dou, H. and B. J. T. J. o. T. T. I. Zuo. 2015. Effect of sodium carbonate concentrations on the degumming and regeneration process of silk fibroin. The Journal of The Textile Institute 106: 311–19.

Duan, Xiaodong, Christopher McLaughlin, May Griffith and Heather Sheardown. 2007. Biofunctionalization of collagen for improved biological response: scaffolds for corneal tissue engineering. Biomaterials 28: 78–88.

Fallahiarezoudar, Ehsan, Mohaddeseh Ahmadipourroudposht, Noordin Mohd Yusof and Ani Idris. 2015. Influence of process factors on diameter of core (γ-Fe_2O_3)/shell (polyvinyl alcohol) structure magnetic nanofibers during co-axial electrospinning. International Journal of Polymeric Materials and Polymeric Biomaterials 64: 15–24.

Farasatkia, A. and M. Kharaziha. 2021. Robust and double-layer micro-patterned bioadhesive based on silk nanofibril/GelMA-alginate for stroma tissue engineering. Int J Biol Macromol 183: 1013–25.

Farasatkia, A., M. Kharaziha, F. Ashrafizadeh and S. Salehi. 2021. Transparent silk/gelatin methacrylate (GelMA) fibrillar film for corneal regeneration. Mater Sci Eng C Mater Biol Appl 120: 111744.

Farokhi, Mehdi, Fatemeh Mottaghitalab, Saeed Samani, Mohammad Ali Shokrgozar, Subhas C. Kundu, Rui L. Reis et al. 2018. Silk fibroin/hydroxyapatite composites for bone tissue engineering. Biotechnology Advances 36: 68–91.

Fedic, R., M. Zˇurovec and F. e. J. J. o. B. C. Sehnal. 2003. Correlation between fibroin amino acid sequence and physical silk properties. Journal of Biological Chemistry 278: 35255–64.

Frauchiger, Daniela A., Rahel D. May, Ezgi Bakirci, Adel Tekari, Samantha C. W. Chan, Michael Wöltje et al. 2018. Genipin-enhanced fibrin hydrogel and novel silk for intervertebral disc repair in a loaded bovine organ culture model. Journal of Functional Biomaterials 9: 40.

Gholipourmalekabadi, Mazaher, Ali Samadikuchaksaraei, Alexander M. Seifalian, Aleksandra M. Urbanska, Hossein Ghanbarian, John G. Hardy et al. 2018. Silk fibroin/amniotic membrane 3D bi-layered artificial skin. Biomedical Materials 13: 035003.

Greenhalgh, David G. 2003. Wound healing and diabetes mellitus. Clinics in Plastic Surgery 30: 37–45.

Gu, Kai, Yixuan Tong, Ruixin Mi, Siyan Leng, Hanwen Huang, Jingrong Yao et al. 2024. The production of soluble regenerated silk fibroin powder with high molecular weight and silk protein-based materials. Giant 19: 100313.

Guan, Ying, Feng Sun, Xiaojuan Zhang, Zhibin Peng, Bo Jiang, Min Liang et al. 2020. Silk fibroin hydrogel promote burn wound healing through regulating TLN1 expression and affecting cell adhesion and migration. Journal of Materials Science: Materials in Medicine 31: 1–11.

Guang, Shanyi, Yang An, Fuyou Ke, Dongmei Zhao, Yuhua Shen and Hongyao Xu. 2015. Chitosan/silk fibroin composite scaffolds for wound dressing. Journal of Applied Polymer Science 132.

Gunatillake, P. A., R. Adhikari and N. J. E. C. M. Gadegaard. 2003. Biodegradable synthetic polymers for tissue engineering. Eur Cell Mater 5: 1–16.

Guo, Peng, Pan Du, Peng Zhao, Xue Chen, Chenyang Liu, Yong Du et al. 2021. Regulating the mechanics of silk fibroin scaffolds promotes wound vascularization. Biochemical and Biophysical Research Communications 574: 78–84.

Hao, Zhanzhang, Dingpei Long, Yuli Zhang, Diane Umuhoza, Jing Dai, Zhen Xu et al. 2021. New insight into the mechanism of *in vivo* fibroin self-assembly and secretion in the silkworm, Bombyx mori. International Journal of Biological Macromolecules 169: 473–79.

Harkin, Damien G., Karina A. George, Peter W. Madden, Ivan R. Schwab, Dietmar W. Hutmacher and Traian V. Chirila. 2011. Silk fibroin in ocular tissue reconstruction. Biomaterials 32: 2445–58.

Harkin, D. G., Z. Barnard, Peter Gillies, S. L. Ainscough and A. J. G. Apel. 2004. Analysis of p63 and cytokeratin expression in a cultivated limbal autograft used in the treatment of limbal stem cell deficiency. British Journal of Ophthalmology 88: 1154–58.

Hayes, Anthony Joseph, Clare Elizabeth Hughes, James Robert Ralphs and Bruce Caterson. 2011. Chondroitin sulphate sulphation motif expression in the ontogeny of the intervertebral disc. Eur Cell Mater 21: 1–14.

Heydari Foroushani, Parisa, Erfan Rahmani, Iran Alemzadeh, Manouchehr Vossoughi, Mehrab Pourmadadi, Abbas Rahdar et al. 2022. Curcumin sustained release with a hybrid chitosan-silk fibroin nanofiber containing silver nanoparticles as a novel highly efficient antibacterial wound dressing. Nanomaterials 12: 3426.

Hodgkinson, Tom, Xue-Feng Yuan and Ardeshir Bayat. 2014. Electrospun silk fibroin fiber diameter influences *in vitro* dermal fibroblast behavior and promotes healing of *ex vivo* wound models. Journal of Tissue Engineering 5: 2041731414551661.

Hu, Jingen, Yang Lu, Ling Cai, Kwabena Gyabaah Owusu-Ansah, Gewen Xu, Feilong Han et al. 2017. Functional compressive mechanics and tissue biocompatibility of an injectable SF/PU hydrogel for nucleus pulposus replacement. Scientific Reports 7: 2347.

Huang, Te-Yang, Guo-Shou Wang, Chia-Chen Tseng and Wen-Ta Su. 2019. Epidermal cells differentiated from stem cells from human exfoliated deciduous teeth and seeded onto polyvinyl alcohol/silk fibroin nanofiber dressings accelerate wound repair. Materials Science and Engineering: C 104: 109986.

Huang, Ting, Chunquan Fan, Min Zhu, Yufang Zhu, Weizhong Zhang and Lei Li. 2019. 3D-printed scaffolds of biomineralized hydroxyapatite nanocomposite on silk fibroin for improving bone regeneration. Applied Surface Science 467: 345–53.

Huang, X. W., J. J. Wei, T. Liu, X. L. Zhang, S. M. Bai and H. H. Yang. 2017. Silk fibroin-assisted exfoliation and functionalization of transition metal dichalcogenide nanosheets for antibacterial wound dressings. Nanoscale 9: 17193–98.

Huiwen, Wu, Liang Shuai, Xie Jia, Deng Shihao, Wei Kun, Yang Runhuai et al. 2024. 3D-printed nanohydroxyapatite/methylacrylylated silk fibroin scaffold for repairing rat skull defects. Journal of Biological Engineering 18: 22.

Hwang, B., L. Korsnick, M. Shen, L. Jin, Y. Singh, M. Abdalla et al. 2024. FSTL-1 loaded 3D bioprinted vascular patch regenerates the ischemic heart tissue. iScience 27: 110770.

Ige, O. O., L. E. Umoru and S. Aribo. 2012. Natural products: a minefield of biomaterials. ISRN Mater Sci 2012: 1–20.

Jaipaew, J., P. Wangkulangkul, J. Meesane, P. Raungrut and P. Puttawibul. 2016. Mimicked cartilage scaffolds of silk fibroin/hyaluronic acid with stem cells for osteoarthritis surgery: Morphological, mechanical, and physical clues. Mater Sci Eng C Mater Biol Appl 64: 173–82.

Jiang, Xinquan, Jun Zhao, Shaoyi Wang, Xiaojuan Sun, Xiuli Zhang, Jake Chen et al. 2009. Mandibular repair in rats with premineralized silk scaffolds and BMP-2-modified bMSCs. Biomaterials 30: 4522–32.

Jin, Shue, Xiaoxue Fu, Weinan Zeng, Anjing Chen, Zhenyu Luo, Yubao Li et al. 2023. Chopped fibers and nano-hydroxyapatite enhanced silk fibroin porous hybrid scaffolds for bone augmentation. Journal of Materials Chemistry B 11: 1557–67.

Jo, You-Young and Ji-Hyeon Oh. 2018. New resorbable membrane materials for guided bone regeneration. Applied Sciences 8: 2157.

Ju, Hyung Woo, Ok Joo Lee, Jung Min Lee, Bo Mi Moon, Hyun Jung Park, Ye Ri Park et al. 2016. Wound healing effect of electrospun silk fibroin nanomatrix in burn-model. International Journal of Biological Macromolecules 85: 29–39.

KA, Williams. 1993. The Australian Corneal Graft Registry; 1990 to 1992 report. Aust NZ J Ophthalmol 21: 1–48.

Kadakia, Parin U., Emily A. Growney Kalaf, Andrew J. Dunn, Laurie P. Shornick and Scott A. Sell. 2018. Comparison of silk fibroin electrospun scaffolds with poloxamer and honey additives for burn wound applications. Journal of Bioactive and Compatible Polymers 33: 79–94.

Kaerkitcha, Navaporn, Surawut Chuangchote and Takashi Sagawa. 2016. Control of physical properties of carbon nanofibers obtained from coaxial electrospinning of PMMA and PAN with adjustable inner/outer nozzle-ends. Nanoscale Research Letters 11: 1–9.

Kato, B., G. Wisser, D. K. Agrawal, T. Wood and F. G. Thankam. 2021. 3D bioprinting of cardiac tissue: current challenges and perspectives. J Mater Sci Mater Med 32: 54.

Kim, B. S., J. S. Lee, G. Gao, and D. W. Cho. 2017. Direct 3D cell-printing of human skin with functional transwell system. Biofabrication 9: 025034.

Koh, Leng-Duei, Yuan Cheng, Choon-Peng Teng, Yin-Win Khin, Xian-Jun Loh, Si-Yin Tee et al. 2015. Structures, mechanical properties and applications of silk fibroin materials. Progress in Polymer Science 46: 86–110.

Krishna, Lekshmi, Sagar Nilawar, Murugeswari Ponnalagu, Murali Subramani, Chaitra Jayadev, Rohit Shetty et al. 2020. Fiber diameter differentially regulates function of retinal pigment and corneal epithelial cells on nanofibrous tissue scaffolds. ACS Applied Bio Materials 3: 823–37.

Lawrence, Brian D., Jeffrey K. Marchant, Mariya A. Pindrus, Fiorenzo G. Omenetto and David L. Kaplan. 2009. Silk film biomaterials for cornea tissue engineering. Biomaterials 30: 1299–308.

Lee, Jung Min, Md. Tipu Sultan, Soon Hee Kim, Vijay Kumar, Yeung Kyu Yeon, Ok Joo Lee et al. 2017. Artificial auricular cartilage using silk fibroin and polyvinyl alcohol hydrogel. International Journal of Molecular Sciences 18: 1707.

Lee, Min Chae, Dong-Kyu Kim, Ok Joo Lee, Jung-Ho Kim, Hyung Woo Ju, Jung Min Lee et al. 2016. Fabrication of silk fibroin film using centrifugal casting technique for corneal tissue engineering. Journal of Biomedical Materials Research Part B: Applied Biomaterials 104: 508–14.

Lee, Ok Joo, Hyung Woo Ju, Ji Heui Kim, Jung Min Lee, Chang Seok Ki, Jung-Ho Kim et al. 2014. Development of artificial dermis using 3D electrospun silk fibroin nanofiber matrix. Journal of Biomedical Nanotechnology 10: 1294–303.

Lee, V., G. Singh, J. P. Trasatti, C. Bjornsson, X. Xu, T. N. Tran et al. 2014. Design and fabrication of human skin by three-dimensional bioprinting. Tissue Eng Part C Methods 20: 473–84.

Leroy, A., P. Beaufils, B. Faivre, C. Steltzlen, P. Boisrenoult and N. Pujol. 2017. Actifit® polyurethane meniscal scaffold: MRI and functional outcomes after a minimum follow-up of 5 years. Orthopaedics & Traumatology: Surgery & Research 103: 609–14.

Li, Danrong, Genlong Jiao, Wu Zhang, Xiaofeng Chen, Ruijian Ning and Chang Du. 2016. Hybrid scaffolding strategy for dermal tissue reconstruction: a bioactive glass/chitosan/silk fibroin composite. RSC Advances 6: 19887–96.

Li, Guangfei and Shan Sun. 2022. Silk fibroin-based biomaterials for tissue engineering applications. Molecules 27: 2757.

Li, S., L. Li, C. Guo, H. Qin and X. Yu. 2017. A promising wound dressing material with excellent cytocompatibility and proangiogenesis action for wound healing: Strontium loaded Silk fibroin/Sodium alginate (SF/SA) blend films. Int J Biol Macromol 104: 969–78.

Li, Xiufang, Yan Liu, Jian Zhang, Renchuan You, Jing Qu and Mingzhong Li. 2017. Functionalized silk fibroin dressing with topical bioactive insulin release for accelerated chronic wound healing. Materials Science and Engineering: C 72: 394–404.

Li, Yuwan, Ziming Liu, Yaping Tang, Qinghong Fan, Wei Feng, Changqi Luo et al. 2020. Three-dimensional silk fibroin scaffolds enhance the bone formation and angiogenic differentiation of human amniotic mesenchymal stem cells: a biocompatibility analysis. Acta Biochimica et Biophysica Sinica 52: 590–602.

Liang, Renjie, Rui Li, Weidong Mo, Xianzhu Zhang, Jinchun Ye, Chang Xie et al. 2024. Engineering biomimetic silk fibroin hydrogel scaffolds with "organic-inorganic assembly" strategy for rapid bone regeneration. Bioactive Materials 40: 541–56.

Lin, M. J., M. C. Lu and H. Y. Chang. 2021. Sustained release of insulin-like growth factor-1 from Bombyx mori L. silk fibroin delivery for diabetic wound therapy. Int J Mol Sci 22.

Lin, Y. D., C. Y. Luo, Y. N. Hu, M. L. Yeh, Y. C. Hsueh, M. Y. Chang et al. 2012. Instructive nanofiber scaffolds with VEGF create a microenvironment for arteriogenesis and cardiac repair. Sci Transl Med 4: 146ra09.

Liu, Bin, Yu-wei Song, Li Jin, Zhi-jian Wang, De-yong Pu, Shao-qiang Lin et al. 2015. Silk structure and degradation. Colloids and Surfaces B: Biointerfaces 131: 122–28.

Liu, Haifeng, Hongbin Fan, Yue Wang, Siew Lok Toh and James C. H. Goh. 2008. The interaction between a combined knitted silk scaffold and microporous silk sponge with human mesenchymal stem cells for ligament tissue engineering. Biomaterials 29: 662–74.

Liu, J., L. Yan, W. Yang, Y. Lan, Q. Zhu, H. Xu et al. 2019a. Controlled-release neurotensin-loaded silk fibroin dressings improve wound healing in diabetic rat model. Bioact Mater 4: 151–59.

Liu, Jianghui, Liwei Yan, Wei Yang, Yong Lan, Qiyu Zhu, Hongjie Xu et al. 2019b. Controlled-release neurotensin-loaded silk fibroin dressings improve wound healing in diabetic rat model. Bioactive Materials 4: 151–59.

Liu, Jingbo, Brian D. Lawrence, Aihong Liu, Ivan R. Schwab, Lauro A. Oliveira and Mark I. Rosenblatt. 2012. Silk fibroin as a biomaterial substrate for corneal epithelial cell sheet generation. Investigative Ophthalmology & Visual Science 53: 4130–38.

Long, Yanlin, Xian Cheng, John A. Jansen, Sander G. C. Leeuwenburgh, Jing Mao, Fang Yang et al. 2021. The molecular conformation of silk fibroin regulates osteogenic cell behavior by modulating the stability of the adsorbed protein-material interface. Bone Research 9: 13.

Lujerdean, Cristian, Gabriela-Maria Baci, Alexandra-Antonia Cucu and Daniel Severus Dezmirean. 2022. The contribution of silk fibroin in biomedical engineering. Insects 13: 286.

Luo, Baiwen, Qiu Li Loh, Marcus Thien Chong Wong, Nguan Soon Tan and Cleo Choong. 2014. Bioactivated protein-based porous microcarriers for tissue engineering applications. Journal of Materials Chemistry B 2: 7795–803.

Luo, Dandan, Chenxue Yao, Rui Zhang, Ruibo Zhao, M. Zubair Iqbal, Asim Mushtaq et al. 2021. Silk fibroin/collagen blended membrane fabricated via a green papermaking method for potential guided bone regeneration application: *in vitro* and *in vivo* evaluation. ACS Biomaterials Science & Engineering 7: 5788–97.

Lyu, Y., Y. Liu, H. He and H. Wang. 2023. Application of Silk-Fibroin-Based Hydrogels in Tissue Engineering. Gels 9.

Ma, Peter X. and Ruiyun Zhang. 1999. Synthetic nano-scale fibrous extracellular matrix. Journal of Biomedical Materials Research: An Official Journal of The Society for Biomaterials, The Japanese Society for Biomaterials, and The Australian Society for Biomaterials 46: 60–72.

Madappura, Alakananda Parassini and Srinivas Madduri. 2023. A comprehensive review of silk-fibroin hydrogels for cell and drug delivery applications in tissue engineering and regenerative medicine. Computational and Structural Biotechnology Journal.

Majidi Ghatar, Jilla, Arian Ehterami, Simin Nazarnezhad, Maryam Sadat Hassani, Nariman Rezaei Kolarijani, Solmaz Mahami et al. 2023. A novel hydrogel containing 4-methylcatechol for skin regeneration: *In vitro* and *in vivo* study. Biomedical Engineering Letters 13: 429–39.

Malay, Ali D., Ryota Sato, Kenjiro Yazawa, Hiroe Watanabe, Nao Ifuku, Hiroyasu Masunaga et al. 2016. Relationships between physical properties and sequence in silkworm silks. Scientific Reports 6: 27573.

Malherbe, Ilana, Ron D. Sanderson and Eugene Smit. 2010. Reversibly thermochromic micro-fibres by coaxial electrospinning. Polymer 51: 5037–43.

Mazurek, Łukasz, Mateusz Szudzik, Mateusz Rybka and Marek Konop. 2022. Silk fibroin biomaterials and their beneficial role in skin wound healing. Biomolecules 12: 1852.

Mehrabani, M. G., R. Karimian, B. Mehramouz, M. Rahimi and H. S. Kafil. 2018. Preparation of biocompatible and biodegradable silk fibroin/chitin/silver nanoparticles 3D scaffolds as a bandage for antimicrobial wound dressing. Int J Biol Macromol 114: 961–71.

Meka, Sai Rama Krishna, Vipul Agarwal and Kaushik Chatterjee. 2019. *In situ* preparation of multicomponent polymer composite nanofibrous scaffolds with enhanced osteogenic and angiogenic activities. Materials Science and Engineering: C 94: 565–79.

Meka, Sai Rama Krishna, Shailendra Kumar Verma, Vipul Agarwal and Kaushik Chatterjee. 2018. *In situ* silication of polymer nanofibers to engineer multi-biofunctional composites. ChemistrySelect 3: 3762–73.

Melke, Johanna, Swati Midha, Sourabh Ghosh, Keita Ito and Sandra Hofmann. 2016a. Silk fibroin as biomaterial for bone tissue engineering. Acta Biomaterialia 31: 1–16.

Melke, J., S. Midha, S. Ghosh, K. Ito and S. Hofmann. 2016b. Silk fibroin as biomaterial for bone tissue engineering. Acta Biomaterialia 31: 1–16.

Menon, Gopinathan K. 2002. New insights into skin structure: scratching the surface. Advanced Drug Delivery Reviews 54: S3–S17.

Miguel, Sónia P., Déborah Simões, André F. Moreira, Rosa S. Sequeira and Ilídio J. Correia. 2019. Production and characterization of electrospun silk fibroin based asymmetric membranes for wound dressing applications. International Journal of Biological Macromolecules 121: 524–35.

Millán-Rivero, José E., Carlos M. Martínez, Paola A. Romecín, Salvador D. Aznar-Cervantes, Marina Carpes-Ruiz, José L. Cenis et al. 2019. Silk fibroin scaffolds seeded with Wharton's jelly mesenchymal stem cells enhance re-epithelialization and reduce formation of scar tissue after cutaneous wound healing. Stem Cell Research & Therapy 10: 1–14.

Miyasaka, Kenji C. 1991. The incidence of knee ligament injuries in the general population. Am J Knee Surg 4: 3–8.

Mu, Xuan, Vincent Fitzpatrick and David L. Kaplan. 2020. From silk spinning to 3D printing: Polymer manufacturing using directed hierarchical molecular assembly, Advanced Healthcare Materials 9: 1901552.

Mu, Xuan, Yu Wang, Chengchen Guo, Yamin Li, Shengjie Ling, Wenwen Huang et al. 2020. 3D printing of silk protein structures by aqueous solvent-directed molecular assembly. Macromolecular Bioscience 20: 1900191.

Mukherjee, Sunny, Anand Krishnan, R. K. Athira, Naresh Kasoju and Mahesh K. Sah. 2022. Silk fibroin and silk sericin in skin tissue engineering and wound healing: Retrospect and prospects. In Natural Polymers in Wound Healing and Repair (Elsevier).

Murakami, Kaoru, Hiroshi Aoki, Shingo Nakamura, Shin-ichiro Nakamura, Megumi Takikawa, Motoaki Hanzawa et al. 2010. Hydrogel blends of chitin/chitosan, fucoidan and alginate as healing-impaired wound dressings. Biomaterials 31: 83–90.

Negahdari, Nazanin, Sanaz Alizadeh, Jila Majidi, Mahdi Saeed, Tayyeb Ghadimi, Kobra Tahermanesh et al. 2024a. Heat-treated alginate-polycaprolactone core-shell nanofibers by emulsion electrospinning process for biomedical applications. International Journal of Biological Macromolecules: 133709.

Negahdari, Nazanin, Sanaz Alizadeh, Jila Majidi, Mahdi Saeed, Tayyeb Ghadimi, Kobra Tahermanesh et al. 2024b. Heat-treated alginate-polycaprolactone core-shell nanofibers by emulsion electrospinning process for biomedical applications. International Journal of Biological Macromolecules 275: 133709.

Neo, Puay Yong, Pujiang Shi, James Cho-Hong Goh and Siew Lok Toh. 2014. Characterization and mechanical performance study of silk/PVA cryogels: towards nucleus pulposus tissue engineering. Biomedical Materials 9: 065002.

Nguyen, T. P., Q. V. Nguyen, V. H. Nguyen, T. H. Le, V. Q. N. Huynh, D. N. Vo et al. 2019a. Silk fibroin-based biomaterials for biomedical applications: a review. Polymers (Basel) 11.

Nguyen, Thang Phan, Quang Vinh Nguyen, Van-Huy Nguyen, Thu-Ha Le, Vu Quynh Nga Huynh, Dai-Viet N. Vo et al. 2019b. Silk fibroin-based biomaterials for biomedical applications: a review. Polymers 11: 1933.

Paladini, Federica, Cinzia Di Franco, Angelica Panico, Gaetano Scamarcio, Alessandro Sannino and Mauro Pollini. 2016. *In vitro* assessment of the antibacterial potential of silver nano-coatings on cotton gauzes for prevention of wound infections. Materials 9: 411.

Paladini, Federica and Mauro Pollini. 2022. Novel approaches and biomaterials for bone tissue engineering: a focus on silk fibroin. Materials 15: 6952.

Park, Sang-Hyug, Eun Seok Gil, Biman B. Mandal, Hongsik Cho, Jonathan A. Kluge, Byoung-Hyun Min et al. 2012. Annulus fibrosus tissue engineering using lamellar silk scaffolds. Journal of Tissue Engineering and Regenerative Medicine 6: s24–s33.

Park, Ye Ri, Hyung Woo Ju, Jung Min Lee, Dong-Kyu Kim, Ok Joo Lee, Bo Mi Moon et al. 2016. Three-dimensional electrospun silk-fibroin nanofiber for skin tissue engineering. International Journal of Biological Macromolecules 93: 1567–74.

Park, Ye Ri, Md Tipu Sultan, Hyun Jung Park, Jung Min Lee, Hyung Woo Ju, Ok Joo Lee et al. 2018. NF-κB signaling is key in the wound healing processes of silk fibroin. Acta Biomaterialia 67: 183–95.

Partheniadis, Ioannis, Ioannis Nikolakakis, Ivo Laidmäe and Jyrki Heinämäki. 2020. A mini-review: Needleless electrospinning of nanofibers for pharmaceutical and biomedical applications. Processes 8: 673.

Patil, Priyanka P., Michaela R. Reagan and Raghvendra A. Bohara. 2020. Silk fibroin and silk-based biomaterial derivatives for ideal wound dressings. International Journal of Biological Macromolecules 164: 4613–27.

Patil, Smita and Neetu Singh. 2019. Antibacterial silk fibroin scaffolds with green synthesized silver nanoparticles for osteoblast proliferation and human mesenchymal stem cell differentiation. Colloids and Surfaces B: Biointerfaces 176: 150–55.

Pellegrini, Graziella, Elena Dellambra, Osvaldo Golisano, Enrica Martinelli, Ivana Fantozzi, Sergio Bondanza et al. 2001. p63 identifies keratinocyte stem cells. Proceedings of the National Academy of Sciences 98: 3156–61.

Qi, Yu, Hui Wang, Kai Wei, Ya Yang, Ru-Yue Zheng, Ick Soo Kim et al. 2017. A review of structure construction of silk fibroin biomaterials from single structures to multi-level structures. International Journal of Molecular Sciences 18: 237.

Rahmati, Maryam, David K. Mills, Aleksandra M. Urbanska, Mohammad Reza Saeb, Jayarama Reddy Venugopal, Seeram Ramakrishna et al. 2021. Electrospinning for tissue engineering applications. Progress in Materials Science 117: 100721.

Rama, Paolo, Stefano Bonini, Alessandro Lambiase, Osvaldo Golisano, Patrizia Paterna, Michele De Luca et al. 2001. Autologous fibrin-cultured limbal stem cells permanently restore the corneal surface of patients with total limbal stem cell deficiency1. Transplantation 72: 1478–85.

Ran, Jisheng, Yejun Hu, Huihui Le, Yangwu Chen, Zefeng Zheng, Xiao Chen et al. 2017. Ectopic tissue engineered ligament with silk collagen scaffold for ACL regeneration: A preliminary study. Acta Biomaterialia 53: 307–17.

Rastogi, S. and B. J. E. S. Kandasubramanian. 2020. Progressive trends in heavy metal ions and dyes adsorption using silk fibroin composites. Environmental Science, and Pollution Research 27: 210–37.

Richardson, S. M., R. Knowles, D. Marples, J. A. Hoyland and A. Mobasheri. 2008. Aquaporin expression in the human intervertebral disc. Journal of Molecular Histology 39: 303–09.

Rising, Anna, Helena Nimmervoll, Stefan Grip, Armando Fernandez-Arias, Erica Storckenfeldt, David P. Knight et al. 2005. Spider silk proteins–mechanical property and gene sequence. Zoological Science 22: 273–81.

Rodríguez-Tobías, Heriberto, Graciela Morales and Daniel Grande. 2019. Comprehensive review on electrospinning techniques as versatile approaches toward antimicrobial biopolymeric composite fibers. Materials Science and Engineering: C 101: 306–22.

Roy, Trina, Priti Prasanna Maity, Arun Prabhu Rameshbabu, Bodhisatwa Das, Athira John, Abir Dutta et al. 2018. Core-shell nanofibrous scaffold based on polycaprolactone-silk fibroin emulsion electrospinning for tissue engineering applications. Bioengineering 5: 68.

Sahoo, Sambit, Siew Lok Toh and James C. H. Goh. 2010. A bFGF-releasing silk/PLGA-based biohybrid scaffold for ligament/tendon tissue engineering using mesenchymal progenitor cells. Biomaterials 31: 2990–98.

Sahu, Neety, Shilpa Pal, Sunaina Sapru, Joydip Kundu, Sarmistha Talukdar, N. Ibotambi Singh et al. 2016. Non-mulberry and mulberry silk protein sericins as potential media supplement for animal cell culture. BioMed Research International 2016.

Sanyal, Arka and Sourabh Ghosh. 2024. 3D bioprinting strategies for recapitulation of hepatic structure and function in bioengineered liver: A state-of-the-art review. Current Opinion in Biomedical Engineering 30: 100526.

Schermer, Alexander, Sharon Galvin and Tung-Tien Sun. 1986. Differentiation-related expression of a major 64K corneal keratin in vivo and in culture suggests limbal location of corneal epithelial stem cells. The Journal of Cell Biology 103: 49–62.

Schneider, Karl Heinrich, Benjamin J. Goldberg, Onur Hasturk, Xuan Mu, Marvin Dötzlhofer, Gabriela Eder et al. 2023. Silk fibroin, gelatin, and human placenta extracellular matrix-based composite hydrogels for 3D bioprinting and soft tissue engineering. Biomaterials Research 27: 117.

Schneiderman, Deborah K. and Marc A. Hillmyer. 2016. Aliphatic polyester block polymer design. Macromolecules 49: 2419–28.

Schwab, Ivan R. 1999. Cultured corneal epithelia for ocular surface disease. Transactions of the American Ophthalmological Society 97: 891.

See, Eugene Yong-Shun, Siew Lok Toh and James Cho Hong Goh. 2012. Simulated intervertebral disc-like assembly using bone marrow-derived mesenchymal stem cell sheets and silk scaffolds for annulus fibrosus regeneration. Journal of Tissue Engineering and Regenerative Medicine 6: 528–35.

Sell, Scott A., Michael J. McClure, Koyal Garg, Patricia S. Wolfe and Gary L. Bowlin. 2009. Electrospinning of collagen/biopolymers for regenerative medicine and cardiovascular tissue engineering. Advanced Drug Delivery Reviews 61: 1007–19.

Shen, Weiliang, Xiao Chen, Jialin Chen, Zi Yin, Boon Chin Heng, Weishan Chen and Hong-Wei Ouyang. 2010. The effect of incorporation of exogenous stromal cell-derived factor-1 alpha within a knitted silk-collagen sponge scaffold on tendon regeneration. Biomaterials 31: 7239–49.

Song, Y., H. Wang, F. Yue, Q. Lv, B. Cai, N. Dong et al. 2020. Silk-based biomaterials for cardiac tissue engineering. Adv Healthc Mater 9: e2000735.

Sorg, Heiko, Daniel J. Tilkorn, Stephan Hager, Jörg Hauser and Ursula Mirastschijski. 2017. Skin wound healing: an update on the current knowledge and concepts. European Surgical Research 58: 81–94.

Stepp, Mary Ann and James D. Zieske. 2005. The corneal epithelial stem cell niche. The Ocular Surface 3: 15–26.

Strodtbeck, Frances. 2001. Physiology of wound healing. Newborn and Infant Nursing Reviews 1: 43–52.

Subbiah, Ramesh and Robert E. Guldberg. 2019. Materials science and design principles of growth factor delivery systems in tissue engineering and regenerative medicine. Advanced Healthcare Materials 8: 1801000.

Sun, Fengchao, Dongdong Xiao, Hui Su, Zhiliang Chen, Bijia Wang, Xueling Feng et al. 2023. Highly stretchable porous regenerated silk fibroin film for enhanced wound healing. Journal of Materials Chemistry B 11: 1486–94.

Sun, Wei, Tania Incitti, Claudio Migliaresi, Alessandro Quattrone, Simona Casarosa and Antonella Motta. 2017. Viability and neuronal differentiation of neural stem cells encapsulated in silk fibroin hydrogel functionalized with an IKVAV peptide. Journal of Tissue Engineering and Regenerative Medicine 11: 1532–41.

Sun, Weizhen, David Alexander Gregory, Mhd Anas Tomeh and Xiubo Zhao. 2021a. Silk fibroin as a functional biomaterial for tissue engineering. International Journal of Molecular Sciences 22: 1499.

Sun, W., D. A. Gregory, M. A. Tomeh and X. Zhao. 2021b. Silk fibroin as a functional biomaterial for tissue engineering. International Journal of Molecular Sciences 22: 1499.

Suntivich, Rattanon, Irina Drachuk, Rossella Calabrese, David L. Kaplan and Vladimir V. Tsukruk. 2014. Inkjet printing of silk nest arrays for cell hosting. Biomacromolecules 15: 1428–35.

Tran, H. Q., S. M. S. Shahriar, Z. Yan and J. Xie. 2023. Recent advances in functional wound dressings. Adv Wound Care (New Rochelle) 12: 399–427.

Tsai, Ray Jui-Fang, Lien-Min Li and Jan-Kan Chen. 2000. Reconstruction of damaged corneas by transplantation of autologous limbal epithelial cells. New England Journal of Medicine 343: 86–93.

Varshney, Neelima, Ajay Kumar Sahi, Suruchi Poddar and Sanjeev Kumar Mahto. 2020. Soy protein isolate supplemented silk fibroin nanofibers for skin tissue regeneration: Fabrication and characterization. International Journal of Biological Macromolecules 160: 112–27.

Verdonk, Peter C. M., Marie E. E. Van Laer and René Verdonk. 2008. Meniscus replacement: from allograft to tissue engineering. Sport-Orthopädie-Sport-Traumatologie-Sports Orthopaedics and Traumatology 24: 78–82.

Vert, M., S. Li, G. Spenlehauer and P. J. J. o. m. s. M. i. m. Guérin. 1992. Bioresorbability and biocompatibility of aliphatic polyesters. Journal of Materials Science: Materials in Medicine 3: 432–46.

Vidya, M. and Senthilkumar Rajagopal. 2021. Silk fibroin: a promising tool for wound healing and skin regeneration. International Journal of Polymer Science 2021: 9069924.

Wang, Mengmeng, Ying Wang, Peng Pan, Xueping Liu, Wenjing Zhang, Cheng Hu et al. 2023. A high molecular weight silk fibroin scaffold that resists degradation and promotes cell proliferation. Biopolymers 114: e23554.

Wang, Qi, Yutian Feng, Anqi Wang, Yuelei Hu, Yannan Cao, Jingjing Zheng et al. 2024. Innovations in 3D bioprinting and biomaterials for liver tissue engineering: Paving the way for tissue-engineered liver. iLIVER 3: 100080.

Wang, Richard N., Jordan Green, Zhongliang Wang, Youlin Deng, Min Qiao, Michael Peabody et al. 2014. Bone Morphogenetic Protein (BMP) signaling in development and human diseases. Genes & Diseases 1: 87–105.

Wang, Yongzhong, Darya D. Rudym, Ashley Walsh, Lauren Abrahamsen, Hyeon-Joo Kim, Hyun S. Kim et al. 2008. *In vivo* degradation of three-dimensional silk fibroin scaffolds. Biomaterials 29: 3415–28.

Wang, Yu-Chi, Meng-Yi Bai, Wan-Yuan Hsu and Mu-Hsien Yu. 2018. Evaluation of a series of silk fibroin protein-based nonwoven mats for use as an anti-adhesion patch for wound management in robotic surgery. Journal of Biomedical Materials Research Part A 106: 221–30.

Wani, Shahid Ud Din, Surya Prakash Gautam, Zulfkar Latief Qadrie and H. V. Gangadharappa. 2020. Silk fibroin as a natural polymeric based bio-material for tissue engineering and drug delivery systems—A review. International Journal of Biological Macromolecules 163: 2145–61.

Wen, Dan-Liang, De-Heng Sun, Peng Huang, Wen Huang, Meng Su, Ya Wang et al. 2021. Recent progress in silk fibroin-based flexible electronics. Microsystems & Nanoengineering 7: 35.

Withanage, Sinith, Artemii Savin, Valeria Nikolaeva, Aleksandra Kiseleva, Marina Dukhinova, Pavel Krivoshapkin et al. 2021. Native spider silk-based antimicrobial hydrogels for biomedical applications. Polymers 13: 1796.

Wu, Cenhao, Genlin Wang, Qi Yan, Jeffrey C. Wang, Huilin Yang and Jun Zou. 2019. Silk fibroin regulates osteoconduction of hydroxyapatite in rat spine fusion model. Journal of Hard Tissue Biology 28: 341–48.

Wu, Hao, Kaili Lin, Cancan Zhao and Xudong Wang. 2022. Silk fibroin scaffolds: A promising candidate for bone regeneration. Frontiers in Bioengineering and Biotechnology 10.

Wu, Ziquan, Zhulong Meng, Qianjin Wu, Delu Zeng, Zhengdong Guo, Jiangling Yao et al. 2020. Biomimetic and osteogenic 3D silk fibroin composite scaffolds with nano MgO and mineralized hydroxyapatite for bone regeneration. Journal of Tissue Engineering 11: 2041731420967791.

Xiao, Wenqian, Xiaohang Qu, Jiale Li, Lin Chen, Yunfei Tan, Kejiang Li et al. 2019. Synthesis and characterization of cell-laden double-network hydrogels based on silk fibroin and methacrylated hyaluronic acid. European Polymer Journal 118: 382–92.

Xu, G., L. Gong, Z. Yang and X. J. S. M. Liu. 2014. What makes spider silk fibers so strong? From molecular-crystallite network to hierarchical network structures. Soft Matter 10: 2116–23.

Xu, Xinyuan, Xingyu Chen and Jianshu Li. 2020. Natural protein bioinspired materials for regeneration of hard tissues. Journal of Materials Chemistry B 8: 2199–215.

Yan, Shuqin, Qiusheng Wang, Zeeshan Tariq, Renchuan You, Xiufang Li, Mingzhong Li et al. 2018. Facile preparation of bioactive silk fibroin/hyaluronic acid hydrogels. International Journal of Biological Macromolecules 118: 775–82.

Yang, Peilang, Di Wang, Yan Shi, Mingzhong Li, Min Gao, Tianyi Yu et al. 2020. Insulin-containing wound dressing promotes diabetic wound healing through stabilizing HIF-1α. Frontiers in Bioengineering and Biotechnology 8: 592833.

Yao, Q., Q. H. Lan, X. Jiang, C. C. Du, Y. Y. Zhai, X. Shen et al. 2020. Bioinspired biliverdin/silk fibroin hydrogel for antiglioma photothermal therapy and wound healing. Theranostics 10: 11719–36.

You, Zhi-Rong, Ming-Hsien Hu, Ho-Yi Tuan-Mu and Jin-Jia Hu. 2016. Fabrication of poly (glycerol sebacate) fibrous membranes by coaxial electrospinning: Influence of shell and core solutions. Journal of the Mechanical Behavior of Biomedical Materials 63: 220–31.

Yu, D.G., C. Branford-White, K. White, N. P. Chatterton, L. M. Zhu, L. Y. Huang et al. 2011. A modified coaxial electrospinning for preparing fibers from a high concentration polymer solution. Express Polymer Letters 5.

Yu, Miao, Rui-Hua Dong, Xu Yan, Gui-Feng Yu, Ming-Hao You, Xin Ning et al. 2017. Recent advances in needleless electrospinning of ultrathin fibers: from academia to industrial production. Macromolecular Materials and Engineering 302: 1700002.

Yu, PiJun, JiAn Guo, JunJie Li, Xiao Shi, LuPing Wang, Wei Chen et al. 2017. Repair of skin defects with electrospun collagen/chitosan and fibroin/chitosan compound nanofiber scaffolds compared with gauze dressing. Journal of Biomaterials and Tissue Engineering 7: 386–92.

Zakaria, Nadia, Carina Koppen, Viggo Van Tendeloo, Zwi Berneman, Andrew Hopkinson and Marie-José Tassignon. 2010. Standardized limbal epithelial stem cell graft generation and transplantation. Tissue Engineering Part C: Methods 16: 921–27.

Zhang, Cen, Fengqin Feng and Hui Zhang. 2018. Emulsion electrospinning: Fundamentals, food applications and prospects. Trends in Food Science & Technology 80: 175–86.

Zhang, Huijie, Renchuan You, Kun Yan, Zhentan Lu, Qunmei Fan, Xiufang Li et al. 2020. Silk as templates for hydroxyapatite biomineralization: A comparative study of Bombyx mori and Antheraea pernyi silkworm silks. International Journal of Biological Macromolecules 164: 2842–50.

Zhang, Jiamei, Lingshuang Wang, Cheng Xu, Yingui Cao, Shengsheng Liu, Rui L. Reis et al. 2024. Transparent silk fibroin film-facilitated infected-wound healing through antibacterial, improved fibroblast adhesion and immune modulation. Journal of Materials Chemistry B 12: 475–88.

Zhang, W., L. Chen, J. Chen, L. Wang, X. Gui, J. Ran et al. 2017. Silk fibroin biomaterial shows safe and effective wound healing in animal models and a randomized controlled clinical trial. Adv Healthc Mater 6.

Zhang, Y., H. Yang, H. Shao and X. J. B. R. I. Hu. 2010. Antheraea pernyi silk fiber: a potential resource for artificially biospinning spider dragline silk. BioMed Research International 2010.

Zhao, Jun, Zhiyuan Zhang, Shaoyi Wang, Xiaojuan Sun, Xiuli Zhang, Jake Chen et al. 2009. Apatite-coated silk fibroin scaffolds to healing mandibular border defects in canines. Bone 45: 517–27.

Zheng, H. and B. Zuo. 2021. Functional silk fibroin hydrogels: preparation, properties and applications. J Mater Chem B 9: 1238–58.

Zhou, C. Z., F. Confalonieri, M. Jacquet, R. Perasso and Z. G. Li, J. J. P. S. Janin. 2001. Silk fibroin: structural implications of a remarkable amino acid sequence. Proteins: Structure, Function, and Bioinformatics 44: 119–22.

Zhou, J., C. Cao, X. Ma, L. Hu, L. Chen and C. J. P. D. Wang. 2010. *In vitro* and *in vivo* degradation behavior of aqueous-derived electrospun silk fibroin scaffolds. Polymer Degradation, and Stability 95: 1679–85.

Zhu, HuiMin, Wei Liu, Qi Yin, SiWei Li, ChangYong Wang, Lun Song et al. 2020. Silk fibroin sponge combined with cell-derived ECM for tissue-engineered 3D functional neural tissues. Science China Technological Sciences 63: 2113–22.

Zuluaga-Vélez, Augusto, Adrián Quintero-Martinez, Lina M. Orozco and Juan C. Sepúlveda-Arias. 2021. Silk fibroin nanocomposites as tissue engineering scaffolds—A systematic review. Biomedicine & Pharmacotherapy 141: 111924.

CHAPTER 12

Dextran Composites as Tissue Engineering Scaffolds

Noushin Ezati,[1,*] *Ebad Asadi*[2] and *Mehrnoush Ezati*[3]

1. Introduction

Polysaccharides are known as outstanding biomaterials owing to some considerable advantages such as suitable immunoactivity, promising bioactivities, natural abundance, and flexibility for chemical modification for tissue engineering purposes. Moreover, owing to significant similarities of biochemical characteristics of the extracellular matrix of the live body to the polysaccharides, the non-toxic nature of degradation remains of polysaccharide-based scaffolds and the notable mechanical and chemical characteristics have expanded the applications of these materials in tissue engineering such as bone/cartilage regeneration, cardiac regeneration, neural regeneration, and skin regeneration (Ahmed Omar, Amédée et al., 2022).

The discovery of dextran can be dated back to 1874. Dextran is a α-glucan composed of linear α-1-6, linked D-glucopyranose residuals. It's component has a non-toxic nature and its promising biocompatibility has extended its application in the biomaterial field. The remarkable properties of dextran derivatives in the vascular thrombosis decrement, anti-inflammation effects, and vascularization have made it a potential candidate for tissue regeneration (Jalaja, Kumar et al., 2014).

Dextrans (DEX), produced industrially from sucrose, are synthesized by microorganisms that secrete dextran sucrases. These enzymes can create various types of linkages, resulting in dextrans with varying degrees of branching. The most widely used bacterial strains are Leuconostoc mesenteroides NRRL B512F and

[1] Department of Chemistry, Science and Research Branch, Islamic Azad University, Tehran, Iran.
[2] Department of Chemistry, Amirkabir University of Technology, Tehran, Iran.
[3] Department of Microbiology, Faculty of Biological Sciences, Alzahra University, Tehran, Iran.
* Corresponding author: s416em@gmail.com

DOI: 10.1201/9781003518792-12

Leuconostoc mesenteroides NRRL B1299. Strain B512F yields dextran primarily composed of α-(1-6) linkages (95%), with a smaller proportion of α-(1-3) linkages (5%). In contrast, strain B1299 produces dextran with a more complex branching pattern, including 63% α-(1-6), 27% α-(1-2), and 8% α-(1-3) linkages. The specific microorganisms employed significantly impact the physical and chemical properties of the resulting dextran.

The widespread use of dextrans in diverse applications makes their production a continuous industrial challenge. Ongoing research focuses on developing commercially viable production processes that allow for control over the structure of these branched molecules. Recent work by Sarwat et al. (Sun and Mao, 2012) demonstrated that Leuconostoc mesenteroides CMG713 can produce high molecular weight dextran without any branching, highlighting the potential for further optimization in dextran production. Dextrans are gaining increasing attention in the biomedical field due to their desirable properties. They exhibit good biocompatibility, are non-toxic, and possess hydrophilicity, meaning, they readily interact with water. Additionally, dextrans are naturally degraded by dextranases, enzymes present in organs like intestine, kidneys, lungs, and spleen. These properties have led to a long history of dextran use in medical and pharmaceutical applications. In the medical field, dextrans have proven valuable as blood plasma extenders, blood flow improvers, components of ophthalmic solutions, and agents for preserving human tissues during transplantation. Sulphated dextran serves as an alternative to heparin in anticoagulant therapy. Furthermore, dextrans are used to coat materials such as quantum dots and iron or gadolinium nanospheres, enhancing their application in medical imaging (Wang, Dijkstra et al., 2016).

2. Dextran Chemical Structure

Dextran (DEX) (Figure 1), a naturally occurring polysaccharide produced by bacteria, is composed of α-(1→6) linked D-glucopyranose residues with a small percentage of side chains linked at positions α-(1→2), α-(1→3), or α-(1→4). This linear, uncharged polymer is available in a range of molecular weights and undergoes enzymatic degradation in the spleen, liver, and colon. Based on the preparation method, branches comprising α-(1→2), α-(1→3) and α-(1→4) linkages may be present and molecular weights may change from 5 to as high as 500 kDa.

Crosslinked DEX hydrogel beads have found extensive use as low protein-binding matrices in column chromatography and as microcarriers in cell culture technology. DEX's resistance to protein adsorption and cell adhesion has made it a valuable material in soft tissue engineering (Shibusawa, 1999).

Porous DEX hydrogels can be produced through crosslinking reactions targeting the hydroxyl groups present in the α-1,6-linked D-glucose residues. Each repeat unit of the polymer contains three hydroxyl groups, with reactivity decreasing in the order of C2 > C4 > C3. Extensive research has explored various chemical modifications, leading to DEX derivatives with customized physicochemical and functional properties (Cunningham and Hodgson, 1992).

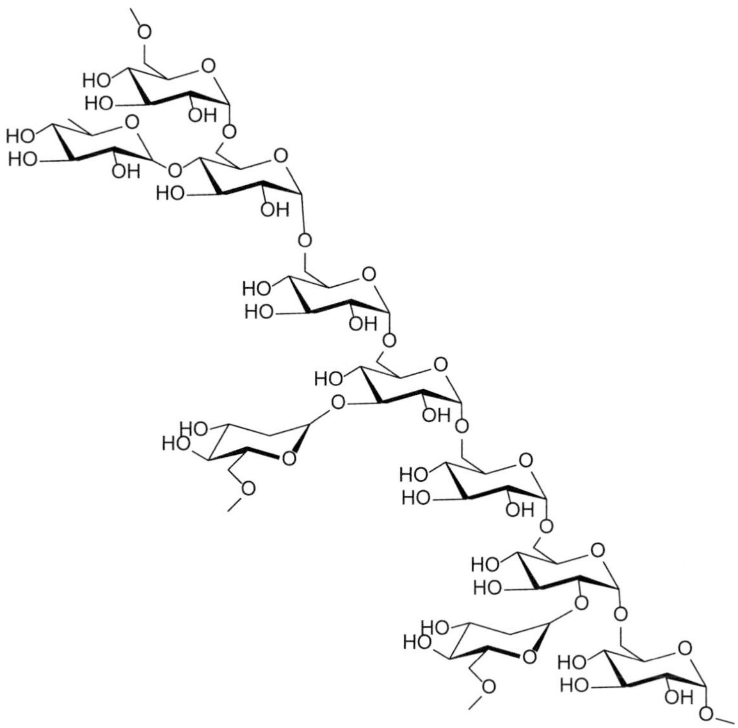

Figure 1. Dextran structure.

Chemical Modification of Dextran

Dextran's high density of hydroxyl groups allows for a wide range of chemical modifications. This discussion focuses on the chemical modification of dextran through oxidation which extends the range of the dextran usage as biomaterial. Periodate oxidation of dextran is a catalysis-free, aqueous reaction that yields a purified product through simple dialysis and freeze-drying. Originally described by Malaprade in 1928, this method was initially used to characterize and elucidate polysaccharide structure by analyzing degradation products. Dextran's glucose residues contain vicinal diols, presenting two oxidizable bonds within the same residue. Oxidation of either bond generates an aldehyde at C3, which is also susceptible to periodate oxidation due to the presence of hydroxyl groups at C2 or C4, leading to double oxidation of the same residue. This process can be monitored by a drop in pH due to the release of formic acid at C3 (Jeanes and Wilham, 1950; Maia, Ferreira et al., 2005). Periodate oxidation of dextran is a harsh reaction that can cause chain degradation and alter dextran's physicochemical properties. The oxidized dextran becomes less soluble in water and more viscous than regular dextran. This increased viscosity can impair injectability and miscibility in certain formulations. The degree of oxidation (DO) also affects the glass transition temperature as determined by dynamic mechanical thermal analysis. Controlling the DO in dextran is crucial for its final properties and applications. The DO significantly impacts the final characteristics of this macromonomer, making its accurate determination essential for effectively designing hydrogels, nanoparticles, surface coatings, and drug immobilization systems (Portalska, Teixeira et al., 2014).

In a surprising discovery during investigations on the enzymatic modification of polysaccharides in organic solvents, it was observed that certain enzymes can perform a double transesterification reaction, resulting in hydrogel formation. From a set of eleven enzymes, protease from *Bacillus subtilis* (Proleather FG-F) and lipases

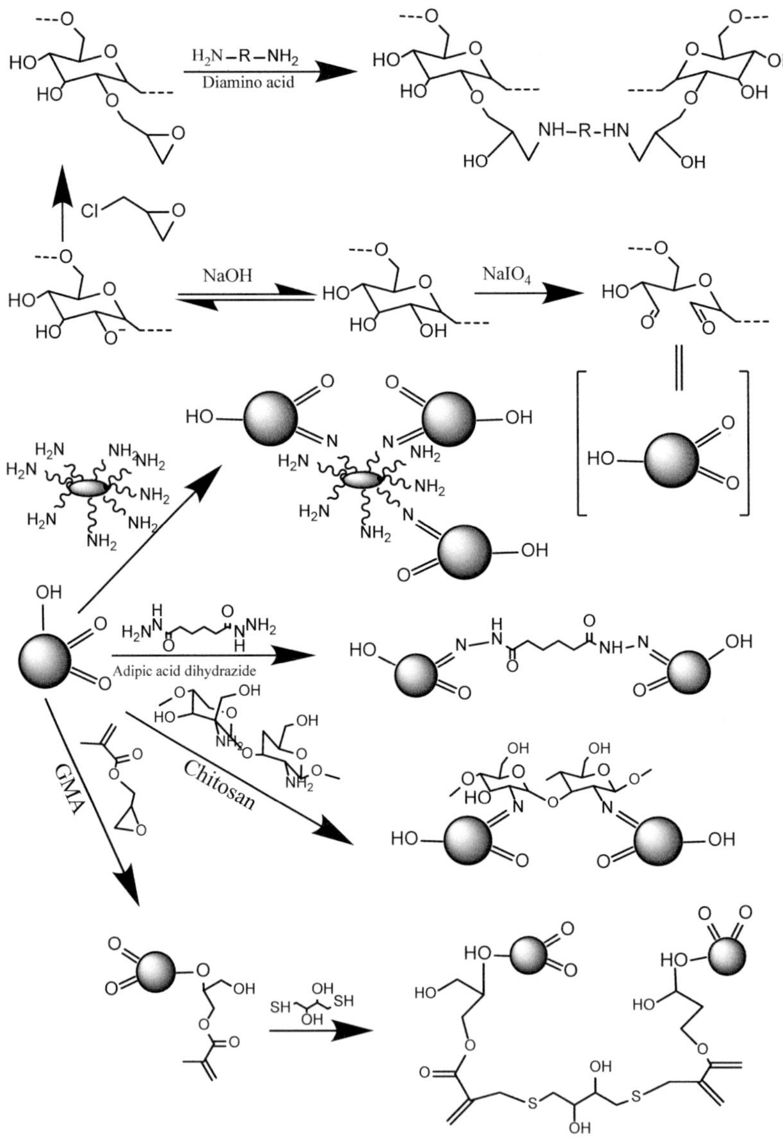

Figure 2. Scheme showing the crosslinking approaches for oxidized DEX.

from Candida rugosa and Pseudomonas cepacia (Lipase AY and PS) were found to catalyze the transesterification of dextran (Mw 70,000 Da) with divinyl adipate (DVA), producing a gel product. After 48 hours, the conversions obtained using Proleather FG-F, Lipases AY and PS were 71.4%, 62.6%, and 58.4%, respectively. NMR experiments confirmed that DVA molecules attached to the dextran backbone at positions 2, 3, and 4. However, the overlapping of dextran bands prevented precise determination of the ratio of positional isomers. This finding contrasts with previous results obtained with enzymatic transesterification of dextran with VA, where hydroxyls at positions 2 and 3 were the only attachment points. This highlights that the nature of the acylating agent plays a crucial role in enzyme regioselectivity (Maia, Evangelista et al., 2014). The crosslinking approaches for oxidized DEX is depicted in Figure 2.

3. Advantages of Dextran Comparing to Other Polysaccharides

Unlike polysaccharides like chitosan, alginate, and hyaluronic acid, which possess diverse functional groups (e.g., amine and amide) that support cell attachment, dextran solely comprises hydroxyl groups, limiting its inherent cell-binding capabilities. While dextran lacks inherent functionality, its structure allows for the precise control and introduction of desired functionalities through chemical modification. This contrasts with biopolymers like chitosan, which possess amine groups that can be chemically modified to alter their functional characteristics. Dextran's unique structure, with three hydroxyl groups per repeat unit, allows for the creation of novel dextran derivatives with diverse functionalities without compromising its primary properties. The reactivity of these hydroxyl groups varies, with C2 being the most reactive, followed by C4 and then C3. The degree of substitution, which refers to the number of replaced hydroxyl groups per unit, significantly influences the properties of dextran derivatives. This value can be determined from the 1H NMR spectrum. Through precise chemical modification, dextran can be tailored for various applications, offering a versatile platform for the development of novel biomaterials (Sun and Mao, 2012).

4. Physicochemical Properties of Dextran and its Derivatives

Dextran stands out as a biopolymer with several advantageous properties. It exhibits excellent solubility in various solvents like water, DMSO, ethylene glycol, and glycerol, attributed to its α-1, 6 glycosidic bonds, which enhance chain mobility. This solubility contributes to its biocompatibility, making it a suitable material for biomedical applications.

Dextran is a biocompatible material, meaning it is non-toxic and well-tolerated by the body. It is metabolized by digestive enzymes, unlike synthetic polymers which can accumulate as toxic degradation products. Subcutaneous implantation studies and tissue response evaluations have confirmed the *in vivo* biocompatibility of dextran-based delivery systems, indicating a lack of toxic or immune response. Dextran is unique in its resistance to breakdown by common amylases like salivary amylase and malt amylase. It is specifically depolymerized by dextranase, an enzyme presents in the lumen of the large intestine, liver, spleen, and kidney. This selective degradation mechanism contributes to its safety and efficacy as a biomaterial (Hu, Lu et al., 2021).

4.1 Esters of Dextran

While synthetic polymers can be readily manipulated into various shapes, native polysaccharides lack this inherent versatility. To tailor their physicochemical properties for applications like nanocarrier construction, chemical modification of polysaccharides has become a widely employed strategy. Dextran is an ideal starting material for designing such structures. Its narrow molecular weight distribution and abundance of active hydroxyl groups make it highly amenable to chemical modification. Various glycoconjugates have been synthesized through reactions including esterification, etherification, and oxidation. Esterification, often considered cleaner and safer than other methods, involves reacting the hydroxyl groups of dextran with carboxyl groups from carboxylic acids or amino acids. This reaction can be achieved through direct ester formation or the carbonate ester method (Stanciu, Belei et al., 2020).

Li and colleagues (2017) synthesized a dextran ester by incorporating lysine, an alkaline amino acid, onto the dextran backbone (Figure 3). This esterification significantly impacted the hydrogen bonding within the dextran structure, leading to reduced crystallization and a change in morphology observed through field emission scanning electron microscopy (FESEM) (Hornig and Heinze, 2007). The dextran ester exhibited a looser, porous, and rough surface compared to the denser, smoother structure of native dextran. This altered structure resulted in enhanced melting behavior due to weakened hydrogen bonds. Furthermore, the amino groups

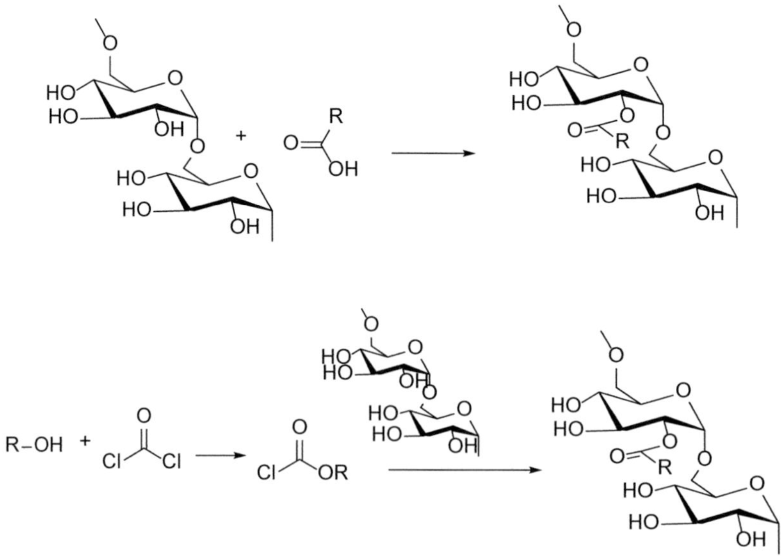

Figure 3. Synthesis of dextran ester.

within the long polymer chain contributed to strong flocculation performance in acidic conditions by interacting with negatively charged substances (Hu, Lu et al., 2021). In another study, Liebert et al. (Liebert, Wotschadlo et al., 2011) synthesized a long-chain fatty ester of dextran using in situ activation of the carboxylic acid with iminium chloride. The resulting dextran ester, with its favorable melting behavior and biocompatibility, was able to form stable films on materials like glass and nitride titanium implants, effectively preventing chronic inflammatory stimuli. This dextran derivative holds significant promise for diverse applications in the biomedical field.

4.2 Ethers of Dextran

Dextran's versatility extends to its modification through etherification, a process that grafts ether-type groups onto its backbone. This alteration significantly influences dextran's physical and chemical properties, including its solubility, ionic characteristics, hydrophilicity-lipophilicity balance, and resistance to enzymatic breakdown (Pahimanolis, Vesterinen et al., 2010). For example, amphiphilic ethers, which spontaneously form micelles in water, are valuable for encapsulating hydrophobic bioactive compounds or drugs. This is achieved by introducing lipophilic groups, like phenoxy groups or long alkyl chains, onto the dextran structure, fine-tuning its hydrophilic-lipophilic balance. Notably, poly(ethylene-glycol)-alkyl dextran ethers have emerged as promising nanocarriers for oral drug delivery (Figure 4). They protect nutrients and drugs from degradation in the gastrointestinal tract, extend their residence time, and enhance absorption efficiency and bioavailability (Fuentes, Segura et al., 2004). Furthermore, the excellent emulsifying properties of amphiphilic ethers make them effective surfactants (Hu, Lu et al., 2021).

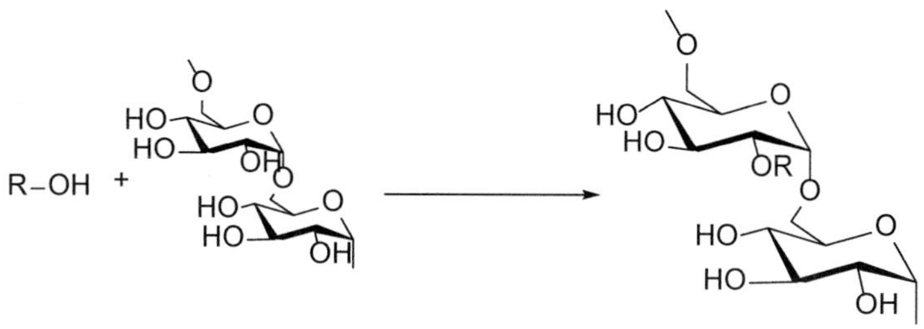

Figure 4. Synthesis of dextran ether.

4.3 Dialdehyde Dextran

Dialdehyde dextran, a reactive derivative of dextran, is produced through periodate oxidation. This derivative possesses aldehyde groups, enabling the attachment of bioactive molecules containing terminal NH_2 groups, such as proteins and enzymes (Figure 5). This conjugation is facilitated by a Schiff base reaction between the amino group and the aldehyde group. Protein-dialdehyde dextran conjugates have proven valuable as carrier molecules in immunodetection. They offer an advantage over hapten-protein conjugates produced using the carbodiimide coupling method, as they avoid the issue of immune cross-reactions. Dialdehyde dextran has also been

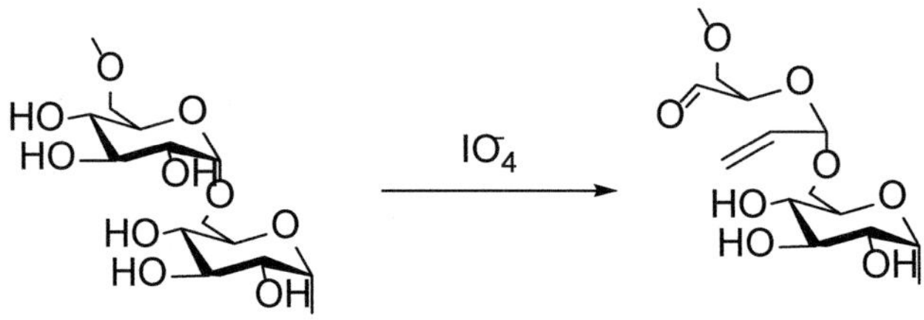

Figure 5. Dialdehyde dextran.

successfully employed as a non-toxic, macromolecular cross-linker to reinforce the nanostructure of biopolymer-based nanocarriers. By forming covalent bonds, it enhances the stability and passage of these carriers through the gastrointestinal tract (Fuentes, Segura et al., 2004; Hu, Lu et al., 2021).

5. Enzymatic Cross-Linking of Dextran-Tyramine

Dextran-tyramine (Dex-TA) conjugates can be crosslinked enzymatically using horseradish peroxidase and hydrogen peroxide. This method has been successfully applied to create hydrogels for encapsulating chondrocytes, demonstrating the potential of Dex-TA hydrogels as 3D scaffolds for cartilage tissue engineering. Swelling and degradation studies revealed that Dex-TA hydrogels linked by urethane bonds exhibit good stability under physiological conditions, with less than 25% weight loss over 5 months. In contrast, DG-TA hydrogels, containing ester-containing diglycolic groups, undergo complete degradation within 4 to 10 days. These dextran-based hydrogels showed promise for use in injectable systems for various biomedical applications, including tissue engineering and protein delivery (Jin, Moreira Teixeira et al., 2010; Zhao, Liu et al., 2020).

6. Dextran-Based Scaffold

Dextran, an exopolysaccharide produced by lactic acid bacteria, is synthesized using sucrose as a substrate. Dextran's chemically reactive hydroxyl groups allow for modification with different functional groups, leading to the formation of spherical, tubular, and 3D network structures. Biodegradable dextran-based scaffolds serve as bioactive carriers for various protein biomolecules, enabling effective controlled release and tissue regeneration. Dextran-derived hydrogels have also shown potential as bioartificial cardiac tissue matrices (BCT) for *in-vitro* cardiac tissue regeneration (Kang, Yang et al., 2022).

Dextran hydrogel scaffolds are particularly advantageous for soft tissue engineering due to their high resistance to protein adsorption and cell adhesion. This property allows for the design of scaffolds with specific recognition sites. It has been demonstrated that the synthesis of a crosslinked network of dextran hydrogels through radical methacrylate group polymerization. The immiscibility between the dextran matrix and poly(ethylene glycol) in aqueous solutions led to the formation of macroporous, beaded-wall morphology scaffolds that effectively manage liquid-liquid phase separations, enhancing cell penetration and nutrient diffusion (Samrot, Sathiyasree et al., 2023).

Dextran-based scaffolds, while acting as bioactive carriers for numerous protein biomolecules, are inherently biodegradable but can be more costly and have lower bioavailability. The development of novel functional scaffolds for efficient controlled release and tissue regeneration remains a significant research area with promising therapeutic potential for many diseases. Dextran-based scaffolds are naturally biodegradable and serve as bioactive carriers for a wide range of protein biomolecules. However, the reconstruction of the *in-vitro* microenvironment with appropriate signaling cues for large-scale tissue regenerative scaffolds remains a significant challenge in regenerative medicine (Jin, Teixeira et al., 2010).

7. Dextran Hydrogels

Dextran hydrogel is extensively explored for preparation of 3D scaffolds in regeneration of tissue, by main focus on their capability in imitation of the mechanical characteristics and water content of tissues (Wang, Li et al., 2017). Dextran hydrogels can be formed through two distinct methods: physical and chemical crosslinking. Both methods utilize the hydroxyl groups present on the α-1,6-linked D-glucose residues of dextran. Physical crosslinking involves derivatizing dextran with L-lactic acid and D-lactic acid oligomers. Chemical crosslinking, on the other hand, can be achieved through several approaches: covalent crosslinking with bifunctional isocyanates, glutaraldehyde, or by partially oxidizing dextran hydroxyl groups to aldehydes, which are then crosslinked with gelatin (Kim, Won et al., 1999; Luanda and Badalamoole, 2023).

Another approach involves synthesizing dextran hydrogels via radical polymerization of derivatized dextran containing polymerizable groups. Edman et al. introduced polymerizable moieties to dextran by derivatizing it with glycidyl acrylate under aqueous conditions. Similar studies have employed various polymerizable groups, including (meth)acrylates, allyl isocyanate, maleic anhydride, and vinyl acrylate using chemo-enzymatic procedures. Hennink and co-workers developed a method to control the degree of substitution (DS) of methacrylated groups on dextran using glycidyl methacrylate (GMA), hydroxyethyl methacrylate (HEMA), or lactate-HEMA. Hydrogels obtained with dextran derivatized with glycidyl methacrylate (Dex-GMA) exhibit stability under physiological conditions and have been shown to be biocompatible *in vivo*. In contrast, hydrogels of Dex-HEMA or Dex-lactate-HEMA are susceptible to hydrolysis (van Dijk-Wolthuis, Franssen et al., 1995; van Dijk-Wolthuis, Kettenes-Van Den Bosch et al., 1997; van Dijk-Wolthuls and Tsang, 1997).

Dextran hydrogel scaffolds are particularly promising for soft tissue engineering applications due to their resistance to protein adsorption and cell adhesion. This property enables the design of scaffolds with specific sites for cell recognition, similar to poly(ethylene glycol) (PEG). While dextran hydrogels have been investigated as drug delivery vehicles, their use as scaffolds for tissue engineering remains relatively unexplored (Lévesque, Lim et al., 2005). This may be attributed to previously synthesized dextran hydrogels having pore sizes too small for efficient cell penetration and regeneration. The Table 1 gives an overview of different preparation methods for dextran hydrogels (Edman, Ekman et al., 1980; Ferreira, Gil et al., 2002; Maire, Logeart-Avramoglou et al., 2005; Zhang, Tang et al., 2005; Reichert, 2006).

Table 1. Preparation methods for dextran hydrogels.

	Dextran derivatives	**Cross linking method**
Chemical crosslinking	Dex-MA Dex-HEMA	Radical polymerization: KPS/TEMED
	Functional dextran	STMP
	CM-dextran	EDC/NHS
	Dex-Al or dex-MA copolymerized with NiPAAm	UV polymerization
	Dex-AE copolymerized with PEGDA	UV polymerization
	Dex-MA copolymerized with con A-MA	UV polymerization
	Oxidized dextran + Gelatin	AAD
	Dex-DVA	Enzyme
Physical Crosslinking	Dextran 6000	Crystallization
	Dex-PEG or dex-PPG + Cyclodextrin	Hydrogen bonding
	Dex+PL Cyclodextrins	Hydrogen bonding
	Dex-L-Lactate + Dex-D-Lactate	Stereocomplexes
	Dextran-sulfate + Chitosan	Ionic interactions
	Dex-HEMA-MAA + Dex-HEMA-DMAEMA	Ionic interactions

8. Applications of Dextran-Based Hydrogels

A wide range of materials have been developed and utilized for tissue engineering, including natural molecules, synthetic polymers, inorganic materials, nanoscale materials, and composites. Natural polymers are particularly interesting in biomedical applications because their biochemical properties often resemble those of the human ECM. This similarity makes these natural macromolecular materials more readily recognized and accepted by the body.

Successful tissue engineering relies on the careful selection of suitable substrates or scaffolds that support cell growth and function. These scaffolds play a critical role in promoting cell adhesion, proliferation, and the secretion of extracellular matrix (ECM). Additionally, they can act as vectors for delivering growth factors or cytokines, which are essential for tissue repair and regeneration (Li, Ding et al., 2017).

Polysaccharides are composed of monosaccharides (sugars) linked together by O-glycosidic bonds. Their physical properties, such as solubility and viscosity, are largely determined by their monosaccharide composition, molecular weight, chain shapes, and other factors. Polysaccharides are abundant in nature and can be readily extracted from various sources, including algae, higher plants, microorganisms, and animals. Polysaccharides extracted from different sources possess distinct physicochemical properties due to variations in their chemical structures, functional groups, and electric charges (Jana, Jana et al., 2022).

Polysaccharides offer significant potential as biomaterials due to their numerous advantages, including their natural abundance, biocompatibility, hydrophilicity, water solubility, polyfunctionality, high chemical reactivity, and biodegradability. However, they also present some challenges. For example, most polysaccharides are water soluble and can be oxidized at temperatures above their melting point. Additionally, the acid solubility of chitosan limits its use in aqueous environments, hindering its broader biomedical applications. These limitations can be addressed by combining polysaccharides with other natural or polymeric materials or by employing chemical modifications. The presence of various functional groups, such as carboxyl, hydroxyl, and amino groups, within the polysaccharide structure allows for tailoring their chemical properties through modifications (Jana, Jana et al., 2022).

Numerous studies have demonstrated the promise of polysaccharides as biomaterials in repairing tissues such as cartilage, bone, cardiac tissue, and skin (Jin, Shi et al., 2021). They are frequently utilized as bio-scaffolds in tissue engineering. Dextran, a family of neutral polysaccharides, is primarily composed of α-(1→6) linked D-glucose backbones. The precise structure, including linkages and branching proportions, is determined by the specific type of bacteria that produces it (Neishlos, Novikova et al., 2004).

Dextran's versatility in biomedical applications stems from its biocompatibility, biodegradability, and tunable molecular weight. This biomacromolecule's volume expansion property makes it suitable as a deswelling agent in cornea transplantation, while its molecular weight similarity to albumin allows it to serve as a blood and plasma substitute. Dextran and its derivatives are crucial components in tissue engineering scaffolds. In cardiovascular applications, PVA/dextran hybrid hydrogels have shown promise as myocardial patches due to their robust mechanical properties, stability, and biocompatibility. These hydrogels can encapsulate and release antioxidants, further enhancing their therapeutic potential. Moreover, PVA/dextran grafts have been developed for MSC-based therapies and anticoagulation in preclinical models. These grafts exhibit comparable or even superior patency rates compared to traditional grafts, making them a viable option for vascular reconstruction (Ramasundaram, Saravanakumar et al., 2022; Sharma and Jagadeesh Chandra Bose, 2024).

The versatility of dextran extends to its modifiable nature. Introducing methacrylic groups allows for crosslinking capabilities through UV or ionic initiation. This property has led to the development of innovative biomaterials such as gelatin and Dex-MA fibers, created through electrospinning and photo crosslinking techniques. The composition of Dex-MA within gelatin/Dex-MA fibers directly influences their physical and biological characteristics. *In vitro* studies have shown that these fibers promote cell adhesion and proliferation of L929 cells, indicating their potential in tissue engineering applications (Maia, Evangelista et al., 2014).

Another innovative approach utilizes oxidized dextran (OD) as a novel crosslinker for blood-derived cryogel macroporous scaffolds. These scaffolds, constructed using platelet lysate (PL) and OD, exhibit mechanical strength that is directly proportional to the concentration of OD. *In vivo* studies using rat models demonstrated that these scaffolds degrade within 90 to 240 days. Furthermore, *in vitro* cytotoxicity studies revealed that PL/OD cytogel scaffolds possess the ability to support chondrogenic differentiation of human adipose-derived stem cells (hASCs) in the presence of chondro-inductive factors. This promising property underscores their potential in regenerative medicine applications (Biswal and Swain, 2021; Jin, Shi et al., 2021).

9. Bioartificial Cardiac Tissues

Bioartificial cardiac tissues (BCTs) hold significant promise as a potential solution for repairing and replacing damaged myocardium (Ye and Black, 2011). Creating these functional tissues often involves embedding cells within a three-dimensional (3D) scaffold that mimics the native extracellular matrix (ECM). Hydrogels, with their viscoelastic and customizable properties, are ideal candidates for 3D matrix formation. While both synthetic and natural hydrogels have been explored for cardiac tissue regeneration, natural hydrogels are preferred due to their inherent biocompatibility and biodegradability. Anionic polysaccharides like hyaluronic acid and alginate have already proven useful in heart tissue formation, but the potential of the neutral polysaccharide dextran as a cardiac tissue scaffold remains largely unexplored (DeForest and Anseth, 2012; Dahlmann, Krause et al., 2013).

Dextran, a linear, non-ionic polysaccharide derived from bacteria, exhibits low cell-adhesive properties in its native form. To enhance cell-scaffold interactions, integrin-specific molecules like the arginine-glycine-aspartic acid (RGD) sequence can be incorporated into the dextran backbone. Cyclic RGD peptides, with their enhanced selectivity and affinity to α-1,3 integrin, offer a promising approach for improving cell adhesion (Shachar, Tsur-Gang et al., 2011). Previous studies have demonstrated the positive impact of RGD motifs on heart tissue formation and neovascularization of implants. This research explores the potential of a modified dextran-based matrix as a novel scaffold material for bioartificial cardiac tissue generation. It represents the first successful encapsulation of human induced pluripotent stem cell-derived cardiomyocytes (hiPSC-CMs) and human foreskin fibroblasts (HFFs) within a modified dextran hydrogel under physiological conditions, leading to the formation of contractile tissue. Notably, both the formation and maturation of cardiac tissue occurred within the dextran hydrogel, regardless of the presence of the cyclic pentapeptide RGD. These findings suggest that dextran-based hydrogels are suitable materials for cardiac tissue engineering (Banerjee, Szepes et al., 2021).

10. Bone Tissue Engineering

Dextran and pullulan, natural hydrophilic polysaccharides, stand out as promising biomaterials for bone tissue engineering (BTE). BTE approaches typically rely on scaffolds combined with cells and/or bioactive molecules to create the environment necessary for bone formation. The design of scaffolds that mimic the mechanical and structural features of the bone extracellular matrix is paramount. Ideal biomaterials for bone scaffolding must support cell viability and proliferation, promote osteo differentiation and angiogenesis, facilitate host integration, and, when required, provide load-bearing capacity (Ahmed Omar, Amédée et al., 2022).

Several studies have investigated the bone regeneration capabilities of dextran-derived biomaterials by combining this polysaccharide with natural or synthetic polymers. For instance, dextran-coated polyvinyl formal (PVF) sponges have been developed to enhance the osteogenic potential of PVF, which suffers from poor osteogenic ability due to its fibrous construction. These dextran-coated sponges, with their high water-holding capacity, significantly improved bone formation in a rat femoral defect model. Dextran-based hydrogels loaded with mineral particles have also demonstrated enhanced osteogenic properties. Dextran-based hydrogels combined with polyacrylamide and loaded with or without hydroxyapatite (HA) particles showed significant improvement in bone regeneration when implanted in a rabbit femoral defect model, with the presence of HA leading to even greater bone formation (Fang, Li et al., 2019).

Further research has explored the combination of dextran with chitosan and strontium-doped mineral particles to optimize bone regeneration. Nanohydroxyapatite particles doped with increasing molar ratios of strontium (0%, 50%, and 100%) were incorporated into a dextran-chitosan mixture and implanted in a rat calvarial defect model. The group with 100% strontium-doped nanohydroxyapatite showed the most significant bone regeneration, while the dextran-based hydrogel without strontium-nanohydroxyapatite exhibited the lowest regeneration rate. Finally, an injectable hydrogel composed of dextran and poly (L-glutamic acid), inspired by mussel adhesion, was designed to create an adhesive, self-healing biomaterial with osteogenic properties. This scaffold, incorporating HA and bisphosphonate, demonstrated significant enhancement of bone regeneration compared to controls when implanted in a rat calvarial defect model. These studies highlight the versatile potential of dextran-based biomaterials in bone tissue engineering, paving the way for the development of innovative solutions for bone regeneration and repair (Ahmed Omar, Amédée et al., 2022; Alves, Simão et al., 2023).

11. Conclusion

Dextran-based hydrogels and scaffolds represent a highly versatile and promising class of biomaterials for regenerative medicine and biomedical applications. Their natural biocompatibility, biodegradability, and capacity for chemical modification enable the development of tailored structures suitable for tissue engineering, drug delivery, and regenerative therapies across various tissues such as cartilage, bone, cardiac, and vascular systems. Advances in synthesis methods, including enzymatic crosslinking and click chemistry, alongside structural modifications like oxidation and esterification, have enhanced their stability, functionality, and bioactivity. Despite these significant strengths, challenges remain in optimizing pore architecture and micro-environmental conditions to facilitate large-scale tissue regeneration. Overall, dextran's inherent properties and modifiability position it as a valuable platform for future innovations in regenerative medicine, with ongoing research poised to address current limitations and expand its clinical potential.

References

Ahmed Omar, N., J. Amédée, D. Letourneur, J.-C. Fricain and M. Fenelon. 2022. Recent advances of pullulan and/or dextran-based materials for bone tissue engineering strategies in preclinical studies: a systematic review. Frontiers in Bioengineering and Biotechnology 10: 889481.

Alves, P., A. F. Simão, M. F. Graça, M. J. Mariz, I. J. Correia and P. Ferreira. 2023. Dextran-based injectable hydrogel composites for bone regeneration. Polymers 15(23): 4501.

Banerjee, S., M. Szepes, N. Dibbert, J.-C. Rios-Camacho, A. Kirschning, I. Gruh et al. 2021. Dextran-based scaffolds for *in-situ* hydrogelation: Use for next generation of bioartificial cardiac tissues. Carbohydrate Polymers 262: 117924.

Biswal, A. and S. K. Swain. 2021. Dextran and pullulan-based hybrid materials for tissue engineering applications. Polysaccharide-Based Nanocomposites for Gene Delivery and Tissue Engineering, Elsevier: 131–154.

Cunningham, J. and H. Hodgson. 1992. Microcarrier culture of hepatocytes in whole plasma for use in liver support bioreactors. The International Journal of Artificial Organs 15(3): 162–167.

Dahlmann, J., A. Krause, L. Möller, G. Kensah, M. Möwes, A. Diekmann et al. 2013. Fully defined *in situ* cross-linkable alginate and hyaluronic acid hydrogels for myocardial tissue engineering. Biomaterials 34(4): 940–951.

DeForest, C. A. and K. S. Anseth. 2012. Advances in bioactive hydrogels to probe and direct cell fate. Annual Review of Chemical and Biomolecular Engineering 3(1): 421–444.

Edman, P., B. Ekman and I. Sjöholm. 1980. Immobilization of proteins in microspheres of biodegradable polyacryldextran. Journal of Pharmaceutical Sciences 69(7): 838–842.

Fang, J., P. Li, X. Lu, L. Fang, X. Lü and F. Ren. 2019. A strong, tough, and osteoconductive hydroxyapatite mineralized polyacrylamide/dextran hydrogel for bone tissue regeneration. Acta Biomaterialia 88: 503–513.

Ferreira, L., M. H. Gil and J. S. Dordick. 2002. Enzymatic synthesis of dextran-containing hydrogels. Biomaterials 23(19): 3957–3967.

Fuentes, M., R. L. Segura, O. Abian, L. Betancor, A. Hidalgo, C. Mateo et al. 2004. Determination of protein-protein interactions through aldehyde-dextran intermolecular cross-linking. Proteomics 4(9): 2602–2607.

Hornig, S. and T. Heinze. 2007. Nanoscale structures of dextran esters. Carbohydrate Polymers 68(2): 280–286.

Hu, Q., Y. Lu and Y. Luo. 2021. Recent advances in dextran-based drug delivery systems: From fabrication strategies to applications. Carbohydrate Polymers 264: 117999.

Jalaja, K., P. A. Kumar, T. Dey, S. C. Kundu and N. R. James. 2014. Modified dextran cross-linked electrospun gelatin nanofibres for biomedical applications. Carbohydrate Polymers 114: 467–475.

Jana, S., S. Jana and A. J. Domb. 2022. Polysaccharide-based biomaterials: delivery of therapeutics and biomedical applications. Royal Society of Chemistry.

Jeanes, A. and C. Wilham. 1950. Periodate oxidation of dextran. Journal of the American Chemical Society 72(6): 2655–2657.

Jin, M., J. Shi, W. Zhu, H. Yao and D.-A. Wang. 2021. Polysaccharide-based biomaterials in tissue engineering: a review. Tissue Engineering Part B: Reviews 27(6): 604–626.

Jin, R., L. S. Moreira Teixeira, P. J. Dijkstra, Z. Zhong, C. A. van Blitterswijk, M. Karperien et al. 2010. Enzymatically crosslinked dextran-tyramine hydrogels as injectable scaffolds for cartilage tissue engineering. Tissue Engineering Part A 16(8): 2429–40.

Jin, R., L. M. Teixeira, P. Dijkstra, C. Van Blitterswijk, M. Karperien and J. Feijen. 2010. Enzymatically-crosslinked injectable hydrogels based on biomimetic dextran–hyaluronic acid conjugates for cartilage tissue engineering. Biomaterials 31(11): 3103–3113.

Kang, D.-S., S.-Y. Yang and C.-Y. Lee. 2022. Fabrication of innocuous hydrogel scaffolds based on modified dextran for biotissues. Carbohydrate Research 522: 108699.

Kim, S., C. Won and C. Chu. 1999. Synthesis and characterization of dextran-based hydrogel prepared by photocrosslinking. Carbohydrate Polymers 40(3): 183–90. https://doi.org/10.1016/s0144-8617(99)00053-3.

Lévesque, S. G., R. M. Lim and M. S. Shoichet. 2005. Macroporous interconnected dextran scaffolds of controlled porosity for tissue-engineering applications. Biomaterials 26(35): 7436–7446.

Li, T., X. Ding, L. Tian and S. Ramakrishna. 2017. Engineering BSA-dextran particles encapsulated bead-on-string nanofiber scaffold for tissue engineering applications. Journal of Materials Science 52: 10661–10672.

Liebert, T., J. Wotschadlo, P. Laudeley and T. Heinze. 2011. Meltable dextran esters as biocompatible and functional coating materials. Biomacromolecules 12(8): 3107–3113.

Luanda, A. and V. Badalamoole. 2023. Past, present and future of biomedical applications of dextran-based hydrogels: A review. International Journal of Biological Macromolecules 228: 794–807.

Maia, J., M. B. Evangelista, H. Gil and L. Ferreira. 2014. Dextran-based materials for biomedical applications. Res Signpost 37661: 31–53.

Maia, J., L. Ferreira, R. Carvalho, M. A. Ramos and M. H. Gil. 2005. Synthesis and characterization of new injectable and degradable dextran-based hydrogels. Polymer 46(23): 9604–9614.

Maire, M., D. Logeart-Avramoglou, M.-C. Degat and F. Chaubet. 2005. Retention of transforming growth factor β1 using functionalized dextran-based hydrogels. Biomaterials 26(14): 1771–1780.

Neishlos, A., E. Novikova, B. Passet and A. Moskvin. 2004. IR and NMR study of the structure of dextran polyaldehyde. Russian Journal of Applied Chemistry 77: 128–130.

Pahimanolis, N., A.-H. Vesterinen, J. Rich and J. Seppala. 2010. Modification of dextran using click-chemistry approach in aqueous media. Carbohydrate Polymers 82(1): 78–82.

Portalska, K. J., L. M. Teixeira, J. C. Leijten, R. Jin, C. van Blitterswijk, J. de Boer et al. 2014. Boosting angiogenesis and functional vascularization in injectable dextran–hyaluronic acid hydrogels by endothelial-like mesenchymal stromal cells. Tissue Engineering Part A 20(3-4): 819–829.

Ramasundaram, S., G. Saravanakumar, S. Sobha and T. H. Oh. 2022. Dextran sulfate nanocarriers: design, strategies and biomedical applications. International Journal of Molecular Sciences 24(1): 355.

Reichert, J. M. 2006. Trends in US approvals: new biopharmaceuticals and vaccines. Trends in Biotechnology 24(7): 293–298.

Samrot, A. V., M. Sathiyasree, S. B. A. Rahim, R. E. Renitta, K. Kasipandian, S, Krithika Shree et al. 2023. Scaffold using chitosan, agarose, cellulose, dextran and protein for tissue engineering—A review. Polymers 15(6): 1525.

Shachar, M., O. Tsur-Gang, T. Dvir, J. Leor and S. Cohen. 2011. The effect of immobilized RGD peptide in alginate scaffolds on cardiac tissue engineering. Acta Biomaterialia 7(1): 152–162.

Sharma, S. and K. Jagadeesh Chandra Bose. 2024. Biomedical applications of dextran. Biopolymers for Biomedical Applications: 219–251.

Shibusawa, Y. 1999. Surface affinity chromatography of human peripheral blood cells. Journal of Chromatography B: Biomedical Sciences and Applications 722(1-2): 71–88.

Stanciu, M. C., D. Belei, E. Bicu, C. G. Tuchilus and M. Nichifor. 2020. Novel amphiphilic dextran esters with antimicrobial activity. International Journal of Biological Macromolecules 150: 746–755.

Sun, G. and J. J. Mao. 2012. Engineering dextran-based scaffolds for drug delivery and tissue repair. Nanomedicine 7(11): 1771–1784.

van Dijk-Wolthuis, W., O. Franssen, H. Talsma, M. Van Steenbergen, J. Kettenes-Van Den Bosch and W. Hennink. 1995. Synthesis, characterization, and polymerization of glycidyl methacrylate derivatized dextran. Macromolecules 28(18): 6317–6322.

van Dijk-Wolthuis, W., J. Kettenes-Van Den Bosch, A. Van Der Kerk-Van Hoof and W. Hennink. 1997. Reaction of dextran with glycidyl methacrylate: an unexpected transesterification. Macromolecules 30(11): 3411–3413.

van Dijk-Wolthuls, W. and S. Tsang. 1997. A new class of polymerizable dextrans with hydrolyzable groups: hydroxyethyl methacrylated dextran with and without oligolactate spacer. Polymer 38(25): 6235–6242.

Wang, R., P. J. Dijkstra and M. Karperien. 2016. Dextran. Biomaterials from Nature for Advanced Devices and Therapies: 307–319.

Wang, X., Z. Li, T. Shi, P. Zhao, K. An, C. Lin et al. 2017. Injectable dextran hydrogels fabricated by metal-free click chemistry for cartilage tissue engineering. Materials Science and Engineering: C 73: 21–30.

Ye, K. Y. and L. D. Black. 2011. Strategies for tissue engineering cardiac constructs to affect functional repair following myocardial infarction. Journal of Cardiovascular Translational Research 4: 575–591.

Zhang, R., M. Tang, A. Bowyer, R. Eisenthal and J. Hubble. 2005. A novel pH-and ionic-strength-sensitive carboxy methyl dextran hydrogel. Biomaterials 26(22): 4677–4683.

Zhao, H., M. Liu, Y. Zhang, J. Yin and R. Pei. 2020. Nanocomposite hydrogels for tissue engineering applications. Nanoscale 12(28): 14976–95.

CHAPTER 13

Pectin Based Scaffolds for Tissue Engineering

Dipti Soni,[1,#,*] *Khushboo Jain*,[2,#] *Jyotsana Panwar*,[2] *Manjinder Kour*,[3] *Girdhar Pal Singh*,[4] *Ritu Tomar*[4] and *Renu Rathore*[4]

1. Introduction

Tissue engineering (TE) is a multidisciplinary subject that aspires to create functional structures to repair, maintain, and improve damaged tissues and organs. The concept of replacing a broken body part using natural materials was presented and even implemented over 4000 years ago. However, living tissues were not engineered until recently, resulting in the discipline of tissue engineering. Tissue engineering is a new discipline that holds great promise for replacing damaged tissues by combining three key components: cells, scaffolds, and biomechanical or biochemical signals. Applied tissue engineering necessitates a thorough understanding of the fundamental sciences guiding cells, their microenvironment, and the signals that influence their behavior (Mhanna and Hasan, 2017). Smart biomaterials, new stem-cell sources, sophisticated three-dimensional bioprinting, vascular engineering, advanced bioreactors, organoids, and microfluidics-based physiological platforms are examples of emerging trends and technologies that are having a positive impact on this sector. The current obstacles in this subject include the availability of reliable cell sources, optimal bioinks, lowering the immunogenicity of TE scaffold, engineering vasculature and innervation in bioengineered tissues, commercialization, and

[1] Department of Chemistry, Faculty of Science, Janardan Rai Nagar Rajasthan Vidyapeeth (Deemed to be University), Udaipur 313002, Rajasthan, India.
[2] Department of Biotechnology, Faculty of Science, Janardan Rai Nagar Rajasthan Vidyapeeth (Deemed to be University), Udaipur 313002, Rajasthan, India.
[3] Department of Chemistry and Biochemistry, Montana State University, Bozeman, Montana, USA.
[4] Department of Chemistry, Bhupal Nobles' University, Udaipur, Rajasthan, India.
# Equal contribution.
* Corresponding author: diptisoni7381@gmail.com

DOI: 10.1201/9781003518792-13

regulatory impediments. Tissue engineering applications frequently involve the use of three-dimensional (3D) scaffolds to provide an appropriate microenvironment for the integration of cells or growth factors in order to restore damaged tissues or organs. These scaffolds are designed to imitate the actual *in vivo* microenvironment, where cells interact and respond in response to mechanical cues from their surroundings.

As a consequence, the scaffolds' material qualities play an important role in regulating cellular response and fate. These 3D scaffolds are often very porous, with linked pore networks that aid in nutrition and oxygen diffusion and waste elimination. Until now, synthetic materials have not been as effective as native tissues, especially in small diameter applications. The advancement of tissue engineering raised the prospect of a living, biological graft that could match the functional features of native vessels. In the last decades, various materials, such as polymers, ceramics, and metals, have been studied to fabricate scaffolds, and a variety of techniques have been investigated in this area, including a top-down approach that uses biodegradable polymeric scaffolds or "Materiomics", which employs high throughput screening technology to understand cell-material interactions. Fibrous structures are excellent as tissue engineering scaffolds because their shape is comparable to that of the extracellular matrix. The structural optimization and surface modification of the scaffolds could improve cell adhesion and proliferation. The majority of materials used in tissue engineering have been produced as scaffolds. Scaffolds are made by foaming that nearly fit the specifications for a good scaffold. These include HA foam gel-cast, bioactive glass foam, and biodegradable polymer foam/bioactive glass composites. Nonetheless, no material or manufacturing method has met all of the specifications for an ideal scaffold. Scaffolds created using Rapid Prototyping (RP) and Solid Freeform Fabrication (SFF) technologies have highly ordered microstructures and may be easily fabricated to complex shapes.

A variety of biodegradable materials have been proposed for the creation of three-dimensional porous scaffolds for tissue engineering. Natural polymers are one of the most appealing choices, owing to their resemblance to the extracellular matrix, chemical adaptability, high biological performance, and inherent cellular connections. This chapter will highlight the use of pectin as a building block for bidimensional (2D) and three-dimensional (3D) scaffolds in TE applications. Including the features of pectin this chapter also summarizes advantages and disadvantages of scaffolds made through biopolymers.

2. Scaffolds in Tissue Engineering and its Role

A scaffold is a structure that offers temporary mechanical support and a guiding template for cells during the synthesis of new tissue. Featuring the intended shape, architecture, and functionalities. At the same time, the scaffold biodegrades, allowing new tissue to develop. Notably, the biodegradability of the scaffold distinguishes it from permanent implants. The total biodegradation of the scaffolds eliminates the need for subsequent surgical treatments to remove or, eventually, replace them. The scaffold can be placed directly into the damaged region to promote tissue regeneration *in vivo*. Otherwise, prior to implantation, the scaffold can be initially cellularized with cells obtained from the patient, and then cultivated *in vitro* to produce tissues that would eventually be implanted into the defect to restore its function. The properties of scaffold are described in Figure 1.

The qualities of the scaffold are heavily influenced by the biomaterial formulations utilized in its manufacture (Kim, 2000). Thus, selecting the appropriate biomaterial formulation is critical for promoting tissue regeneration in a controlled manner while avoiding any undesirable side effects (e.g., cytotoxicity, apoptosis, carcinogenesis). The most often applied biomaterial formulations in TE are mostly based on as described in Figure 2.

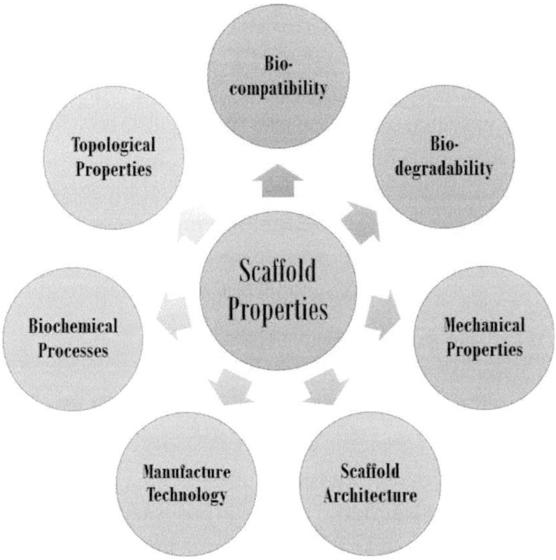

Figure 1. Properties of scaffold.

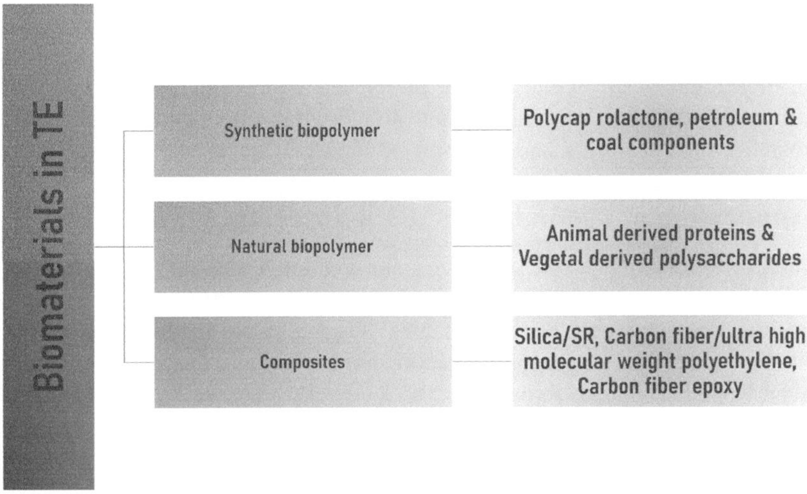

Figure 2. Types of biomaterials used in tissue engineering.

3. Pectin: The Super Biomaterial in TE

Depending on the kind of plant and cell type it develops in, pectin has a varied structure. The size, acetylation type, methyl group content, degree of esterification, and other variables that are controlled by the differences between the galacturonic acid that leads to the homogalacturonan chain and the side chain type of the rhamnogalacturonan-1 vary depending on the source of the pectin (Ross et al., 2012). Rhamnogalacturonan-1 is primarily found in the branching portions of the pectin polysaccharide, which constitute the main carbohydrate chains. Interestingly, Homogalacturonan forms the linear fragment of the polysaccharide, and the chain occasionally produces the component that rhamnogalacturonan-1 normally consists of. The source from which it is obtained and the conditions under which it is extracted determine the type of pectin polysaccharide (Reilly et al., 2010). No matter how diverse the pectin polysaccharide is, the structure is still considered canonical. Even though pectin was discovered more than 200 years ago, its composition's design is still not fully understood. In the core structure of the pectin polymer, alpha-1, 4-linked D-galacturonic acid and alpha-1, 2-L-rhamnose units alternate. The regulation of the polysaccharide's effect on cytokine production by the pectin system demonstrates that the polysaccharide's elemental properties are linked to its capacity to affect cellular and biological conditions.

Pectin is commercially obtained from the residual part of the plant materials in three stages that are;

i. *Pre-Processing*: (Cleaning, Drying and Maceration)
ii. *Processing*: (Conventional extraction methods, Enzyme assisted extraction, Microwave assisted extraction, Ultra sound assisted extraction, Subcritical water extraction, Pressurized CO_2 and deionized water).
iii. *Post-Processing*: (Neutralization, Precipitation, Centrifugation, Filtration).

The yield of the pectin depends on various variables such as solid-to-liquid ratio, acid strength, temperature, extraction duration and precipitation method. These variables also affect the qualities of the pectin in terms of degree of esterification (DE), molecular weight, composition, purity and color. This purified pectin can be efficiently used for drug delivery and tissue engineering through gel beads or microspheres, 3D scaffolds, and membranes (Figure 3).

4. Application of Pectin in Tissue Engineering

One use for formulations based on pectin is the creation of two-dimensional hydrogel patches for wound care. These patches operate as an antibacterial barrier to stop future infections and give cells mechanical support while they build new tissue. Additionally, the film's hydrophilic pectin molecules have the ability to combine with the wound's fluids to form a soft gel. The gel's presence makes it possible to keep the wound wet (Table 1). This promotes the healing process by assisting in the removal or control of secretions from the injured tissue. By including bioactive substances like medications (like antibiotics) and/or growth factors inside the pectin

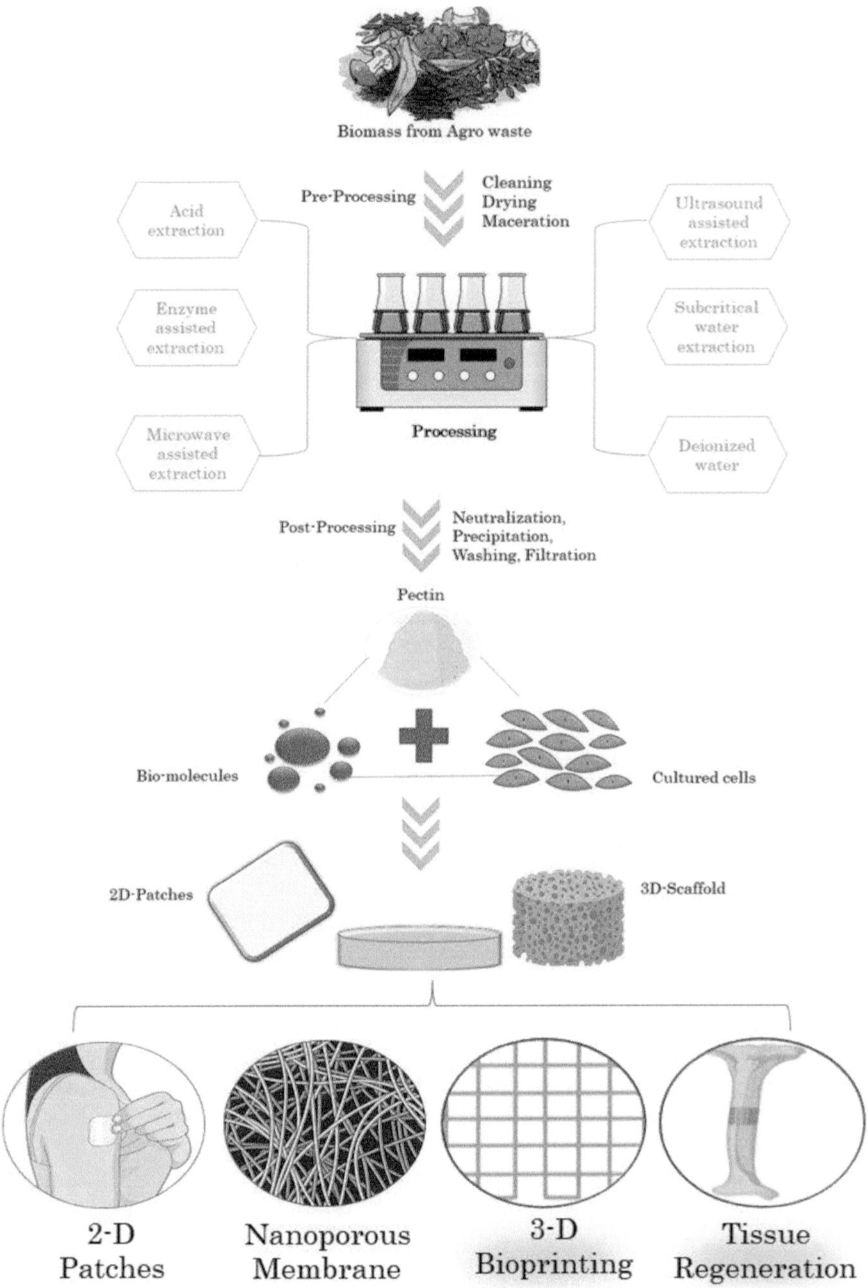

Figure 3. Processing of pectin for development of scaffolds and their application in tissue engineering.

Table 1. Illustration of different pectin sources and their application in tissue engineering.

S. No.	Pectin Source	Method	Application	Reference
1	Citrus peels	Cross-linked with GPTMS, Freeze-dried	3D bioprinting, Tissue engineering	Lapomarda et al., 2020
2	Citrus peels, Chitosan, Alginate	Freeze-drying	Tissue engineering	Archana et al., 2013
5	Chitosan, Pectin	Hydrogel preparation	Tissue engineering, *In vitro* modelling	Morello et al., 2023
6	Citrus peels, Chitin, Nano $CaCO_3$	Lyophilization	Tissue engineering, Drug delivery	Kumar et al., 2013
7	Citrus peels	Modified with RGD peptide, Injection molding	Bone tissue engineering	Munarin et al., 2011
8	Citrus peels, Chitosan	Freeze-drying	Bone tissue engineering	Coimbra et al., 2011
9	Calcium phosphate, Pectin	Cement preparation	Bone tissue engineering	Zhao et al., 2016
10	Silk, Pectin	Dialysis against methanol	Tissue engineering, Hydrogels with high mechanical properties	Numata et al., 2014
11	Chitosan, Pectin, Cellulose nanocrystals	Hydrogel preparation	Cartilage tissue regeneration	Ghorbani et al., 2020
13	Bitter gourd (pectin)	Green template synthesis, Composite films	Bone tissue engineering	Shebi and Lisa, 2018
14	Gelatin methacryloyl (GelMA), Pectin-g-PCL	IPN and semi-IPN hydrogels	Tissue engineering, Regenerative medicine	Fares et al., 2018
15	Apple and Citrus	Preparation of calcium pectate (CaP) beads	Study of chemical characteristics, gelation properties, and biological application	Zhao et al., 2018
16	Periodate-oxidized	Combination with gelatin, analysis of swelling, reswelling, SEM, FTIR	Controlled drug delivery systems, wound dressing	Garrido et al., 2019
17	Pectin	Internal gelation with calcium carbonate, pH modulation, rheological study	Injectable hydrogels for cell, drug, or bioactive molecule delivery	Moreira et al., 2014
18	Pectin	Electrospinning with gelatin, PCL, PVA	Wound healing applications, improved hydrophilicity, degradability, and antibacterial activity	Hosseini et al., 2022
19	Chitosan and Pectin	Electrospinning	Skin tissue scaffolds with improved mechanical properties and faster drug release	Lin et al., 2013

Table 1 contd. ...

...Table 1 contd.

S. No.	Pectin Source	Method	Application	Reference
20	Pectin	Electrospinning with periodate oxidation, crosslinking	Vascular cell differentiation from mesenchymal stem cells, potential in tissue engineering	Li et al., 2019
21	Pectin/PAA/PRF and Pectin/PAA/SIM	3D printing and electrospinning	Skin tissue regeneration, enhanced angiogenesis, and wound healing	Tavakoli et al., 2024
23	Pectin and PVA	Electrospinning with rectorite (REC)	Composite nanofabrics with improved thermal stability and cell viability	Ye et al., 2016
24	Pectin and Derivatives	Review of various pectin-based scaffolds	Tissue engineering applications, focusing on biocompatibility, biodegradability, and permeability	Pandey et al., 2023
25	Chitosan, PLLA, Pectin	Freeze-drying and crosslinking with EDC/ NHS	Cartilage tissue regeneration with enhanced mechanical strength and chondrocyte attachment	Mallick et al., 2018
26	PVDF and Pectin	Electrospinning with benzalkonium chloride (BAC)	Drug delivery systems with controlled release profile	Hamzah et al., 2021
27	Low-methoxyl citrus pectin	UV photocrosslinking with peptide crosslinkers and adhesive ligands; lyophilization	Skin tissue engineering	Pereira et al., 2018
28	Sugar beet pectin (SBP)	Crosslinked by visible light (405 nm) in the presence of tris(bipyridine) ruthenium (II) chloride hexahydrate and sodium persulfate; 3D bioprinting	Liver and other soft tissue engineering	Mubarok et al., 2023
29	Pectin/strontium/ hydroxyapatite	Solution-based chemical technique	Bone regeneration	Akshata et al., 2022
30	Collagen/ polyurethane/pectin	Semi-interpenetration process	Bone regeneration	Guzmán-Chávez et al., 2022
31	Pectin/PVA	Freezing–thawing	Bone regeneration	Hu et al., 2022
32	Poly(L-lactide-co-caprolactone) (PLCA)/ pectin	Scaffolds functionalized with pectin	*In vitro* and *in vivo* bone regeneration	Suliman et al., 2022

patches, the healing of the injured tissue can be stimulated even more (Munarin et al., 2012).

Direct delivery of these molecules to the injured region over an extended period of time can help promote the growth of new tissue and lower the risk of infection. The two main forms of pectin-based patches are non-porous films and porous membranes; as explained below;

a) *Pectin-based films*: These are often two-dimensional, flexible, non-porous substrates with the capacity to hold an excessive amount of water in their matrix. "Solvent Casting" is one method employed in the creation of these films. The following is the solvent casting process:
b) *Nanoporous membranes*: These are derived from pectin are predominantly fabricated via electrospinning. This method enables the creation of highly porous and flexible patches by subjecting pectin-based polymer solutions to an external electric field (Zhao et al., 2016; Shebi and Lisa, 2018).

A typical electrospinning setup comprises four parts mainly;

i. A syringe pump containing the polymer solution,
ii. A metallic needle through which the solution is dispensed,
iii. A high voltage power supply (ranging from several tens of kilovolts),
iv. A grounded collector, often a metal plate.

When the polymer solution is extruded through the needle, the intense electric forces between the needle and the collector cause the solution to stretch and form fibers with diameters ranging from nanometers to micrometers (Bhardwaj and Kundu, 2010). These fibers are then deposited onto the collector, where they form a nonwoven fibrous membrane following the complete evaporation of the solvent (Figure 4).

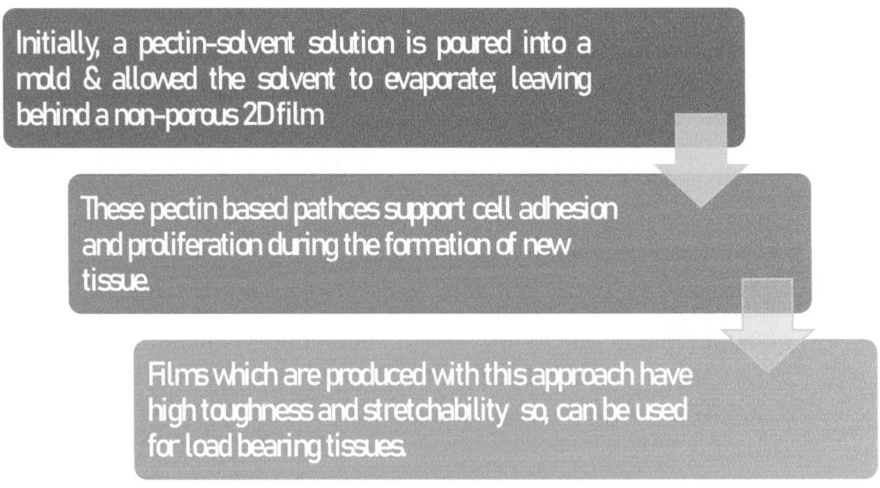

Figure 4. Preparation of 2-D pectin films through solvent casting method.

Pectin-based nano-fibers find application for the preparation of films/structures that can be potentially used as patches for wound healing of soft tissues (e.g., vascular tissue, retinal tissue). In addition, drugs (such as antibiotics) and particles (such as argentum ions for antibacterial purposes) can be successfully loaded in these structures obtaining patches for a local and controlled release of drugs directly into the wound (Chen et al., 2018; Alipour et al., 2019; Ahadi et al., 2019; Li et al., 2019).

5. 3D Pectin Scaffolds

Pectin-based formulations can be advanced to produce three-dimensional (3D) scaffolds that replicate the intricate structures of biological tissues. These 3D pectin-based scaffolds are primarily created as porous sponges or through 3D bioprinting techniques.

5.1 Porous Sponges

Sponges, like foams, have an interconnected network of pores. This form of architecture promotes cell penetration and scaffold colonization while also assuring appropriate nutrition transport to cells within the scaffold. Furthermore, a highly porous scaffold with open and linked pores is crucial because it allows nutrients and waste products to diffuse throughout the scaffold (Loh and Choong, 2013). The high hydrophilicity of pectin molecules, along with interconnected porosity, allows these sponges to entrap a vast volume of water, resulting in a 3D hydrogel-based environment that can imitate the natural extracellular matrix (Figure 5).

These sponges sustain wet environments, which speeds up wound healing because of their hydrophilicity. Sponge scaffolds have been utilized to regenerate several tissues, including cartilage, bone, and skin (Türkkan et al., 2018; Yu et al., 2019; Cui et al., 2020; Lapomarda et al., 2020). The technique of producing pectin-based sponges is described in Figure 6.

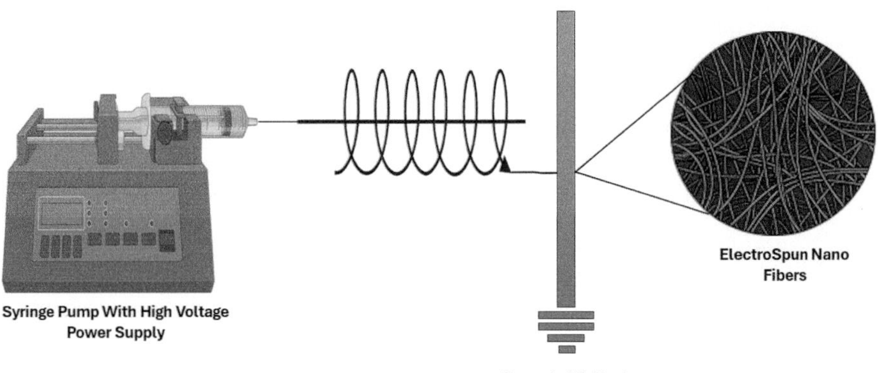

Figure 5. Preparation of nanoporous membranes through electro-spun technique.

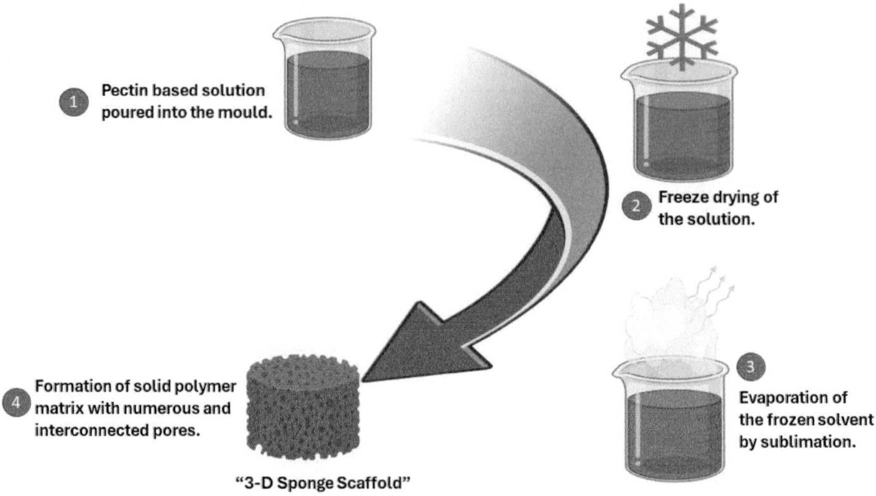

Figure 6. Preparation of 3-D Scaffold in tissue engineering.

5.2 3-D Bioprinting

Pectin-based scaffolds with intricate shapes have predominantly been created using extrusion-based bioprinting techniques. This method is highly favoured in tissue engineering due to its straightforwardness and its ability to handle a wide range of biomaterials, cells, and biomolecules (Archana et al., 2013). An extrusion-based bioprinter typically comprises a moveable cartridge filled with the biomaterial formulation, known as 'ink,' and a movable deposition stage. Prior to the bioprinting process, the scaffold's design can be developed using computer-aided design (CAD) software or based on medical images from sources like computed tomography (CT) scans or magnetic resonance imaging (MRI). The 3D model of the scaffold is then sliced into bioprinting paths using computer-aided manufacturing (CAM) software and converted into a printable code file known as 'G-code.' During the bioprinting process, the ink is extruded onto the deposition stage according to the predefined paths in the G-code, building up the scaffold layer by layer (Figure 7). For example, pectin-based scaffolds with more complex shapes such as a human ear and nose shape for cartilage tissue regeneration were successfully obtained.

6. Advantages of Pectin as Scaffolds in Tissue Engineering

- **Regulation of Immune Response:** Pectin plays a crucial role in managing the body's immune response to foreign materials or transplanted cells. Instead of eliminating the immune response, pectin acts as an immunomodulator to regulate and enhance the efficiency of applications requiring immune system modulation. Research has shown that pectin can reduce inflammatory reactions by promoting anti-inflammatory cytokines and decreasing proinflammatory cytokine levels.

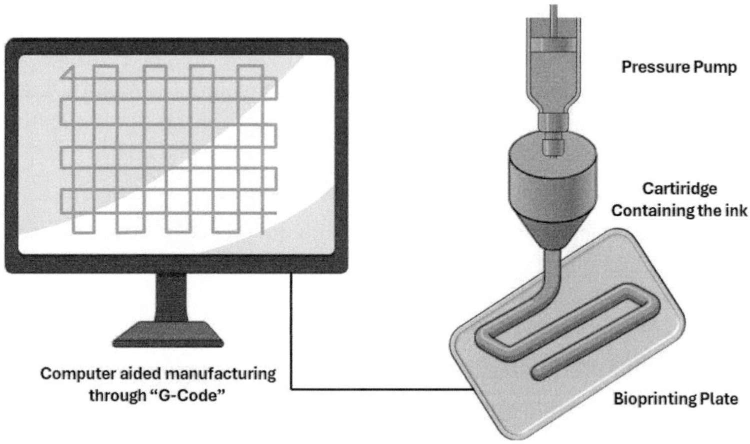

Figure 7. 3-D Bioprinting.

- **Structural Versatility:** Pectin's complex and diverse structure provides it with excellent modification capabilities compared to other biopolymers. This structural complexity allows for a wide range of applications in biomedical fields (Morello et al., 2023).
- **Hydrophilicity and Water Retention:** As a hydrophilic natural polymer, pectin can absorb and retain significant amounts of water, which contributes to its swelling properties. This characteristic is beneficial for forming hydrogels and composite materials through crosslinking and other methods.
- **Smart Composite Formation:** Pectin can be used to create smart composites with various sensitivities—such as to physical stimuli (light, temperature, electricity), chemical stimuli (pH, redox, glucose), and biological stimuli (enzymes). These properties are advantageous for controlled delivery systems of bioactive compounds.
- **Biodegradability and Biocompatibility:** Pectin-based materials are biodegradable and biocompatible, making them suitable for use in medical applications (Numata et al., 2014).
- **Widespread Availability:** Due to its extensive availability, pectin has become a significant focus in the research and development of nature-based biomedical and healthcare technologies.

7. Disadvantages of Pectin as Scaffold in Tissue Engineering

- **Rapid Degradation:**

  Pectin degrades relatively quickly in biological environments, which can lead to premature scaffold breakdown before the tissue has sufficiently regenerated. Degradation can produce acidic by-products that may cause local inflammation or affect the pH balance of the surrounding tissue.

- **Limited Cell Adhesion:**

  Pectin lacks specific cell-binding sites (like RGD peptides found in other materials), which can result in poor initial cell attachment and proliferation. To enhance cell adhesion, pectin scaffolds often require surface modification or the addition of other biomolecules, increasing complexity and cost.

- **Difficulty in Mimicking Native Tissue Properties:**

  Pectin's natural properties make it difficult to precisely tailor the scaffold's mechanical and chemical properties to closely mimic those of specific native tissues. Creating intricate, three-dimensional structures that are needed to mimic the extracellular matrix of tissues can be challenging with pectin.

- **Swelling and Stability Issues:**

  Pectin-based hydrogels can absorb large amounts of water, leading to excessive swelling, which may disrupt the scaffold's shape and negatively impact its mechanical properties. Pectin scaffolds can lose their integrity in the presence of fluids, leading to challenges in maintaining scaffold shape and function over time.

- **Crosslinking Challenges:**

  To improve mechanical stability and control degradation rates, pectin scaffolds often require crosslinking. However, crosslinking agents can be toxic or alter the scaffold's biocompatibility. Achieving uniform crosslinking throughout the scaffold can be difficult, leading to inconsistent properties and performance.

- **Potential for Immunogenicity:**

  Pectin is derived from plant sources, which can vary in composition and purity, potentially leading to immunogenic responses or inconsistent scaffold performance.

- **Processing and Manufacturing Challenges:**

  Creating pectin scaffolds with consistent properties and desirable architecture can be complex and require sophisticated techniques, which may not be easily scalable or cost-effective.

8. Conclusion

Pectin scaffolds have gained attention in tissue engineering for their natural origin, biocompatibility, and biodegradability, making them suitable for a range of biomedical applications. Despite these advantages, their use is limited by inherent drawbacks such as poor mechanical strength, rapid degradation rates, insufficient cell adhesion, and stability challenges in physiological environments. These factors can hinder their performance, particularly in applications requiring long-term structural support or specific mechanical properties. To address these limitations, modifications such as crosslinking, blending with other polymers, or functionalization with bioactive molecules are often necessary. Such enhancements can improve mechanical properties, control degradation rates, and promote better cell interactions. Moving forward, continued research and innovation in scaffold design and material modification are essential to fully leverage the potential of pectin scaffolds in tissue engineering. By overcoming current challenges, pectin-based scaffolds could become more versatile and effective tools in the regeneration of a wide variety of tissues.

9. Future Scope of Work

The future of pectin scaffolds in tissue engineering holds promising potential for significant advancements in their design, functionality, and application, addressing current limitations and broadening their use in regenerative medicine. One crucial area for development is enhancing the mechanical properties of pectin scaffolds. This can be achieved by creating hybrid scaffolds that combine pectin with other polymers, nanoparticles, or reinforcing agents to improve their mechanical strength, making them more suitable for load-bearing applications such as bone and cartilage regeneration. Another important focus is on controlling the degradation rates of pectin scaffolds. By employing advanced crosslinking techniques or chemical modifications, researchers can fine-tune the degradation profiles to ensure that scaffold longevity aligns with the tissue healing process. Improving cell adhesion and bioactivity is also a key area for future research. Incorporating bioactive molecules, peptides, or growth factors into pectin scaffolds can enhance cell adhesion, proliferation, and differentiation, thus promoting more effective tissue regeneration. Additionally, exploring innovative fabrication techniques such as 3D bioprinting and electrospinning could lead to the development of more complex and patient-specific scaffold architectures, improving their integration and functionality *in vivo*. Expanding the applications of pectin scaffolds beyond traditional uses is another exciting avenue. Tailoring these materials for specific purposes like drug delivery systems, wound healing, or as carriers for cell therapy could significantly broaden their biomedical applications. Finally, translating these advancements into clinical practice will require scaling up production, ensuring reproducibility, and conducting thorough preclinical and clinical trials to confirm the safety and efficacy of pectin scaffolds in real-world medical scenarios.

References

Ahadi, F., S. Khorshidi and A. Karkhaneh. 2019. A hydrogel/fiber scaffold based on silk fibroin/oxidized pectin with sustainable release of vancomycin hydrochloride. Eur. Polym. J. 118: 265–274. DOI: 10.1016/j.eurpolymj.2019.06.001.

Akshata, C. R., G. Harichandran and E. Murugan. 2023. Effect of pectin on the crystallization of strontium substituted HA for bone reconstruction application. Colloids and Surfaces B: Biointerfaces 226: 113312.

Alipour, R., A. Khorshidi, A. F. Shojaei, F. Mashayekhi and M. J. M. Moghaddam. 2019. Skin wound healing acceleration by Ag nanoparticles embedded in PVA/PVP/pectin/Mafenide acetate composite nanofibers. Polymer Testing 79(5): 106022. DOI: 10.1016/j.polymertesting.2019.106022.

Archana, D., L. Upadhyay, R. Tewari, J. Dutta, Y. B. Huang and P. Dutta. 2013. Chitosan-pectin-alginate as a novel scaffold for tissue engineering applications. Indian J. Biotechnol. 12: 475–482.

Bhardwaj, N. and S. C. Kundu. 2010. Electrospinning: A fascinating fiber fabrication technique. Biotechnol. Adv. 28(3): 325–347. DOI: 10.1016/j.biotechadv.2010.01.004.

Chen, S., S. Cui, H. Zhang, X. Pei, J. Hu, Y. Zhou et al. 2018. Cross-linked pectin nanofibers with enhanced cell adhesion. Biomacromolecules 19(2): 490–498. DOI: 10.1021/acs.biomac.7b01605.

Coimbra, P., P. Ferreira, H. C. de Sousa, P. Batista, M. A. Rodrigues, I. J. Correia et al. 2011. Preparation and chemical and biological characterization of a pectin/chitosan polyelectrolyte complex scaffold for possible bone tissue engineering applications. Int. J. Biol. Macromol. 48(1): 112–118.

Cui, Y., Q. Wu, J. He, M. Li, Z. Zhang and Y. Qiu. 2020. Porous nano-minerals substituted apatite/chitin/pectin nanocomposites scaffolds for bone tissue engineering. Arab. J. Chem. 13(10): 7418–7429. DOI: 10.1016/j.arabjc.2020.08.018.

Fares, M. M., E. Shirzaei Sani and R. Portillo Lara. 2018. Interpenetrating network gelatin methacryloyl (GelMA) and pectin-g-PCL hydrogels with. Biomaterials.

Garrido, C. A., M. Vargas and J. F. Alvarez-Barreto. 2019. Auto-cross-linking hydrogels of hydrogen peroxide-oxidized pectin and gelatin for applications in controlled drug delivery. International Journal of Polymer Science (1): 9423565.

Ghorbani, M., L. Roshangar and J. Soleimani Rad. 2020. Development of reinforced chitosan/pectin scaffold by using cellulose nanocrystals as nanofillers: An injectable hydrogel for tissue engineering. Eur. Polym. J. 130: 109697. https://doi.org/10.1016/j.eurpolymj.2020.109697.

Guzmán-Chávez, M. L., J. A. Claudio-Rizo, M. Caldera-Villalobos, D. A. Cabrera-Munguía, J. J. Becerra-Rodríguez and N. Rodríguez-Fuentes. 2022. Novel bioactive collagen-polyurethane-pectin scaffolds for potential application in bone regenerative medicine. Applied Surface Science Advances 11: 100317.

Hamzah, M. S. A., C. Ng, J. A. Sukor, K. S. M. Ali and N. H. M. Nayan. 2021. Electrospinning, preparation and characterization of polyvinylidene fluoride/pectin electrospun loaded with benzalkonium chloride as a drug reservoir. Journal of Mechanical Engineering (JMechE) 18(1): 39–51.

Hosseini, S. A., S. J. Hoseini, V. R. Askari, R. Salarinia, A. Ebrahimzadeh-Bideskan, F. Tara et al. 2022. Pectin-reinforced electrospun nanofibers: Fabrication and characterization of highly biocompatible mats for wound healing applications. Journal of Drug Delivery Science and Technology 77: 103916.

Hu, Z., J. Cheng, S. Xu, X. Cheng, J. Zhao, Z. W. K. Low et al. 2022. PVA/pectin composite hydrogels inducing osteogenesis for bone regeneration. Materials Today Bio 16: 100431.

Kim, B.-S., C. E. Baez and A. Atala. 2000. Biomaterials for tissue engineering. World J. Urol. 18(1): 2–9. DOI: 10.1007/s003450050002.

Kumar, P. T., C. Ramya, R. Jayakumar, S. K. Nair and V. K. Lakshmanan. 2013. Drug delivery and tissue engineering applications of biocompatible pectin-chitin/nano $CaCO_3$ composite scaffolds. Colloids Surf. B Biointerfaces 106: 109–116. https://doi.org/10.1016/j.colsurfb.2013.01.048.

Lapomarda, A., A. De Acutis, I. Chiesa, G. M. Fortunato, F. Montemurro, C. De Maria et al. 2020. Pectin-GPTMS-based biomaterial: Toward a sustainable bioprinting of 3D scaffolds for tissue engineering application. Biomacromolecules 21(2): 319–327. DOI: 10.1021/acs.biomac.9b01332.

Li, N., F. Xue, H. Zhang, H. J. Sanyour, A. P. Rickel, A. Uttecht et al. 2019. Fabrication and characterization of pectin hydrogel nanofiber scaffolds for differentiation of mesenchymal stem cells into vascular cells. ACS Biomater. Sci. Eng. 5(12): 6511–6519. DOI: 10.1021/acsbiomaterials.9b01178.

Lin, H. Y., H. H. Chen, S. H. Chang and T. S. Ni. 2013. Pectin-chitosan-PVA nanofibrous scaffold made by electrospinning and its potential use as a skin tissue scaffold. Journal of Biomaterials Science, Polymer Edition 24(4): 470–484.

Loh, Q. L. and C. Choong. 2013. Three dimensional scaffolds for tissue engineering applications: Role of porosity and pore size. Tissue Eng. Part B Rev. 19(6): 485–502. DOI: 10.1089/ten.teb.2012.0437.

Mallick, S. P., B. N. Singh, A. Rastogi and P. Srivastava. 2018. Design and evaluation of chitosan/poly(L-lactide)/pectin-based composite scaffolds for cartilage tissue regeneration. International Journal of Biological Macromolecules 112: 909–920.

Mhanna, R. and A. Hasan. 2017. Introduction to tissue engineering. *In*: Hasan, A. (ed.). Tissue Engineering for Artificial Organs. https://doi.org/10.1002/9783527689934.ch1.

Moreira, H. R., F. Munarin, R. Gentilini, L. Visai, P. L. Granja, M. C. Tanzi et al. 2014. Injectable pectin hydrogels produced by internal gelation: pH dependence of gelling and rheological properties. Carbohydrate Polymers 103: 339–347.

Morello, G., G. De Iaco, G. Gigli, A. Polini and F. Gervaso. 2023. Chitosan and pectin hydrogels for tissue engineering and *in vitro* modeling. Gels 9: 132. https://doi.org/10.3390/gels9020132.

Mubarok, W., K. C. M. L. Elvitigala, T. Kotani and S. Sakai. 2023. Visible light photocrosslinking of sugar beet pectin for 3D bioprinting applications. Carbohydrate Polymers 316: 121026.

Munarin, F., M. C. Tanzi and P. Petrini. 2012. Advances in biomedical applications of pectin gels. Int. J. Biol. Macromol. 51(4): 681–689. DOI: 10.1016/j.ijbiomac.2012.07.002.

Numata, K., S. Yamazaki, T. Katashima, J. A. Chuah, N. Naga and T. Sakai. 2014. Silk-pectin hydrogel with superior mechanical properties, biodegradability, and biocompatibility. Macromol. Biosci. 14(6): 799–806. https://doi.org/10.1002/mabi.201300482.

Pandey, A., P. K. Sharma, R. Malviya and K. Rahate. 2023. Pectin/pectin derivatives as potential scaffolds for the tissue engineering applications. The Natural Products Journal 13(6): 63–78.

Pereira, R. F., C. C. Barrias, P. J. Bártolo and P. L. Granja. 2018. Cell-instructive pectin hydrogels crosslinked via thiol-norbornene photo-click chemistry for skin tissue engineering. Acta Biomaterialia 66: 282–293.

Reilly, G. C. and A. J. Engler. 2010. Intrinsic extracellular matrix properties regulate stem cell differentiation. J. Biomech. 43(1): 55–62. DOI: 10.1016/j.jbiomech.2009.09.009.

Ross, A. M., Z. Jiang, M. Bastmeyer and J. Lahann. 2012. Physical aspects of cell culture substrates: Topography, roughness, and elasticity. Small 8(3): 336–355. DOI: 10.1002/smll.201100934.

Shebi, A. and S. Lisa. 2018. Pectin mediated synthesis of nano hydroxyapatite-decorated poly(lactic acid) honeycomb membranes for tissue engineering. Carbohydr. Polym. 201: 39–47.

Suliman, S., A. Mieszkowska, J. Folkert, N. Rana, S. Mohamed-Ahmed, T. Fuoco et al. 2022. Immune-instructive copolymer scaffolds using plant-derived nanoparticles to promote bone regeneration. Inflammation and Regeneration 42(1): 12. https://doi.org/10.1186/s41232-022-00196-9.

Tavakoli, M., M. H. Al-Musawi, A. Kalali, A. Shekarchizadeh, Y. Kaviani, A. Mansouri et al. 2024. Platelet-rich fibrin and simvastatin-loaded pectin-based 3D printed-electrospun bilayer scaffold for skin tissue regeneration. International Journal of Biological Macromolecules 265(Pt 1): 130954. https://doi.org/10.1016/j.ijbiomac.2024.130954.

Türkkan, S., D. Atila, A. Akdağ and A. Tezcaner. 2018. Fabrication of functionalized citrus pectin/ silk fibroin scaffolds for skin tissue engineering. J. Biomed. Mater. Res. Part B Appl. Biomater. 106(7): 2625–2635. DOI: 10.1002/jbm.b.34079.

Ye, X., Y. Zhan, T. Li, X. Shi, H. Deng and Y. Du. 2016. Pectin-based composite nanofabrics incorporated with layered silicate and their cytotoxicity. International Journal of Biological Macromolecules 93: 123–130.

Yu, N., X. Wang, F. Ning, C. Jiang, Y. Li, H. Peng et al. 2019. Development of antibacterial pectin from Akebia trifoliata var. australis waste for accelerated wound healing. Carbohydr. Polym. 217: 58–68. DOI: 10.1016/j.carbpol.2019.03.071.

Zhao, L., J. Li, L. Zhang, Y. Wang, J. Wang, B. Gu et al. 2016. Preparation and characterization of calcium phosphate/pectin scaffolds for bone tissue engineering. RSC Adv. 6: 10. 1039/C6RA07800A.

Zhao, S., F. Yang, Y. Liu, D. Sun, Z. Xiu, X. Ma et al. 2018. Study of chemical characteristics, gelation properties and biological application of calcium pectate prepared using apple or citrus pectin. International Journal of Biological Macromolecules 109: 180–187.

Index

For Product Safety Concerns and Information please contact our
EU representative GPSR@taylorandfrancis.com Taylor & Francis
Verlag GmbH, Kaufingerstraße 24, 80331 München, Germany